The Emerging Optical Network Comprehensive Report

Presented by the
International Engineering Consortium

ISSN: 0886-229X
ISBN: 0-933217-97-8

International Engineering Consortium
549 West Randolph Street, Suite 600
Chicago, Illinois 60661-2208, USA
+1-312-559-4100 voice • +1-312-559-4111 fax
publications@iec.org • www.iec.org

About the International Engineering Consortium

The International Engineering Consortium (IEC) is a nonprofit organization dedicated to catalyzing positive change in the information industry and its university communities. Since 1944, the Consortium has provided high-quality educational opportunities for industry professionals, academics, and students. In conjunction with the industry, the IEC has developed free, on-line, Web-based tutorials. The IEC conducts industry-university programs that have substantial impact on curricula. It also conducts research and develops publications, conferences, and technological exhibits that address major opportunities and challenges of the information age. More than 70 leading, high-technology universities are currently affiliated with the Consortium. The industry is represented through substantial corporate support and through the involvement of many thousands of executives, managers, and professionals.

Through the University Program, sponsoring organizations provide grants for full-time faculty members and their students to attend IEC Forums. The generous contributions of the following sponsoring organizations make this valuable program possible. For more information about the program, please call +1-312-559-4103 or send an e-mail to *universityprogram@iec.org*.

This past year, the University Program has provided more than 400 grants to professors to attend the educational opportunities offered at IEC Forums. Because of their participation in IEC Forums, professors created or updated 600 university courses; more than 10,000 students were impacted by these new and upgraded courses, and more than 12,000 students benefited from improvements made in university laboratories. Since its inception in 1984, the University Program has enhanced the education of more than 500,000 students worldwide.

Canadian Sponsors

European Sponsors

These leading high-technology universities participate in the University Grant Program:

The University of Arizona
Arizona State University
Auburn University
University of California at Berkeley
University of California, Davis
University of California, Santa Barbara
Carnegie Mellon University
Case Western Reserve University
Clemson University
University of Colorado at Boulder
Columbia University
Cornell University
Drexel University
École Nationale Supérieure des Télécommunications de Bretagne
École Nationale Supérieure des Télécommunications de Paris
École Supérieure d'Électricité
University of Edinburgh
University of Florida

Georgia Institute of Technology
University of Glasgow
Howard University
Illinois Institute of Technology
University of Illinois at Chicago
University of Illinois at Urbana/Champaign
Imperial College of Science, Technology and Medicine
Institut National Polytechnique de Grenoble
Instituto Tecnológico y de Estudios Superiores de Monterrey
Iowa State University
KAIST
The University of Kansas
University of Kentucky
Lehigh University
University College London
Marquette University
University of Maryland at College Park

Massachusetts Institute of Technology
University of Massachusetts
McGill University
Michigan State University
The University of Michigan
University of Minnesota
The University of Mississippi
University of Missouri-Columbia
University of Missouri-Rolla
Technische Universität München
Universidad Nacional Autónoma de México
North Carolina State University at Raleigh
Northwestern University
University of Notre Dame
The Ohio State University
Oklahoma State University
The University of Oklahoma
Oregon State University

Université d'Ottawa
The Pennsylvania State University
University of Pennsylvania
University of Pittsburgh
Polytechnic University
Purdue University
The Queen's University of Belfast
Rensselaer Polytechnic Institute
University of Southampton
University of Southern California
Stanford University
Syracuse University
University of Tennessee, Knoxville
Texas A&M University
The University of Texas at Austin
University of Toronto
VA Polytechnic Institute and State University
University of Virginia
University of Washington
University of Wisconsin-Madison

Consortium-Affiliated Universities

The International Engineering Consortium is affiliated with the following leading high-tech universities:

International Affiliated Universities

Ecole Supérieure d'Électricité

Ecole Nationale Supérieure des Télécommunications de Bretagne

Ecole Nationale Supérieure des Télécommunications de Paris

Imperial College of Science, Technology and Medicine

Institut National Polytechnique de Grenoble

Instituto Tecnológico y de Estudios Superiores de Monterrey

KAIST

McGill University

Technische Universität München

The Queen's University of Belfast

Universidad Nacional Autónoma de México

Université d'Ottawa

University of Edinburgh

University of Glasgow

University of Southampton

University College London

University of Toronto

The IEC's Media Sponsors and Partners program was established as a means for the Consortium and the leading industry publications to provide reciprocal support for the many service programs throughout the year. The IEC thanks these publications for their efforts in keeping the industry up to date on the Consortium's events, publications, and Web-based educational programs.

Media Sponsors

Publications participating as Media Sponsors provide substantial support and exposure to the Consortium's overall programs, and are intimately involved in specific events and publications as Official Publishing Sponsors.

America's Network

201 East Sandpointe Avenue, Suite 600
Santa Ana, CA 92707-8700
714-513-8400
www.americasnetwork.com

Interactive Week
THE INTERNET'S NEWSPAPER

100 Quentin Roosevelt Boulevard
Garden City, NY 11530
516-229-3700
www.interactiveweek.com

Telephony

One IBM Plaza, Suite 2300
Chicago, IL 60611
312-595-1080
www.internettelephony.com

Broadband week
NETWORKS, APPLICATIONS & CONTENT

8878 South Barrons Blvd
Highlands Ranch, CO 80129
303-470-4800
www.broadbandweek.com

LIGHTWAVE

98 Spit Brook Road
Nashua, NH 03062-5737
603-891-0123
www.light-wave.com

Wireless
WEEK

8878 South Barrons Boulevard
Highlands Ranch, CO 80129-2345
303-470-4875
www.wirelessweek.com

CED

600 South Cherry Street, Suite 400
Denver, CO 80222
303-393-7449
www.cedmagazine.com

RCR Wireless news

777 East Speer Boulevard
Denver, CO 80203-4214
303-733-2500
www.rcrnews.com

XCHANGE

3300 North Central Avenue, Suite 2500
Phoenix, AZ 85012
480-990-1101
www.x-changemag.com

Communications News
Solutions for Today's Networking Decision Makers

2500 Tamiami Trail North
Nokomis, FL 34275
941-966-9521
www.comnews.com

tele.com
strategic context for service providers

3 Park Ave, 30th Floor
New York, NY 10016
212-592-8400
www.teledotcom.com

Media Partners

Publications participating as Media Partners provide general support to the Consortium's overall programs.

BROADBAND solutions

www.broadbandsolution.com

[The Net Economy]

www.theneteconomy.com

sounding board

www.soundingboardmag.com

UPSTART

www.upstartmag.com

BUSINESS COMMUNICATIONS REVIEW

www.bcr.com

Outside Plant

www.ospmag.com

telecom

www.telecomasia.net

comm verge

www.commvergemag.com

Private & Wireless Broadband MAGAZINE

www.privatebroadband.com

Telecom BUSINESS
IDEA ■ STRATEGY ■ OPPORTUNITY / WWW.TELECOMBUSINESS.COM

www.telecombusiness.com

Global Telephony

www.globaltelephony.com

PHONE+

www.phoneplusmag.com

TELECOMMUNICATIONS MAGAZINE

www.telecommagazine.com

Table of Contents

Part I: Metro-Area Optical Networks

Part V: OSS and Network Management

Table of Contents by Author

Acronym Guide

2B1Q	two binary, one quaternary
3G	third-generation
3GPP	third-generation partnership program
4B3T	four binary, three ternary
4F/BDPR	four-fiber bidirectional dedicated protection ring
4F/BSPR	four-fiber bidirectional shared protection ring
AAA	authentication, authorization, and accounting
AAL–x	ATM adaptation layer–x
ABC	activity-based costing
ABR	available bit rate
AC	alternating current OR authentication code
ACD	automatic call distributor
ACL	access control list
ACLEP	adaptive code excited linear prediction
ACR	alternate carrier routing
ADM	add/drop multiplexer OR asymmetrical digital multiplexer
ADPCM	adaptive differential pulse code modulation
ADS	add/drop switch
ADSI	analog display services interface
ADSL	asymmetric digital subscriber line
AFE	analog front end
AIN	advanced intelligent network
ALI	automatic location identification
AM	amplitude modulation
AMI	alternate mark inversion
AMPS	advanced mobile phone service
AN	access network
ANI	automatic number identification
ANSI	American National Standards Institute
ANX	Automotive Network Exchange
AOL	America Online
AON	all-optical network
AP	access provider OR access point
APC	automatic power control
API	application programming interface
APON	ATM passive optical network
APS	automatic protection switching
ARCNET	attached resource computer network
ARI	assist request instruction
ARP	address resolution protocol
ARPANET	Advanced Research Projects Agency Network
AS	autonomous system
ASC	Accredited Standards Committee
ASCII	American Standard Code for Information Interchange
ASE	amplified spontaneous emission
ASIC	application-specific integrated circuit

ASIP	application-specific instruction processor
ASON	automatically switched optical network
ASP	application service provider
ASR	automatic service request
ASSP	application-specific standard part
ATC	automatic temperature contol
ATIS	Alliance for Telecommunications Industry Solutions
ATM	asynchronous transfer mode
ATP	analog twisted pair
ATU–C	ADSL transmission unit–CO
ATU–R	ADSL transmission unit–remote
AUI	attachment unit interface
AVI	audio video interleaved
AWG	arrayed waveguide grating OR American Wire Gauge
AYUTOS	as-yet-un-thought-of services
B-Box	breakout box
BCSM	basic call state model
BDCS	broadband digital cross-connect system
BDPR	bidirectional dedicated protection ring
BE	border element
BER	bit-error rate
BERT	bit error–rate test
BGP	border gateway protocol
BH	busy hour
BI	bit rate independent
BICC	bearer independent call control
BID	bit rate identification
BIP	bit interactive parity
B–ISDN	broadband ISDN
BLEC	building local-exchange carrier
BLES	broadband loop emulation services
BLSR	bidirectional line-switched ring
BML	business management layer
BOC	Bell operating company
BOND	back-office network development
BOSS	broadband operating system software
BPSK	binary phase shift keying
B–RAS	broadband–remote access server
BRI	basic rate interface
BSA	business services architecture
BSPR	bidirectional shared protection ring
BSS	base station system OR business support system
BTS	base transceiver station
CA	call agent
CAC	call admission control OR carrier access code OR connection admission control
CAD	computer-aided design
CAGR	compound annual growth rate
CAM	computer-aided manufacture
CAMEL	customized application of mobile enhanced logic
CAP	competitive access provider OR carrierless amplitude and phase modulation
CAT	conditional access table

CATV	cable television
C-band	conventional band
CBDS	connectionless broadband data service
CBR	constant bit rate
CBT	core-based tree
CC	control component
CCITT	Consultative Committee on International Telegraphy and Telephony
CCK	complementary code keying
CD	chromatic dispersion OR compact disc
cDCF	conventional dispersion compensation fiber
CDD	content delivery and distribution
CDDI	copper-distributed data interface
CDMA	code division multiple access
CDN	control directory number
CDPD	cellular digital packet data
CDR	call detail record OR clock and data recovery
CD–ROM	compact disc–read-only memory
CEI	comparable efficient interface
CEO	chief executive officer
CERT	computer emergency response team
CES	circuit emulation service
CESID	caller emergency service ID
CEV	controlled environment vault
CFO	chief financial officer
CFP	contention free period
CGI	common gateway interface
CHCS	composite health care system
CHN	centralized hierarchical network
CIC	circuit identification code
CIO	chief information officer
CIP	classical IP over ATM
CIR	committed information rate
CLASS	custom local-area signaling services
CLE	customer-located equipment
CLEC	competitive local-exchange carrier
CLR	circuit layout record
CM	cable modem
CM&B	customer management and billing
CMIP	common management information protocol
CMISE	common management information service element
CMOS	complementary metal oxide semiconductor
CMRS	commercial mobile radio service
CMTS	cable modem termination system
CNAM	calling name (also defined as "caller I.D. with name" and simply "caller I.D.")
CO	central office
CODEC	coder-decoder
COI	community of interest
COPS	common open policy service
CORBA	common object request broker architecture
CORE	council of registrars
CoS	class of service

COT	central office terminal
COTS	commercial off-the-shelf
COW	cellsite on wheels
CP	connection point
CPAS	cellular priority access service
CPC	calling-party category (also calling-party control OR calling-party connected)
CPE	customer-premises equipment
CPI	continual process improvement
CPL	call processing language
CPN	calling-party number
CPU	central processing unit
CRC	cyclic redundancy check OR cyclic redundancy code
CR–LDP	constraint-based routing–label distribution protocol
CRM	customer-relationship management
CRV	call reference value
CS	client signal
CS–1	capability set 1
CSA	carrier serving area
CSCW	computer supported collaborative work
CS–IWF	control signal interworking function
CSMA/CA	Carrier Sense Multiple Access/Collision Avoidance
CSP	communications service provider
CSR	customer-service representative
CSU	channel service unit
CSV	circuit-switched voice
CT–2	cordless telephony generation 2
CTI	computer telephone integration
CTIA	Cellular Telecommunications Industry Association
CTM	Center for Telecommunications Management
CWD	centralized wavelength distribution
CWIX	cable and wireless Internet exchange
DAC	digital access carrier
DACS	digital access cross-connect system
DAM	DECT authentication module
DAMA	demand assigned multiple access
DARPA	Defense Advanced Research Projects Agency
DAVIC	Digital Audio Video Council
dB	decibel(s)
dBrn	decibels above reference noise
DBS	direct broadcast satellite
DC	direct current
DCC	data communications channel
DCF	dispersion compensation fiber OR discounted cash flow
DCLEC	data competitive local-exchange carrier
DCN	data communications network
DCOM	distributed component object model
DCS	digital cross-connect system OR distributed call signaling
DCT	discrete cosine transform
DDN	defense data network
DDS	dataphone digital service
DECT	Digital European Cordless Telecommunication

demarc	demarcation point
DFB	distributed feedback
DGD	differentiated group delay
DGFF	dynamic gain flattening filter
DHCP	dynamic host configuration protocol
DiffServ	differentiated services
DIN	digital information network
DIS	distributed interactive simulation
DITF	Disaster Information Task Force
DLC	digital loop carrier
DLCI	data-link connection identifier
DLEC	data local-exchange carrier
DLR	design layout report
DM	dense mode
DMD	dispersion management device
DMS	digital multiplex system
DMT	discrete multitone
DN	distinguished name
DNS	domain name server
DOC	department of communications
DOCSIS	data over cable service interface specifications
DOD	Department of Defense
DOJ	Department of Justice
DOS	disk operating system
DOT	Department of Transportation
DPC	destination point code
DPE	distributed processing environment
DPT	dial pulse terminate
DS–x	digital signal level x
DSAA	DECT standard authentication algorithm
DSC	DECT standard cipher
DSF	dispersion-shifted fiber
DSL	digital subscriber line
DSLAM	digital subscriber line access multiplexer
DSLAS	DSL–ATM switch
DSP	digital signal processor
DSSS	direct sequence spread spectrum
DSU	data service unit OR digital service unit
DTMF	dual-tone multifrequency
DTV	digital television
DVB	digital video broadcast
DVC	dynamic virtual circuit
DVD	digital video disc
DVMRP	distance vector multicast routing protocol
DVOD	digital video on demand
DWDM	dense wavelength division multiplexing
DXC	digital cross-connect
E911	enhanced 911
EAI	enterprise application integration
EBITDA	earnings before interest, taxes, depreciation, and amortization
EC	electronic commerce

ECD	echo-cancelled full-duplex
ECRM	echo canceller resource module
EDA	electronic design automation
EDF	erbium-doped fiber OR electronic distribution frame
EDFA	erbium-doped fiber amplifier
EDGE	enhanced data rates for GSM evolution
EDI	electronic data interchange
EDSX	electronic digital signal cross-connect
EFT	electronic funds transfer
EJB	Enterprise Java Beans
ELAN	emulated local-area network
EM	element manager
EMI	electromagnetic interference
EML	element management layer
EMS	element management system OR enterprise messaging server
E–O	electrical-to-optical
EO	end office
EoA	Ethernet over ATM
EOC	embedded operations channel
EPD	early packet discard
ERP	enterprise resource planning
ESCON	enterprise system connection
ETC	establish temporary connection
ETSI	European Telecommunications Standards Institute
EU	European Union
FBG	fiber Bragg grating
FCAPS	fault, configuration, accounting, performance, and security
FCC	Federal Communications Commission
FDA	Food and Drug Administration
FDD	frequency division duplex
FDDI	fiber distributed data interface
FDF	fiber distribution frame
FDM	frequency division multiplexing
FDMA	frequency division multiple access
FDS–1	fractional DS–1
FE	extended framing
FEC	forward error correction
FEPS	facility and equipment planning system
FEXT	far-end crosstalk
FHSS	frequency hopping spread spectrum
FICON	fiber connection
FITL	fiber-in-the-loop
FM	fault management OR frequency modulation
FOC	firm order confirmation
FOT	fiber-optic terminal
FOTS	fiber-optic transmission system
FP	Fabry-Perot [laser]
FPGA	field programmable gate array
FPLMTS	future public land mobile telephone system
FR	frame relay
FRAD	frame-relay access device

FSAN	full-service access network
FSN	full-service network
FT	fixed-radio termination
FT1	fractional T1
FTP	file transfer protocol
FTTB	fiber-to-the-building
FTTC	fiber-to-the-curb
FTTCab	fiber-to-the-cabinet
FTTEx	fiber-to-the-exchange
FTTH	fiber-to-the-home
FTTN	fiber-to-the-neighborhood
FWM	four-wave mixing
FX	foreign exchange
GbE	Gigabit Ethernet (also GE)
Gbps	gigabits per second
GDIN	global disaster information network
GE	Gigabit Ethernet (also GbE)
GEO	geosynchronous Earth orbit
GETS	government emergency telecommunications service
GFF	gain flattening filter
GFR	guaranteed frame rate
GKMP	group key management protocol
GMPCS	global mobile personal communications services
GMPLS	generalized MPLS
GNP	gross national product
GOCC	ground operations control center
GPIB	general-purpose interface bus
GPRS	general packet radio service
GPS	global positioning system
GR	generic requirement
GSA	Global Mobile Suppliers Association
GSM	Global System for Mobile Communications
GSMP	generic switch management protocol
GTT	global title translation
GUI	graphical user interface
GVD	group velocity dispersion
HCC	host call control
HD	home domain
HDLC	high-level data-link control
HDML	hand-held device markup language
HDSL	high–bit rate DSL
HDT	host digital terminal
HDTV	high-definition television
HDVMRP	hierarchical distance vector multicast routing protocol
HEC	head error control OR header error check
HEPA	high-efficiency particulate arresting
HFC	hybrid fiber/coax
HLR	home location register
HN	home network
HOM	high-order mode
HomePNA	Home Phoneline Networking Alliance (also HomePNA2)

HomeRF	Home Radio Frequency Working Group
HPC	high probability of completion
HPO	high-performance option
HQ	headquarters
HSCSD	high-speed circuit-switched data
HSD	high-speed data
HSIA	high-speed Internet access
HTML	hypertext markup language
HTTP	hypertext transfer protocol
HW	hardware
IAD	integrated access device
IAM	initial address message
IAS	integrated access service or Internet access server
IAST	integrated access, switching, and transport
IBC	integrated broadband communications
IC	integrated circuit
ICD	Internet call diversion
ICL	intercell linking
ICMP	Internet control message protocol
ICP	integrated communications provider
ICW	Internet call waiting
IDE	integrated development environment
IDF	intermediate distribution frame
IDL	interface definition language
IDLC	integrated digital loop carrier
IDSL	integrated services digital network DSL
IEC	International Engineering Consortium OR International Electrotechnical Commission
IEEE	Institute of Electrical and Electronics Engineers
I-ERP	integrated enterprise resource planning
IETF	Internet Engineering Task Force
IFITL	integrated [services over] fiber-in-the-loop
IFMA	International Facility Managers Association
IFMP	Ipsilon flow management protocol
IGMP	Internet group management protocol
IGP	interior gateway protocol
IGRP	interior gateway routing protocol
IGSP	independent gateway service provider
IHL	Internet header length
IIS	Internet Information Server
ILA	in-line amplifier
ILEC	incumbent local-exchange carrier
ILMI	interim link management interface
IMA	inverse multiplexing over ATM
IMRP	Internet multicast routing protocol
IMT	International Mobile Telecommunications OR intermachine trunk
IMTC	International Multimedia Teleconferencing Consortium
IN	intelligent network
INAP AU	INAP adaptation unit
INAP	intelligent network application part
INE	intelligent network element
INM	integrated network management

INMD	in-service, nonintrusive measurement device
INT	[point-to-point] interrupt
InterNIC	Internet Network Information Center
IntServ	integrated services
IOF	interoffice facility
IP	Internet protocol
IPBX	Internet protocol private branch exchange
IPDC	Internet protocol device control
IPDR	Internet protocol data record
IPe	intelligent peripheral
IPG	intelligent premises gateway
IPO	initial public offering OR Internet protocol over optical
IPoA	Internet protocol over ATM
IPQoS	Internet protocol quality of service
IPSec	Internet protocol security
IPTel	IP telephony
IPv6	Internet protocol version 6
IP–VPN	Internet protocol–virtual private network
IPX	Internet package exchange
IR	infrared
IS	information service
IS–IS	intermediate system to intermediate system
ISAPI	Internet server application programmer interface
ISC	integrated service carrier
ISDF	integrated service development framework
ISDN	integrated services digital network
ISDN–BA	ISDN basic access
ISDN–PRA	ISDN primary rate access
ISM	industrial scientific medical OR industrial, scientific, and medical
ISO	International Standards Organization
ISOS	integrated software on silicon
ISP	Internet service provider
ISUP	ISDN user part
IT	information technology OR Internet telephony
ITSP	Internet telephony service provider
ITU	International Telecommunications Union
ITU–T	ITU–Telecommunication Standardization Sector
IVR	interactive voice response
IVRU	Interactive voice response unit
IWF	interworking function
IXC	interexchange carrier
JAIN	Java APIs for integrated networks
JCAT	Java coordination and transactions
JCC	JAIN call control
JDMK	Java dynamic management kit
JMAPI	Java management application programming interface
JSCE	JAIN service creation environment
JSLEE	JAIN service logic execution environment
JVM	Java virtual machine
kbps	kilobits per second
kHz	kilohertz

L2F	Layer-2 forwarding
L2TP	Layer-2 tunneling protocol
LAC	L2TP access concentrator
LAN	local-area network
LANE	local-area network emulation
LATA	local access and transport area
L-band	long band
LCD	liquid crystal display
LCP	link control protocol
LCUG	Local Competition User Group
LD	laser diode OR long distance
LDAP	lightweight directory access protocol
LDP	label distribution protocol
LDS	local digital service
LE	local exchange OR line equipment
LEAF®	large-effective-area fiber
LEC	local-exchange carrier
LED	light-emitting diode
LEO	low Earth orbit
LEOS	low Earth-orbiting satellite
LES	loop emulation service
LIDB	line information database
LLC	logical link control
LMDS	local multipoint distribution system
LMP	link management protocol
LMS	loop management system OR loop monitoring system
LNNI	LANE network-to-network interface
LNP	local number portability
LNS	L2TP network server
LOL	loss of lock
LOS	line of sight OR loss of signal
LPF	low-pass filter
LRN	local routing number
LSA	link state advertisement
LSMS	local service management system
LSO	local service office
LSP	label-switched path
LSR	line service request OR leaf setup request OR label-switched router
LT	line terminator OR logical terminal
LTE	lite terminating equipment
LUNI	LANE user network interface
LX	local exchange
MAC	media access control
MADU	multiwave add/drop unit
MAN	metropolitan-area network
MAP	mobile applications part
MAS	multiple application selection
MBAC	measurement-based admission control
MBGP	multicast border gateway protocol
MBone	multicast backbone
Mbps	megabits per second

MCC	mobile country code
MCU	multipoint control unit
MDF	main distribution frame
MDSL	multiple DSL
MDU	multiple dwelling unit
MEGACO	media gateway control
MEMS	micro-electromechanical system
MExE	mobile execution environment
MFJ	modified final judgment
MG	media gateway
MGC	media gateway controller
MGCP	media gateway control protocol
MHz	megahertz
MIB	management information base
MIME	multipurpose Internet mail extensions
MIPS	millions of instructions per second
MIS	management information system
MITI	Ministry of International Trade and Industry (in Japan)
MLT	mechanized loop testing
MM	Mobility Management
MMDS	microwave multipoint distribution system
MMPP	Markov-Modulated Poisson Process
MNC	mobile network code
MOM	message-oriented middleware
MON	metropolitan optical network
MOP	method of procedure
MOS	mean opinion score
MOSFP	multicast open shortest path first
MOU	minutes of use
MP x	MPEG Layer-x
MPEG	Moving Pictures Experts Group
MPI	message passing interface
MPLambdaS	multiprotocol lambda switching
MPLS	multiprotocol label switching
MPOA	multiprotocol over ATM
MPOE	multiple point of entry
MPOP	metropolitan point of presence
MPP	massively parallel processor
MRSP	mobile radio service provider
MSC	mobile switching center
MSF	Multiservice Switch Forum
MSIN	mobile station identification number
MSNAP	multiple services network access point
MSO	multiple system operator
MSS	multiple-services switching system
MSSP	mobile satellite service provider
MTA	message transfer agent
MTP	message transfer part
MTTR	mean time to repair
MTU	multiple tenant unit
MVL	multiple virtual line

MZI	Mach-Zender Interferometer
NAFTA	North America Free Trade Agreement
NANC	North American Numbering Council
NANP	North American Numbering Plan
NAP	network access point
NARUC	National Association of Regulatory Utility Commissioners
NAS	network access server
NASA	National Aeronautics and Space Administration
NAT	network address translation
NATA	North American Telecommunications Association
NBN	node-based network
NCF	National Communications Forum
NCP	network control protocol
NCS	national communications system
NE	network element
NEBS	network-equipment building standards
NEL	network element layer
NEXT	near-end crosstalk
NGCN	next-generation converged network
NGDLC	next-generation digital loop carrier
NGF	next-generation fiber
NGN	next-generation network
NHRP	next-hop resolution protocol
NI	network interface
NIC	network interface card
NID	network interface device
NIU	network interface unit
NML	network management layer
NMS	network management system
NNI	network-to-network interface
NOC	network operations center
NOMAD	national ownership, mobile access, and disaster communications
NPAC	Number Portability Administration Center
NP–REQ	number-portable request query
NPV	net present value
NRC	Network Reliability Council
NRIC	Network Reliability and Interoperability Council
NRRI	National Regulatory Research Institute (The Ohio State University)
NRSC	Network Reliability Steering Committee
NRZ	non–return to zero
NS/EP	national security and emergency preparedness
NSAP	network service access point
NSAPI	Netscape server application programming interface
NSDB	network and services database
NSP	network service provider
NSTAC	National Security Telecommunications Advisory Committee
NT	network termination
NTN	network terminal number
NTSC	National Television Standards Committee
NVP	network voice protocol
NZ–DSF	nonzero dispersion-shifted fiber

OADM	optical add/drop multiplexer
OAM	operations, administration, and maintenance
OAM&P	operations, administration, maintenance, and provisioning
OBF	Ordering and Billing Forum
OBLSR	optical bidirectional line-switched ring
OC–x	optical carrier–level x
OCBT	ordered core-based protocol
OCD	optical concentration device
OCh	optical channel
OCR	optical character recognition
OCU	office channel unit
OCX	open compact exchange
ODBC	open database connectivity
ODSI	optical domain services interface
OE	optical-to-electrical
O–E	optical-to-electrical
O–EC	optical–electrical converter
OECD	Organization for Economic Cooperation and Development
OEM	original equipment manufacturer
O–E–O	optical-to-electrical-to-optical
OEXC	opto-electrical cross-connect
OFC	Optical Fiber Conference
OFDM	optical frequency division multiplexing
OIF	Optical Internetworking Forum
OLA	optical line amplifier
OLAP	on-line analytical processing
OLI	optical link interface
OLT	optical line termination
OLTP	on-line transaction processing
OMC	Operations and Maintenance Center
OMG	Object Management Group
OMS	optical multiplex section
OMS SW	optical multiplex section switch
OMSSPRING	optical multiplex section shared protection ring
ONA	open network architecture
ONE	optical network element
ONI	optical network interface
ONT	optical network termination
ONTAS	optical network test access system
ONU	optical network unit
OP	optical path
OPS	operator provisioning station
OPTIS	overlapped PAM transmission with interlocking spectra
OPXC	optical path cross-connect
ORB	object request broker
ORT	operational readiness test
OS	operating system
OSC	optical supervisory panel
OSD	on-screen display
OSGI	open services gateway initiative
OSI	open systems interconnection

OSMINE	operations systems modification of intelligent network elements
OSN	optical-service network
OSNR	optical signal-to-noise ratio
OSP	outside plant
OSPF	open shortest path first
OSS	operations support system
OTM	optical terminal multiplexer
OTN	optical transport network
OUI	optical user interface
OWSR	optical wavelength switching router
OXC	optical cross-connect
PABX	private automatic branch exchange
PACA	priority access channel assignment
PACS	picture archiving communications system
PAL	phase alternate line
PAM	pulse amplitude modulation
PANS	pretty amazing new services
PBN	point-to-point–based network
PBX	private branch exchange
PC	personal computer
PCI	peripheral component interconnect
PCM	pulse code modulation
PCN	personal communications network
PCR	peak cell rate
PCS	personal communications service
PDA	personal digital assistant
PDH	plesiochronous digital hierarchy
PDN	public data network
PDP	policy decision point
PDU	protocol data unit
PFD	phase-frequency detector
PHB	per-hop behavior
PHY	physical layer
PIC	predesignated interexchange carrier OR primary interexchange carrier
PICS	plug-in inventory control system
PIM	protocol-independent multicast
PIN	personal identification number
PINT	PSTN and Internet Networking [IETF working group]
PINTG	PINT gateway
PKI	public key infrastructure
PLC	planar lightwave circuit OR product life cycle
PLCP	physical layer convergence protocol
PLL	phase locked loop
PLMN	public land mobile network
PLOA	protocol layers over ATM
PM	performance monitoring
PMD	physical-medium dependent OR polarization mode dispersion
PMP	point-to-multipoint
PNNI	private network-to-network interface
PnP	plug and play
PO	purchase order

PODP	public office dialing plan
POET	partially overlapped echo-cancelled transmission
POF	plastic optic fiber
POH	path overhead
PON	passive optical network
POP	point of presence
POS	packet over SONET OR point of service
PosReq	position request
POT	point of termination
POTS	plain old telephone service
PP	point-to-point
PPD	partial packet discard
PPP	point-to-point protocol
PPPoA	point-to-point protocol over ATM
PPPoE	point-to-point protocol over Ethernet
PPTP	point-to-point tunneling protocol
PP–WDM	point-to-point–wavelength division multiplexing
PRI	primary rate interface
PSAP	public safety answering point
PSC	Public Service Commission
PSD	power spectral density
PSDN	public switched data network
PSID	private system identifier
PSN	public switched network
PSPDN	packet switched public data network
PSTN	public switched telephone network
PTE	path terminating equipment
PTN	personal telecommunications number service
PTP	point-to-point
PTT	Post Telephone and Telegraph Administration
PUC	public utility commission
PVC	permanent virtual circuit
PVM	parallel virtual machine
PVN	private virtual network
PWS	planning workstation
QAM	quadrature amplitude modulation
QOE	quality of experience
QoS	quality of service
QPSK	quaternary phase shift keying
RADIUS	remote authentication dial-in user service
RADSL	rate-adaptive DSL
RAM	remote access multiplexer
RAS	remote access server
RBOC	regional Bell operating company
RCP	remote call procedure
RCU	remote control unit
RDC	regional distribution center
RDSLAM	remote DSLAM
RF	radio frequency
RFC	request for comment
RFI	request for information

RFP	request for proposal
RFQ	request for quotation
RGU	revenue-generating unit
RHC	regional holding company
RIAC	remote instrumentation and control
RIP	routing information protocol
RISC	reduced instruction set computing
RJ	registered jack
RM	resource management
ROBO	remote office/branch office
ROI	return on investment
ROW	right of way
RPC	remote procedure call
RPF	reverse path forwarding
RSVP	resource reservation protocol
RSVP–TE	resource reservation protocol–traffic engineering
RT	remote terminal
RTCP	real-time conferencing protocol
RTOS	real-time operating system
RTP	real-time transport protocol
RxTx	receiver/transmitter
RZ	return to zero
SAM	service access multiplexer
SAN	storage-area network
SAP	service access point
SAR	segmentation and reassembly
SBS	stimulated Brillouin scattering
SCAN	switched-circuit automatic network
SCCP	signaling connection control part
SCE	service creation environment
SCF	service control function
SCM	service combination manager OR station class mark OR subscriber carrier mark
SCN	service circuit node
SCP	service control point
SCR	sustainable cell rate
SCSP	server cache synchronization protocol
SCTP	simple computer telephony protocol OR simple control transport protocol
SD	selective discard
SDA	separate data affiliate
SDB	service design bureau
SDC	service design center
SDF	service data function
SDH	synchronous digital hierarchy
SDN	software-defined network
SDP	session description protocol
SDRP	source demand routing protocol
SDSL	symmetric DSL
SDV	switched digital video
SEC	Securities and Exchange Commission
SEE	service execution environment
SEP	signaling end point

SET	secure electronic transaction
SFA	sales force automation
SFD	start frame delimiter
SFGF	supplier-funded generic element
SG	signaling gateway
SGCP	simple gateway control protocol
SGSN	serving GPRS support node
SHLR	standalone home location register
SHV	shareholder value
SIBB	service-independent building blocks
SIC	service initiation charge
SICL	standard interface control library
SID	silence indicator description
SIM	subscriber identity module OR service interaction manager
SIP CPL	SIP call processing language
SIP	session initiation protocol
SKU	stock-keeping unit
SL	service logic
SLA	service-level agreement
SLC	subscriber line carrier
SLEE	service logic execution environment
SLIC	subscriber line interface circuit
SM	sparse mode
SMC	service management center
SMDI	simplified message desk interface
SMDS	switched (multiple-) Megabit data service
SME	small-to-medium enterprise
SMF	single-mode fiber
SML	service management layer
SMP	service management point
SMS	service management system OR short message service
SMSC	short messaging service center
SMTP	simple mail transfer protocol
SN	service node
SNA	service node architecture OR systems network architecture
SNAP	subnetwork access protocol
SNMP	simple network management protocol
SNR	signal-to-noise ratio
SOA	service order activation
SOAC	service order analysis and control
SOCC	satellite operations control center
SOHO	small office/home office
SON	service order number
SONET	synchronous optical network
SOP	service order processor
SP	signaling point OR service provider
SPC	stored program control
SPE	synchronous payload envelope
SPF	shortest path first
SPIRITS	Service in the PSTN/IN Requesting Internet Service [working group]
SPIRITSG	SPIRITS gateway

SPM	self-phase modulation OR subscriber private meter
SQL	structured query language
SRF	special resource function
SRP	space reusing protocol
SRS	stimulated Raman scattering
SS7	signaling system 7
SSE	service subscriber element
SSG	service selection gateway
SSL	secure sockets layer
SSM	service and sales management
SSMF	standard single-mode fiber
SSP	service switching point
STE	section terminating equipment
STM	synchronous transfer mode
STN	service transport node
STP	shielded twisted pair OR signal transfer point OR spanning tree protocol
STS	synchronous transport signal
SVC	switched virtual circuit
SW	software
SWAN	storage wide-area network
SWAP	shared wireless access protocol
SWOT	strengths, weaknesses, opportunities, and threats
SYN	IN synchronous transmission
TALI	transport adapter layer interface
TAT	terminating access trigger OR termination attempt trigger
TBD	to be determined
Tbps	terabits per second
TC	tandem connect
TCAP	transactional capabilities application part
TCB	transfer control block
TCIF	Telecommunications Industry Forum
TCM	time compression multiplexing
TCO	total cost of ownership
TCP	transmission control protocol
TCP/IP	transmission control protocol/Internet protocol
TC–PAM	trellis coded–pulse amplitude modulation
TDD	time division duplex
TDM	time division multiplex
TDMA	time division multiple access
TDMDSL	time division multiplex digital subscriber line
TDOA	time difference of arrival
TDR	time domain reflectometer OR transaction detail record
TE	traffic engineering
TEAM	transport element activation manager
TEM	telecommunications equipment manufacturer
TIA	Telecommunications Industry Association
TIMS	transmission impairment measurement set
TINA	Telecommunications Information Networking Architecture
TINA-C	Telecommunications Information Networking Architecture Consortium
TIPHON	Telecommunications and Internet Protocol Harmonization over Networks
TIWF	trunk interworking function

TL1	transaction language 1
TLS	transparent LAN service OR transport-layer security
TLV	tag length value
TMF	TeleManagement Forum
TMN	telecommunications management network
TMO	trans-metro optical
TN	telephone number
TNO	telecommunications network operator
TO&E	table of organization and equipment
ToS	type of service
TP	twisted pair
TPM	transaction processing monitor
TPS–TC	transmission control specific–transmission convergence
TR	tip and ring
TRA	technology readiness assessment
TSB	telecommunication system bulletin
TSI	time slot interchange
TSP	telecommunications service provider
TTC	Telecommunications Technology Committee
TTCP	test TCP
TTL	transistor-transistor logic
TTS	TIRKS® table system
UADSL	universal ADSL
UAK	user-authentication key
UAWG	Universal ADSL Working Group
UBR	unspecified bit rate
UCS	uniform communication standard
UDP	user datagram protocol
UM	unified messaging
UML	unified modeling language
UMTS	Universal Mobile Telecommunications System
UNE	unbundled network element
UNH–IOL	University of New Hampshire–Interoperability Laboratory
UNI	user network interface
UPC	usage parameter control
UPI	user personal identification
UPSR	unidirectional path-switched ring
URI	uniform resource identifier
URL	uniform resource locator
USB	universal serial bus
USTA	United States Telephone Association
UTOPIA	Universal Test and Operations Interface for ATM
UTS	universal telephone service
VAD	voice activity detection
VAN	value-added network
VASP	value-added service provider
VBNS	very–high-speed backbone network service
VBR	variable bit rate
VBRnrt	variable bit rate non–real-time
VBRrt	variable bit rate real time
VC	virtual circuit

VCC	virtual channel connection
VCI	virtual channel identifier
VCLEC	voice CLEC
VCO	voltage-controlled oscillator
VCR	videocassette recorder
VD	visited domain
VDM	value delivery model
VDSL	very-high–data rate DSL
VeDSL	voice-enabled DSL
VHS	video home system
VITA	virtual integrated transport and access
VLAN	virtual local-area network
VLR	visitor location register
VLSI	very–large-scale integrated
VM	virtual machine
VoADSL	voice over ADSL
VoATM	voice over ATM
VOD	video on demand
VoDSL	voice over DSL
VoFR	voice over frame relay
VoIP	voice over IP
VP	virtual path
VPDN	virtual private dial network
VPI	virtual path identifier
VPN	virtual private network
VPR	virtual path ring
VPRN	virtual private routed network
VRE	Virtual Radiology Environment
VRU	voice response unit
VSAT	very-small–aperture terminal
VSM	virtual services management
VSN	virtual service network
VT	virtual tributary
VToA	voice traffic over ATM
VXML	voice extensible markup language
W3C	World Wide Web Consortium
WAN	wide-area network
WAP	wireless application protocol
WB DCS	wideband DCS
WCDMA	wideband CDMA
WCT	wavelength converting transponder
WDCS	wideband digital cross-connect
WDM	wavelength division multiplexing
WECA	wireless Ethernet compatibility alliance
WEP	wired equivalent privacy
WFA	work and force administration
WFQ	weighted fair queuing
WIN	wireless intelligent network
WLAN	wireless local access network
WLL	wireless local loop
WML	wireless markup language

WS	work station
WTA	wireless telephony application
WVPN	wireless VPN
WWW	World Wide Web
XC	cross-connect
XD	extended distance
xDSL	digital subscriber line
XML	extensible markup language
XNS	Xerox network system
XPM	cross-phase modulation
XPS	cross-point switch
XT	crosstalk
XTP	express transport protocol
Y2K	Year 2000

Requirements and Solutions for Reconfigurable Metro WDM Networks

E. Almström, P. Evaldsson, and C. P. Larsen
Ericsson

S. Hubendick, S. Larsson, and C. Wickman
Formerly with Telia, currently with Wavium

Background

The successful introduction of wavelength division multiplexing (WDM) into the long-distance networks has boosted the expectation for the even larger metropolitan market. When comparing metro and long-haul network requirements, a few differences can be identified: The fiber distance and capacity requirements are relaxed, but service transparency and cost efficiency are more important in the metro network [1, 2]. The deregulation of the telecom market has increased the requirements of fast provisioning of bandwidth within and between the metro networks. Moreover, new networks and service providers can quickly get large coverage within the limited metropolitan area.

The switched optical metro network has been forecast to be a multibillion-dollar market within the next few years. One of the reasons is that the introduction of new dynamic broadband applications, usually encapsulated by Internet protocol (IP), will evolve first in the metropolitan area. The questions to be answered, then, are the following: Can the WDM layer offer the IP layer more than only capacity increase? If so, how should the two layers interwork? This paper describes the functionality and requirements of the metro network, and it reports the results from a metro IP/WDM testbed in the Stockholm area.

Compared to the access network, which is characterized by, mainly, a hub traffic pattern with its large variety of services and formats, the core network of the metropolitan area has higher requirements on accommodating varying traffic patterns and capacity. Still, the transparency requirements on the metro core are higher than on the long-distance network. A common service is interconnection of enterprise networks—e.g., local-area network (LAN) extensions, which are characterized by demands for high reliability and security. Other services such as remote storage and remote servers will generate a massive increase in traffic in the coming years. Common for all these services is that the growth relies on a relatively low-cost and service-transparent network.

Selective and simple protection—if any—is often preferred in the access network, while shared bulk protection or even restoration is often used in the core. In some cases the network protection is limited to provide only network redundancy through diversity. Optimization for the transport resources of a limited number of wavelengths and nodes in the metro core network are justified by the fact that a minor reconfiguration will have a large impact on network performance [3]. At the same time, the cost of excess bandwidth to support the network flexibility can be kept low because of short distances between limited numbers of nodes (compared to long distances, which would require amplifiers, better transmitters, etc.). Reconfigurability demands faster provisioning and increased manageability of the network. Applications and functions, which can be supported by a reconfigurable metro network are, for example, ring-interconnect, bandwidth-brokers, optical multiplex section shared protection ring (OMSSPRING)/optical bidirectional line-switched ring (OBLSR), preemption of low-priority channels, fast provisioning of virtual private networks (VPNs), multicast, and traffic engineering [4].

The requirements can be summarized as follows:

- A network that can handle mainly IP services cost-efficiently, but that can also transmit and support other services and formats, such as enterprise systems connection (ESCON), fiber channel, pure bandwidth, and synchronous optical network (SONET)/synchronous digital hierarchy (SDH)–framed services
- A technology that can handle the capacity increase, including functionality to handle the dynamic traffic behaviors and the requirement to establish bandwidth connections quickly

Technology Solutions

The architecture and topology of a metro network depends on the services offered by the network provider. It is important to find the best combination of technologies to offer these services. The solution of choice will be the one that offers these services at the lowest cost, not only in terms of hardware but also concerning operation and maintenance. In the following text, considerations concerning optical and data network elements (NEs), interfaces, and functions are examined for an IP–centric metropolitan network.

Optical Network
A high degree of reconfigurability and transparency are required in the metro core network. These features can typically be traded against each other, and for today's networks no viable technology exists that fully combines high reconfigurability/flexibility with high transparency [5], even though they are the main goals eventually (see *Figure 1*). However, two very different technology options, or perhaps rather "design philosophies," exist that may complement each other rather than compete, namely an optical cross-connect (OXC)–based network and a broadcast-and-select network. Before further explanation it should be mentioned that promising complementary technologies to these categories exist. There are coarse WDM (direct modulated, uncooled, and low wavelength accuracy) and time division multiplex (TDM) integration into WDM NEs, enabling WDM systems to be cost-efficient at lower speeds and shorter distances. A variety of hybrid designs can, of course, be envisioned.

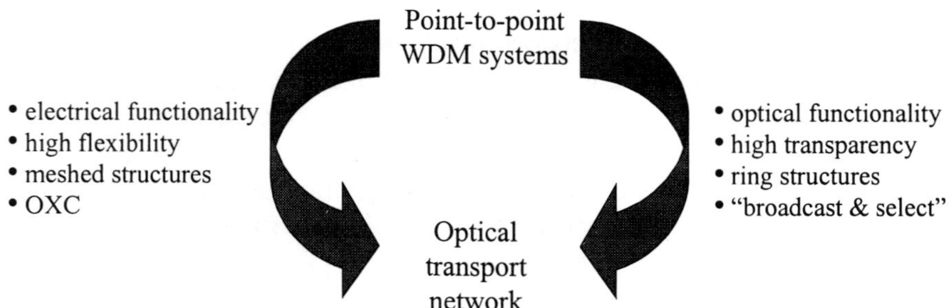

Figure 1: *Two Paths toward the Ultimate Goal of High Transparency and High Flexibility in the Optical Transport Network (OTN)*

A broadcast-and-select–type network is one in which there is no wavelength selection within the network and where information from one node is accessible at all other nodes. The routing elements are typically passive splitters/combiners whereby the optical multiplex section (OMS) is handled (as opposed to the individual channels) (see *Figure 2*). Channel selection will not be before the receiver. One advantage of such an approach is a very high degree of transparency. In addition, it is static from a transmission point of view, but it still offers dynamic connectivity. The only practical technological implementation is using all-optical elements, which is very cost-effective for high-speed signals. Such all-optical, broadcast-and-select networks with inherent resilience functionality have been demonstrated [6]. One drawback, however, is that, because there is no electrical signal, there is no regeneration of the signal beyond simple amplification, causing accumulation of various impairments of the analogue optical signal.

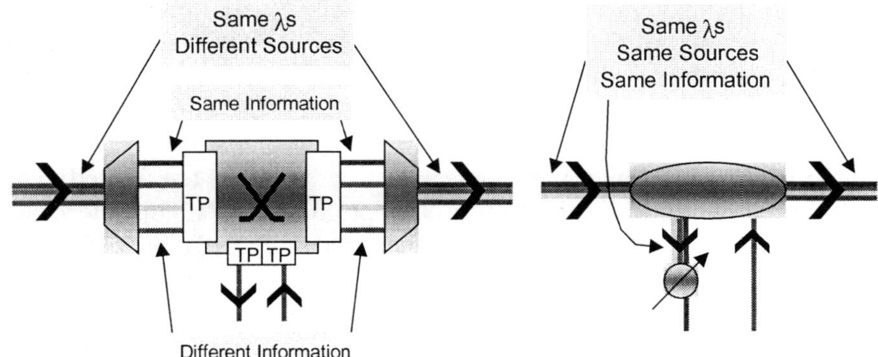

Figure 2: *An Optical Add/Drop Multiplexer (OADM) in an Opaque Network and an All-Optical, Broadcast-and-Select Network*
The figure shows that an "OXC–based" OADM allows wavelength reuse, while a broadcast-and-select OADM is fully transparent. Note that the OXC with either an optical or an electrical switchcore is interfaced by transponders. The different technologies have both been successfully tested in field trials in the Stockholm metropolitan area [7, 8].

In an opaque OXC network all routing is done at wavelength level, and due to accumulation of impairments and costs at electrical termination, there is often electrical processing. It should be noted that the term "OXC" is used as a rather broad term in this paper: It has optical interfaces (transponders) to the point-to-point connections, but the switching can be done electrically or optically. Optical switch cores will typically use low-cost optics at 1300 nm, while electrical switch cores—despite being less transparent—will typically process whole channels at native bit rates. In *Table 1* the broadcast-and-select solution is compared with an OXC–based network, while *Figure 2* show schematics of OADMs (in this case, equal to a two-port OXC) representing the two technologies. As indicated, many factors must be considered when choosing between an OXC–based network and a broadcast-and-select network, but in general, if the node density is high, if the transparency requirements are moderate, and if the traffic pattern is very mesh and dynamic, the OXC solution is usually preferred.

Feature	Broadcast and Select	OXC
Reconfigurability	*Lower*	*High*
Survivability	*Self-healing rings*	*Restoration*
Transparency	*High*	*Lower*
Provisioning	*Intrinsically very fast*	*Potentially fast*
Wavelength efficiency	*Low*	*High*
Routing granularity	*OMS*	*Wavelength*
Architecture	*Tree, ring*	*Meshed*

Table 1: *Comparison between Broadcast-and-Select and OXC–Based Networks*

When comparing the cost of an OXC comprising optical switches and egress transponders (used for channel equalization, wavelength conversion, channel monitoring and, optionally, for the regeneration of the signals) with an electrical switch interfaced by wavelength selective transceivers, it can be shown that as long as the line rate is moderate (fewer than about 3 Gbps), the electrical switch alternative will be the more cost-effective solution. When the bit rate increases, either the cost or the building practice will limit the benefits of the electrical switch solution. If the transponder is removed from the optical switch alternative, the cost will go down, but then obstacles such as unbalanced channels and channel performance and identification must be considered [9].

Data Network

SONET/SDH is traditionally used in transparent networks, and in LANs the preferred choice has been Ethernet. Metro networks are in an intermediate position between these network parts, so both protocols should be considered as possible to use. It is important to have a clear view of the requirements of the metro network when making this choice. It is also important to see in what ways WDM can relieve demands put on higher-layer protocols by implementing the functionality directly in the physical layer.

SONET/SDH is commonly used in router interconnection where packets are framed in a high-level data-link control (HDLC) frame according to the point-to-point protocol (PPP). This interface is usually called packet over SONET (POS) [10]. POS uses either concatenated SONET/SDH frames or channelized interfaces. Both realizations provide monitoring, alarm, and protection functionality of standard SONET/SDH, but the channelized interfaces are compatible with legacy SONET/SDH NEs. SONET/SDH provides low overhead and good manageability. However, the hierarchy of SONET/SDH was developed for circuit-switched telephone networks that are not well adapted for the highly variable packet size of today's IP traffic. Instead, the statistical multiplexing in the IP/Ethernet packet-switched layer, together with the increased bandwidth offered by the WDM layer, is more adapted to the IP services. An IP over SONET metro network can be realized either with IP routers directly interconnected with these POS interfaces or by using routers at the edge of a SONET. The result is either a Layer-3 (packet-switched) or Layer-1 (circuit-switched) network with fundamentally different properties. It is clear that the equipment costs for the SONET case are lower, and the restoration speed is faster. However, the flexibility of a packet-based transport network is lost.

An alternative approach would be to extend the Ethernet technology from the LAN into the metropolitan area as not only a framing technology for running IP over fiber, but as a Layer-2 switching protocol. It does provide service differentiation through the eight priority classes in Ethernet (IEEE 802.1P). It also benefits from the virtual LAN standard (VLAN, IEEE 802.1Q) to provide secure tunnels in the network. This has some implications due to the plug-and-play features inherent in Ethernet switches, which means that they adapt dynamically to topological changes of the network. To learn the topology of the network switches, one must broadcast to all ports except to the one where the packet arrived. This implies that if there are closed loops in the network, so-called broadcast storms could evolve. To avoid this, the switches use the spanning tree protocol (STP, IEEE 801.2D) to form a loop-free tree structure in the network where unused links are disabled. The disabling of links and the formation of a tree topology gives a poor utilization of network resources. However, the STP provides a way of self-healing networks in Layer 2, but the recovery time is long—from a few seconds to several minutes. If not, the rapid STP [11] is implemented, which then decreases the time to around one second.

Today the highest line speed standardized for Ethernet is Gigabit Ethernet (GbE). Until the 10 GbE (IEEE 802.3ae) standard is available, the utilization of the fiber is lower compared to SONET/SDH (optical carrier [OC]–192/synchronous transfer mode [STM]–64). There are, however, some available products that multiplex up to eight GbE channels on one wavelength to increase utilization without following Ethernet standardization. The functionality to handle a bundle of Ethernet links as one exists through link aggregation (IEEE 802.3ad).

Control Plane

To automate the optical path setup and to introduce more efficient protection schemes, an IP–based control plane should be considered for the optical layer. In this perspective, it is the intent to use as much as possible of the existing IP–based signaling and routing protocols, as is also the case for multiprotocol label switching (MPLS). However, the circuit-switching characteristics of the optical network are clearly introducing some major differences. For instance, once a circuit has been established, the deployed resources are not available to other traffic. For the label-switched path (LSP) no resources need to be allocated, except for storing the label mappings, until the traffic is flowing.

Figure 3 illustrates the four fundamental parts needed to perform automated path setup in an OXC network. First, the OXCs must establish adjacency, meaning that they must learn how their ports are interconnected to their neighbors. This can either be manually configured or automated through a messaging protocol. The result of this procedure is a table in each OXC containing a map of port interconnections. Second, there is the global distribution of the network topology/state information. For this the routing protocols—i.e., open shortest path first (OSPF) and intermediate system–to–intermediate system (IS–IS)—must be extended to support the distribution of optical layer–specific information. New tag length values (TLVs) are being defined [12]. Third, the path calculation must support constraint-based routing, meaning that it must go beyond simply selecting the "shortest path first." For example, it must be able to calculate disjoint paths for protection and to consider a longer path, with available resources. Initially, the path calculation does not run on every node but is implemented in a centralized location. Fourth, the signaling part is used to set up, maintain, modify, and tear down optical channels. The signaling protocol must support the concept of constraint-based paths. For this reason constraint-based routing–label distribution protocol (CR–LDP) and resource reservation protocol–traffic engineering (RSVP–TE) extensions defined for MPLS will be modified to take additional optical network constraints into account.

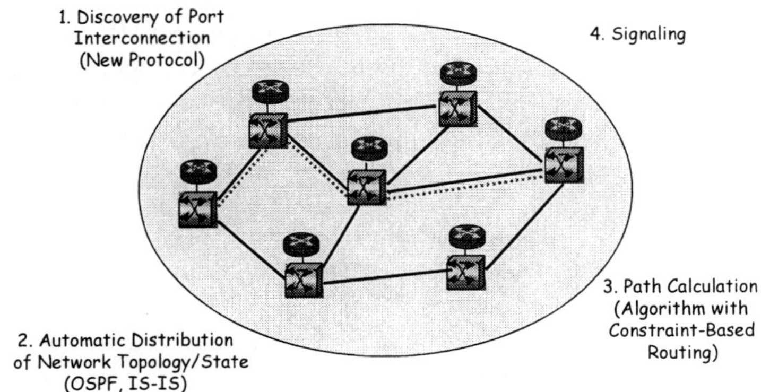

Figure 3: Four Main Functions to Enable Automated Provisioning

The Winchester Testbed

A field trial has been established in the Telia network in Stockholm with a mix of commercially available products and technologies developed within the project. The Winchester network is a three-layer network (physical, link, and network layers), with the focus on the network control plane.

Infrastructure

This test network has four nodes, which are connected by a fiber pair in a 70 km metro ring. Each transmission section is configured as WDM point-to-point links. The terminal WDM equipment is an eight-channel system with 200 GHz channel spacing. This system has a preamplifier and, optionally, a booster amplifier, depending on the losses of the fiber link. Even though the fiber lengths are moderate

(10–20 km) local office connectors and cabling cause high power losses, and usually both booster and preamplifiers are used. In each node an OXC connects the optical signals between the WDM equipment and the router switches[1], which have two interfaces. Thus, a large degree of flexibility in logical network configuration and protection is possible. In the experimental network four wavelengths are dedicated for IP traffic on GbE, one mainly used as a control channel, two used as a demonstrator network to the ACTS DEMON[2] project, and one to carry digital video. The physical fiber structure is a ring topology, but arbitrary connections can be set up in the network with the use of the OXCs (see *Figure 4*).

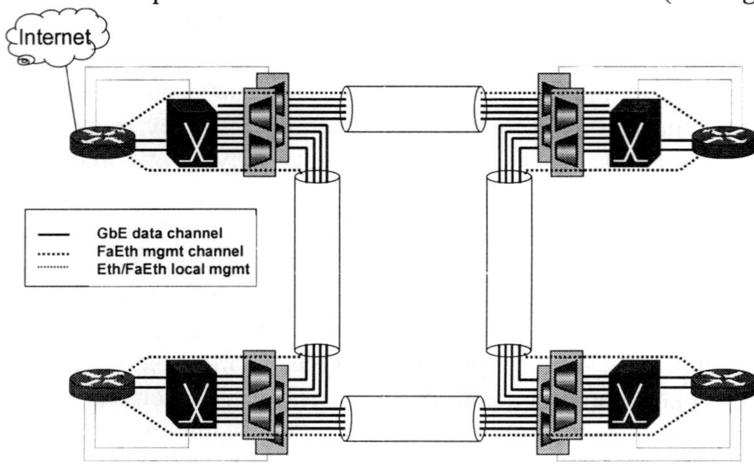

Figure 4: *Physical Architecture, Including WDM Terminals, OXCs, Routers, and Router Switches*

Data Connectivity in the Winchester Testbed
The data network in the testbed is based on router switches, and it is connected to the reconfigurable WDM network via GbE interfaces. The router switches, which can connect enterprise and residential networks via Ethernet switches, can create (IEEE 802.1Q) and protect (IEEE 802.1D) VLANs. To get external IP access to the metro network, the router switches can be connected to dual gigarouters, which are equipped with MPLS and boarder gateway protocol (BGP) functionality. The Gigarouters can interface the incumbent transport network via POS interfaces (see *Figure 5*).

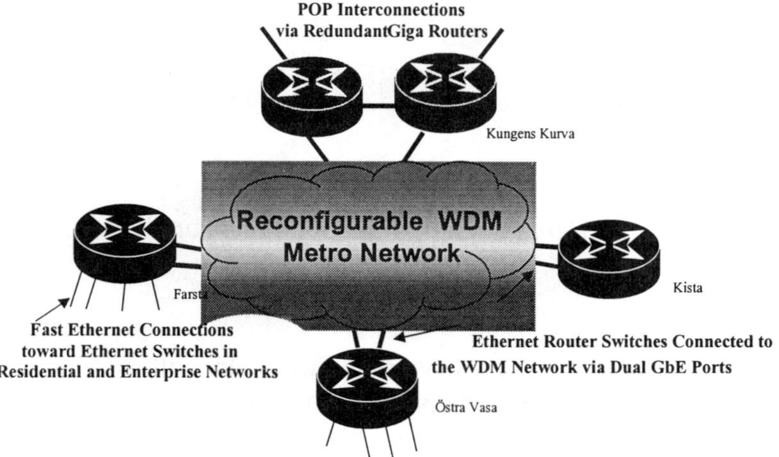

Figure 5: *Data Network Interconnected via OXCs and WDM Terminals in the Four-Node Testbed*

---

[1] Router switches are here defined as NEs, which have both IP and Ethernet functionality. Their Ethernet functions are usually preferred from a performance perspective.
[2] ACTS DEMON is a European research project focusing on interconnection of optical subnetworks.

What is not shown in the figure, but is easily supported, is wavelength connectivity to ESCON, fiber channel, video services, static metro WDM rings, and direct metro external wavelength interconnections, for example.

Optically Interfaced Electrical Cross-Connect
The OXC in *Figure 6*, which is developed within the project, consists of 1R 1300 nm transceivers (RxTx) for the optical-to-electrical-to-optical (O–E–O) conversion and an electrical 2R GaAs cross-point switch (XPS). Multirate clock and data recovery (CDR) circuits, located between the transceivers and the switch, provide 3R functionality. Here, only CDR circuits for 1.25 Gbps are used—corresponding to GbE. The switch state is supervised by monitoring the output/input ports and is configured by a local element manager (EM). The signal quality can also be monitored on the protocol level by periodically scanning all input ports (using the XPS multicast functionality) to a specific output port connected to a high-speed line card on the EM. Packets are then processed by conventional protocol-monitoring software. Hereby, the status of the transmission equipment can be monitored, and flow identification (verifying the wavelength channel connections) can be made. The local EM is connected to other EMs via a low-speed interface on the local router. The cascadability of OXCs in 2R mode is limited by timing jitter [13]. A conservative assumption is to limit the number of OXC hops to the system bandwidth (here 1.25 Gbps) divided by the application bandwidth [13], e.g., a 200 Mbps ESCON connection could be established over six OXC hops (1250/200). To avoid this jitter limitation the multirate clock recovery circuit must be used.

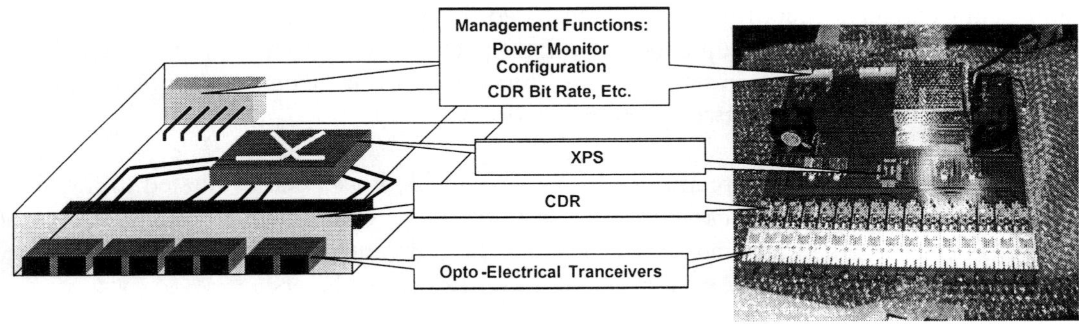

Figure 6: *The 16-Port OXC Circuit Board with Optional CDR*

Data Communication Network
The control plane of the OXC can include a routing function. In contrast to routers, this function can be located on a separate piece of equipment, with an interface to the element agent of the OXC. This means that the control channel can be transported separately from the client data channels. This is in contrast to the case for routers, where the signaling and data channels are carried over the same physical links. There will be one network topology for the control channel and one for the data carried on the optical channels. Consequently, separate redundancy must be considered for the control and client data channels.

There are several options to realize the data communications network (DCN) for the control channel. It can be a completely separate network, it can use the optical supervisory channel (OSC), it can use one of the wavelength channels, or if SONET/SDH framing is available, for example, it may use part of the overhead for the DCN. This field trial uses GbE framing and considers the use of both part of an in-band wavelength (for 10/100 Mbps Ethernet) and a separate physical network. Usually, OSCs in metro networks are avoided to minimize excess optical losses (from the 1510 nm couplers).

Control Plane

The control plane is a critical part of any network. It handles fault, configuration, accounting, performance, and security (FCAPS), which are complex tasks in a typical, flexible metro network. Combining the IP, Ethernet, and optical layers of a network to improve the network efficiency and minimize the interference between layers can potentially complicate things even further. In the Winchester test network some of these issues are addressed and solutions are evaluated.

By reducing the responsibility for a central management system, the complexity of managing the network is reduced, and network survivability is improved. Also, intermanagement between the layers should be easier to implement with a distributed paradigm in all layers. However, connection-oriented optical-path services will be available in the near future, and thus it is infeasible to allow single nodes to make independent decisions about routing. In the Winchester network, paths are allocated from the edge of the network (from source to destination node). A connection manager resides in each edge node and handles restoration in a fault situation. There exists a central network manager for bootstrap reasons, but after initialization and distribution of network topology, the network will act independently.

Fault and Performance Monitoring

In the case of an OTN for IP, it may be beneficial to correlate IP–related performance statistics with optical performance measurements or simply to draw conclusions on the state of an optical path based on Layer-3 performance statistics. Burst errors or high packet loss may indicate problems in the optical network and could trigger a reconfiguration event. Also, by identifying IP flows between routers, the network can be optimized by setting up end-to-end light paths between routers through which are heavy flows, thus reducing transit routing in the IP layer. This will result in better utilization of the IP routing capacity and will reduce delay and network load in the IP layer.

A tool to ensure that service-level agreements (SLAs) are fulfilled has been evaluated in the Winchester network, where both the WDM alarms and the IP performances were integrated using simple network management protocol (SNMP) traps and syslog. This enables guaranteed bandwidth to be provided more easily and correlation of the alarm and performance from the different layers.

Restoration and Protection Resilience

By using timed-out refresh packets, the IP layer has functionality for rerouting in the event of failed links. Layer 2—GbE in this case—can, by counting invalid code groups for the 8B/10B line-code, estimate the bit-error rate (BER) and declare failed links in only a few 10-bit groups or errored frames. However, it would be beneficial to do restoration in the optical layer for several reasons, one of which is speed. Handling a link failure in an IP backbone requires updating of the routing tables of a number of routers, which takes time and which may introduce instabilities into the network.

To avoid this, the optical layer in the Winchester network can restore the optical path fast enough that the IP layer need not take action. By precalculating restoration paths, this service is achieved within a few milliseconds in case of cable break.

Distributed and Automated Provisioning

There exist two main approaches for enabling automation of the functionality that a reconfigurable optical layer can provide to an IP network. One assumes separate control planes for IP and for the optical layer (overlay model). By gathering statistical information from routers on traffic load and flows, the optical-layer manager may decide to rearrange the logical network topology. It is up to the IP routing protocol to adapt to the changes in topology. The other approach is to integrate the IP and the optical-layer control plane (peer model). Then the IP routers could request or reallocate optical path resources when needed. Alternatively, the optical-layer manager could make the decision but inform affected routers of the change in time for them to update routing tables (or do this for them).

In the Winchester network, the first approach is currently implemented. The architecture can support path provisioning from routers, but this has not been tested as yet. The possibility of migrating toward a truly integrated IP/WDM network platform is under investigation.

Control Plane Architecture
Node-element agents control and survey optical-node operation (via a small hardware control application) and report faults. A central manager provides user operation and maintenance, keeps track of configurations and connections, and restores failed connections. It delegates this responsibility to the distributed agents, which use signaling for network configuration changes (see *Figure 7*) [14].

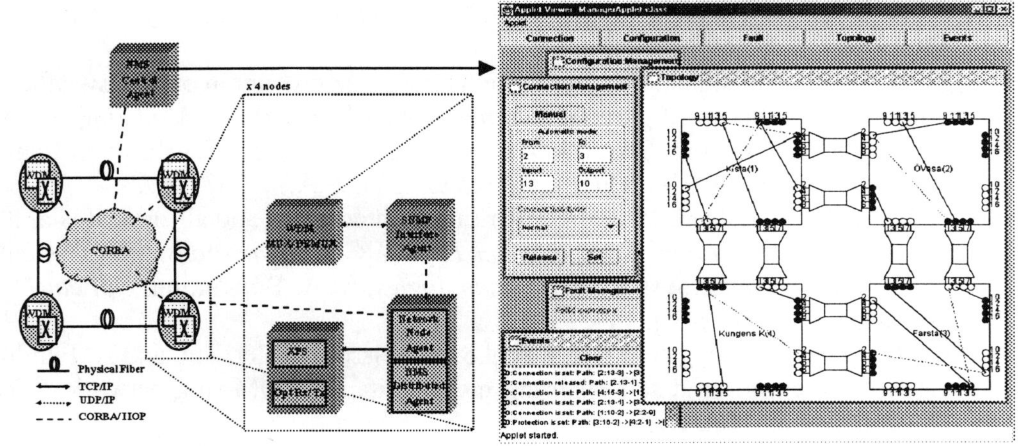

*Figure 7: The Physical Network with NEs and Corresponding Management Applications,
Including a Screen Dump of the User Interface
All agents communicate via a common object request broker architecture (CORBA)
distributed platform.*

This distributed architecture gives *flexibility* (addition of Java objects is simple, and CORBA provides flexibility in choice of programming language), *robustness* (agents can crash while the rest of the control system still will work and provide state information to agents that restarts), and *scalability* (central agents can control separate parts of the network as well as the entire network).

Summary

A reconfigurable transport metro network can achieve major network profits, such as improved network resilience and utilization and faster bandwidth provisioning by slightly increasing the excess network capacity.

GbE–framed IP packets are a cost-efficient way to provide fault and performance information sufficiently in the metro network for the main traffic. The WDM technology as a complement still enables a transparent transport for several data formats.

All-optical, broadcast-and-select transport networks, compared to OXC–based networks that utilize O–E conversion, can provide a high degree of transparency. On the other hand, medium-sized, very dynamic networks are usually more optimized for OXC solutions, provided that the interface costs are optimized with, for example, coarse WDM. Both technologies have been successfully tested in field trials in the Stockholm area.

A metro testbed, the Winchester network, has been created in Stockholm that can utilize network functions of the WDM, Ethernet, and IP layers, depending on the service and the connection requirements.

The optical layer has been extended with routing and signaling functionality to achieve 1:N protection with preemption and fast provisioning of optical circuits. The next step will be to add and evaluate the path provisioning from the routers to the optical network for enabling clients to establish optical circuits in the network.

References

1. Chen, Y., et al. Jan.–Mar. 1999. Metro optical networking. *Bell Labs Tech. J.* 4 #1: 163–185.
2. Redifer, G. B. Sept. 1999. DWDM in the Metro Marketplace—does it really cost in? *NFOEC* 1: 239.
3. Larsson, S. N, and S. Hubendick. Feb. 2000. Reduction of hop-count in packet-switched networks using wavelength reconfiguration. *Optical Network Design and Modeling*, 1004–1024.
4. Evaldsson, P., and E. Almström. Sept. 1999. Evaluating the technical feasibility of transporting IP and ATM through WDM. *NFOEC* (Chicago, IL, U.S.A.) 1:282–289.
5. Almström, E., and C. P. Larsen. 1999. Interworking with domains and all-optical islands. *11th Tyrrhenian International Workshop on Digital Communications, The Optical Network Layer: Management, Systems and Technologies*, in A. Bononi, ed. *Optical Networking*. London: Springer-Verlag, 14–25.
6. Goobar, E., J. Sandell, S. Wingstrand, K. Käyhkö, K. Nilsson, L. Persson, and A. Lundberg. Sept. 1999. Fully engineered self-restoring metropolitan DWDM ring network. *ECOC* (Nice, France), session WeD1.
7. Gillner, L., J. Karlsson, E. Goobar, A. Lundberg, S. Magnusson, S. Melin, and Y. Nilsson. Mar. 2000. Field trial of a European self-healing WDM optical network. *OFC* (Baltimore, MD, U.S.A.), paper TuQ1.
8. Larsson, S., S. Hubendinck, H. Carldén, C. Wickman, and E. Almström. Mar. 2000. Wavelength label switched IP backbone: Architecture and field trial. *OFC* (Baltimore, MD, U.S.A.), paper ThE4.
9. Almström, E., and S. Johansson. Sept. 1998. An optical cross-connect prototype. *NFOEC* (Orlando, FL, U.S.A.).
10. IETF RFC 2615, PPP over SONET/SDH.
11. IEEE Draft P802.1w/D4, Rapid spanning tree protocol, Jan. 27, 2000
12. Kompella, K., et al. Feb. 2000. Extension to IS–IS/OSPF and RSVP in support of MPL(ambda)S. *IETF draft-kompella-mpls-optical-00.txt.*
13. Almström, E., and S. Larsson. Sept. 1999. Experimental comparison between optical and electrical switches for transparent networks. *ECOC* (Nice, France), session MoA2.
14. Hubendick, S., and J. Söderqvist. Sept. 2000. Management of connections services in dynamic optical networks. Accepted for publication at *ECOC* (Munich, Germany).

Optical Networking in the Metro Area

Chris Bowers
Senior Member, Technical Staff
SBC Technology Resources

Introduction

There is much speculation about when dense wavelength division multiplexing (DWDM) will displace synchronous optical network (SONET). However, DWDM is not popular in metro networks for several reasons. The long-haul space and metro space are completely different. In the long-haul, DWDM reduces the number of regenerators and fibers, as shown in *Figure 1*.

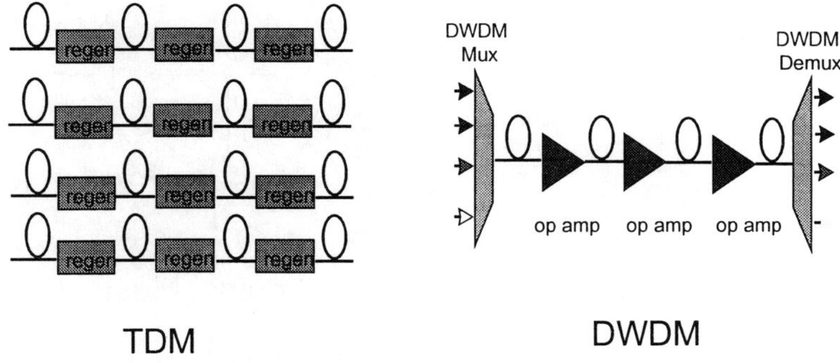

Figure 1: *Use of DWDM Minimizes Number of Regenerators and Fibers in Long-Haul*

It is not only fiber; by having an optical amplifier perform the amplification on many different wavelengths at the same time, four regenerators can be replaced with, for example, one optical amplifier, as shown in *Figure 2*. In reality, this replaces between 30 to 100 regenerators with one optical amplifier.

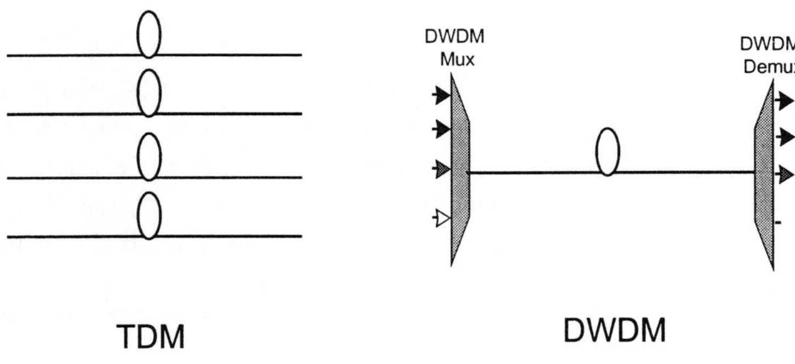

Figure 2: *In Metro, DWDM Minimizes Fibers*

DWDM clearly cuts costs in long-haul networks, and removing even more regenerators is the motivation for new ultra-long-haul DWDM products.

The Metro Experience

Metro networks have very different physical characteristics from long-haul networks. Typical distances are fewer than 50 km, which can easily be reached with long-reach optics. The cost of the DWDM equipment for most transport is, typically, more than the cost of additional fiber. There will certainly be applications for which it is prohibitively expensive to install more fiber, for which right-of-way is not possible, or for which the carrier has only a few existing fibers. On the other hand, it may simply be more expedient to put in DWDM when the service must be activated. However, as a general, ubiquitous deployment, DWDM does not currently cost in for metro networks.

Optical versus Electrical Add/Drop

For DWDM and optical networking to be more than simply niche products when they come to metro, they must provide more than fiber gain. The obvious thing to try, in view of SONET's add/drop multiplexer (ADM) role as the workhorse of metro transport networks, is to make optical ADMs (OADMs), as shown in *Figure 3*, and set up rings based on wavelengths that make logical connections across the rings, instead of the time slots in the time division multiplex (TDM) case.

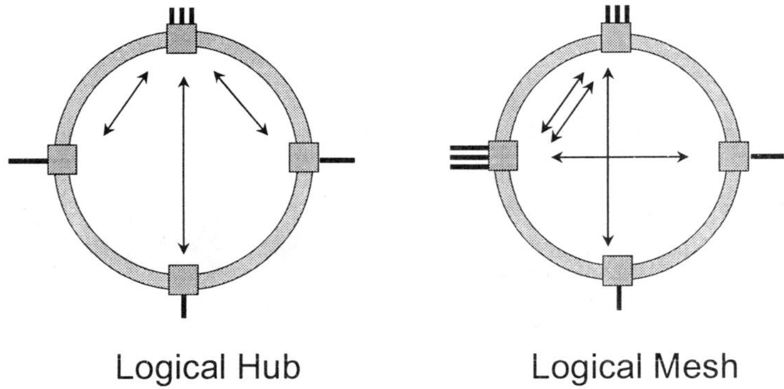

Figure 3: Optical Add/Drop Rings

Many vendors make these products. They allow the carrier inexpensively to drop a large amount of bandwidth in the form of wavelengths by dropping individual wavelengths with an interference filter. This is much less expensive than going into an electrical cross-connect (XC) of equivalent bandwidth and having to detect the light at a very high rate, shuffle around the time slots, and drop out the time slot of interest. *Figure 4* illustrates the differences between optical and electrical add/drops. OADMs afford high bandwidth dropped out at low cost.

However, the cost is not that low because even though the interference filter itself is very inexpensive, all the light must be at the right wavelength for the interference filter to drop it out correctly. Expensive wavelength-specific optics must be used. Also, the loss budgets and engineering rules involved are further disadvantages to this approach. With a TDM–based ADM, the optical signal is detected at every node. Engineering is required only span by span, and one time slot does not affect another time slot. With the OADM, as shown in *Figure 4*, dropping a wavelength at a node can slightly affect the power of the wavelengths that pass through the node. The whole ring must be engineered at one time, which detracts from its flexibility.

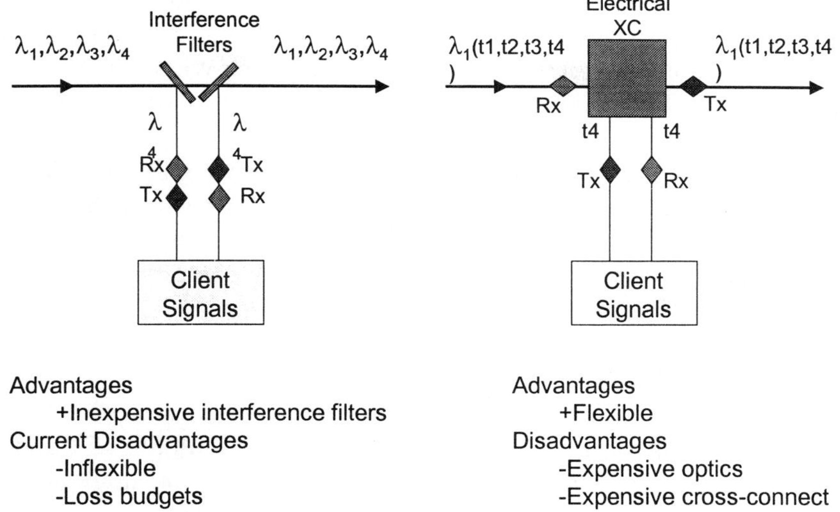

Figure 4: Optical versus Electrical Add/Drop

Also, electrical, TDM–based ADMs are remotely reconfigurable. For example, one can sit down at a PC and decide to drop time slot three instead of time slot four, hit "return," and it is done. So far, that is not the case with OADMs. When, then, will all of this be possible? It will take a "trigger condition."

The Trigger Condition for Metro Optical Networking—OC–48c

This move to OADM will depend on a trigger condition within the metro network—namely, a predominance of optical carrier (OC)–48c and higher-rate port cards and data interfaces on asynchronous transfer mode (ATM) switches and Internet protocol (IP) routers that demand it. Currently, with an OC–12c or below interface, it makes a reasonable amount of sense to put it into a SONET ADM and go through the traditional infrastructure. With an OC–48c interface and above, planners will simply put it on dark fiber, as shown in *Figure 5*. One can go from an ATM switch or IP router from one side of town to the other with long-reach optics.

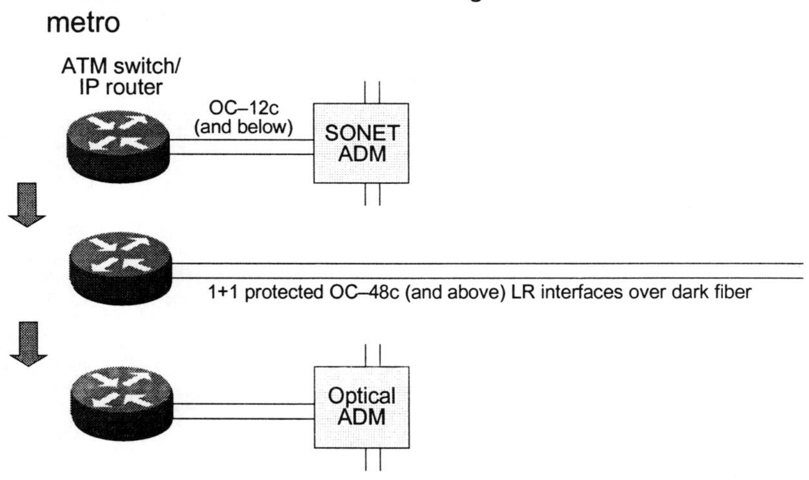

Figure 5: Trigger for Metro Optical Networking

The data equipment also comes with one-plus-one protected interfaces for OC–48c. Diverse fiber routes can provide good protection against fiber cuts. It may not be a glamorous approach, but it will suffice until things get too messy and unmanageable. Hopefully, the industry will enter at the right time with the right solutions to manage that bandwidth. For example, in the center scenario of *Figure 5,* the ATM switch (router) provides performance monitoring and protection switching, but it may not do a good job; or, if it does a good job, there might be three or four different routers or ATM switches. It is unreasonable to hope that every vendor will do it the same way. Thus, no centrally manageable system exists to handle the bandwidth. Management is left to every ATM switch and every IP router. Technology is badly needed to organize and manage the bandwidth. That is where the OADM operating at OC–48c granularity of bandwidth will really take off in metro.

Drivers for OC–48c Interfaces in Metro

What carriers want from the optical layer is everything they have with SONET and more in terms of protection, performance monitoring, and, in particular, being able to configure an add/drop remotely at any node. It will be difficult to convince people to take a step backwards in functionality. Bandwidth demands may make this backstepping necessary, and dark fiber may be the answer in cases where connections must be patched through. Carriers also want in-service bandwidth upgradability. Of course, this must all be cost-competitive. What, then, are the drivers for getting these OC–48c interfaces in the metro?

One internal driver for the telecommunications industry is digital subscriber line (DSL). Enterprise data and application service providers are obviously large bandwidth customers. Point-to-point data applications will also play a part connecting, for example, redundant data centers with fiber connection (FICON), enterprise systems connectivity (ESCON), or Gigabit Ethernet and taking advantage, in that case, of DWDM for protocol transparency. DWDM might be used more than was previously expected in the metro, especially when fiber is in short supply. These are the challenges in moving optical into the metro networks that must be addressed expediently.

Bringing Broadband IP Services to Small Businesses in Metropolitan Areas

John H. Davis
Chief Technology Officer and Director
Allied Riser Communications

In the past, lack of continuity between the application layer and the data structure has been one of the key issues facing telecommunications services businesses. For incumbent service providers that must contend with the legacy systems of their past, current market pressures are forcing major redesign of systems to allow more timely delivery of services and applications, which are highly dependent upon customer data. For new entrants, an opportunity exists to build businesses that are "data centric" instead of "network centric." The engineers who invented digital signal (DS)–1, DS–2, and DS–3 long before synchronous optical network (SONET) was conceived thought mainly about the network, not about the applications that would be built upon it. Thus, most of the data required to develop a customer-focused service was buried in the applications and not readily available. The importance of the relationship between the data architecture, the application layer, and the underlying supporting infrastructure became clear.

Just as the technical relationships have evolved to today's awareness of the importance of customer data, the emergence of the so-called "building-centric" telecom players has created an important transformation in point of view regarding customers. Briefly, this contrast is one of "inside-out" rather than the more traditional "outside-in" thinking. To apply inside-out thinking, one starts at the desktop, determines the needs of the customer, and then develops applications and network resources around those needs. Outside-in thinking starts from the network and tries to figure out how to segment customer needs and somehow penetrate the "last mile." As trivial as this idea may sound, it represents a fundamental difference in the way one thinks about the problem that is being addressed.

This paper is a case study of one company's experience in bringing broadband Internet protocol (IP)–based services to small to medium-sized businesses in metropolitan areas.

ARC

Allied Riser Communication (ARC) was founded on the premise that it would build its business processes around a common database of customer information (currently a lightweight directory access protocol [LDAP] database) and would not deploy legacy circuit-switched networks. ARC also premised the availability of all-optical networks (AONs) and has endeavored to ensure that the "first mile" elements of the network are optical.

ARC is a new-entrant service provider that uses optical networks to supply small businesses with ultrabroadband, IP–based data, video, and voice communications services over owned and operated broadband facilities in commercial office buildings. ARC goes into commercial buildings, drills a 4 inch hole through a utility riser, and fills that with fiber, usually multimode dedicated fiber, which can supply every user in that building. The switches and routers go in the basement, which becomes the building point of presence (POP).

ARC aggregates the traffic in a building and uses a hub-and-spoke architecture homed on a metropolitan POP (MPOP), usually co-located with a wide-area network (WAN) partner. For example, today ARC is in 23 Level 3 co-located sites. From the MPOP, ARC goes to the WAN using backbone transport capabilities of its many partners (see *Figure 1*).

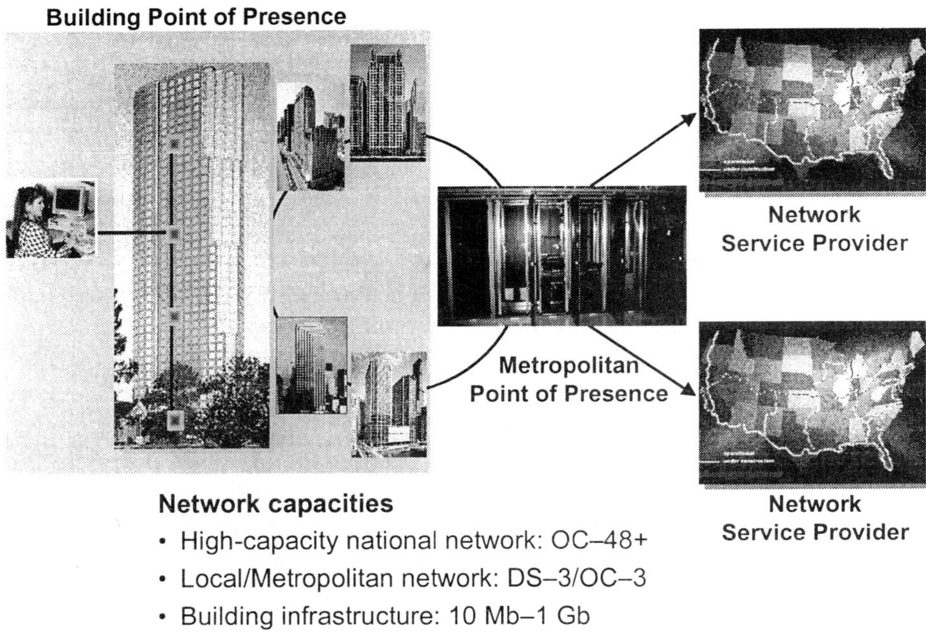

Figure 1: ARC's LightSpeed Network

To these WAN partners, ARC appears as points of substantial traffic aggregation; to the application service providers (ASPs), ARC appears as an effective broadband distribution channel. ARC averages 6 to 10 weeks to build out the buildings that it has under contract. As an example, the Sears Tower in Chicago, a 3.5 million square foot building with 110 floors, was recently completed in about five weeks.

However, getting access to the backbone network can take 20 weeks with current industry-wide DS–3 provisioning intervals. This reality has presented ARC with a major access issue. ARC has begun to light its own fiber in the streets because of the delay. The company has competitive local-exchange carrier (CLEC) rights in every market in which it is involved, so this is a cost-effective and practical solution in many cases. In other cases, ARC uses fixed wireless radio and evaluates the use of wireless optics to break the access bottleneck.

ARC targets mainly small to medium-sized enterprises in the buildings that it has under contract. Interestingly, only six service initiation charge (SIC) codes are needed to describe the activities of more than 48 percent of the tenants in the buildings. This fact represents an opportunity for ARC to work with ASPs to deliver highly targeted vertical applications with major impact.

Key Issues Faced

Some choices the company had to make included deciding on its services architecture, making technological choices for buildings and access, creating a strategic product roadmap, and planning for capital efficiency in light of the choices mentioned.

Services Architecture

ARC's services architecture choice was to become an IP–based broadband service provider, deploying both in-house and ASP–based applications in a new space that is neither enterprise- nor network-based. ARC has a branded presence with all of the desktops served going through the company's myARC™ portal. ARC is not an exchange carrier and does not use its infrastructure to provide services to other people. Currently, ARC offers only basic broadband services.

The myARC™ portal pulls together provisioning and maintenance for end users. These users can activate services themselves through the myARC™ portal. The portal works with an LDAP integrated data architecture. The LDAP database has a customer record of 48 entries on each desktop on the system and is the foundation of the company.

Another aspect of ARC's services architecture might be termed "managed abundance." ARC has large data pipes with managed quality of service (QoS). ARC provisions 10-megabit, fully symmetric pipes to users. The infrastructure itself is equipped with 100 foreign exchange (FX) transceivers and can be upgraded to gigabit capability with a simple swap of electronic plug-ins. ARC provides a platform for innovation with open interfaces and standards-based applications.

ARC uses traffic-shaped virtual LANs (VLANs), and, in fact, when competitors offer a lower–bit rate digital subscriber line (DSL) solution, ARC can offer double the bandwidth for about the same price. That can be accomplished by traffic shaping at the routers in these buildings.

Technology Choices

There is a dizzying array of choices in access technology, distribution and long-haul technology, customer interfaces, and applications protocols. ARC has chosen three technologies for its main thrust. First, the company chose IP. Second, Ethernet will dominate ARC's physical connectivity. Finally, fiber will be ARC's medium of choice in the future.

In addition, wireless, copper, and wireless-optical technologies will play an important supplementary role in filling out connectivity. Those options may be important in breaking the access bottleneck.

Strategic Product Roadmap

ARC's first products are connectivity products that supply the market with access to the ARC LightSpeed Network™. The next suite of products is bandwidth-enabled applications. ARC is working, both in-house and with ASPs, to deliver bandwidth-consumptive applications on-line. By making things possible that were not possible before, ARC increases demand on its network. Finally, ARC is moving toward managed services. The myARC™ portal pulls all this together.

ARC will have open standards. Dozens of potential suppliers have had their products evaluated in an ARC lab where the suppliers are helped to meet the company's architectural requirements. This testing and collaboration process is a part of what will become a more formal process in the future.

Capital Efficiency Issues

Unlike the very robust understanding that has guided generations of engineers in the design and capacity planning for voice telecom systems, the fundamentals of packet data are not well understood, and this leads to inefficient network asset utilization. Even though WANs have abundant capacity, local networks are still constrained. Market demand for high bandwidth among small- and medium-sized-business users is still developing.

One particular issue in capital efficiency is the fact that Poisson modeling does not work. This traditional engineering approach of the circuit-switched voice world does not apply in packet-switched data networks, and the attempt to apply it causes inefficient utilization of capital in many of today's IP networks (see *Figure 2*).

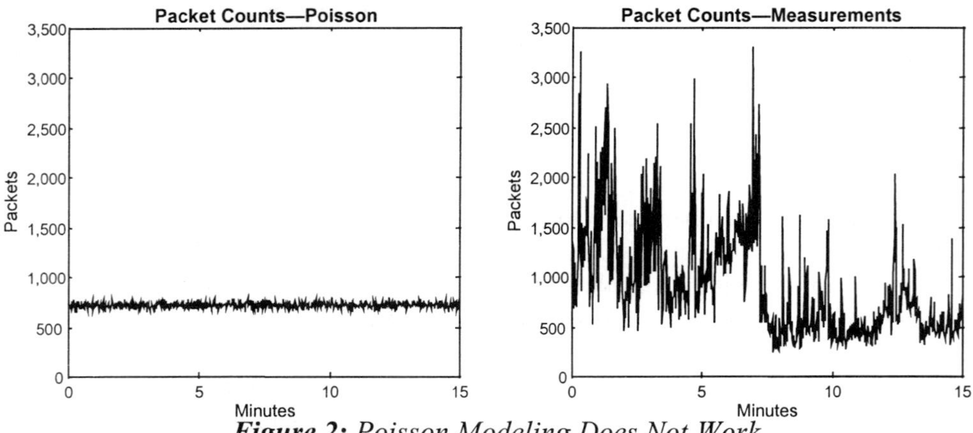

Figure 2: *Poisson Modeling Does Not Work*

ARC, by contrast, is pioneering a planning method using a Markov-Modulated Poisson Process (MMPP) model to model its traffic accurately and to allow efficient asset planning (see *Figure 3*).

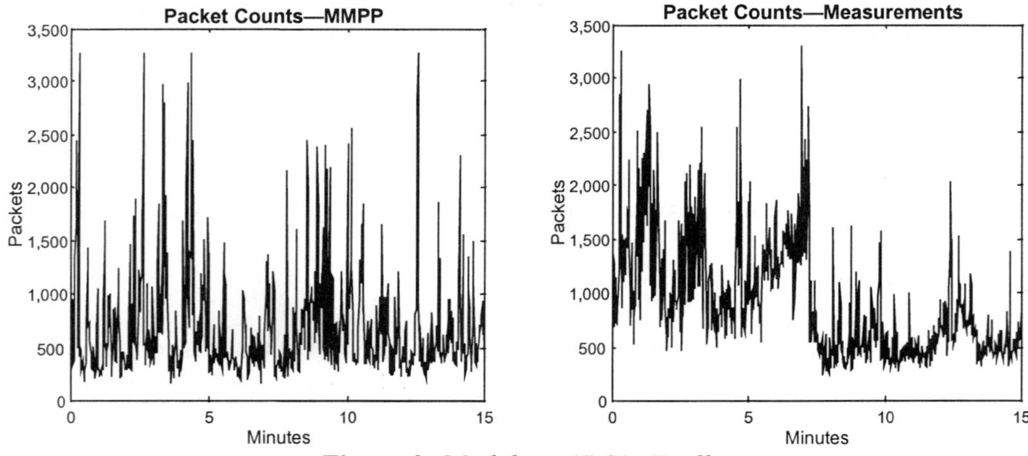

Figure 3: *Modeling ARC's Traffic*

The Future of Broadband

DSL will be a short-term play in the broadband arena because of the fundamental physical limitations of the structures of wire-pair cables. Cable has different problems and will face ongoing economic problems. In addition, cable has a fundamental topological problem: the more branches that are added to a hub, the more traffic congestion results. That traffic congestion forces the operator to put out more nodes, and more nodes cause the economics to spiral downward. Cable works very well at about 500 homes per hub, but below that, the economics are not favorable. Many cable networks are below that number of homes per hub.

Wireless will play an important complementary role in the future. Ultimately, however, AONs will prevail for business and consumers.

Fiber Solutions for All-Optical Networks for Metro

Matthew G. Estep
Product Line Manager–Metro
Corning

The future of technology and telecommunications is only now becoming visible. The dramatic changes brought about by fiber optics, the Internet, and the opening of new markets worldwide are about to be succeeded by a new revolution as technology continues to develop, and demand for technology-based services continues to grow exponentially. This paper concerns today's networks and tomorrow's all-fiber networks. It also provides insight into the fiber backbones of today's networks and into the design of today's networks that are built with a fiber level for tomorrow's all-optical network (AON). The technology behind the next revolution will be optical. That optical future is closer than many might think, and its implications may reach farther than many realize.

This paper will first touch briefly on market bandwidth needs for the next few years. It will then provide a brief discussion of the evolution of AON technology. More detail will be provided on fiber solutions for all-optical networking, including technical reasons to choose one fiber over another, the economic realities behind each of those choices, and where each choice would put service providers in terms of providing all-optical networking in the future.

The Market Driver: Internet Usage

Internet usage will be the driving force behind bandwidth demand growth over the next 10 years. Based on information extracted from the Pioneer Consulting metro dense wavelength division multiplexing (DWDM) report, the number of Internet users worldwide is expected to grow 39 percent annually.

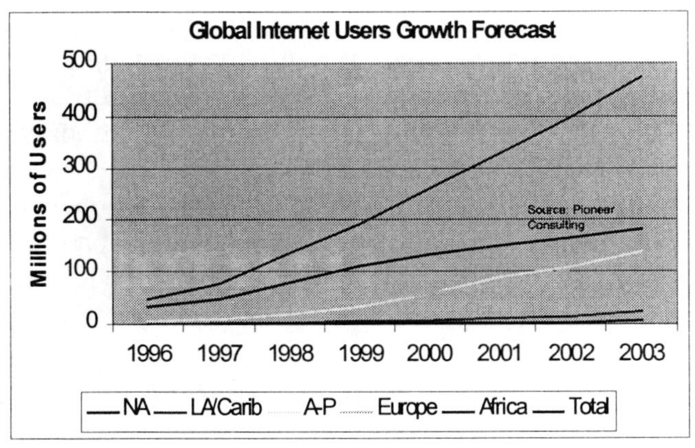

Figure 1: The Market Driver: Internet Usage

Internet usage is migrating away from educational, research, and government uses toward e-commerce and business-to-business activities. Telecommuting and e-commerce are driving Internet service providers (ISPs) to expand and improve services. The growth of data traffic relative to voice traffic requires re-engineering of the telecommunications infrastructure worldwide. Thus, the potential growth in bandwidth demand at a sample metropolitan central office (CO) is expected to be in the vicinity of 40 percent (see *Figure 1*).

Evolution Paths for Metro Networks

One way in which providers are considering re-engineering their networks is with DWDM deployment. DWDM puts multiple channels on a single fiber, thereby expanding the capacity of that fiber from one channel to 16 or 32. In fact, the upgrade path for many systems today could lead to hundreds of channels on a single fiber. North America drives the growth in investment in DWDM equipment, and while growth in Europe is slower, people are moving toward DWDM technology. The reasons for this move include the issue of fiber exhaust and the inability to deploy more fiber in a specific city or city build as well as the level of flexibility that DWDM brings to a network (see *Figure 2*).

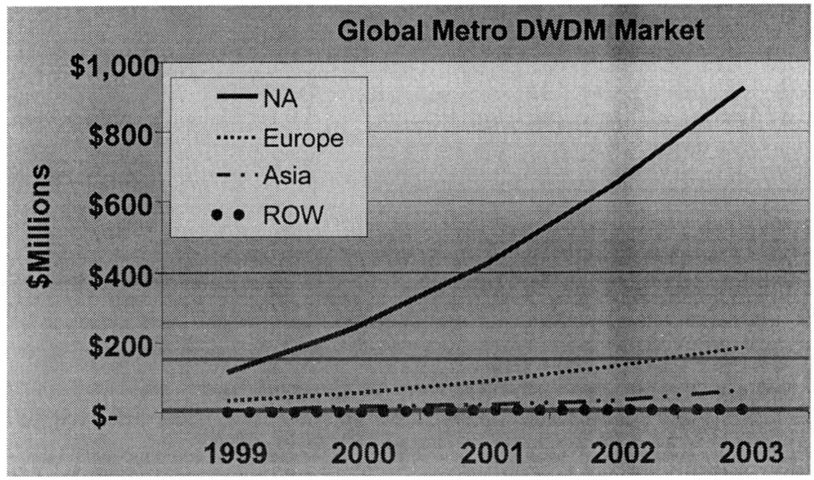

Figure 2: Metro DWDM Anticipated Growth

As DWDM is increasingly deployed and as more fiber is deployed to handle bandwidth demand in the future, there is an opportunity for the networks of today and the future to interconnect and to make a revolutionary change in how transmission of optical signals occurs. This change is the advent of the AON. Components, such as erbium-doped fiber amplifier (EDFA) components and optical cross-connects (OXCs), are being introduced at the optical layer and provide clear optical path transmission without having to convert from an optical-to-electrical (O–E) signal. In the future, increasingly more of these types of components will be developed and deployed to effect this evolutionary change to the AON.

This migration will begin with 1310 nm point-to-point synchronous optical network (SONET) rings, which will live in a hybrid environment with 1550 nm wavelength division multiplexing (WDM) systems. Many people are deploying these WDM components because they need to increase bandwidth without having to deploy more fiber. It is quick, easy, and flexible. Depending on how much it costs to lay a new network or fiber, it can even be less expensive. However, service providers must make long-term strategy decisions. Are they building networks that will be primarily WDM at 1550, or are they looking at WDM only to add bandwidth in areas of fiber exhaust?

If the second proposition is true, SONET can be utilized as a primary, long-term strategy. SONET is, in its own way, configurable, handling Internet protocol (IP) and asynchronous transfer mode (ATM) traffic, and it can be upgraded to optical carrier (OC)–48 or OC–192. It is also a known quantity; if a SONET is deployed, operators will know what they are getting. On the other hand, there exist problems with flexibility. A SONET system provides fixed connections, so to upgrade one OC–192 node, all must be upgraded along with all of the transponders.

In contrast, a WDM scenario provides extreme flexibility. WDM can be used to transmit IP and ATM exactly as with a SONET system. At first, WDM will be fixed and opaque, like the SONET system. Wavelengths will be preprovisioned and set, but as the opto-electronics become more optical and less electronic, the WDM system will become dynamically reconfigurable. Connection costs will drop considerably, and provisioning will become much faster. That is the direction many networks are headed today. The obvious drawback of the WDM system is that it is all new, but its benefits are likely to outweigh this initial problem. WDM may be dramatically increased without pulling additional fiber. A 144-strand cable might have 16 wavelengths on each strand. The bottom line for both of these scenarios is that network efficiencies increase, while operation costs are lowered (see *Figure 3*).

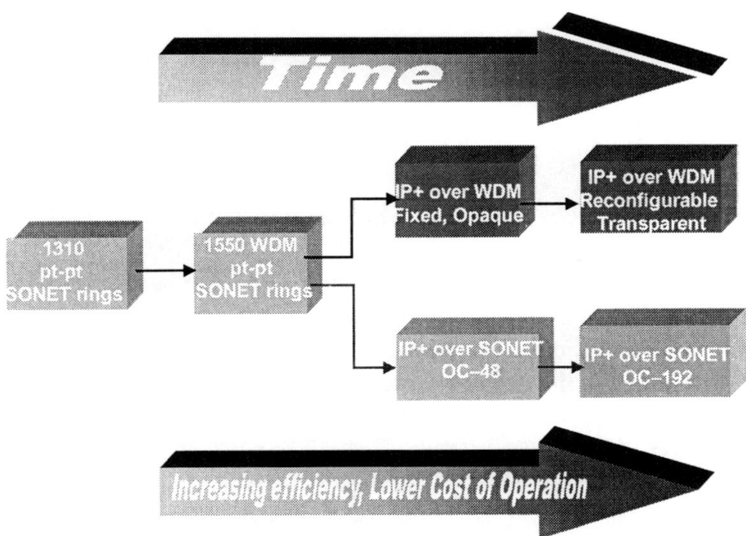

Figure 3: Evolution Paths for Metro Networks

Interconnected Metro Ring Structure Comparison

In comparing the differences in various network configurations, it is instructive to compare an opaque, electronically switched network to a transparent, optically switched network. Consider a generic metro ring structure with a large, interoffice feeder ring of some circumference. Connected to that ring are six to eight nodes of access rings (see *Figure 4*).

Each of the squares in *Figure 4* indicates an access node, perhaps a computer terminal hooked into the network. This is an amplified system, and the amplifiers are indicated by triangles in the figure. At each interconnect node an optical-to-electrical-to-optical (O–E–O) conversion occurs, as is typical of an opaque network. Though this type of network provides the security of a ring-type system, each O–E–O conversion introduces opportunity for signal degradation. System component requirements increase at each node and additional transponders can be costly. In other words, although an electronically switched network can deliver service transparency, the signals cannot flow nonstop through each ring and between them.

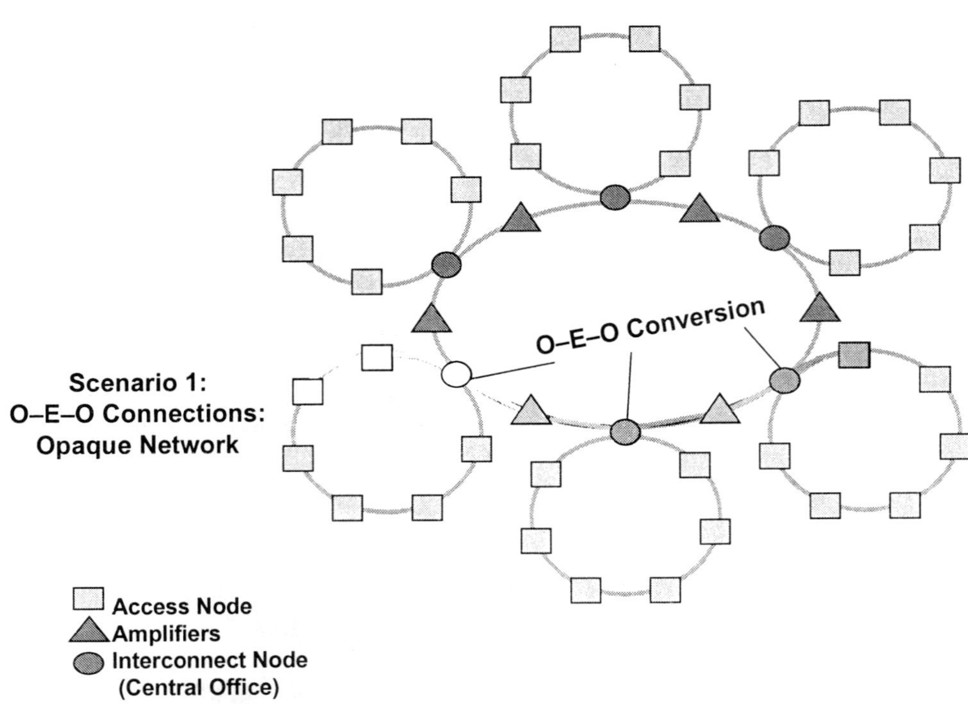

Figure 4: Scenario 1—O–E–O Connections, Opaque Network

In the transparent AON, on the other hand, the electronic interface is replaced with an all-optical interface at each node, and there is a clear path and seamless transition for signals. A signal starting at one access node is not converted back to an electrical signal until it arrives at the next access node. In the all-optical environment, electronic interchange is limited to where it is really needed—at the access nodes where the signal either begins or ends at an electronic device, such as a computer (see *Figure 5*).

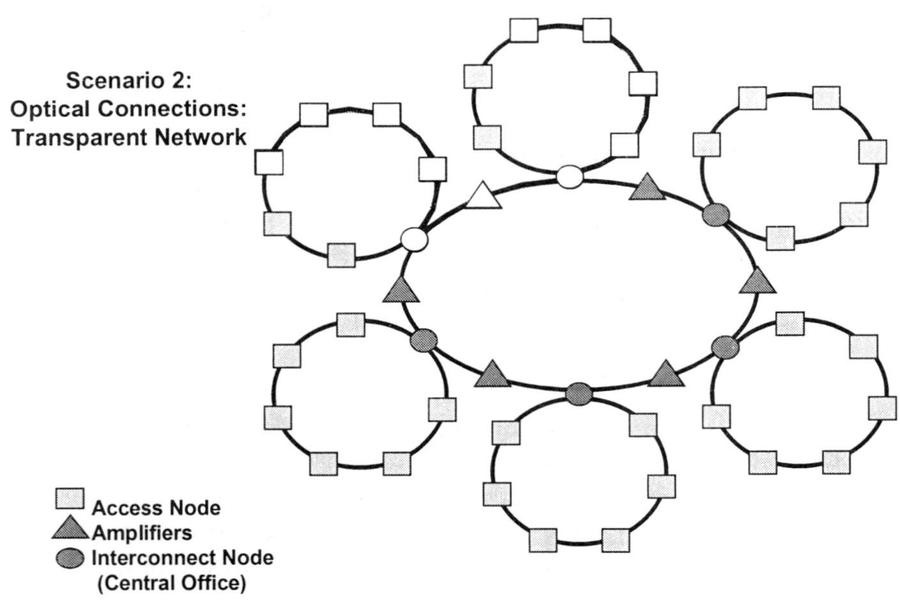

Figure 5: Scenario 2—Optical Connections, Transparent Network

Optical path lengths can be increased significantly in this system to provide easy reconfiguring and flexible network upgrading. There are benefits for both the operations group that deploys and maintains the network and for the customers who seek quick turn-up of their services.

Fiber Attribute Impact on AON

How do various attributes of fiber affect AON system design criteria such as optical path length, amplification, laser type, bit rate, channel spacing, and bit-error rate (BER) level, wavelength band, and cable design?

As previously mentioned, the lack of electrical conversions at each node in ring architectures means that optical path lengths can be increased. With increased path lengths, it is possible to exceed the dispersion limit of a standard single-mode fiber (SSMF) quite readily. This issue will be discussed in greater detail later in this paper. Attenuation is also an issue. Attenuation is the distance an optical signal can travel before having to be re-amplified or regenerated. These two parameters must be remembered.

The next issue is amplification—that is, whether or not to amplify the optical network. The laser type used is also very important. There are many different types of lasers, and each has its own pros and cons. Relative cost is one obvious factor in deciding between lasers. The bit rate at which the network will operate is another important consideration that must be decided in advance for future operation. Channel spacing and the BER level that will be tolerated are other AON system-design criteria that must be considered. All of these criteria are affected by fiber attributes such as the effective area of the fiber, about which more detail will be provided later in the paper. The dispersion of the specific fiber is another relevant attribute, as are the overall attenuation of the fiber, specified attenuation, and nonlinearities.

Wavelength band must be discussed in light of the desired operating wavelength. Single-channel at 1310 nm is widely used today, but with WDM and amplification components becoming much less costly, moving into the 1550 nm window is common.

Cable design involves choosing the type of cable to install, and it is possible from a fiber perspective to design a fiber that is less bend-sensitive and, thus, more cable-friendly.

Various Fiber Applications

On the wavelength continuum from 850 nm to 1625 nm, 850 nm is the playground of multimode fibers. Those low–data rate fibers typically service premises or local-area networks (LANs). At 1310 nm, SSMFs play a big role, because they were designed for that wavelength. In the 1550 nm window, nonzero (NZ)–dispersion-shifted fiber (DSF) is found. At the 850 nm wavelength, the type of transponder typically found is a light-emitting diode. At 1310 nm, single-channel systems are used (see *Figure 6*).

The conventional (C)-band runs from 1530 nm to 1545 nm red and from 1545 nm to 1565 nm blue. DWDM plays a large part in C-band. There is little deployment in the long (L)-band from 1565 nm to 1625 nm, but there is much discussion about when the industry will move into that region, as it is currently an untapped region for the use of fiber.

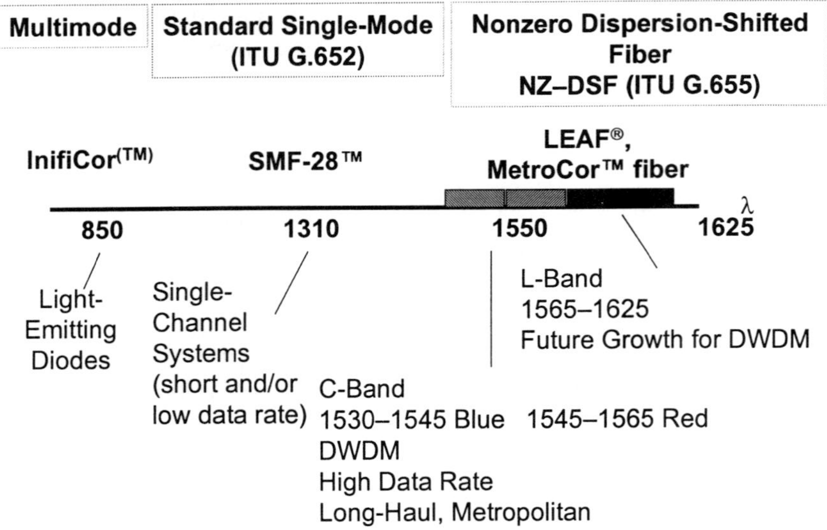

Figure 6: Various Fiber Applications

Fiber Types Vary in Functionality

SSMF has a zero dispersion wavelength near 1310 nm, and it is optimized for 1310 nm transmission. It is G.652–compliant and includes fibers characterized as "low water peak."

In NZ–DSF, the zero-dispersion wavelength is shifted outside the wavelengths in the EDFA window, so it is optimized for operating multiple wavelengths at 1550 nm. There are positive and negative dispersion varieties as well as fibers with large effective areas.

SSMF

SSMF is best for 1310 nm transmission because the zero-dispersion wavelength, or the point where the least amount of dispersion for transmission occurs, is about 1312 nm. However, the attenuation at that level is quite high, about 0.35 to 0.40 dB/km, so greater amplification is required to transmit. There are dispersion problems to address before one can take advantage of the lower attenuation region. Dispersion increases significantly in the 1550 nm window (see *Figure 7*).

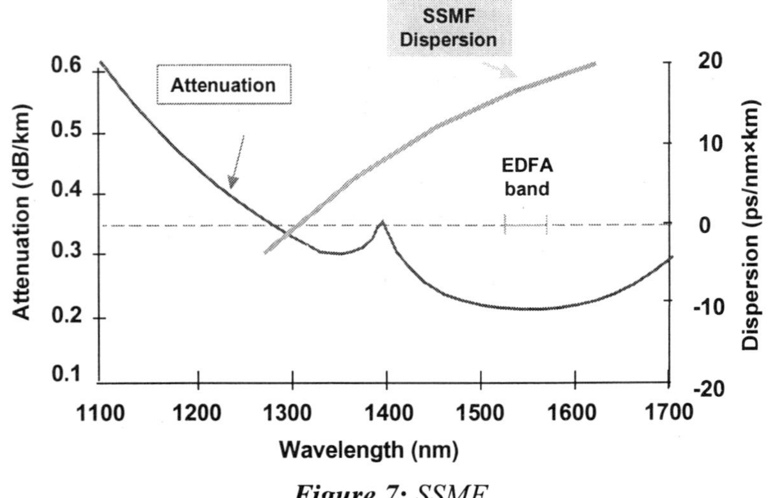

Figure 7: SSMF

SSMF, 1310 nm fiber systems are typically single-channel. Fabry-Perot (FP) lasers, which have little dispersion tolerance and are designed for lower data rates and shorter distances, are used and are inexpensive relative to high–data rate lasers. Directly modulated lasers are used for higher data rates, such as 2.5 Gbps, and longer distances. These lasers are also called distributed feedback (DFB) lasers. They have more reproducible wavelength control and a smaller spectral width, which leads to higher dispersion tolerance and the ability to travel longer distances over the fiber.

SSMF systems have high dispersion in the C-band and L-band and require greater compensation in the AON.

Coarse WDM Systems

Coarse WDM systems are the opposites of DWDM systems. Here, the wavelengths are widely spaced and are measured in nanometers. Coarse WDM systems can potentially transmit these widely spaced wavelengths across the entire 1300 nm to 1600 nm band. So-called "low water peak" fibers claim advantages in these systems. These are SSMFs that have the water peak removed, and the result is lower attenuation in the 1400 nm region. These fibers are specified for WDM systems, but this transmission system depends on the availability of low-cost components in that 1400 nm window. The true cost of these components remains unknown, as only C-band and L-band amplifiers exist. Thus, it is difficult to evaluate this fiber as a solution for the future network.

Considerations revolving around SSMF attenuation versus the distance limitation for specific wavelengths are important in deciding on network design. At 1310 nm, attenuation is a limiting factor for SSMF. In the 1550 nm window, the limiting factor becomes dispersion (see *Figure 8*).

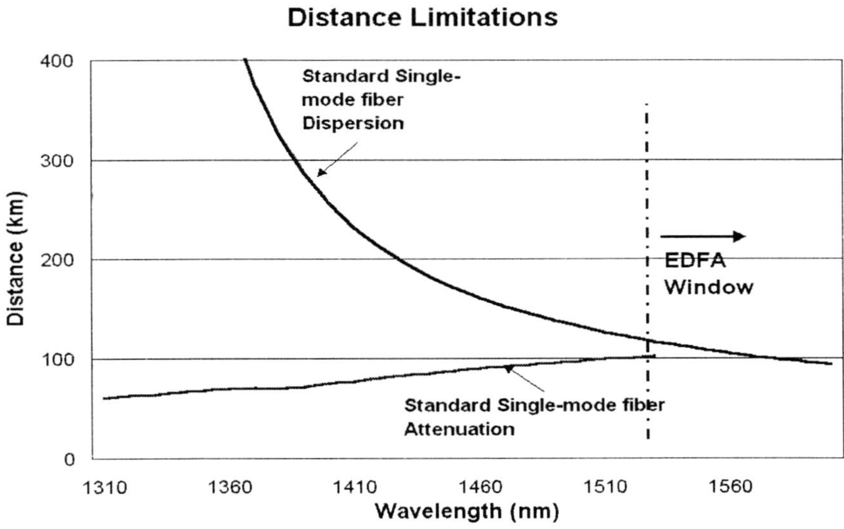

Figure 8: SSMF–DWDM Systems

A pure SSMF solution will limit the wavelengths where data can be most effectively transmitted.

NZ–DSF

Attenuation and dispersion have already been examined for SSMF. *Figure 9* shows these two attributes for NZ–DSF.

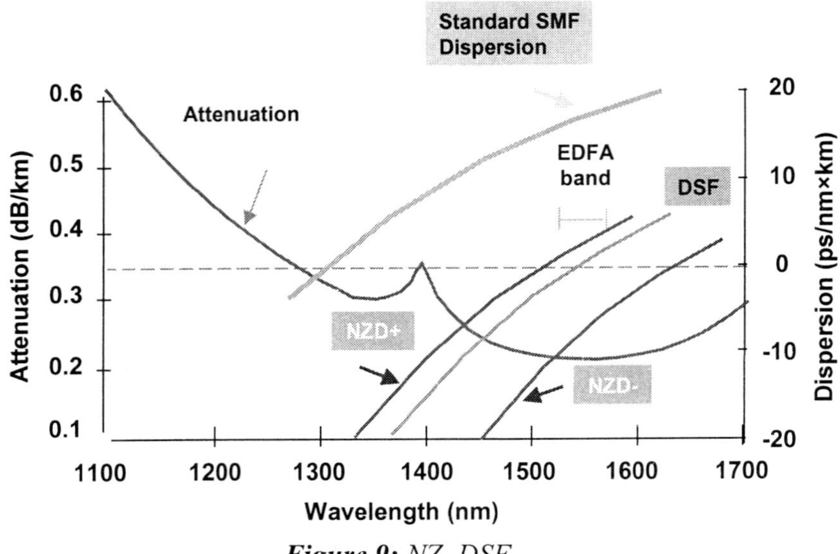

Figure 9: NZ–DSF

Years ago, a fiber was optimized with a zero-dispersion wavelength in the EDFA band. NZ–DSFs have since been developed that are dispersion-shifted outside the EDFA band. These fibers have been optimized for low dispersion in the EDFA window. There are specific benefits to these fibers.

The positive-dispersion NZ–DSF is designed for long-haul, high–data rate, multichannel systems. Typically these systems cover more than 400 km and are OC–192–compatible. Because of the low dispersion in the EDFA band, compensation modules are not required here—or fewer compensation modules are required—and the level of amplification can be reduced. These fibers, also available in large effective areas, reduce power intensity and nonlinear effects not mitigated by low dispersion. This allows even longer distances, and it lowers costs significantly. These fibers are specifically designed for long-haul networks and make good sense for AONs, as dispersion compensation is not required. They can handle the longer optical paths.

Managing Nonlinear Effects

Managing nonlinear effects in an NZ–DSF network can be a limiting factor. Nonlinear effects can influence data transmission or the BER at very long distances and at high data rates. To address those problems, network operators can decrease either the power or the length of the network. Both solutions obviously affect overall network performance. Another solution is to increase the effective area of the fiber.

One way to address these nonlinear problems is to increase the effective area of fiber with large-effective-area NZ–DSF. In a patented design, the dopants in the glass are varied to spread the light radially, reducing power peaks and nonlinearities. A comparison of a standard NZ–DSF and a large-effective-area fiber (LEAF®) demonstrates this effect (see *Figure 10*).

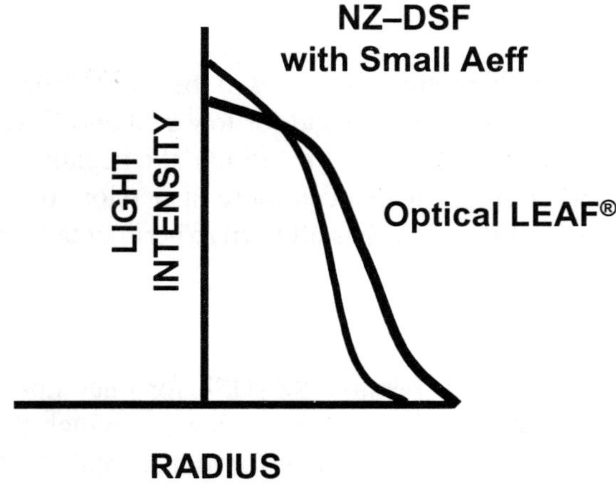

Figure 10: *Patented Design of LEAF®*

Decreasing the overall power intensity at the core of the fiber and thus decreasing the nonlinearities can be compared to reducing water pressure in a hose by increasing the diameter of the hose.

Negative-Dispersion NZ–DSF

Negative-dispersion NZ–DSFs are a recent development. They have been used for some time in submarine networks, but now they are being applied to metropolitan networks. These fibers are designed for distances of 80 to 400 km. Although they can be used at shorter distances, the true advantages of these fibers appear at about 80 km. They are optimized for operation at OC–48 as the primary transmission signal, but OC–192 transmission is well within their grasp. The other advantage of this kind of NZ–DSF is the low dispersion in the EDFA band, which makes compensation modules unnecessary and reduces the need for amplification. Negative dispersion also improves the performance of the directly modulated DFB lasers.

As mentioned before, attenuation at 1310 nm limits the distance signals can be transmitted by SSMF. However, for MetroCor™ fiber or negative DSF, dispersion around 1550 nm is not a criterion of concern relative to the standard single-mode case. It is designed to have very low dispersion and to allow transmission over much longer distances without dispersion compensation (see *Figure 11*).

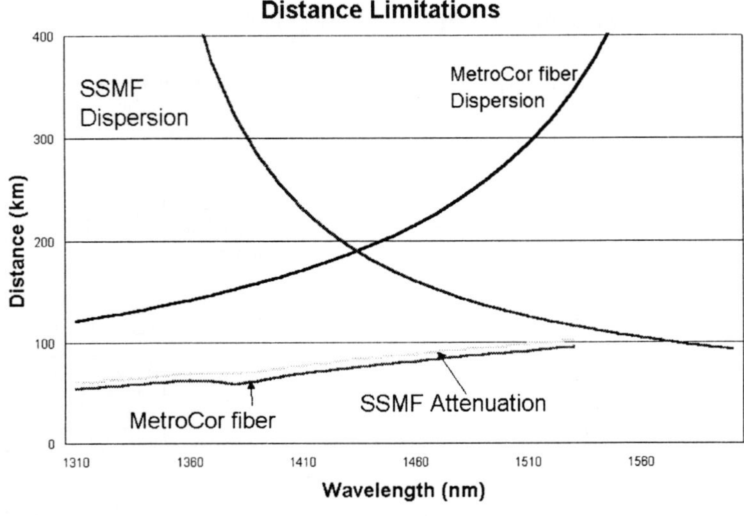

Figure 11: *Negative-Dispersion NZ–DSF Optimized for EDFA*

2.5 Gbps DWDM Systems

The 2.5 Gbps DWDM next-generation metro systems will be very densely spaced, 100 to 200 GHz wavelength systems operating in the C-band and L-band for low cost and flexibility. NZ–DSFs offer longer uncompensated reach as a result of lower dispersion. In addition, the negative-dispersion NZ–DSFs, through interaction with directly modulated lasers, provide even more dispersion tolerance, offering extended reach and allowing the use of cheaper transmitters. This has all been experimentally verified.

SSMF versus MetroCor™ Fiber

How does SSMF compare with negative-dispersion NZ–DSF for uncompensated signal quality with 40 channels at 2.5 Gbps over 330 kilometers? The Q value, or level at which the signal may be comfortably transmitted, typically must be around 9.5 dBQ to allow for an acceptable BER. SSMF drops well below that line, but the negative-dispersion DSF experiences very little effect as a result of longer signal transmission (see *Figure 12*).

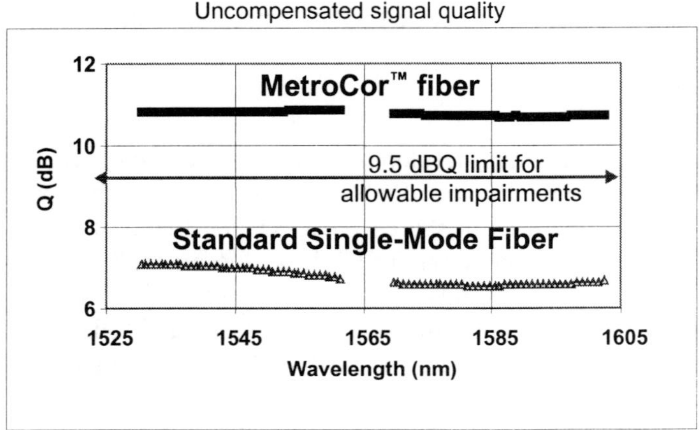

Figure 12: 40 Channel, 2.5 Gbps, 330 km Results

While dispersion compensation could be added to an SSMF system to boost the signal above the 9.5 dBQ line, it would add cost and complexity and may decrease the level of reliability.

MetroCor™ Fiber Performance on Metro System

The power penalty performance relative to fiber distance of commercially available metro components is illustrated in *Figure 13*.

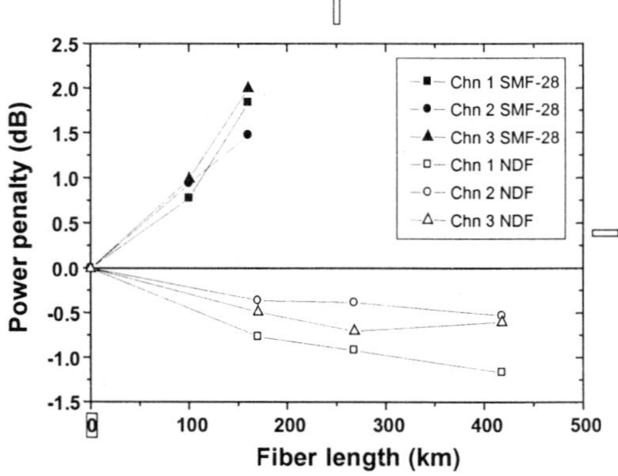

Figure 13: MetroCor™ Fiber Performance on Metro System

The power penalty should be kept below 2 dB for standard transmission systems. The open symbols in *Figure 13* indicate the SSMF, while the solid symbols are the negative-dispersion NZ–DSF. The power penalty increases substantially with distance in the SSMF system, while for the negative-dispersion NZ–DSF the power penalty actually decreases with distance.

The upgrade path to 10 Gbps OC–192 transmission for these negative-dispersion NZ–DSFs is evident when maximum uncompensated reach is compared. In the C-band, maximum compensated reach for MetroCor™ fiber is 170 km, as compared to 83 km for SSMF. In the L-band, maximum uncompensated reach is 205 km for MetroCor™ fiber and 74 km for SSMF. The value of these very–low-dispersion fibers—specifically, negative-dispersion NZ–DSFs—is obvious.

In 10 Gbps DWDM systems, maximum uncompensated distances are limited by dispersion tolerance of externally modulated lasers to around 60 km in erbium bands for SSMF. Positive- or negative-dispersion NZ–DSF can extend reach to 150 to 200 km without dispersion compensation. Nonlinearities may limit channel spacing or usable transmission bands. For longer system length, all fibers require dispersion compensation. Positive-dispersion NZ–DSF with large effective area has an advantage in this area.

Uncompensated system reach for SMF–28™ and AllWave™ fibers is 90 km for C-band and 60 km for L-band. This compares to 250 km (C-band) and 150 km (L-band) for positive-dispersion NZ–DSF and 200 km (C-band) and 250 km (L-band) for negative-dispersion NZ–DSF.

Summary

All-optical networking places unique demands on the fiber that is put into the network: high data rates, protocol transparency, and longer optical path lengths. Longer optical path lengths can greatly reduce operating costs. SSMF dispersion characteristics limit the fiber for application in specific scenarios. Dispersion-optimized fibers provide cost and complexity benefits. Low negative dispersion fibers can extend the useful range of less expensive lasers and further reduce costs of AON in metro applications.

Optical Access Can Provide Flexibility in Architectures

Anand Parikh
Cofounder, Vice President of Product Marketing and Business Development
Appian Communications

This paper will discuss what optical access, which involves a different kind of architecture, may mean. It will also discuss service management, operational costs, and the intelligence required in the optical edge. Access is the next wave. The industry believes in fiber-based access. Some issues for the intelligent optical edge include the optical-access architecture and what it needs to drive the life cycle costs down and to provide superior service-management capabilities to deliver, activate, and manage new services from this all-optical network (AON). It is all about services. Bandwidth is an enabler. In the end, services are needed, and bandwidth is a means to get to the services.

Oceans of Bandwidth

Look at the last couple of years both in the carrier network—the core as well as the metro—and on the enterprise side of the network. Clearly there are oceans of bandwidth enabled by dense wavelength division multiplexing (DWDM) technology in the core of the carrier network. One hundred Mbps Ethernet and Gigabit Ethernet are commonplace in the enterprise networks. In fact, 10 Gbps Ethernet is a standard that is currently under development. The bandwidth in the core of the network is growing very rapidly. In the enterprise networks for end users, there are new applications enabled by the Internet, such as business-to-business e-commerce and applications service provider (ASP) outsourcing. Ethernet technology is becoming ubiquitous in the enterprise networks. However, when looking at the connection points between these oceans of bandwidth on the two sides, one encounters time division multiplexing (TDM) sipping straws.

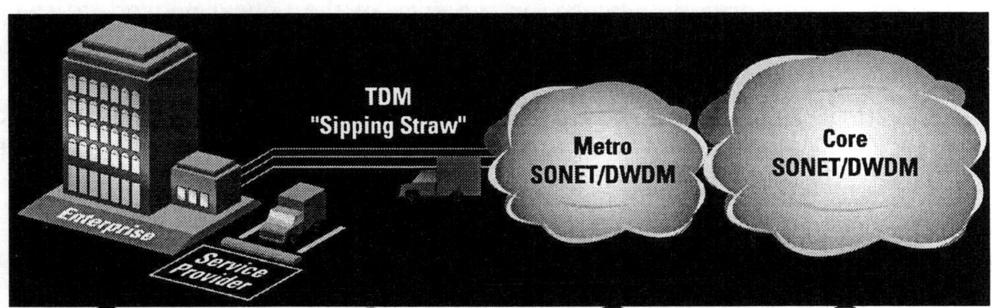

Figure 1: Access—A Sad Story

In many cases, service providers connect business buildings and office parks in metropolitan areas over optical fiber loops so that fiber enters directly into the building. However, the services that the company receives from those fibers are multiples of T1 (1.5 Mbps). Looking at the access piece of this optical network, it becomes apparent that, unless there is a different way of using the fiber that is already coming into the building to offer services beyond T1s, a provider really cannot give the end user the advantage of all this capacity being deployed in the network. The issue with TDM architecture in the last inch of the last mile where the fiber terminates is that, first of all, the bandwidth hierarchy is very rigid. When an end user wants more bandwidth from the service provider, the service provider must roll the truck to the building to enable or provision new services.

Appian Communications ordered a second T1 recently, but it had to wait 14 weeks to get the second T1 provisioned from the synchronous optical network (SONET) add/drop multiplexer (ADM) inside the building, despite the fact that there are two optical carrier (OC)–3s terminating in the building. It involved multiple truck rolls because Appian has an Internet service provider (ISP) that leases a T1 from the local-loop provider. Not only did the local-loop provider have to provision another T1 and run a copper wire from its wiring closet, but the ISP had to come out to the building to upgrade the router and enable inverse multiplexing. With all of the optical capacity deployed in the core network, two OC–3s coming into the building, and the capacity in the local-area network (LAN), it remains a hassle to upgrade a simple service from one T1 to two T1s. In this case, the service provider could not get revenues for the second T1 for 14 weeks—one whole quarter.

Network Bottleneck

ISPs can appreciate the implications of the TDM sipping straws that are in the network, terminating this optical network. In the AON, it is necessary to remove this final big bottleneck (see *Figure 2*). The TDM technology must go if the AON is used to deliver high-speed data services. Again, focus is on the business end users, on not residential or co-location facilities, or on ISPs as customers.

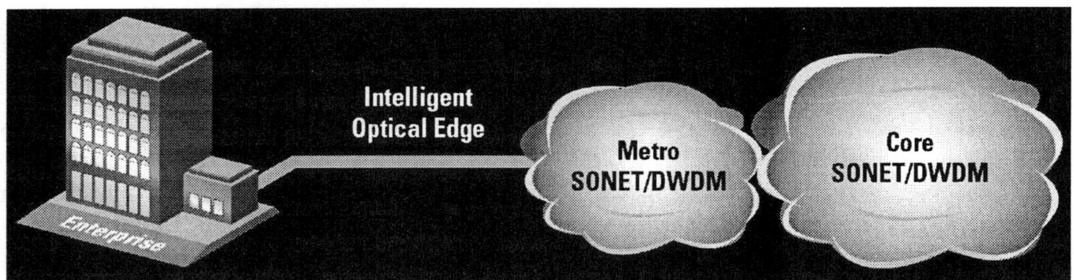

Figure 2: The Next Wave of the Optical Evolution: Remove the Final Bottleneck

It is important to know how to drive intelligence in the edge of the optical network to remove this last bottleneck and to offer tunable bandwidth so that it is possible to provision a T1 service and then remotely tune up the bandwidth as the end user needs more bandwidth—all without rolling the truck. A company should be able to activate the next level of service instantly and remotely without waiting 14 weeks. And 14 weeks is relatively short. A financial services company has a digital signal (DS)–3 network that it wants to upgrade to OC–3. The company's service provider said that eight months would be necessary to upgrade the company's virtual private network (VPN) from DS–3 to OC–3. The company must now decide whether or not to go directly to OC–48, as it does not know how much bandwidth it will need in eight months.

To overcome this problem, suppose there were intelligence at the edge of the network in a platform that could reside at the edge of the existing SONET transmission infrastructure. DWDM could offer services to multiple users inside the multitenant buildings or office parks. This could integrate into the existing infrastructure both from a transmission point of view as well as the point of presence (POP) switching and routing point of view.

Existing Infrastructure

It is a key requirement for service providers not to have to throw away the infrastructure that is already in place. That is true for the transmission infrastructure as well as the routing and switching infrastructure. If there were a platform that could provide connections to multiple customers, end users, and businesses and if that platform could carry traffic over to the optical network intelligently and could communicate directly to the optical transmission nodes (SONET or DWDM) and to switches and routers such as asynchronous transfer mode (ATM), frame relay (FR), or Internet protocol (IP), new services based on existing and emerging networks could be deployed.

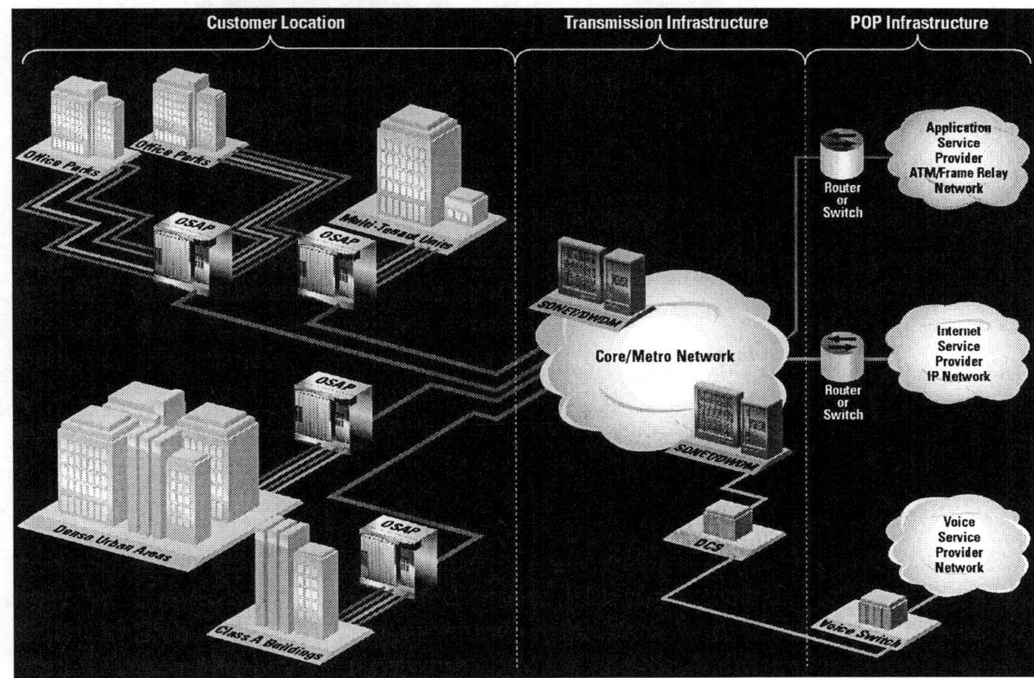

Figure 3: Support Current and Evolving Networks

The access or the edge is all about where LAN and wide-area network (WAN) meet from an enterprise network's point of view. Today's LAN is predominantly IP over Ethernet, running over copper or fiber. There may be a 100-Megabit Ethernet LAN backbone that could be copper or fiber or a Gigabit Ethernet backbone in the LAN that is fiber. The current WAN infrastructure, however, has IP running over whatever different Layer-2 technology may be out there, such as FR, ATM, or point-to-point protocol (PPP) over SONET or DWDM. The industry believes that IP over wavelength will happen, but will it happen tomorrow? Next year? The year after? A key point in an AON from the access-architecture point of view is to be able to work with the existing infrastructure (which is IP over frame/ATM or PPP over SONET) as well as IP straight over wavelength.

Capabilities

To facilitate that need, the platform at the intelligent optical edge that sits between the enterprise LAN and the service-provider network must have several things. First, it must be able to integrate at the transport level over the existing transmission infrastructure, which means that it must be able to connect into OC–3, OC–12, and OC–48 and to drive multiple wavelengths directly (see *Figure 4*). Today, it is SONET. Tomorrow, it might be a DWDM–based network. At the same time, it must have the capability to interwork with any routing and switching infrastructure in the POP. There is no reason why Ethernet, which is ubiquitous in the LAN environment and is experiencing the steepest cost decline of any networking technologies, cannot be used as the universal services interface for data services.

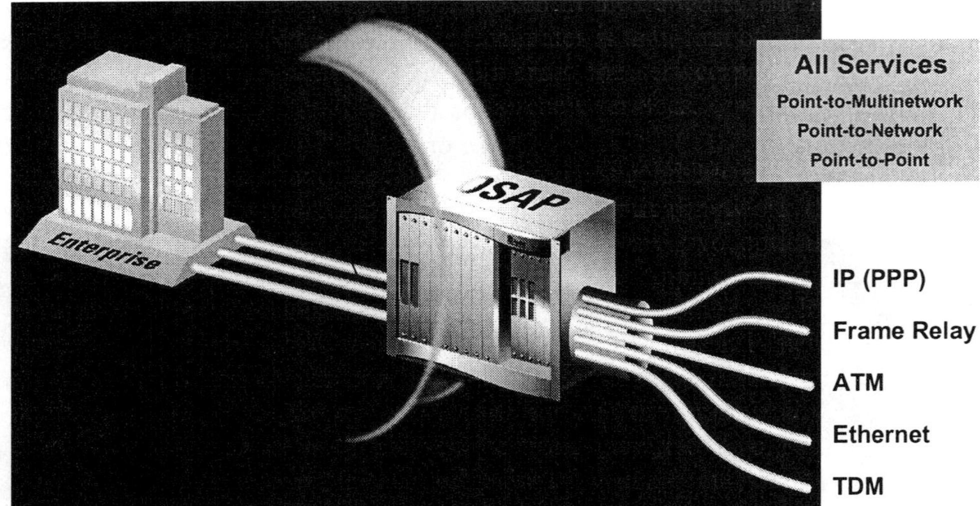

Figure 4: Multiservice Interworking

What if this kind of intelligent optical-edge platform also had the capability to use Ethernet technology for that universal services interface on which all data services could be delivered, regardless of whether the POP infrastructure is IP, FR, or ATM? At the same time, service providers also derive 70 to 80 percent of revenues today from either T1 voice services or T1 or T3 private-line services. Therefore, this intelligent optical edge must support not only the Ethernet technology for data services but also T1 and T3 TDM services that could be viewed as a pass-through capability for this kind of platform.

Delivering data services over Ethernet is simple, but guaranteeing a certain amount of bandwidth on that Ethernet interface is not as easy. If a provider were to replace those TDM sipping straws in the last mile or the last inch of the last mile with packet technology and use packet technology to create this intelligent-optical edge, it would also need to be able to provide the same kind of bandwidth guaranteed by TDM technology. Appian calls it *packet private-line service*. The key is to be able to provide a guaranteed bit-rate service on the Ethernet interfaces so that a provider could come in from the optical network, terminate in this kind of platform, and provide the data services using Ethernet technology as well as private-line services with these guarantees. This would be done using the same Ethernet technology.

Because data is bursty, because the utilization of bandwidth tends to be bursty, and because the peak times in enterprise networks tend to be between 8:00 am and 10:00 am, at noon, and at 5:00 pm, it would be convenient to offer guaranteed bit rate and the ability to burst at a high rate for those periods when high bandwidth is needed.

This is possible only because there is abundant bandwidth or capacity deployed in the optical core. If that were not the case, this would make no sense.

Service Guarantees

At the same time, because real-time applications will be driven over an IP infrastructure, voice over IP (VoIP), or streaming video, it is also necessary to offer class of service (CoS) guarantees. Thus, this intelligent optical edge must deal with not only data services, but also real-time applications, such as voice, riding on the same infrastructure and feeding into the same optical infrastructure.

Service management is all about services (see *Figure 5*). How can a company provide on-demand services and end-to-end provisioning using this intelligence in the optical edge (see *Figure 6*)? Soft, tunable bandwidth, the ability for end users to burst instantaneously, and the ability for end users to increase the delivered guaranteed bandwidth are good, but if a provider cannot signal the core of the network to provision more bandwidth at the same time, there may be a bottleneck.

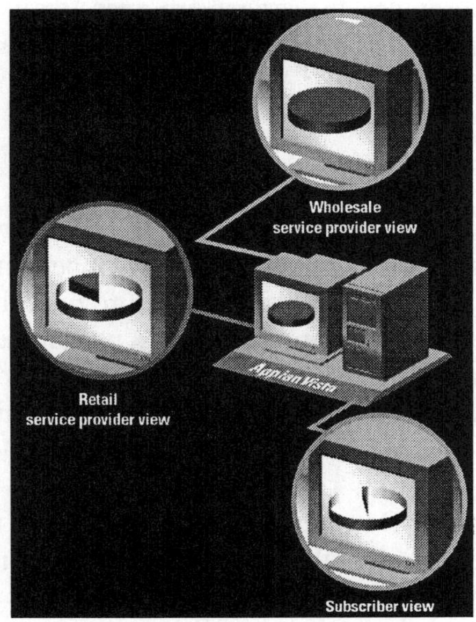

Figure 5: It's All about Services

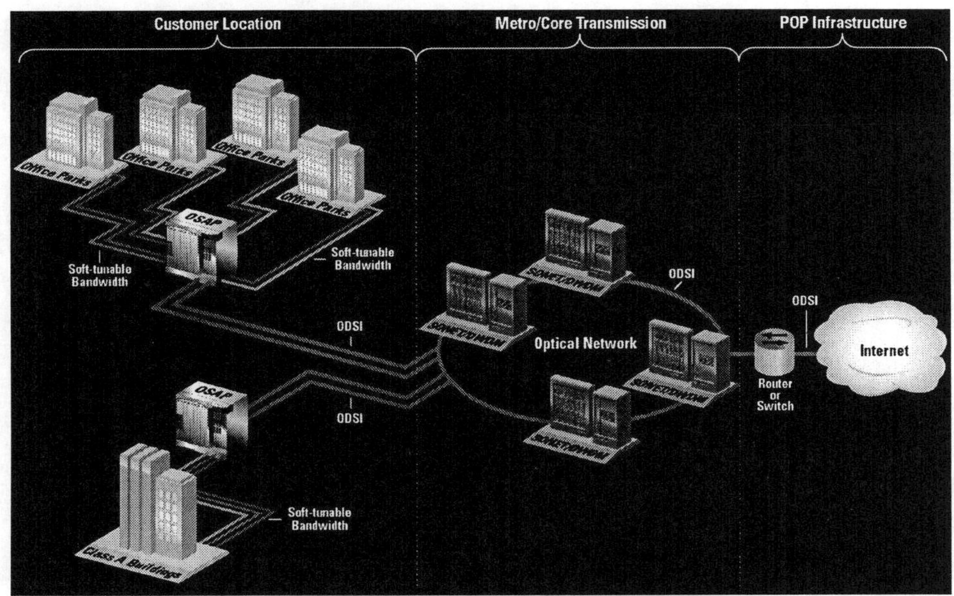

Figure 6: On-Demand, End-to-End Provisioning

The intelligent optical edge must be able to signal back to the optical network using either optical domain services interface (ODSI) or another protocol as defined by Optical Internetworking Forum (OIF). When the bandwidth requirement from the aggregation of the users exceeds the bandwidth provisioned between the intelligent optical edge and the optical core, the edge platform would signal the core platform to provision more bandwidth. If the industry can achieve this kind of signaling mechanism, it would be possible to create a scenario in which the end users themselves can ask for more bandwidth, and it would be delivered all the way through—not only within the service-provider core, but also all the way down to the customer premises. That is what the industry hopes to enable.

Service management is essential. If Appian were able to get its Internet access service over a 100-Megabit Ethernet interface from the service provider, the service provider would not have to make Appian wait for 14 weeks for a service upgrade. The service provider would not have to lose revenue for those 14 weeks. How services are managed is a key issue.

Customer Network Management

Customer network management is the customer's ability to view the resources and services that are delivered by the service provider. When looking a step further, it becomes necessary to have the partition services management capability so that the wholesale service provider that owns the facilities can view all the resources in that network. The retail service provider has a partial view of resources and facilities used and services offered. Then, of course, the subscriber—the end user—can view the subscribed services, their parameters and performance. It is necessary to have both this partition-services-management capability in the intelligent-optical edge and the capability for customer self-service.

It would be nice if, in November and December when e-commerce is at its peak during the holiday season, customers could turn up their bandwidth based on performance monitoring. Customer self-service thus becomes a key issue in the intelligent optical edge—and dynamic bandwidth provisioning becomes possible with the specification efforts led by ODSI or OIF.

Consider, for example, the financial services company previously mentioned. It might have an operations center in Boston, a trading center in Chicago, a call center in Dallas, and a backup center in Atlanta (see *Figure 7*). It has a VPN from a service provider that connects those buildings, and, of course, it has an

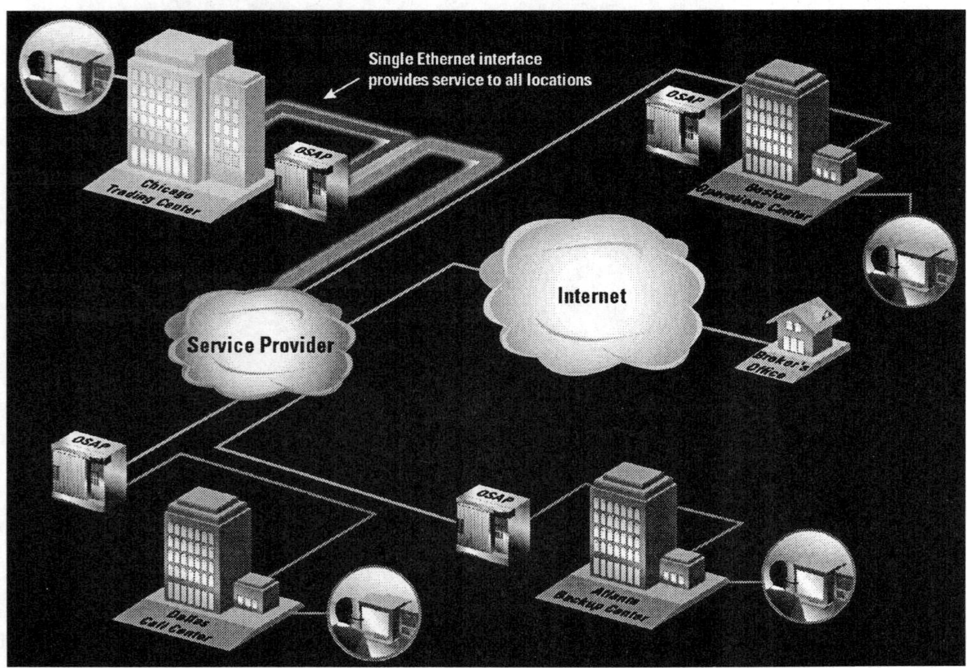

***Figure 7:** Business-to-Business Application—Financial Services Company*

Internet connection for its on-line trading transactions. The company would want a capability that allows it to tweak the bandwidth between its different sites depending on the time of day. For example, at 10:00 am when institutional trading is at a peak, it would want more bandwidth between the operation center, the trading center, and the call center (see *Figure 8*). During lunchtime, when individuals get on the Internet to check their stocks and then trade, the company needs more bandwidth between its private network and the Internet (see *Figure 9*). In that example, the intelligent optical edge can deliver higher bandwidth to the endpoints and the servers, and something like ODSI could be used to signal the optical network to provision more bandwidth to the edge devices.

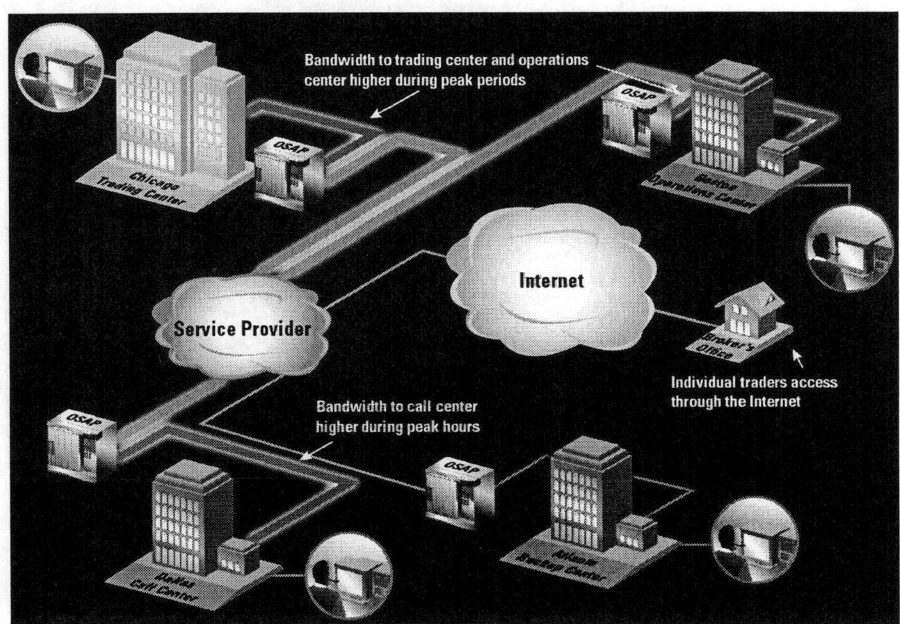

Figure 8: 10:00 am

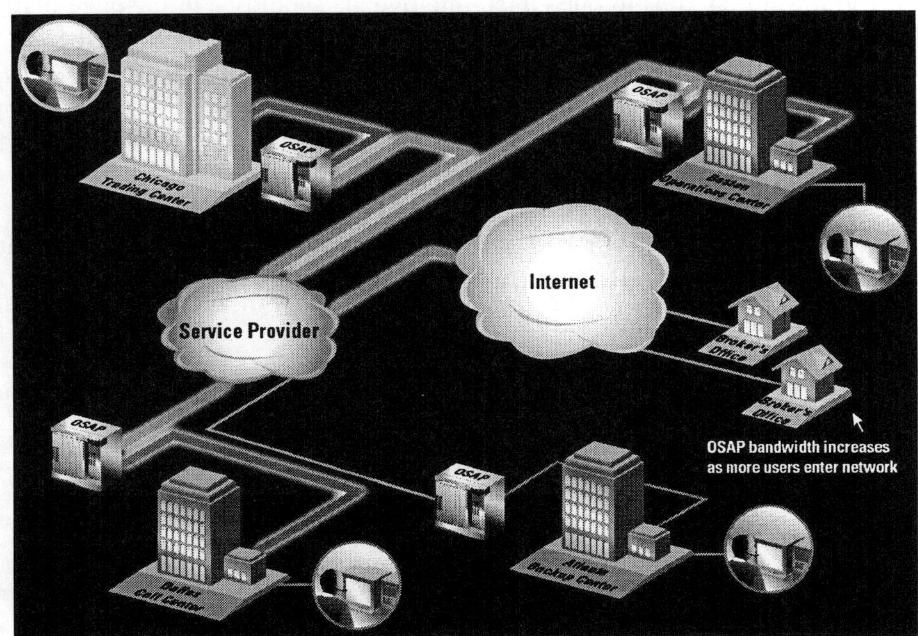

Figure 9: Lunchtime

At 7:00 pm, when it is backup time, the company wants more bandwidth only between the trading center and the backup center (see *Figure 10*). Bandwidth to other locations is turned down. The beauty of this system is that the company pays for bandwidth only when it uses it. The bandwidth is available or could be borrowed from the optical network only when it is needed, as opposed to having it there all the time.

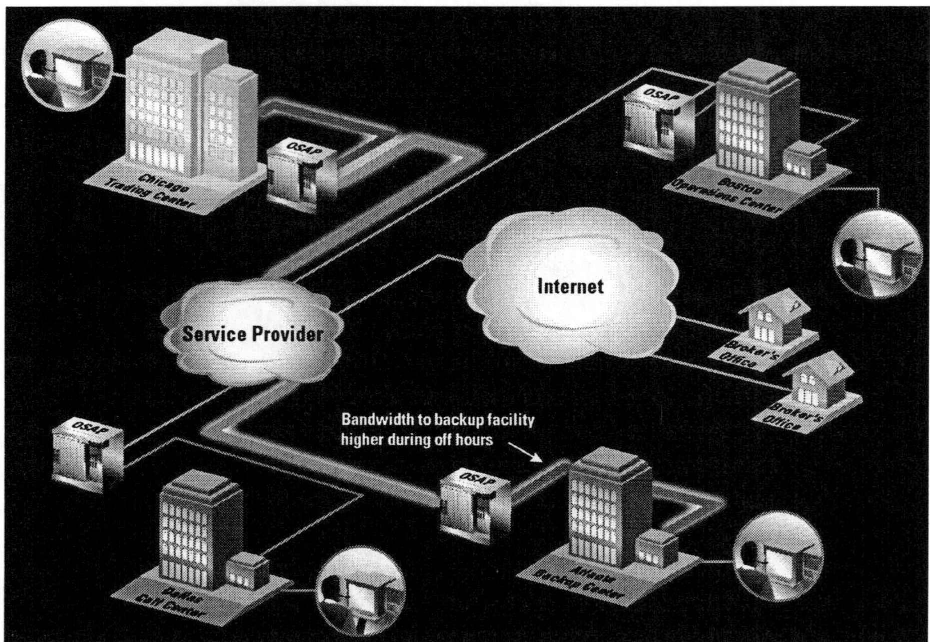

Figure 10: 7:00 pm

Summary

To work properly, the access in the AON should, first of all, remove that rigid bandwidth hierarchy. The TDM in the last inch of the last mile should be removed and replaced with packet switching, on-demand soft, tunable bandwidth from 64 kbps to one Gbps. Intelligence should be in the edge platform to enable the soft tunability of bandwidth to be turned up and down on demand. Truck rolls should be eliminated, as each truck roll costs anywhere between $500 and $1,000. In New York City, it costs $1,000. This becomes exceedingly expensive. Furthermore, the true realization of end-to-end provisioning is to shrink service delivery times from months to minutes. It is not only within the network—it is all the way to the end user.

Having bandwidth in the optical network is wonderful, but the only way to harness that power is by empowering end users (the customers) and by placing certain intelligence at the edge of the network.

Transparent LAN Services over Metropolitan Optical Networks

Luc Roy
Director, Solutions Marketing
Alidian Networks

Introduction

Corporations are starving for bandwidth, but to boost their current bandwidth they must pay high prices and often purchase more than they require. To the end user, native transport at native speeds eliminates the need for dedicated circuits, data conversion equipment, and managing wide-area network (WAN) protocols.

For service providers, an optical-service network (OSN) as a metropolitan-area network (MAN) removes the complexity of supporting transparent local-area network services (TLSs) by natively transporting Ethernet and other protocols over a single wavelength and by providing feature-rich TLSs. Best of all, an OSN offers the service at 1/10 the price of current TLS offerings in North America.

TLS

TLSs provide native local-area network (LAN) connectivity to multiple sites over the MAN and the WAN. Connectivity has traditionally been over Ethernet, token ring, and fiber distributed data interface (FDDI), but today's TLS customers seek fast Ethernet and Gigabit Ethernet services.

What Is the Pain?
Since the early 1990s, TLS has fallen short of end-user and service-provider expectations.

To end users, Ethernet networking is much cheaper, much faster (56 kbps versus 10 Mbps), and far less complicated than WAN networking. TLS is not cost-effective—end users pay astronomical prices ($3,000/month/Ethernet + usage).

To service providers, TLS means excessive bandwidth, network complexity, and configuration chaos. To relieve the bandwidth glut over the existing time division multiplex (TDM) infrastructure, service providers tended to overlay asynchronous transfer mode (ATM) or Ethernet networks. The rigidity of the TDM infrastructure and the overhead of ATM conversion did not offer the flexibility, performance, and cost-effectiveness sought by service providers. *Figure 1* demonstrates today's implementation of TLS.

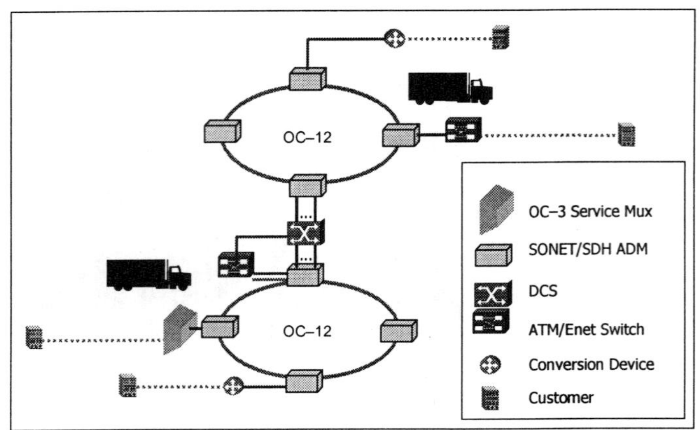

Figure 1: Today's Service-Provider Implementation of TLS

Why Is TLS an Important Service?

End users, medium/large enterprises, and specialized service providers alike have an insatiable need for bandwidth. Internet and corporate traffic is skyrocketing as more information is made available and as new Internet-based applications make it easier to harvest the information. Applications are developed on the belief that they will run over a LAN. When these applications meet the WAN, many modifications to both the applications and the network are required (e.g., enterprise resource planning [ERP] time-out). Censorware.org measured 61 billion bytes (text only) added on the Internet in 24 hours on February 28, 2000.

For service providers, an efficient and feature-rich TLS offering would mean an unprecedented, high-margin revenue stream from a data service.

TLS Requirements

End-User Requirements

1. High-speed Ethernet access—100 and 1,000 Mbps are mandatory, and 10 Gbps Ethernet must be planned. Legacy interfaces such as token ring and FDDI are no longer in demand because they are distance-restrictive and are slower and more expensive than Ethernet.
2. Multipoint topologies—Point-to-point (two sites interconnected) configurations represent a handful of the TLS opportunities. The TLS value proposition is shared media over the MAN without significant performance impact.
3. Privacy via virtual private networks (VPNs)—VPNs may be irrelevant to service providers (because it is a given) but to end users, VPN support is critical. Because the TLS is shared with other end users (and competitors alike), how VPNs are supported must be addressed.
4. Access rate control—Rate control is key to making TLS affordable. The average peak period of a corporate LAN utilizes about 30 percent of an Ethernet network. For many customers, more than a 30 percent rate control is overkill.
5. Traffic differentiation via quality of service (QoS)—End users will demand that they specify traffic discrimination, especially during congestion and when the rate control is exceeded. QoS is the ideal mechanism for communicating which packets can (in severe cases) and cannot be discarded. End users would specify the priority on their network devices using 802.1p or Internet protocol (IP) differentiated services (DiffServ).
6. Service-level agreements (SLAs) and flexible billing—More customers request SLAs in the form of measured end-to-end availability, average response time, volume, jitter variations, and packet loss. Billing flexibility based on bytes transferred and time of day is required to meet end-user preferences.

7. Cost-effectiveness—Aggressive pricing compared to existing services (e.g., digital signal [DS]–1 clear channel, DS–3 ATM, etc.) must be extremely attractive. Historically, service providers did not tackle the cost barriers of Ethernet over the MAN; in response, most end users preferred DS–3 instead of 10 Mbps because of the disparate pricing (DS–3 is one-third the price and five times the performance!).

Service-Provider Requirements

Service-provider requirements must be considered both at the access and at the core. From an access perspective, the solution cannot deviate from the end-user requirements, as specified in the previous section. From a core perspective, three simple requirements must be met:

1. Lower the cost per bit. Data traffic surpassed voice/TDM traffic on the public infrastructure in 1999, but data represents only 20 percent of the revenue due to the current infrastructure's implementation costs for data services. An efficient core network is fundamental in providing a cost-effective TLS service.
2. Eliminate truck rolls. With the explosion of bursty and asynchronous data services, synchronous optical network (SONET)/synchronous digital hierarchy (SDH) is ill-suited for today's new data services. A dense wavelength division multiplexing (DWDM) metro solution that burns up a wavelength per service will become costly and operationally challenging within one year. A data optimized–metro solution is mandatory.
3. Offer multiple services. TLS is one service, but the metro network must be leveraged for other services such as ATM, packet over SONET (POS), digital subscriber line access multiplexer (DSLAM) aggregation, TDM, storage-area networking, high-speed Internet access, and other emerging data applications.

The TLS Solution—An OSN Solution

The ideal solution for TLS will need to bring service intelligence to the next-generation MAN. Bridging service intelligence with the MAN means adopting a new solution that melds access and transport functionality and that offers service delivery, breadth, and awareness within a platform that scales intelligently. A new breed of DWDM network—an OSN—is required.

The OSN delivers the ideal solution for TLS. It melds access and transport functionality, offering point-to-multipoint topologies that perform exactly like a secured, switched Ethernet network while transporting other technologies such as ATM. *Figure 2* shows the features—specifically for TLS—that the OSN must deliver.

1. Access control
2. QoS
3. VPN support
4. Multipoint topologies
5. Integrated digital cross-connect system (DCS) and data switching
6. Multiprotocol and multiple-user wavelengths
7. Topology flexibility with SONET reliability
8. New management tools
9. Future direction

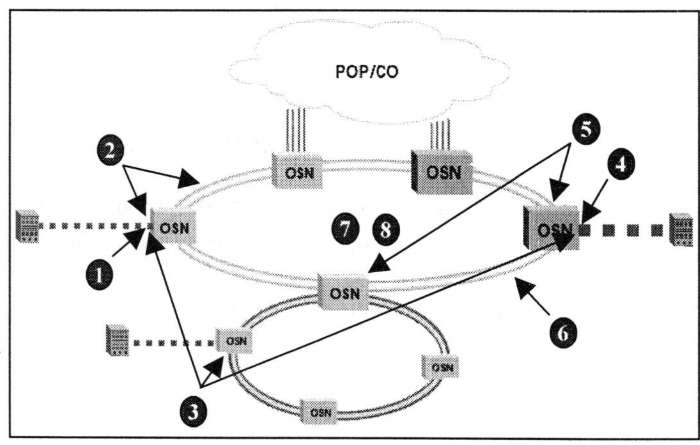

Figure 2: *Features an OSN Must Deliver*

Access Control

The OSN must support fast Ethernet (100 Mbps) and Gigabit Ethernet (1 Gbps), as per Institute of Electrical and Electronics Engineers (IEEE) 802.3 carrier sense multiple access (CSMA)/carrier detect (CD) (International Standards Organization [ISO]/International Electrotechnical Commission [IEC] 8802-3), IEEE 802.3u (fast Ethernet), and IEEE 802.3z (Gigabit Ethernet). The 10 Gbps Ethernet interface should be made readily available once the standard is completed (in approximately the first half of 2001). In support of long distances, the OSN should provide foreign exchange (FX) ports for fast Ethernet that can extend up to 2 km. Gigabit Ethernet can extend up to 5 km with the use of local exchange (LX) ports, or up to 50 km using the extended distance (XD) interface, all in single mode. Customer-located equipment is no longer required because all Ethernet traffic remains native end to end. Optical repeaters are only needed when distance specifications are exceeded.

The OSN's TLS should support all network protocols over Ethernet, such as IP, Internet packet exchange (IPX), AppleTalk, DECnet, Xerox network system (XNS), systems network architecture (SNA), NetBIOS, Banyan Vines, and more.

Each Ethernet interface must support a tunable access rate (average and burst), providing a better economy of scale for end users. The average rate represents a committed bandwidth that an end user elected at sign-up. The burst rate represents the allowable excess traffic in addition to the average rate. These two parameters provide exceptional protection to both the end user and the service provider to ensure that the network is properly designed, even during severe bursting conditions. *Figure 3* illustrates the average rate feature.

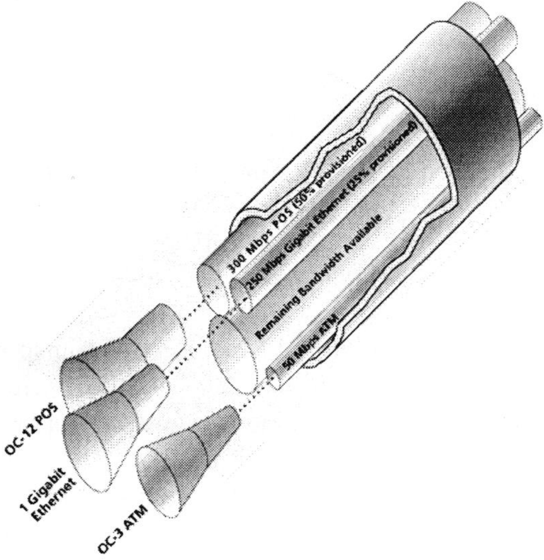

Figure 3: *Average Rate Control in Action*

Each rate control (average and burst) must be measured on a per-byte basis (within a predefined time period) instead of on a packet basis (like most other implementations), as shown in *Figure 4*. Byte-level measurements prevent queue starvation, which often occurs when receiving large Ethernet frames. Each Ethernet interface should support average and burst rates at increments of 10 Mbps.

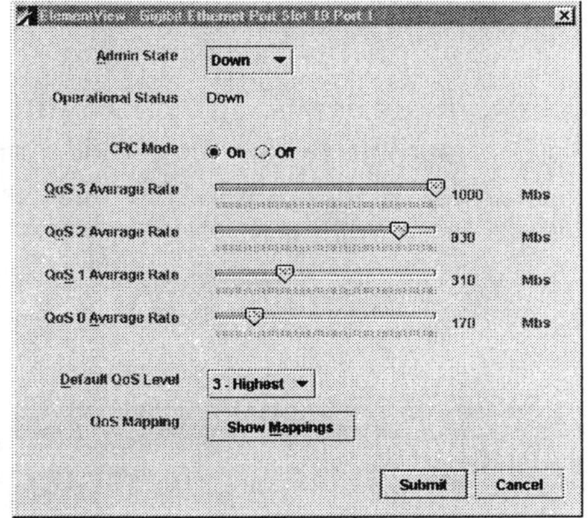

Figure 4: *Example QoS Configuration Graphical User Interface (GUI)*

The OSN access rate control system maintains high customer satisfaction while protecting the network against bandwidth starvation. All rate parameters must be configurable by the service providers to achieve the desired service performance. The OSN access rate control system would perform equally for all OSN data services (ATM and POS). For service providers that oversubscribe their network, the OSN's QoS features protect priority traffic.

QoS

QoS is extremely important for four reasons: differentiation of customer-service levels, treatment discrimination during normal operation, differentiation of traffic when exceeding average and burst rate, and differentiation of traffic when exceeding burst rate during congestion.

The OSN treats QoS by recognizing a combination of standards: 802.1p (Ethernet), type of service (ToS)/DiffServ (IP), multiprotocol label switching (MPLS), and virtual paths (ATM). For enterprise subscribers, 802.1p and DiffServ will be the predominant QoS standards used; for competetive local-exchange carrier (CLEC) subscribers, MPLS and virtual paths may also be used (see *Figure 5*).

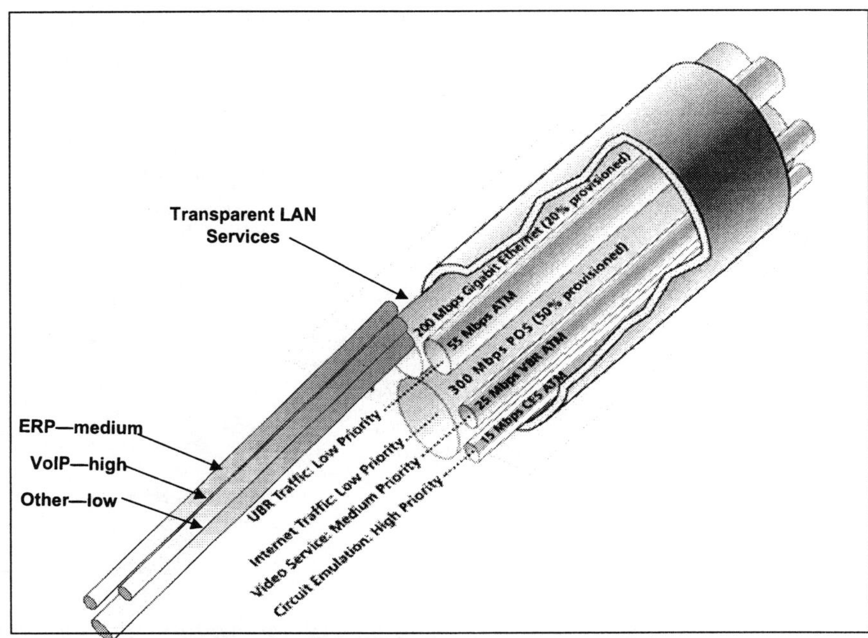

Figure 5: *OSN's Granular QoS Capabilities*

VPN Support

No TLS will be successful unless privacy can be guaranteed to every end user. The OSN offers high-performance and reliable VPNs configurable by service providers using virtual LAN (VLAN) standards 802.1Q or MPLS. End users can further configure multiple VLANs (within the OSN VPN) with 802.1Q, MPLS, IP security (IPSec), or Layer-2 tunneling protocol (L2TP). The OSN does not rewrite or change any customer-specified VPN or VLAN. As demonstrated in *Figure 6*, customer-initiated VPNs are maintained across the entire OSN.

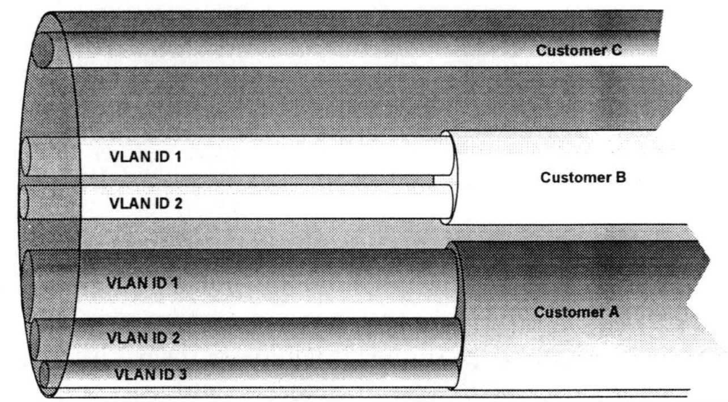

Figure 6: VLAN Capabilities

Multipoint Topologies

The OSN must deliver point-to-multipoint and point-to-point topologies. To allow multipoint topologies, the OSN allows the aggregation of many ingress ports into a single egress port (and vice versa), exactly like a native LAN switch. In comparison, some DWDM solutions can offer only point-to-point solutions because the traffic is multiplexed over a single wavelength. Even though point-to-point solutions can still provide TLS, most end users will require more than two sites to be interconnected.

In multipoint topologies, traffic must be sent efficiently to all other ports participating in the same VPN. With the OSN, the packet should remain on the ring until all participating nodes have received the packet. The last participating node to receive the packet drops the packet. This is possible through the OSN's embedded multicast mechanism. *Figure 7* demonstrates the aggregation and multicast abilities of the OSN.

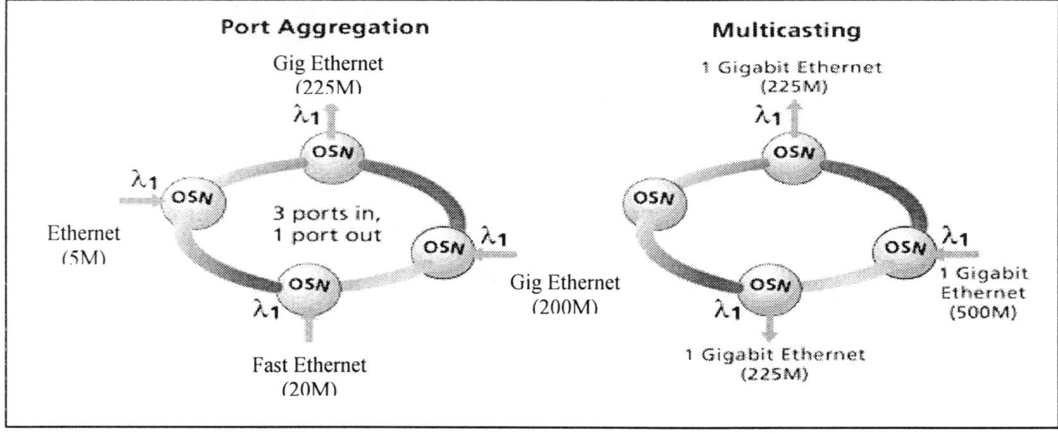

Figure 7: The OSN's Efficient Use of the Bandwidth in Multipoint Topologies

Integrated DCS and Data Switching

Traditional SONET/SDH networks imply many complexities when configuring or modifying a simple circuit end to end. These complexities range from identifying the available bandwidth end to end, to requiring external products such as cross-connects (XCs) and data switches to provide the desired connectivity.

The OSN lets individual services within a wavelength be protected, restored, or added/dropped at any metro node to create efficient hub and meshed topologies. Additions or modifications, as well as switching services between wavelengths, are accomplished just in time without interfering with existing services. This is achieved by switching from one wavelength to another (WaveSwitch). WaveSwitch eliminates the need for XCs or the one-wavelength-per-drop-point requirement of metro DWDM solutions.

With WaveSwitch, the OSN also eliminates the need for Ethernet switches. When a destination is on a separate wavelength, the OSN dynamically switches the traffic from one wavelength to another, automatically. *Figure 8* illustrates the WaveSwitch functionality.

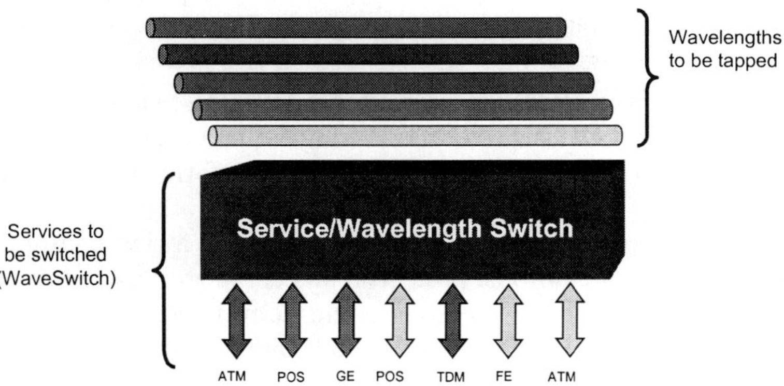

Figure 8: *WaveSwitch*

With an OSN, the service provider simply specifies the endpoints; the OSN takes care of addressing. Customer-specific configurations such as VPN and QoS are managed through the OSN's element (EMS) and network (NMS) management systems.

Multiprotocol and Multiple-User Wavelengths

The OSN allows multiple protocols and multiple users to share the same wavelength. Packets are carried in their native protocol, avoiding conversion to common protocol or rigid time slot. Through VPNs, multiple customers are integrated over the same wavelength without any intrusion.

For TLS, the OSN performs similarly to an Ethernet switch. Channels are not reserved end to end, but network paths are identified and switched accordingly through an internal identifier. Multiprotocol traffic need not be converted to a single common protocol, such as ATM, for transmission purposes. Conversion adds more than 10 percent overhead to the network, reducing available bandwidth by 10 percent and adding unwarranted complexity.

Topology Flexibility with SONET Reliability

The OSN should support ring, linear add/drop, and point-to-point physical topologies, as shown in *Figure 9*. Additionally, the network's underlying logical topology should be separated from the physical environment. Flexible support of TLS and other traffic flows requires the ability to carry hub and mesh logical topologies simultaneously across a common physical medium.

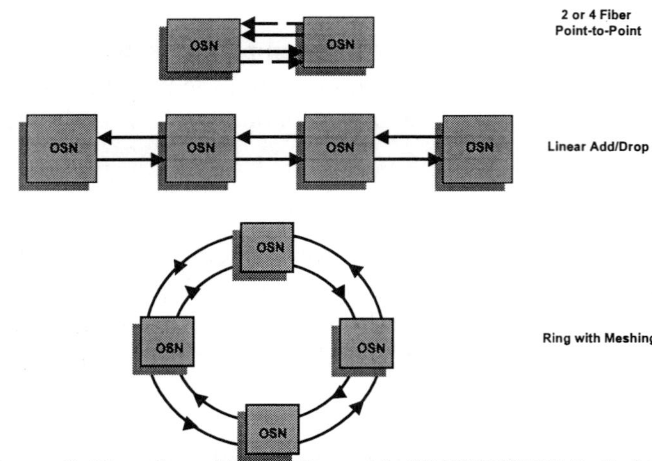

***Figure 9:** Topology Flexibility with SONET/SDH Reliability*

As more businesses and mission-critical applications utilize the public infrastructure for transport, customer demand for reliability increases. All the protection and restoration capabilities inherent in a SONET/SDH infrastructure are maintained with the OSN.

With SONET/SDH, the OSN could offer extensive error monitoring to allow speedy detection of failures before they degrade to serious levels. This error monitoring method is called bit interleave parity-8 (BIP–8). The OSN also provides rapid fault isolation and end-to-end performance monitoring.

Some ideal restoration capabilities come from a ring topology. In a ring configuration, the OSN could deliver two types of restoration capabilities: protection switching and uni/bidirectional traffic transport. With the protection switching technique, the OSN could deliver two modes: line-protection switching (1:1) and path protection (1+1). The OSN supports both bidirectional line-switched ring (BLSR) and unidirectional path-switched ring (UPSR).

New Management Tools

Because of the revolutionary advantages to the OSN, new management tools must be considered, especially for the planning of data traffic over a DWDM network to complement the traditional fault, configuration, accounting, performance, and security (FCAPS) management. The network management solution designed to facilitate the deployment of the OSN for any environment should consist of three main components:

1. An element management platform
2. OSN agents on each network element (e.g., simple network management protocol [SNMP] management information base [MIB])
3. A comprehensive planning tool that simplifies the integration of multiple services over a DWDM metro network.

The EMS should be a comprehensive, standards-based element management platform for the entire OSN family. The EMS supports the telecommunications management network (TMN) model and provides full FCAPS management functionality.

Strict adherence to industry standards facilitates OSN integration into existing operations support systems (OSSs) and NMSs such as MegaSys Telenium and Nortel integrated network management (INM). The OSN and the EMS would utilize SNMP, transaction language 1 (TL1), and hypertext transfer protocol (HTTP) management protocols. Common object request broker architecture (CORBA) support could also be provided to enhance further operational simplicity; operations systems modification of intelligent network elements (OSMINE) and NMA can also be required, especially by incumbent local-exchange carriers (ILECs).

Fault Monitoring and Alarm Handling

The OSN/EMS must provide an extensive monitoring and alarm management facility that can be readily integrated into any existing OSS. Network elements (NEs) may be configured to filter alarms to manage alarm/trap storms that accompany major failures better. Alarm correlation at the EMS enables operators to isolate failures and problems rapidly. Traps should be directed to multiple sites.

Configuration

The EMS should provide a simple point-and-click connection management to enable just-in-time, flow-through provisioning.

Performance Monitoring Enables Service-Level Management

OSN/EMS must deliver comprehensive performance statistics needed by diverse users throughout the organization. Connection utilization and error statistics are provided to enable verification of SLAs. Port statistics provide the basis for capacity planning and usage-based billing, and SONET–level statistics enhance fault isolation and analysis. Performance reports or statistics can be exported in circuit-switched voice (CSV) format to any applications. Performance statistics can be delivered to subscribers in support of SLAs.

Performance alarms may also be used to flag peak or excessive usage and underutilized interfaces, elevating customer satisfaction to new standards.

Accounting

The OSN supports per-connection monitoring at multiple layers in the stack, which enables service providers to offer value-added billing services billed by bytes transferred, packets transferred, and time of day.

New Planning Tool

The new planning tool facilitates the design and analysis of the physical, logical, and optical connectivity of the network. Preplanning is especially critical for DWDM networks where the logical and optical topologies are intertwined. The planning tool delivers in-depth analyses such as bandwidth utilization reports, optical loss analysis, and equipment lists. The network configuration based on the operational OSN can be imported at any time for what-if and growth planning. *Figure 10* demonstrates a snapshot of the new planning tools required for an OSN.

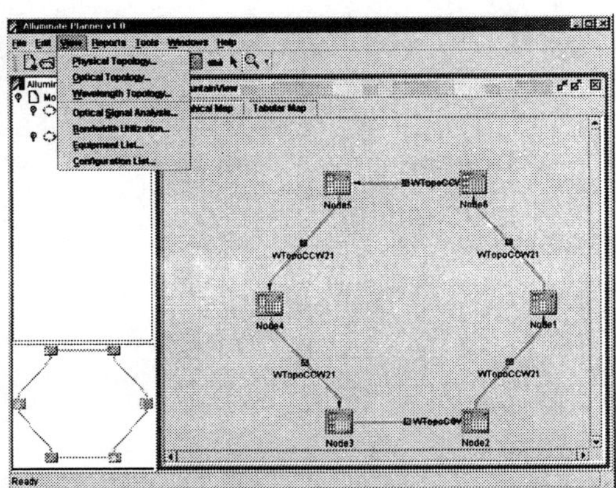

Figure 10: Sample of a Planning Tool

Business Opportunity

TLS has never been a high-revenue opportunity for service providers because of complexity and costs due to TDM's rigid multiplexing hierarchy and to the multiple layers of technology required to offer TLS. Recently, RHK predicted that data services will increase by 2,000 percent by 2005. To tap the revenue opportunity of TLS, a new approach must be considered to allow a flexible and affordable service.

Today's TLSs across North America range from $3,000 per month to $7,000 per month, with an additional $400 per month for every 10 Mbps committed information rate (CIR). Alternatively, a DS–3 clear channel sells for approximately $1,400 per month, an optical carrier (OC)–3 sells for $1,800 per month, and a leased lambda costs approximately $500 per month per mile. Unfortunately, subscribers typically pay higher prices and purchase more capabilities than they need.

An OSN can take a completely different approach to broadband delivery to offer DWDM scale with service breadth, delivery, and awareness. The following TLS business case demonstrates the significant savings attributed to an OSN. Pricing was based on Alidian Networks' OSN 4200.

Basic Assumptions and Requirements
- Interconnect three fast Ethernet LANs
- All sites have Layer 2 (L2) Ethernet switches with single-mode LX connectors.
- All sites are within one mile of the service provider's point of presence (POP).

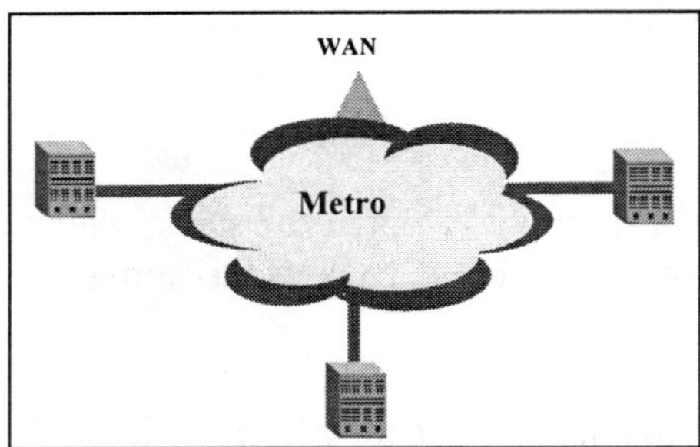

Today's TLS Offers
Assumptions:

100 Mbps access *(source: AT&T)*	$7,320/month
10 Mbps CIR *(source: AT&T)*	$401/month
Fiber cost *(source: Robertson Stephens)*	$500/month/fiber/mile

Calculations (two-year service):

3 * 100 Mbps access (@ $7,320/month)	$527,040
1 * 100 Mbps CIR (@ $401/month/10 Mbps CIR)	$96,240
2 * 50 Mbps CIR (@ $401/month/10 Mbps CIR)	$192,480
Two-year lease for fiber (one mile per site)	$36,000
Total two-year customer cost:	**$851,760**

Leased Lambda Scenario
Assumptions:

- Leased dark lambda (*source: Robertson Stephens*) $500/month/fiber/mile
- Fast Ethernet optical link extender *(estimated)* $1,000/device

Calculations:
- Six fast Ethernet link extenders $6,000
- Two-year lease for dark lambda (3 * 20 miles) $720,000
- **Total two-year customer cost:**   **$726,000**

Note: Subscriber will require two ports per site.

OSN TLS
Assumptions:

- Local fiber access $500/mo/fiber/mile
- Alidian fast Ethernet port cost $156/month/port
    - Includes fully redundant 4200 total cost
    - Includes installation and maintenance
    - Amortization over 24 months
    - Four-lambda system (upgradeable)
    - All prices are list
    - Configured for maximum access rate at 100 percent (worst case)
    - Service provider up charge (* 3) $468/month/port

Calculations:

- Dark fiber cost (one mile * three sites for 24 months) $36,000
- TLS (3 * 24 months) $33,696
- **Total two-year customer cost:**   **$69,696**

Bottom Line
- Subscriber cost (TLS + fiber access) $968/month/site
    - No additional ports or product required on-site
- Subscriber savings $32,586/month
- Service provider margins (50 percent) $1,452/month

Of course, the service provider can offer more profitable and flexible TLS plans based on demand, rate (fast Ethernet access with rates ranging from 10 Mbps to 100 Mbps or Gig access with rates ranging from 100 Mbps to 1 Gbps), and reliability (protected lambda versus unprotected). Multiple tenant units (MTUs) would not incur the cost of the dark fiber.

The cost per bit of an OSN (with protection) is less than 1/10 (on a single wavelength!) than an add/drop multiplexer (ADM) (without including the DCS and muxes!). Further savings will increase from the multiwavelength capability of the OSN up to 32 wavelengths over a single fiber pair (reducing cost to 1/320).

Summary

Over the past three years, a significant shift has occurred in the type of traffic traversing public networks, from voice to data. Data traffic has become the primary driver for dramatic increases in bandwidth demand. In the next three years, data is forecast to comprise 80 percent of traffic, but at the same time, only 20 percent of revenues. Rapid growth in bandwidth demand coupled with a shift in the type of traffic and revenue model creates several new challenges for service providers existing in an increasingly competitive market.

End users have demanded TLS for years, but most end users could not justify the pricing. Ethernet is the norm within the enterprise, and end users, small to large, are looking at leveraging their Ethernet installed base with TLS that extends to the MAN and even the WAN.

With a true OSN, service providers would be able to get wavelength efficiency in DWDM networking with service-level intelligence and scalability from 10 Mbps to 10 Gbps Ethernet, while increasing their profits. In addition, end users should be able to buy a 100 Mbps Ethernet (fast Ethernet) TLS today for the same price they paid for a T1 circuit of yesterday.

OSN will return the revenue model to service providers. The economics of metro DWDM will change the way service providers regard the technology—as a potential revenue generator—by using DWDM as a multiservice platform.

Next Wave of Optical Networking: The Metro Arena

William C. Szeto
Vice President, Technology and Architecture
Iris Labs

This paper will address the lessons learned when implementing a real-life network and will concentrate on a study of the optical networking companies.

Much has been learned since the industry began to put fiber in the ground. For example, it was learned that fiber characteristics could be tracked very well by taking a trace every six months. Each trace can be compared to the previous six months' to see if anything has changed. Sometimes the earth moves and minor changes occur. A periodic study of these changes enables prediction and prevention of changes in the fiber. Problems with groundhogs and other animals also arise sometimes. Groundhogs like to eat fiber, but if a fiber is put in a 3.5-inch conduit the groundhog cannot open its mouth wide enough to chew it. These are some of the lessons learned.

One of the more serious issues in planning an optical network has come up since the introduction of dense wavelength division multiplexing (DWDM) systems. Initially, the plan did not provide an end-to-end solution. Large wavelength routers and optical cross-connects (OXCs) are involved in the core network with long-span DWDM systems with more than 100 channels. However, the critical fact that there is another network—the metro and access network—was ignored. Other considerations involve fiber-to-the-home (FTTH) and a fiber-rich metro and access network.

The fibers today are with the core network. There is no doubt that many people are putting fiber in the metro areas. Many start-up companies are putting fiber in buildings' risers. However, that is still several years behind the core network, as is the DWDM for metro networks. Instead of talking about hundreds of channels, like the systems for the core network, the metro DWDM usually involves 18, 24, or 30 channels. Most of the systems are not designed with amplifiers.

Many people follow Internet development. Where will Gigabit Ethernet be used? It will not be at the core network directly but in the buildings where the business users are located, or perhaps someday at consumers' homes. How will that impact the core network? There could be OC–192 running in the core to support these users. However, it is difficult to deliver this kind of bandwidth from the business customers or from the homes to the core network today. The development of one chunk of the network has been forgotten.

Optical Networking

Because of this neglect the core network has abundant capacity today—hundreds and hundreds of channels. The solution moved the bottleneck downstream. An illustration of this process can be seen in the long, tube-like balloons that clowns use to make figures. When these filled balloons are squeezed at one end, the air moves to the other end. In the industry, people took the capacity in the core and squeezed the bottleneck down to the metro or access network. However, the bandwidth demand continues to grow as people talk about 10 Gbps Ethernet. How will that be delivered from a house or an office to the core? That issue still needs a solution.

The current solution does not necessarily put DWDM systems, fibers, and lasers at a user's desk. It actually goes way beyond that. End-to-end solutions to the users are required. What does that mean? End users would decide to have certain features. For example, an end user may have a Gigabit Ethernet network that he would like to be able to manage. Perhaps he wants bandwidth on demand or to order certain quality of service (QoS) features whenever needed. Maybe he wants to increase or decrease service at a certain time of the day. That is currently impossible because no standard exists that allows talking through the network from end to end.

That there are no existing end-to-end standards brings up a second point about the lessons learned. When the optical network was developed and fibers were put in the ground, an important piece was forgotten: standards. Many people say synchronous optical network (SONET) is a standard. Yet who has tried to combine different vendors' SONET equipment in a single network? There are many variations of SONET standards.

There must be an end-to-end standard in the network. The future network has to be able to interoperate with the core network in a common, standardized manner to allow the user to manage and control the network and to deliver the network features from end to end.

How will that be accomplished? A key is to study the standards works being done in today's industry. There are many standards committees, like the T1 committees, the International Telecommunications Union (ITU), the Internet Engineering Task Force (IETF), and many others, which work to create standards to allow the interoperation of future optical networks.

The standards must be monitored closely because they impact carriers and vendors alike. It would be advantageous if one vendor's equipment worked with every other vendor's equipment. A carrier with a network that allows the users to communicate from end to end has an advantage that can really impact business.

Metropolitan Network Transport Requirements*

Scott T. Wilkinson
Director, Kestrel Atlanta
Kestrel Solutions

Advanced multiservice metropolitan networks are characterized by diverse demands and by an unpredictable traffic pattern. In these networks, the transport layer must carry a variety of signals in a scalable, compact, and economic fashion to have an optimal transport network. Factors such as low initial cost, scalability, high capacity, ease of engineering, and simple operations are required to support metropolitan transport networks. The combination of the transport layer with lower-capacity service platforms at the edges results in an optimal flexible, scalable, and manageable solution for metropolitan network needs.

Characteristics of the Multiservice Metropolitan Network

The metropolitan market is typically characterized by both a wide variety of service offerings and a constantly changing network demand profile. Demands for services range from traditional digital signal (DS)–1/DS–3 services to standard synchronous optical network (SONET) offerings such as optical carrier (OC)–3, OC–12, and OC–48 to non–synchronous transfer mode (STM) signals such as D1 video, asynchronous fiber-optic terminal (FOT) signals, enterprise systems connectivity (ESCON), and others. Demands can change rapidly and unpredictably as customers grow, move, or change their networks. Solutions for building multiservice metropolitan networks must be able to cope simply and economically with a variety of signals in a rapidly changing environment.

Whereas long-distance networks have fairly predictable traffic demands (SONET OC–192, OC–48, etc.), metropolitan networks can be required to carry a huge variety of traffic types on a single network (see *Figure 1*). Many customers still require asynchronous electrical signals, such as DS–1 and DS–3, and this demand is not expected to disappear completely anytime soon. Also, non–STM signals—such as D1 video, ESCON, fiber channel, Ethernet, legacy asynchronous FOTs, and others—are not unusual in metropolitan environments alongside more traditional OC–*n* signals. Even among the SONET–based OC–*n* signals, the protocols can be a variety of concatenated (Internet protocol [IP] and asynchronous transfer mode [ATM]) and nonconcatenated signals.

---

* Portions of this paper were presented at NFOEC '00. Some details of the network study have been updated due to constantly changing market economics and element capabilities.

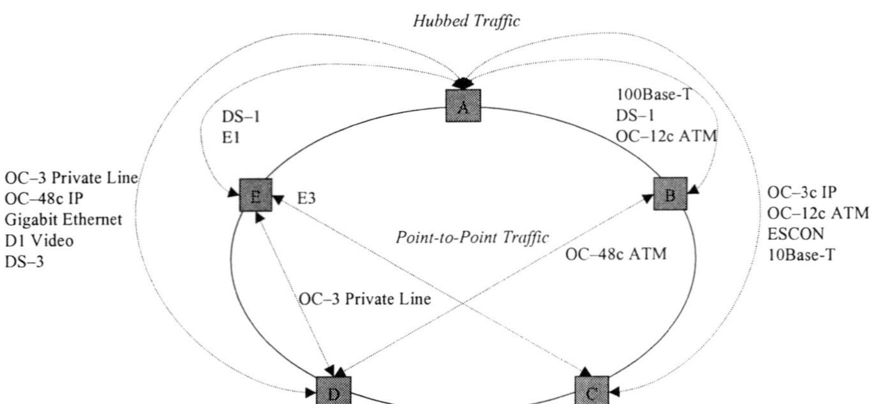

Figure 1: *Metropolitan Network Traffic*

A single *service* platform solution for carrying all possible traffic types in their native form would be complicated and expensive and would require an excess of capability in some locations simply to serve occasional demands in other areas. For example, it can be quite expensive to deploy a platform with a virtual tributary [VT]–level cross-connect [XC] in every location if demand for DS–1s exists in only one location. However, the ability to combine all traffic onto a single *transport* platform can be very beneficial, reducing the number of fibers and the amount of network monitoring required, and increasing the manageability of the network. A transport platform should be able to carry large-capacity signals natively and to leave management of the more granular signals to an external element. In this way, the expense of managing highly granular XCs and multiplexers (such as DS–1s) can be localized onto a service platform in only the locations where such expense is justified.

The one characteristic of metropolitan multiservice networks that makes designing for the future the most difficult is unpredictably changing demands. Customers increase demands or move from one location to another, buildings change tenants (e.g., the computer service center that recently moved in requires much more bandwidth than the bakery that moved out), office parks expand, and the popularity of the Internet creates demand for broadband services in apartment complexes (see *Figure 2*). Time is of the essence in

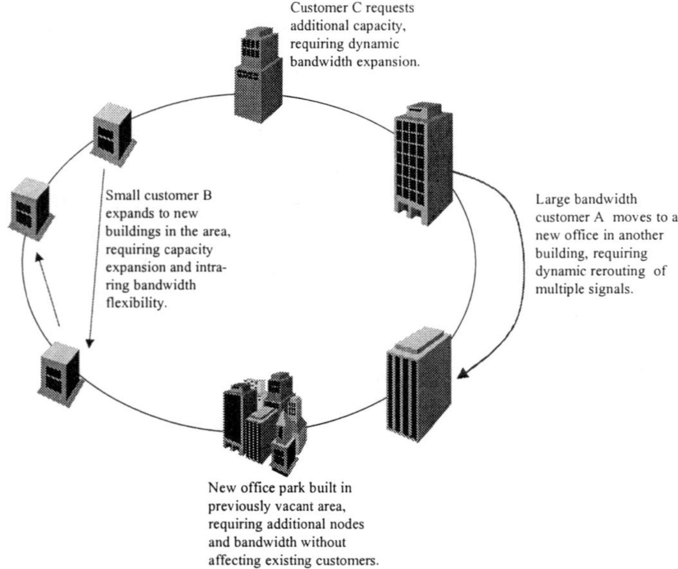

Figure 2: *Customers' Changing Demands on the Metropolitan Multiservice Network*

the metropolitan market, where customers can be lost to the competition if services cannot be provided in a timely fashion. Simply overbuilding the network is not economically viable, but underbuilding a network can create time delays when responding to new service demands, and, therefore, revenue losses. Thus, a solution to metropolitan multiservice traffic demands must be flexible to respond rapidly to changes in demand, and scalable to keep initial costs low, while at the same time providing the ability to increase capacity to support future requirements.

First cost, life cycle cost, scalability, and flexibility are very important in the metropolitan market as carriers increase the capacity of the network incrementally in response to customer demand. First cost enables a short-term return on investment from new revenue streams to pay for the capitalized infrastructure improvements. Concurrent with the low first-cost requirement is the need to keep life-cycle costs down. Ensuring that the proposed solution integrates well with existing methods of operation and administration to minimize engineering, operations, and training costs involves more than simply providing reliable equipment. Integration with existing methods of operation means not only interoperability with existing equipment and management platforms, but also sharing a common "look and feel" with traditional network elements (NEs) to minimize the amount of training that support personnel must have prior to using the equipment.

Scalability implies that capacity can be built out only as it is required without stranding an investment in bandwidth capacity when demand does not materialize. Because the network demands in multiservice metropolitan networks can change dramatically and unpredictably, scalability gives network planners insurance against future changes. Flexibility implies that signals can be added and dropped without restrictions, allowing additional traffic to be terminated at a node or routed to a new location as demands change. This requires a remotely provisionable XC functionality. To reduce costs throughout the network, the XC should be less granular than the subtending service platforms (e.g., OC–n or wavelength-level routing).

Solutions for the Metropolitan Multiservice Network

A pure SONET solution to the multiservice metropolitan network has many beneficial features that have enabled the building of these networks in the past. However, as demand skyrockets, the negative aspects of SONET are starting to create problems for network providers. SONET offers such benefits as ease of management through standard interfaces, a high level of flexibility by including XCs at every node that can manage down to the payload level or below, full performance monitoring and fault isolation at every point in the network, and the ability to handle any electrical or optical signal that follows standard SONET or synchronous digital hierarchy (SDH) standards. However, SONET equipment has poor scalability and difficulty handling non–STM traffic. Because the demand in metropolitan multiservice networks can be very dynamic and unpredictable, network providers are faced with the difficult choice of overbuilding capacity with higher-speed SONET or performing costly and time-intensive upgrades down the line. Because SONET requires that the full high-speed capacity be provided at every node from day one, and because SONET typically requires a highly granular XC at every site—even when such granularity is not required—the start-up cost of a SONET system can be prohibitively high. Therefore, network providers have been reluctant to install large-capacity SONET (e.g., OC–192) in the metropolitan environment. However, a demand for only a few OC–12c circuits can swamp an OC–48 SONET network, and demand for an OC–48c circuit must be serviced as an overlay. In addition, demand for services such as D1 video, ESCON, asynchronous FOT services, and some native data services are difficult to service on a SONET network.

Dense wavelength division multiplexing (DWDM) networks are very well suited for servicing non–SONET signals due to the bit-rate and protocol-independent nature of most optical equipment. DWDM-based networks are also very good at handling larger-capacity signals, such as OC–48c, and they offer transparency so that management of subtending equipment can be accomplished using the already-established management networks.

However, DWDM as the only solution has several serious drawbacks. DWDM cannot efficiently (if at all) handle lower-speed electrical signals such as DS–1 and DS–3. In addition, DWDM networks typically involve complicated engineering rules that make deployment and rearrangement of traffic difficult or impossible, and DWDM networks can be difficult to scale—especially if lower-speed signals are placed onto wavelengths—due to the impact of additional wavelengths on the existing network. Rearranging traffic on a DWDM network can be a very complicated process because strict rules govern the number of nodes that can be passed, the distance between nodes, the total circumference of the ring, the number of wavelengths added and dropped at each node, the bit rates of each signal, and more. Finally, DWDM–based networks are generally more expensive to deploy due to the large number of optical-to-electrical (O–E) and electrical-to-optical (E–O) conversions required in the network. Optical-to-electrical-to-optical (O–E–O) conversions are required for wavelength translation, performance monitoring, and regeneration, and O–E conversions can be the largest source of capital expense in any optical network. In a DWDM–based network, every signal typically requires an O–E conversion at nearly every node (systems without O–E–O conversions at every node are forced to employ optical amplifiers, which add another layer of complexity and expense to any optical network). Because DWDM O–E–O transponders are tuned to specific wavelengths, large numbers of spares can also be required, further driving up the overall cost of a DWDM network.

The optimal solution for the metropolitan multiservice network should combine the versatility of SONET platforms for electrical services with the versatility of DWDM for optical and variable bit rate services. The solution must have a low start-up cost to be economically viable, must be scalable to avoid expensive equipment changeouts as demand rises, and must work with established procedures and management techniques to eliminate training and interoperability issues. In the rapidly changing environment of metropolitan multiservice networks, the solution must be able to rearrange traffic rapidly without resorting to heroic, time-consuming procedures.

Several approaches have been proposed to meet the needs of metropolitan networks by combining the best features of SONET and DWDM. Static multiplexing using time division multiplex (TDM) within DWDM elements increases the bandwidth efficiency of the network and helps to reduce costs by lowering the number of O–E conversions in the network. Combining DWDM capabilities into SONET–type elements is also possible, allowing expansion of the network as demands grow. Some very complex TDMs have been announced recently (often called "God boxes") that propose to serve all demand types on a single, low-cost platform. Several network providers have begun to implement conversion of all traffic to packets for better bandwidth utilization and flexibility. Finally, fundamentally new technologies are being introduced that are designed specifically for metropolitan networks. One of these technologies, optical frequency division multiplexing (FDM), is expounded upon in the following sections.

Optical FDM in a Multiservice Network

Optical FDM was designed specifically for transporting optical signals of any protocol and bit rate, and can be combined with traditional or next-generation SONET or ATM equipment to transport any subtending electrical services required. In many cases, optical FDM offers a cost-effective solution for carrying a variety of signals on a common platform with full flexibility, scalability, performance monitoring, and bit-rate and protocol independence.

Optical FDM is a unique combination of digital signal processing, FDM, and optical modulation. Optical FDM employs state-of-the-art components and forward error correction (FEC), guaranteeing consistent operation over any quality or type of single-mode fiber (SMF), from standard SMF (SSMF), dispersion-shifted fiber (DSF), and nonzero–DSF (NZ–DSF) to older, high–physical-medium dependent (PMD) fiber. Because optical FDM requires all signals to be converted to an electrical format, full regeneration of all incoming signals is performed, and the span budget is guaranteed for all configurations independent of the amount of pass-through traffic. This means that there are no restrictions on the number of nodes through which a signal can pass. By removing restrictions based on fiber type or number of nodes, optical FDM greatly simplifies network design and engineering.

With an optical FDM transport layer, more expensive service-level equipment can be placed at the network edge. This allows network providers to avoid paying for extensive feature sets at every node when, for example, VT–level XCs are required at only a few sites. The backbone transport system offered by the optical FDM solution can transport any high-bandwidth optical signals efficiently without eating up more expensive capacity on a full-services element. Because all signals are converted to an electrical format, a fully flexible XC can be implemented that can be used to reroute traffic dynamically on the ring, while subtending equipment performs further grooming only where needed. The transparency of the optical FDM allows all subtending equipment to be managed from a centralized location without any interoperability concerns.

The scalability of the optical FDM means that network providers can start out with only the bandwidth required at a node on the network and then scale up as necessary—whether the requirement is for 622 Mbps, 2.5 Gbps, 10 Gbps, or an intermediate rate such as 3.1 Gbps (20 OC–3s equivalent) or 8.1 Gbps (52 OC–3s equivalent). While the low start-up cost enables a quick return on investment, the large capacity of the optical FDM (implementations of 10 Gbps and higher are available or proposed) means that in most situations, there will be sufficient capacity to accommodate any unexpected increases in demand. In cases where greater capacity is required, the single wavelength output of optical FDM can be combined with inexpensive passive DWDM to provide significantly higher capacity. In essence, the capacity and scalability of optical FDM provide cheap insurance (a.k.a. "future-proofing") for a network provider in today's unpredictable telecommunications market.

Customer Example Network

The following analysis is based on an actual customer network. The network consisted of five five-node rings with a variety of services unevenly distributed across the rings and the nodes. Services requested varied from large-capacity OC–48c IP circuits to DS–1/E1 circuits to non–SONET D1 video and ESCON circuits. This distribution is typical of a multiservice metropolitan network, where large demands for multiple optical channels can coexist on the same ring with a customer using one or two DS–1s. The total demand across all of the rings is detailed in *Figure 3*.

This network was modeled with four solutions: SONET OC–192, "next-generation" SONET OC–48, DWDM, and optical FDM. The SONET and DWDM solutions each had benefits in transporting some of the services (e.g., OC–48s for DS–1/DS–3 transport), but each also had pitfalls that made them less attractive than an optical FDM transport layer solution.

The overall results from the analysis of the customer metropolitan network are detailed in *Figure 4*. Comparisons were made between the four solutions in the following categories:

- Total cost
- Total cost including fiber
- Common equipment costs
- Total number of managed elements
- Total number of rings
- Spare capacity remaining

Service	Total Demand
ATM OC–12c	5
ATM OC–3c	18
D1 Video	5
DSL	10
ESCON	5
Gigabit Ethernet	5
100B-T Ethernet	5
10B-T Ethernet	5
IP OC–48c	5
IP OC–12c	5
IP OC–3c	27
OC–12	30
OC–3	51
DS–3	30
DS–1	179
E1	9

Figure 3: Total Demand Across Rings

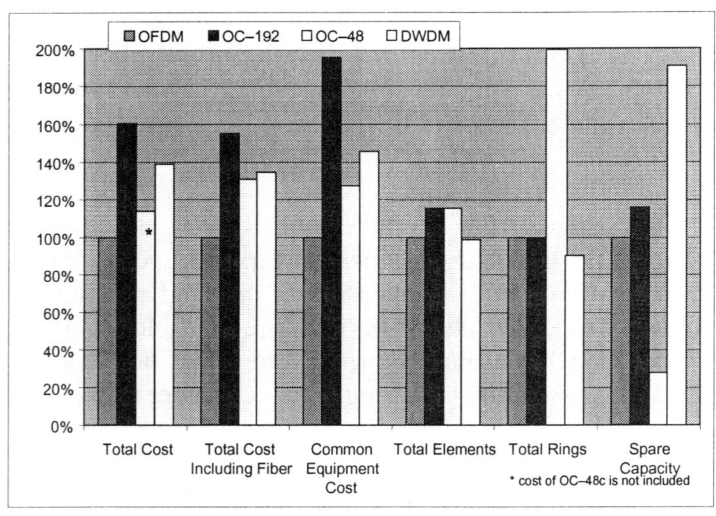

Figure 4: Customer Metropolitan Network Analysis Results

All solutions are normalized to the optical FDM solution, which is represented as 100 percent for all parameters. Total cost included the cost of all equipment required for the solution, including the primary equipment and all subtending equipment. Total cost including fiber is calculated using a price per km of $900 for fiber and is a measure of the total cost of a solution when extra fiber capacity is required. Common equipment costs include only the common equipment for the primary transport part of each solution. This number is representative of the initial cost of each solution, assuming that all of the capacity will not be installed at once. The total number of elements includes all managed elements—primary and subtending—and is an indication of the complexity of each solution. The total rings required is an indication of how many fibers are required for each solution. Finally, spare capacity is the amount of capacity left on all five rings once the network is completely built and is an indication of the ability of each solution to respond to sudden increases in demand. The spare capacity of the DWDM solution is calculated assuming an average of OC–12 per wavelength, which may be larger than the actual average based on real demand.

The SONET OC–192 model had the advantage of 10 Gbps capacity and an assumed ability to carry Gigabit Ethernet as well as all OC–*n* traffic on the same platform. For electrical signals lower than OC–3, external equipment was required—much as would be required in an optical FDM solution. The biggest missing piece for the OC–192 solution, however, was a facility for carrying D1 video and ESCON services. To make some sort of comparison, those signals were assumed to be converted to OC–12s using a theoretical external SONET multiplexer. The SONET OC–192 solution also suffered from a high start-up cost and high continuing costs due to the necessity of purchasing the expensive OC–192 common equipment up front.

The "next-generation" SONET OC–48 case was modeled assuming a single platform with the ability to carry OC–3, OC–12, Ethernet (10/100/1,000), DS–1, DS–3, and E1 traffic. However, the OC–48 solution was unable to handle OC–48c traffic at all and required an additional overlay fiber network whenever OC–48c service was requested. This made the OC–48 solution the largest user of fiber and the most expensive when fiber costs were included. Also, the OC–48 case suffered from the same inability to carry D1 video and ESCON as in the OC–192 case, requiring the theoretical external OC–12 multiplexer. Finally, the OC–48 case left very little spare capacity for expansion. Whereas the OC–192 and optical FDM platforms can carry a full 10 Gbps of information, the OC–48 case required a new ring to be installed every time 2.4 Gbps steps were exceeded. The first cost of the OC–48 case showed some improvement over the OC–192 case (a bit deceptively because the cost of OC–48c traffic was not included), but the first cost of the OC–48 case was still 30 percent higher than the optical FDM solution.

The DWDM solution was well suited to carry non–SONET signals (ESCON and D1 video) as well as SONET OC–*n* signals on the same platform. As in the SONET OC–192 and optical FDM cases, the electrical signals were carried on external multiplexers. The biggest weakness seen in the DWDM solution was lack of scalability and flexibility. As network demands change, a DWDM network has trouble responding quickly because more complicated engineering rules must be considered at each step in the network evolution, a fact that is not well represented by numerical analysis. The large spare capacity of the DWDM network was determined assuming an OC–12 per wavelength, which may not be the case because the current majority of signals are OC–3 or below (with OC–3s on the excess wavelengths, the spare capacity drops to less than 50 percent of the optical FDM solution). Additionally, the strict requirements placed on an optical network means that the spare capacity may not be available in the locations where it is required (unlike SONET or optical FDM solutions). Also, the DWDM solution suffered from higher cost points than the optical FDM solution—both in initial and ongoing cost—because O–E–O conversions were required on every optical signal to meet the wavelength plan. Not included in this analysis is the cost of sparing for a DWDM network, which is also expected to be significantly higher than an optical FDM solution due to unique sparing requirements for each unique wavelength.

The optimal solution for the network model was an optical FDM transport network with "next-generation" SONET deployed at the edges only when required. This solution was by far the most cost-effective:

- More than 60 percent less expensive than the OC–192 case
- More than 35 percent less expensive than the DWDM case
- More than 30 percent less expensive than the OC–48 case when the cost of the additional fiber required is included

The optical FDM solution efficiently combined the flexibility of the SONET cases with the signal-carrying ability of the DWDM case. External multiplexers were deployed only where DS–1, DS–3, or 10/100Base-T services were required. All OC–*n* traffic, regardless of protocol, was carried natively on optical FDM, as were other non–SONET optical signals (D1, ESCON). Management of the subtending equipment—including customer equipment, if required—could be performed from a central location due to the optical FDM's transparency. The initial cost of the optical FDM solution was the lowest of all of the solutions. Rings with lower initial capacity demand could be deployed with a partially filled optical FDM solution (high-speed and low-speed) to reduce first cost, then could easily be upgraded (adding circuit packs) to increase the ring capacity as demand increased. The amount of fiber required in the optical FDM case was comparable to the OC–192 and DWDM cases and was less than 30 percent of the fiber required in the OC–48 case due to the 10 Gbps maximum capacity of optical FDM. The spare capacity in the optical FDM case was also significantly higher than the OC–48 case for the same reason. On the few rings where capacity exceeded 10 Gbps, optical FDM could have been combined with passive DWDM to reduce fiber requirements further if fiber was at a premium. By providing additional spare capacity with a lower start-up cost, the optical FDM solution effectively provided cheap insurance against any future unexpected network growth.

Conclusions

Multiservice metropolitan networks are characterized by a large variety of bandwidth requirements and by an unpredictable demand profile. The optimal solution for these networks must be flexible, scalable, manageable, and economical. By using optical FDM as the transport layer in this metropolitan network, network providers can deploy a platform with the bit-rate and protocol independence of DWDM and the flexibility and ease of deployment of SONET—with the added bonus of low first cost and scalability as the network demand grows. More expensive and complex service-level equipment can then be deployed at the edges where required.

The optical FDM solution offers the lowest first-cost solution and the lowest overall-cost solution. In addition, the optical FDM solution combines the ability of DWDM to handle non–SONET traffic natively with the flexibility and reliability of SONET. The optical FDM solution uses significantly less fiber than a "next-generation" OC–48 solution with much higher spare capacity, costs much less than an OC–192 solution—initially and overall—while retaining the 10 Gbps capacity of an OC–192 solution, and is more flexible and simpler to engineer than a DWDM solution with a much lower cost-sparing requirement. For metropolitan multiservice networks, an optical FDM solution is, in many cases, the optimal transport network architecture.

Optical Access Networking: The Last Mile of the All-Optical Network

Anthony Zona
President and Chief Executive Officer
Quantum Bridge Communications

Introduction

This paper discusses various aspects of all-optical networks (AONs) including the impact of the technological and business landscape changes that the Internet has wrought on the sector. It addresses the effects telecom reforms and innovations in optics have had on carriers, and it outlines strategies and solutions carriers are adopting as they attempt to cope with a rapidly changing environment.

The Last Mile

This paper is concerned primarily with the last mile. Obviously, a carrier hoping to build an AON must go beyond the point of presence (POP) to extend the optical network to more carriers. What are the requirements for access, how do they differ from backbone and metro, and what do those networks look like? The first products introduced are based on passive optical networks (PONs) because of their unique value proposition when they relate to an access plan.

Some changes in the industry facilitate optics deeper and deeper into the network and touch more end carriers. Part of that is simply the learning curve of optical technology, which doubles in performance every nine months and halves in cost. It enables options that were impossible even a few years ago, and as soon as that bandwidth becomes available it is consumed. In that respect it is like the microprocessor and memory; it becomes available in applications, it is filled immediately,and there is call for more. There is no end in sight.

The Internet Business Landscape

Capital markets continue to fuel the boom. They fund carriers and component companies that provide industry observers with insight into what will be possible in the future. More important to carriers is the money that goes into new carrier models. Ambitious competitive local-exchange carriers (CLECs) are aggressive about fiber deployment in the first mile. If they bring fiber to the building, they will own the revenue stream of that building. There is a first-mover opportunity and thus much movement in that area.

The Internet economy drives everything. The Internet model—its content and movement to the edges—is changing the face of telecom. It is a disruption of the old-world model where local access was a loss leader and all the money was made in long distance. Now touching customers, controlling accounts, and pushing services to the edge are paramount.

All-optical technology started in the backbone, in submarine, and in long-haul. It has since moved to interoffice feeder and has begun to move beyond the POP into access. Access is the next wave of optical technology. Copper plant was replaced by an optical and electrical system, and eventually there will be an all-optical environment (see *Figure 1*).

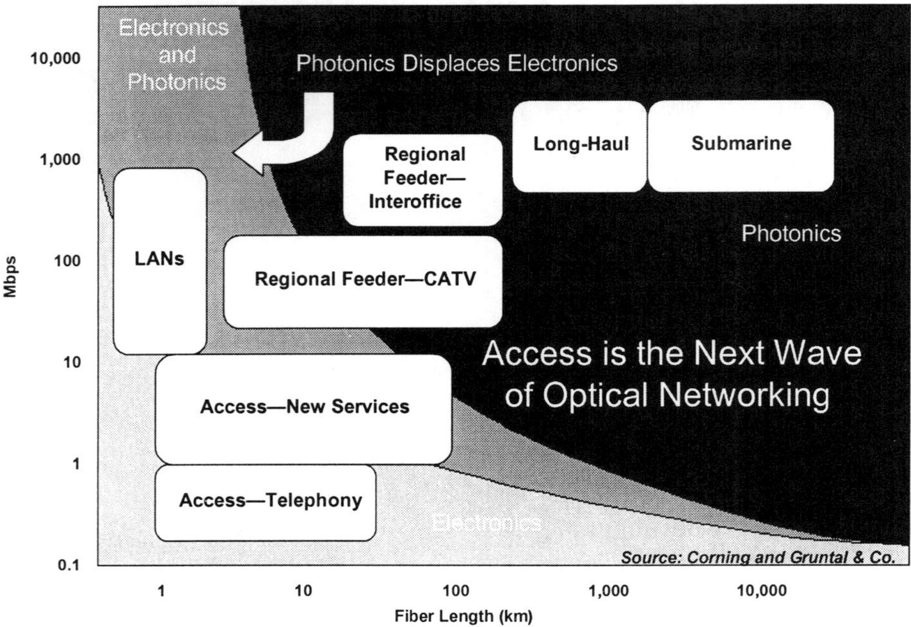

Figure 1: *Waves of Optical Deployment*

Access

What is unique about access? First, it involves many endpoints, unlike the backbone, which has fewer nodes. National Football League (NFL) cities, for example, are interconnected. If a carrier moves toward the edge, the number of endpoints and the variation in churn increase. Obviously, deep into the network, more traffic aggregates and there is less variability. Toward the edge, the carrier has to handle churn and variability in bandwidth demand. There are different barriers to entry. Some topologies make more sense in access than in, say, metro. For instance, rings in access are ineffective when it comes to many endpoints; a ring is appropriate for a few nodes. Thus, three to four nodes on a ring are sufficient in a metropolitan environment, but numerous nodes on infrastructure rings are unnecessary.

PON is a point-to-multipoint system (see *Figure 2*). Carriers receive tremendous fan-out, and they can activate those endpoints when revenue is there. Thus, it is an opportunistic model that resonates with CLECs. When they build a network, they do not know exactly where the customers will be. They have to build the infrastructure and then add customers to the network. Therefore, they need a scalable and flexible network. It must be flexible in terms of distributing bandwidth because of the sheared architecture. They can shear bandwidth among all the endpoints.

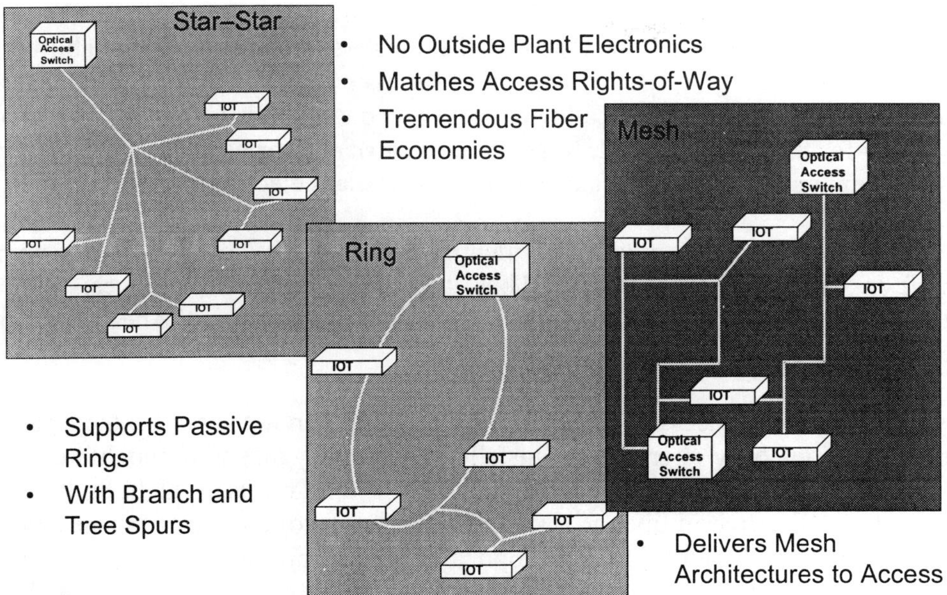

Figure 2: Passive Optical Networking

With this model, carriers can shift capacity to where the band hot spots happen to be as a function of time, time of day, and time of year. It is scalable, passive, and broadband. A carrier can overlay wavelengths on the infrastructure to add capacity. In addition, it can accelerate the fiber in access by leveraging what fiber already exists. Thus, cable companies, multiple service operators (MSOs), CLECs, and competitive access providers (CAPs), that built fiber infrastructures in the late 1980s and 1990s, have limited fiber already in access; they can distribute those limited resources, capture many endpoints, and bring them into the network.

The first incarnation of the network is a branch-and-tree network. That is the most logical design, and it matches the current design of the outside plant in access. In the copper plant, carriers take copper from an edge office and bring it to a feeder distribution interface. They can fan it out again to the edge of the neighborhood and then fan it out to homes. It is a star–star type of architecture.

Carriers can also build rings. They can build a passive ring, which is similar to a unidirectional path switched ring (UPSR) where carriers hub the bandwidth back to the node in the edge office, consolidate it if there is a fiber cut, and then switch directions to allow a counter rotation of the ring. The main advantage of a passive ring is that it offers carriers the ability to deploy branch-and-tree spurs. Thus, they can have both ring architecture and branch-and-tree on the same infrastructure, which allows them to segment the market from the endpoints that want the resiliency of path protection and the endpoints that are unwilling to pay for it.

PON also enables carriers to deliver mesh architectures to access. Without expensive electronics in the outside plant, they can link passive branches and allow the end nodes to decide in which direction to put traffic, depending on the congestion and fiber cut in the network.

In the early 1990s, CAPs decided to strive for what was called the tall, shiny building (TSB). It was an opportunity that represented a high concentration of bandwidth consumption and low variability in terms of aggregate bandwidth. It was also easy to identify—simply fly a helicopter over a city, pick the tallest points and decide that that is where to bring fiber. Carriers took fiber to the basement of the building and installed a synchronous optical network (SONET)–add/drop multiplexer (ADM).

That same opportunity exists with clusters of small or medium-sized businesses that represent high concentrations of bandwidth consumption across the clusters, possess low variability in aggregate demand, and are easy to identify. Carriers can, again, fly over business parks and indicate clusters that represent bandwidth consumption. PON allows carriers economically to drive fiber to one of those buildings and to pay for the network build. It then allows carriers to bring other carriers into the network with only passive couplers in the outside plant and no outside plant electronics.

PON also allows carriers to create pools and identify customers by bandwidth consumption and to manage the shift in bandwidth as necessary. Thus, they can offer retail services, sell parts of their networks, and offer wholesale services simultaneously, managing it all from the head end with partitioned customer control.

Again, there is fiber already in access. CAPs have built metropolitan networks for the last 10 or 15 years. Utilities and municipalities have aggressively installed fiber along major thoroughfares and connecting municipal buildings. Perhaps most aggressive have been the MSOs, which built out their cable infrastructure based on HFC (hybrid fiber/coax). In fact, 76 percent of mid-sized businesses will be within one mile of a fiber source. Therefore, they are optically passed, but not optically served, today. Some would argue that those customers should be brought into the network by leveraging the fiber that is already in place, using passive couplers, and fanning the fiber out (see *Figure 3*).

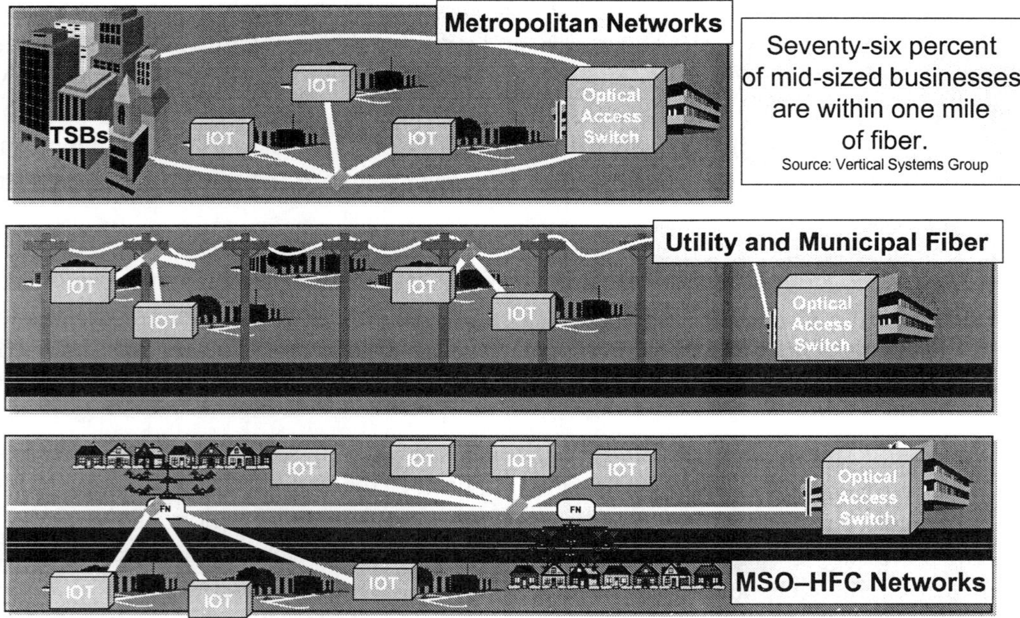

Figure 3: Low Infrastructure Investment

There is also an opportunity born of deregulation and various Federal Communications Commission (FCC) rulings. The FCC ruled that the incumbent local-exchange carriers (ILECs) had to unbundle all the pieces of their network infrastructure, including the fiber that extended to digital loop carriers (DLCs). That created an opportunity for CLECs to access fiber at the DLC, which happened to be a perfect location—close to business clusters. Naturally, the ILECs revolted. The FCC then passed an additional ruling—99-238—that dictated that as long as ILECs deploy packet-based infrastructure they do not need to unbundle their fiber assets. Instead, they must provide wholesale interfaces to that packet-based infrastructure.

This has accelerated fiber deployments by some ILECs based on packet technology instead of on traditional time division multiplexing (TDM) technology. For example, SBC Communications has announced *Project Pronto*, an aggressive look at building out a fiber-based architecture based on next-generation technology—packet-based with fiber both to the building and to the neighborhood gateway infrastructure. The race is on. Both CLECs and ILECs will be aggressive over the next five or 10 years deploying fiber in access. There is activity in both market segments. Communications Industry Researchers (CIR) projects a $600 million opportunity for products this year. An optical access network affords carriers a foundation for retail and wholesale services, broadband and legacy, and the chance to build the foundation of the next-generation e-economy (see *Figure 4* and *Figure 5*).

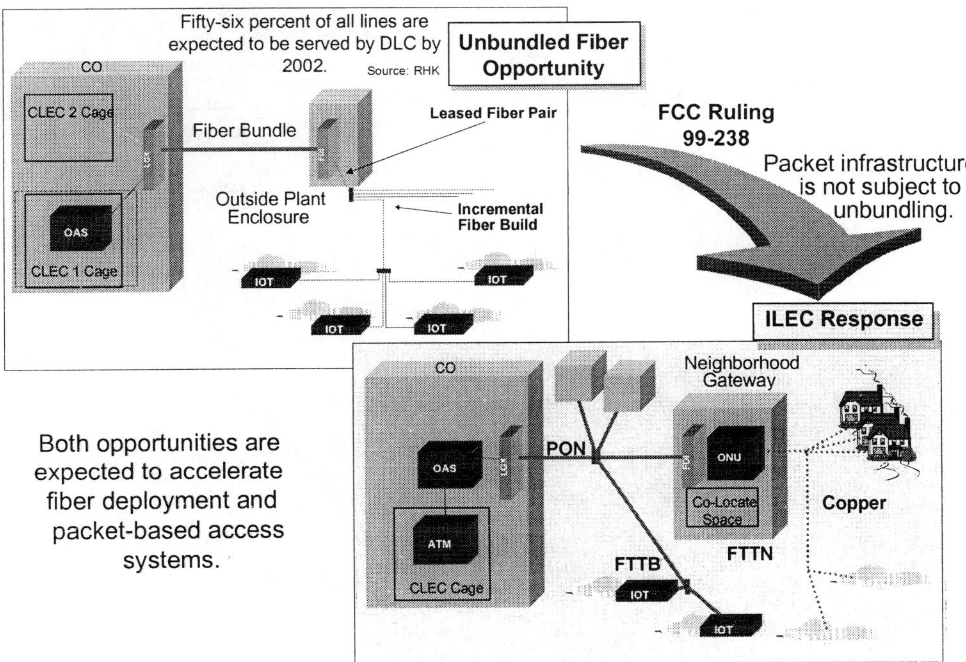

Figure 4: The Leased Fiber Opportunity

- CLEC Options
  - Unbundled or Leased Fiber
  - Leased Bandwidth
  - New Construction
  - Leverage Existing Plant
  - Acquire and Upgrade Plant
- ILEC Competitive Response
  - Avoid Unbundling by Deploying Packet-Based Access Systems per FCC 99-238
    - Rehabilitate Existing Plant
    - New Build

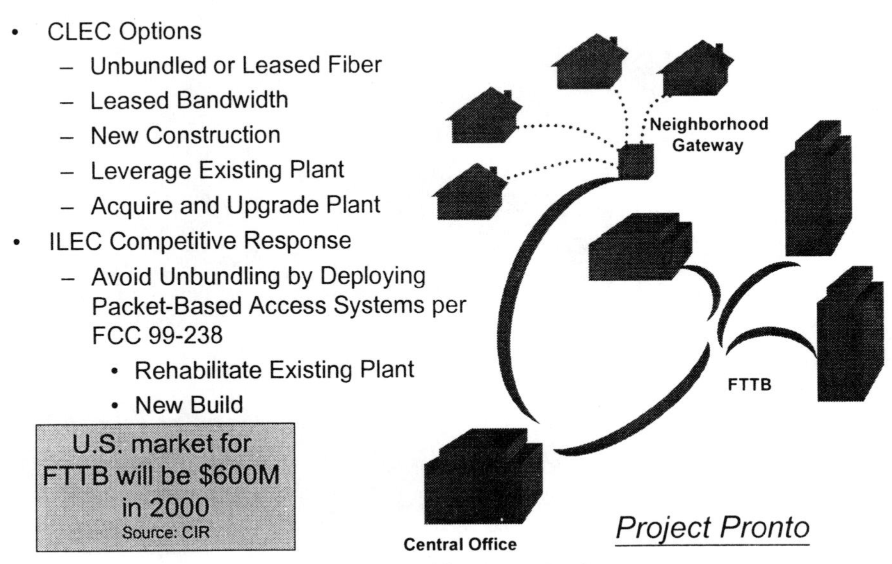

Figure 5: The Race Is On

An optical access network links into the metro or backbone network and provides an economic aggregation point. It provides diverse routes, mesh restoration, and high concentration through which carriers can best utilize northbound wavelengths. It also completes the end-to-end optical network. Fiber is the only optical access technology that will scale and, therefore, the only one that will endure. Although digital subscriber line (DSL) and cable modems (CMs) can move residential carriers from one Mb to 10 Mb, business carriers need 10 times that. Businesses must move quickly from 10 Mb to 100 Mb, and the only way to meet that need is via optical networking (see *Figure 6*).

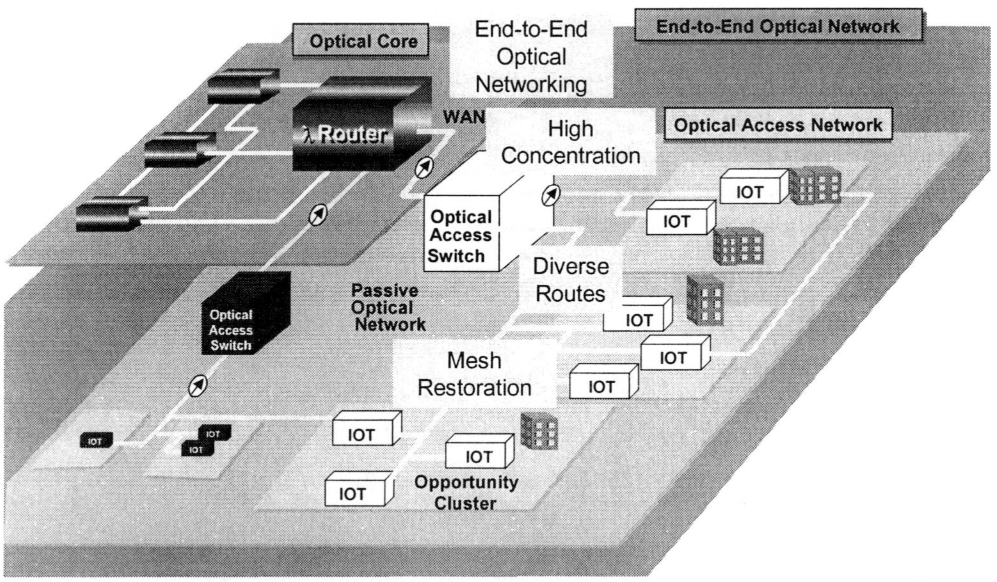

Figure 6: Optical Access Networking

Optical networking is efficient because of its point-to-multipoint solution and because of the way that it leverages the embedded base of fiber that carriers already have. CLECs believe that if they bring fiber to a building they have claimed that building and consequently have a first-mover advantage over the building's revenue stream. That is the basis for the next generation of services in the coming decade.

Migration to an All-Optical Network

John C. Adler
Director of Marketing
Cisco Systems

What are the barriers to migrating to an all-optical network (AON)? Why is the industry going there, what is lacking in the electrical infrastructure, and why must this absence be remedied as quickly as possible? Systems providers moving toward the optical world must address the whys and wherefores of these issues. Topics of concern include the following:

- Optical transparency, which for many years has been the focus of much discussion within the telecommunications industry, both in academia and in conferences
- Scaling questions and their real perspective on systems as well as the strong bearing on the AON, especially when it comes to granularity and the management of packets, cells, and slots
- Optical network service creation

Networks have been built for telephones in the same fashion for the past 124 years. Now, however, systems providers and carriers have a choice: They can build the optical network to behave in the same way as when phones were in buildings and tied to the wall, or they can try to build the optical network in a new way—a way that is more like the data network that is driving optical's growth.

Why, Where, and When?

Why migrate to all-optical? All-optical promises to lower cost—even to drive cost out of the network. That is why wavelength division multiplexing (WDM) was invented. WDM eliminated electrical regenerators in the network and lowered costs substantially. Where should WDM be incorporated in the network? To begin, electronics will be around for a very long time. Electronics is being pushed out to the edge, where it collects information and packetizes it, or puts it in slots, or makes cells out of it, and puts that information on the wavelengths so the optical network can do its job. When will this happen? Some key technology building blocks will be needed to create true AONs and to determine where they belong in the network.

Finally, the question arises as to how those optical networks will be built. A migration question exists here—migration from the centralized, nailed-up, circuit-switched networks, which is how networks have always been built, to distributed, circuit-routed intelligent networks (INs), where the elements themselves understand the topology, and the carrier does not have to understand how the network is built or what topology underlies the service.

The Dream versus the Reality

Besides lowering cost, transparency will free the network from both transmission rate and protocol when building services across the topology. There are no more worries about what specific card types reside inside the system when creating new services—for both rate and protocol. The dependency between the service and the underlying equipment is eliminated. However, at the core of the network lie realities about how core optical networks are built. First, only the highest bit rates really exist. Optical networks today run at 2.5 to 10 Gbps, and 40 Gbps is on the horizon. They are the only data rates that warrant dealing with; a wide plethora of data rates are not independently running through the core network.

Second, there is really only one protocol format: synchronous optical network (SONET). Carriers like SONET as a protocol because it enables provisioning, fault management, detection, and security. It is a very efficient transport mechanism—a sort of sea–land protocol. Sea–land containers are the carriers on the backs of trucks that travel U.S. highways and deliver goods across seas. If a sea–land container is loaded in Australia, a magnetic tag is placed on its corner. The container goes on a truck, which takes it to the harbor. The container goes on a ship and then travels to the United States, where it is placed on a train. Containers of that exact same size, all carrying these universal tags, can be hauled by trucks, trains, and ships that are adapted to receive the container's design. SONET accomplishes the same task. It is a protocol that will be around for a long time, but as a technology, SONET is changing.

Another reality is that everything comes as a trade-off. Nothing is free. The move to transparent networks costs the carrier several things. For example, it severely limits the carrier's ability to do operations, administration, maintenance, and provisioning (OAM&P). A technology does not currently exist that allows an understanding of the quality of the underlying services at the all-optical layer. Things such as signal presence, Q factor, and signal-to-noise ratio (SNR) can be examined, and they are somewhat predictive. However, they are not definitive when the carrier is trying to find things such as bit errors and intermittent problems in the network.

Transparency also abandons the idea of wavelength routing. The current circuit-routing networks have distributed intelligence and use routing protocols to build services. That is much more challenging to do in a completely transparent network, given wavelength blocking and the difficulty in engineering a long-haul network.

Span Engineering

When an optical network is installed, some design is required. An optical network must be engineered because it is not as simple as connecting an optical transmitter to an optical receiver. There are characteristics in the fiber, transmitter, and receiver that must be taken into account. When it comes to affecting optical amplifier spacing, the number of wavelengths and the fiber type and characteristics must be considered. Also, optical path engineering is dependent on the distance between regenerators—i.e., the minimum received signal level, which varies by rate (optical carrier [OC]–48 or OC–192).

An example of a span engineering problem in an AON is shown in *Figure 1*. To the left, between Portland and Los Angeles, a circuit represents the service that engineers are attempting to create.

Engineers perform the analysis and decide how much regeneration is required, based on amplifier spacing, the number of channels, etc., and they provision that circuit. Then there is a cut between Portland and San Francisco, and the circuit is rerouted between Portland and Los Angeles by going through Salt Lake City. Characteristically, the working route between Portland and Los Angeles is about 1,700 km, but between Portland, Salt Lake City, and Los Angeles, the distance is nearly 2,500 km. This is the most simple, parametric change in making this restoration, radically affecting the noise figure or the budget—

the decibel (dB) budget in that optical span. Depending on fiber type and span design, this can be a 6-to-9 dB change in optical span budget.

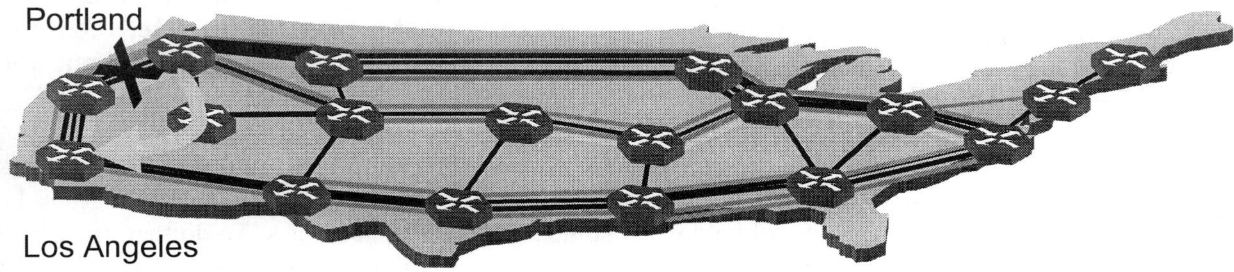

Figure 1: Network Transparency Span Engineering

In a completely transparent network, the engineer has the option of engineering for the worst-case scenario, which looks at every route in the network, simulates failures, tries to figure out where the network would be rerouted, and then builds the network for the worst case, which puts too much cost in the network. Thus, the issue of whether money is really being saved by going transparent again arises. Another possibility is to preplan the restoration routes to contain the problem so that every route need not be engineered twice. These are some of the challenges an engineer faces in the transparent network.

Wavelength Blocking

Wavelength blocking is fairly straightforward. Consider the earlier route, shown again in *Figure 2*. Observe what happens when provisioning new services through this network. Wavelengths are scarce resources, and market value, which any economics class will make clear, depends on scarcity. Scarcity defines demand, pricing, and the size of a market. Furthermore, in the present optical network, a certain window of transmission exists across which wavelengths can be sent, and this window can be sliced only so far to create a comb of wavelengths. It is a finite resource, through which fiber is pumped with today's technology.

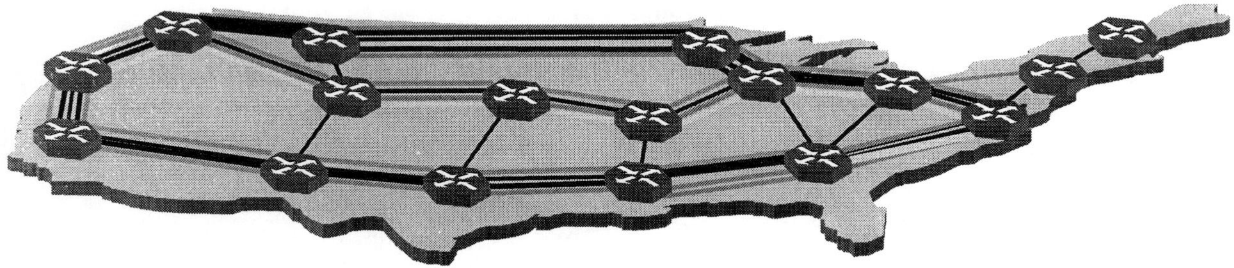

Figure 2: Network Transparency Wavelength Blocking

Depending on the vendor, 120 to 160 wavelengths are available today. As shown in *Figure 2,* the longer the path and the greater the number of hops, the more likely the possibility of a blocking probability in this network when trying to stay on the same route. This references only a transparent case. This system requires affordable, all-optical wavelength conversions. A way to change the color of the light in an all-optical manner must be found. Some of this technology already exists. All-optical wavelength converters are available for sale from some suppliers, although they are expensive. It is a costly proposition, indeed, and the technology is relatively new.

When it comes to transparency, some fundamental pieces are still missing. The first missing piece is a practical, all-optical wavelength conversion. As mentioned, some all-optical wavelength converters are available, but they are limited in terms of how large a bandwidth they can convert. A converter must cover both the conventional (C)- and the long (L)-bands. They have to be very low noise, because noise affects the engineering of spans and the inherent budget and cost considerations. Furthermore, a converter must be very low cost—substantially less than the quickly falling cost of regenerators. Optical-to-electrical-to-optical (O–E–O) rates are falling fast in the network.

After the wavelength conversion is accomplished to manage the number of colors in the network, regeneration still must occur somewhere. There exist ultra-long-haul systems that do between 2,000 km and 4,000 km, but regeneration must still be done at some point for most services. To do that, the carrier must perform a full reamplify, reshape, retime (3R); retiming is critical to control jitter and signal quality along a signal path.

The next optical switching technology that is needed is scale—micro-electromechanical systems (MEMS) technology is currently very hot in the industry. When optical switches first started, they were all, unfortunately, mechanical. They lacked speed, switching in hundreds of milliseconds—which is unacceptable, particularly for restoration cases. Consequently, the industry turned to other technologies, including various semiconductor technologies, thermo-optic, and other varieties of mechanical. Now the tide has turned back to mechanical, as the machines have been miniaturized and the switching speeds improved. It is not known if these will be sufficiently fast, however.

Finally, the lasers must be tuned. Technology is very close at hand for good tunable lasers to reduce systems cost radically. Tunable lasers also help with some of these wavelength-blocking problems, but they do not represent a complete solution.

There has also been much discussion in the industry about islands of transparency, where islands are created in a core network of transparent pieces and other parts of the network remain opaque. This is very analogous to SONET and SONET islands. The carriers simply waited. They did not put SONET islands in until they could build an end-to-end SONET overlay; and then they moved forward with that.

The Advantages of All-Optical Network Transparency

Where does transparency help? It helps in metro networks, as this is where all the service noise is. As shown in *Figure 3*, there are many rates, formats, and different kinds of services.

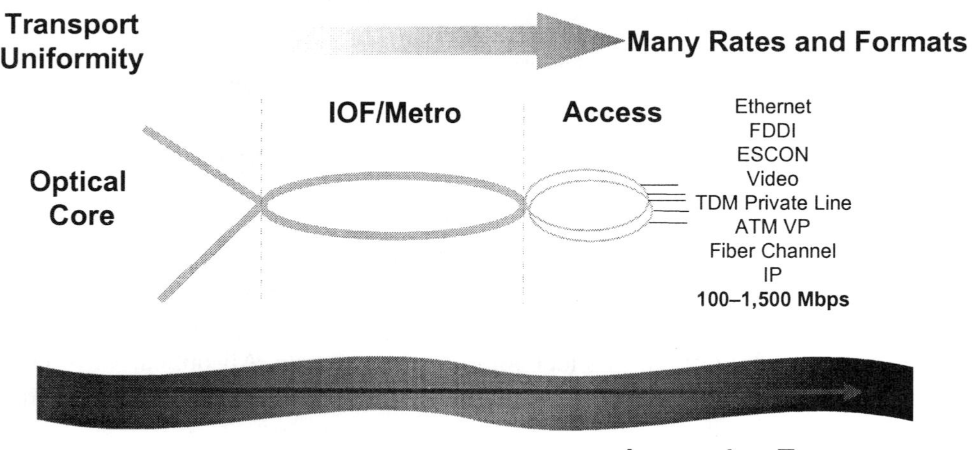

Figure 3: Network Transparency "All-Optical"

Also, that part of the network is physically rather short. Amplifiers are not needed in that kind of environment. A carrier can easily do 50 km or 100 km, depending on how many wavelengths are being pushed, and that is a great role for transparency to play in the network. In the core, where transport uniformity is desired (the sea–land analogy), things are very definitive and easy to monitor, track, and switch.

Scaling and Provisioning

When it comes to all-optical, scaling is an important issue because it affects the question of regularity. In an all-optical world, granularity is the photonic line. If it is 155 Mbps, then that is it; if it is 10 Gbps, that is it. The ability to perform grooming between those rates is sacrificed. *Figure 4* illustrates a case study of one of these core network nodes with a four-way junction and 16 wavelengths coming from each direction to total 64 wavelengths.

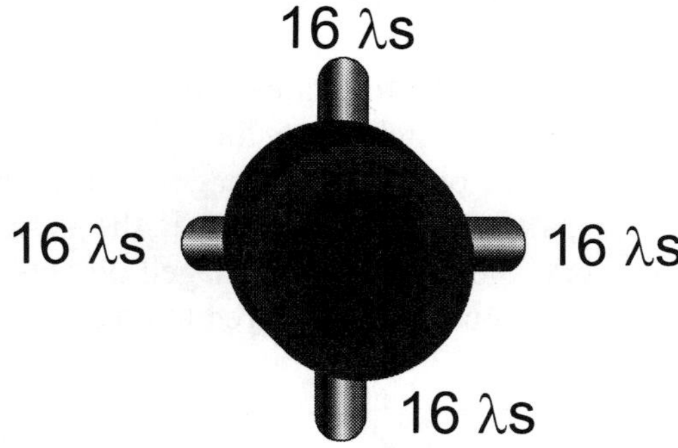

Figure 4: Core Node Scaling Requirements

Consider how quickly this node doubles, and thus how quickly the scale problems faced in the network are realized. If the network is on a doubling rate, somewhere in the six-month range the analysis will predict that this network node, which started in January 1999, will amount to 1,000 wavelengths by January 2001. That is the power of doubling. Doubling statistics must be carefully calculated, although a window of about two and a half years is workable. *Figure 5* illustrates the need for scaling at the dense WDM (DWDM) junction to meet traffic growth.

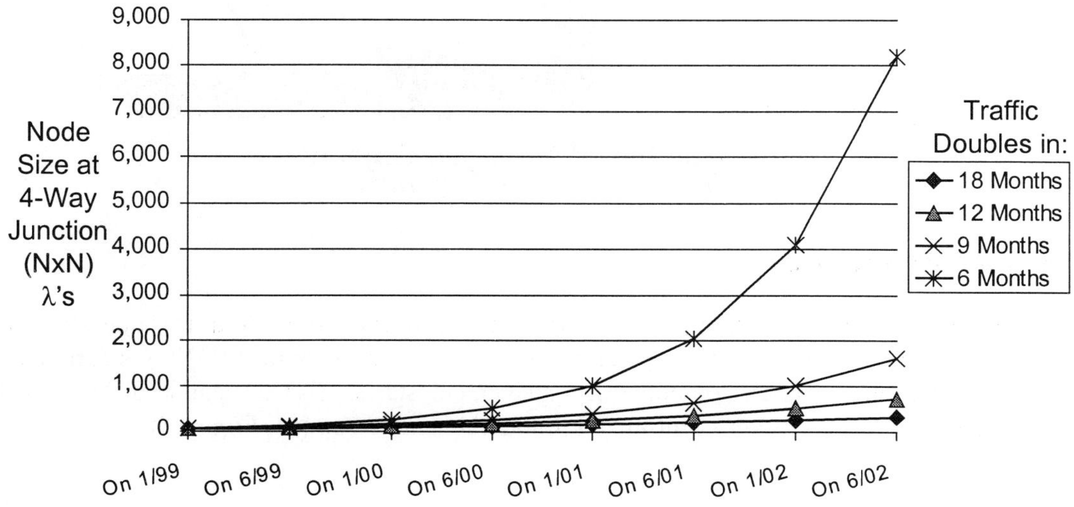

Figure 5: Need for Scale at DWDM Junction

Figure 6 shows the need for wavelength provisioning when going to an optical network and when using wavelength granularity to provision services. The OC–48 provisioning between two locations, in this case Los Angeles and Chicago, and the number of cross-connects necessary between these two places is substantially different.

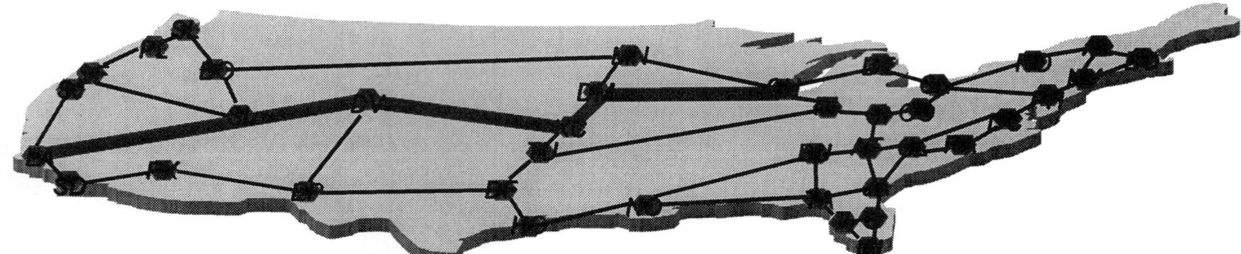

Figure 6: STS–1 and Wavelength Provisioning

This is one reason to start switching close to the line rate in the all-optical world. More specifically, in the three cases shown in *Figure 7*, if traffic is switched for the OC–48s at the synchronous transport signal (STS)–1 rate, this amounts to about 1,200 objects to manage. If it is switched at the wavelength level, transparently or not, the number drops to 72 objects. Then, if the system is switched in a wavelength-routed or an intelligent-optical network, the number of managed objects can be reduced to two. Only the source and the destination of this service must be considered to create this end-to-end path.

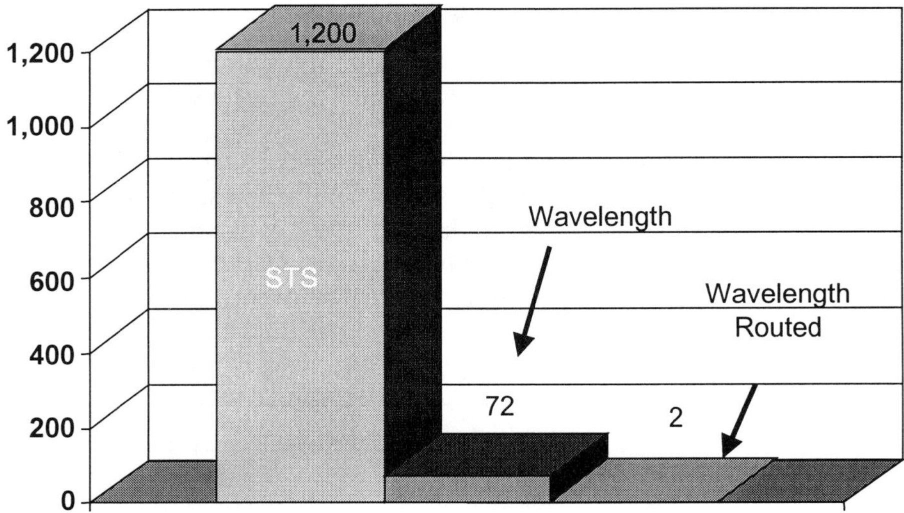

Figure 7: New-World Operations

Working with the Legacy System

In a legacy approach, when building optical networks and migrating them into a new operations paradigm, an automated device such as an automated patch panel or optical cross-connect (OXC) is combined with an enterprise messaging server (EMS) or network management station (NMS) standard database to create services. The network database is very centralized; in fact, it is nailed up (see *Figure 8*).

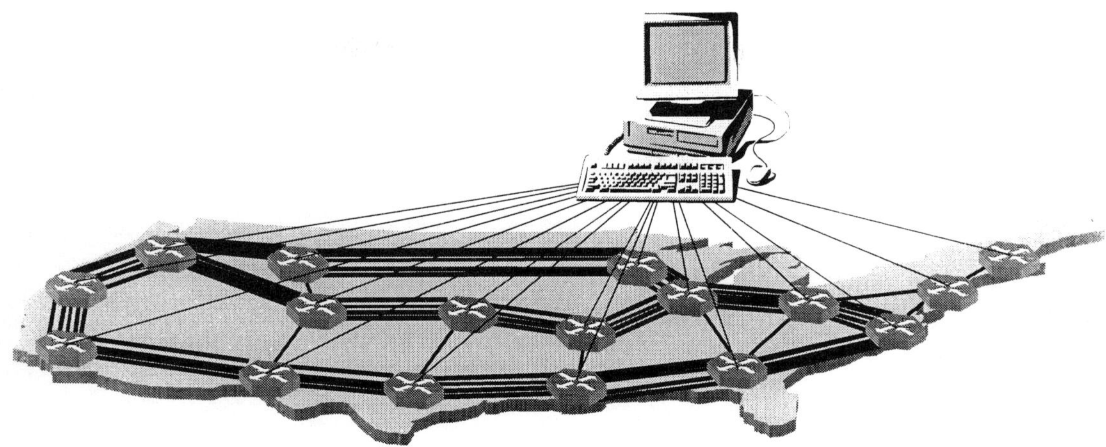

Figure 8: *Legacy Approach*

These devices are automated, but not intelligent. They have no idea what occurs in the network or its topology, and all of the services remain dependent on it—a very legacy approach to creating services in the optical network. The other way to do it is to make the devices truly intelligent so that the NMS system, which does not go away, is radically simplified. All it has to know is the source and destination of the wavelength route. Now the elements are intelligent; a distributed network database that can scale is used. Circuit-routed networks are built across this path with inherent scaling, in which all of the elements reside, and as elements are added more capability is added to the network. This results in *topology-driven* services, not a *service-driven* topology.

Summary

Some technology is needed on this migration path, and this technology is needed in the right places in systems. Electronics will be around for quite a while, and ultra-long-haul technology is needed. DWDM helps, but some wavelength conversion capability is required. Most likely, the cost of O–E–O conversions to perform regeneration should decrease in the future.

Finally, the challenge is to deploy this technology rapidly and to start offering wavelength services because the Internet protocol (IP) routers and the asynchronous transfer mode (ATM) switches are not waiting for the carrier in the optical layer. They are deploying OC–48 and OC–192 interfaces now. They will go to 40 Gbps, and that means that wavelength services at those rates will follow.

The Bandwidth "Big Bang" Meets "The Farm System"

Matthew Bross
Chief Technology Officer
Williams Communications

Introduction

Out of a small, dense mass of matter came an explosion, which resulted in the known physical universe—perhaps one of the only things most people can accept as infinite or boundless. However, out of the small collection of technologies known as the Internet (dense wavelength division multiplexing [DWDM], digital subscriber line [DSL], microprocessing, wireless, and software technologies ad infinitum), has come a similar explosion, with limitless possibilities. Although 70 billion or so light years are visible with the Hubbell telescope, the effort exerted by humankind today to explore that physical space amounts to only about four ten-millionths of a light year. Similarly, all of our telecommunications innovations have barely gotten us out of the driveway in exploring the information universe.

The All-Optical Network as an Enabler for the Future

The all-optical network (AON) is a huge enabler in exploring this information universe, but let us first go back to the 1950s. The cost per byte was approximately zero, with a market size projected into infinity over a period of time. Over a 14-year period, telecommunications costs showed a 1 percent decrease while experiencing a 3 percent increase in the market. Fourteen years represents a fairly mundane growth. This data referred specifically to U.S. long distance, a market elasticity that indicates that, for a 1 percent decrease in the cost to explore that space, there is a 3 percent increase in the interest to explore it. What that means is that companies can drive down their unit costs while exploring a market that is expanding to infinity with an increasing margin opportunity—an exciting possibility.

For example, if a sphere is represented by the information universe as a radius that is driven out first by DSL technology, then by DWDM, and then, finally, by wireless technologies, that sphere starts with a 1.5-inch radius and grows to a 10-inch radius. Each one of these technologies grows tenfold in response to the number of people getting on-line and the speed of DSL and DWDM technologies. When the "sphere" reaches 15 inches, the volume grows 98,000 percent.

Capability = Demand = Volume = Demand = Capability

When we talk about information as if it has volume, we equate it to the volumes in the Library of Congress or to the volumes on a disk drive. We do not think about information itself as having volume. When it comes to the Internet, of course, function is primary. The marketers are interested in which Web sites consumers are visiting so that ads can be placed in those locations. They move a bit of information between the Web site and the consumer. The transporters (the carriers) are very interested parties, because it is their task as go-betweens to make sure that bytes are delivered between marketers and consumers. So, looking at a byte from these orthogonal views, it actually does have volume. As these technologies accelerate, the extent of the information universe and the volume of information filling up that universe is expanding spherically.

No one has been able to indicate exactly what is driving this insatiable demand because, once again, we are trying to put a box around something that is actually infinite if the working cost can be kept sufficiently low. Other industry specialists, such as the cybercoretographers, are mapping this information universe, documenting the indications that this bandwidth big bang has, in fact, happened. Because we are corporeal beings, we do not really look at it in that context; so the density of that universe increases.

Building the Networks and Leveraging New Technologies—Past and Present

Building on these new technologies is much different than building the networks of the past. The job of the chief technical officer is to ensure that a gap never develops between what is technologically possible and what the company is delivering. But how does one really mount an effort to eliminate that gap? Interestingly, an industry compromise exists for all carriers—as the network scales, the rate of technological innovation decelerates. If the situation is reversed, one should accelerate the rate of technological innovation while scaling the network. This is like defying gravity, but some interesting possibilities emerge.

In the first era of building networks, AT&T and technology were synonymous. If it was high-tech, it was AT&T, and vice versa. In the second era, there were 5 or 6 long-haul players, 8 to 12 local players, and 10 or so telecom technology providers. What mattered was getting big fast. Scale mattered. Everything was about scale, and the process worked as follows: A carrier would go to one of these 10 or so telecom technology companies, bring 3 of them into the lab, and evaluate them. Whether the technology worked or not was only marginally important, because these companies had balance sheets that could rule a technology into existence. They could provide 50 of the "other things" while they got this new thing to work. The other attribute of this era was that when a technology was chosen, the carrier was married to it. The carrier worked with a relevant vendor on incrementally improving that technology and applying it to the network over the next decade or so. And again, scale meant everything.

Today, there are thousands of local companies, not 8 or 12. There are hundreds of interexchange carriers (IXCs), and there are thousands of telecom technology companies. No model exists to enable a lab to look at these thousands of technologies. However, two things are guaranteed in today's era of telecommunications:

- First, there will be more failures than in the days of balance-sheet strength, mainly due to the pigheadedness of the entrepreneurs. Failures will occur because of a lack of channel strategy. There are many different reasons why—out of those thousands of technology companies, an increased number of failures will occur. That is a guarantee.
- Second, it is much harder to pick the winning pony when betting on thousands of choices. In the past, picking from an offering of 10 companies was no big deal. It did not really matter. Everybody was under the same heat. Today, out of those thousands of technologies, an order of magnitude exists, like shifts on multiple axes. These shifts, unlike the technologies that leap-frogged before them, now leap-frog in 10X manners; so the old model simply does not work.

So, What to Do?

Carriers should find insurgent technologies—i.e., those technologies that drive disruptions in the carrier's marketplace.

For example, consider a small, new network in the process of launching a technology initiative. At first, this takes very little power. As the network scales and becomes bigger, with margin sought from it, the carrier must struggle to maintain 99.99 percent performance. What happens, then, when it attempts to launch that new technology initiative off a much bigger mass? It sits on the launch pad, it smolders and shakes, and it explodes and kills everybody around it who was important. New technologies are difficult to launch from this network scale; the rate of technological innovation decelerates. Thus, it is necessary to be able to go outside the telecom industry.

The Farm System

Hundreds of technologies must be narrowed to three or four that go through lab implementation. Today, no lab is ready to handle all of these thousands of technologies, so these labs perform a quantitative analysis. The lab gets at the speeds and feeds. It achieves a photograph, a snapshot of what is going on with a particular company. However, the lab does not see how the management team builds out. It cannot see how the channel strategy is developing, as it works with a chosen interval. When bringing technologies into the farm system, the talent, the best-performing athletes are desired. This requires telling the scouts what to look for: In this case, the goal is technologies that move by an order of magnitude or by one or more vectors, and the first vector is service velocity.

Service velocity can be considered in two ways. The first is the speed with which networks can be provisioned, such as a wavelength or a virtual private network (VPN) service. The second is the speed with which new services can be created—i.e., the time it takes to prototype, build, roll out, and bill services in the network. Velocity is the speed with which new products can be created, so the 10X promised improvements are needed by that new technology, and these technologies can come from the incumbent or emerging players. It does not matter. Insurgent technologies come from all places.

The second vector is to move by an order of magnitude of cost to build the network. If each port costs $200, one must attempt to lower the cost to $20 per port. Most companies do not permit such a price because they intend to make a sizable profit.

Cost-owned refers to, for example, taking 20 racks of equipment and reducing them to two racks. If the network used 1000 amperes (amps) of power, it now takes only 100 amps. Moving by an order of magnitude on all three of these axes represents an insurgent technology that can drive a disruption in the marketplace. They must be identified early on and placed in a farm system concept that takes these thousands of technology potentials and applies a set of scouts who view them through a filter. Multiple local and long-haul technologies exist in the optical space, and the best athletes are working on an increasingly competitive play to determine whether all the qualitative attributes of that technology and its company are survivable.

Imagine an athlete coming in to play triple-A ball. When that athlete plays 365 days a year, it seems more like a documentary than a photograph (that single snapshot from the "triple-A ball" lab), which enables a better choice of insurgent "triple-A ball" technology to be applied across the network.

Blending the Technologies and the Markets

The farm system provides the earliest view of new technologies, knowledge of the intimate relationship with the emerging and existing technology sponsors, the most efficient blend of technology, and more than a quantitative analysis. It provides a qualitative analysis, which improves the ability to identify these insurgents. What does it do in terms of being able to accelerate the rate of technological innovation? Internal customers, engineering operations, capacity planning, and carrier customers are more risk-averse. They do not want to see that new technology in there until it is proven. Fiber customers have a different risk profile. Stock investments in S1s are considered foolish, to say the least, because of risk factors. Every risk known to humankind is listed, but there are people who buy that stock because they have a certain risk tolerance—exactly like people who will buy blue-chip stocks. The notion in the farm system is that these technologies are put to work against each other in a real, competitive situation. What results is a 10X movement on multiple axes, a blending-in with the existing network, and a most efficient operation (i.e., low cost, which is what drives that margin out of the business and affords the sustainability over time that it takes to be a player in this space).

Conclusion

If the carrier does not employ this farm system approach, the outlook is bleak. In terms of selecting technologies via the farm system, consider the emergence of major league baseball in the 1920s. Branch Rickey organized the first farm system team in St. Louis. The second team to adopt the system was the New York Yankees. Now, after 80 years of major-league baseball, the strongest franchises in the history of World Series major league baseball are still the New York Yankees and the St. Louis Cardinals—the first two teams to implement the farm system. The industry is now applying these principles to assess emerging optical technologies in the ultra-long-haul. This cannot be done in a lab; it requires a network. The network becomes the lab and hints at the probability of being able to adopt those technologies more quickly than other in-class competitors. The other thing that it does is build a defensible business against the "Pet Rock" or fad networks when it comes to scale. These networks simply get a little bit of fiber, build out a network, and end up relegated. They quickly become a legacy because it takes a lot of assets to run efficiently with fiber, to run co-location, and to have a farm system that builds an efficient, profitable network.

Building Future Optical Networks

Milorad Cvijetic
Chief Technology Strategist, Photonic Networks
NEC America

Everything is going digital, and the growth in data services for businesses and individuals is exploding at an exponential rate. From corporate intranets and local-area networks (LANs) to residential services such as interactive television, the increased demand for advanced network services shows no signs of abating. To the contrary, as shown in *Figure 1*, the variety of available services has only increased with time, and with this traffic explosion has come the increased demand for large-capacity files.

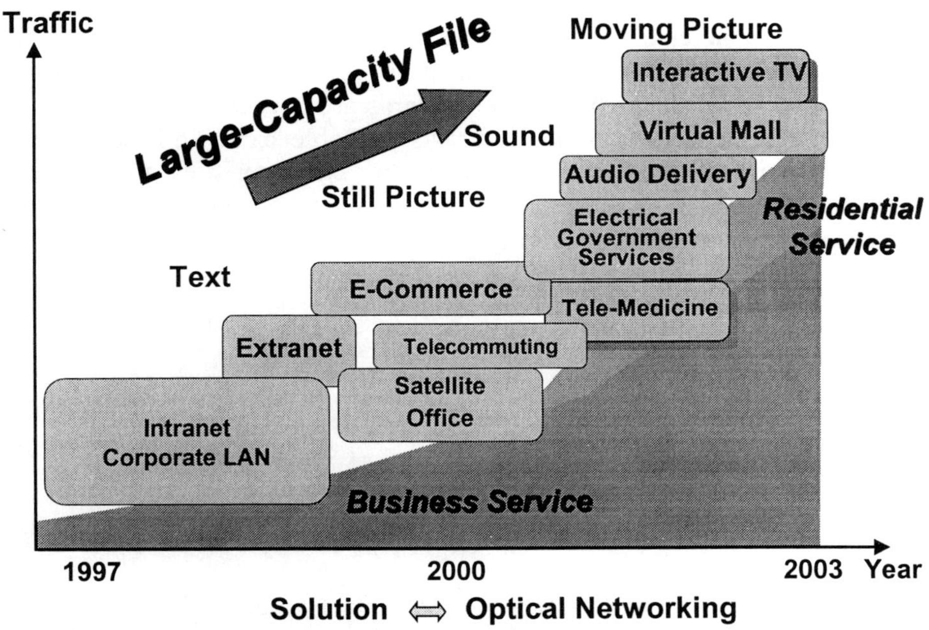

Figure 1: Advanced Network Service

The Solution: Optical Networking

The solution to much of this demand lies in optical networking, and to build an optical network, a transport vehicle is needed. *Figure 2* presents the other building blocks for the optical network, providing a mix-and-match of high-speed bit rates (2.5 Gbps, 10 Gbps, 40 Gbps) and optical networking features. To have a fully optical network, wavelength routing and optical protection are necessary.

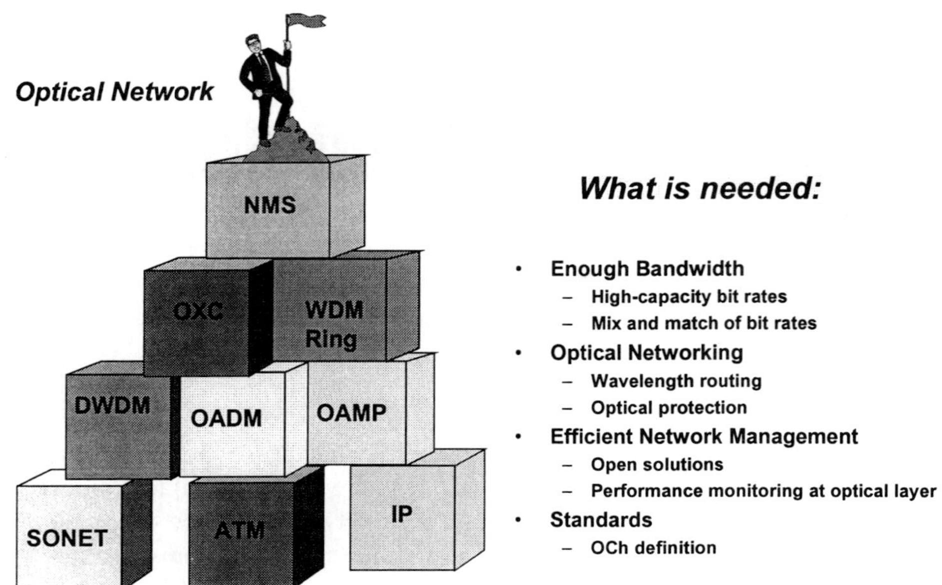

Figure 2: Optical Network Building Blocks

Very efficient network management is also required, as are open solutions with multivendor environment and performance monitoring at the optical level. Finally, to accomplish solutions that are multifunctional and universally acceptable, open standards are needed. These standards are related to both transport and networking issues. *Figure 3* charts probable bandwidth capacity limits over the network. In this simplified analysis, based on optical fiber characteristics, total bandwidth is plotted according to how many channels there are, what space exists between channels, and so forth. The resultant 50 Tbps capacity is a realistic number.

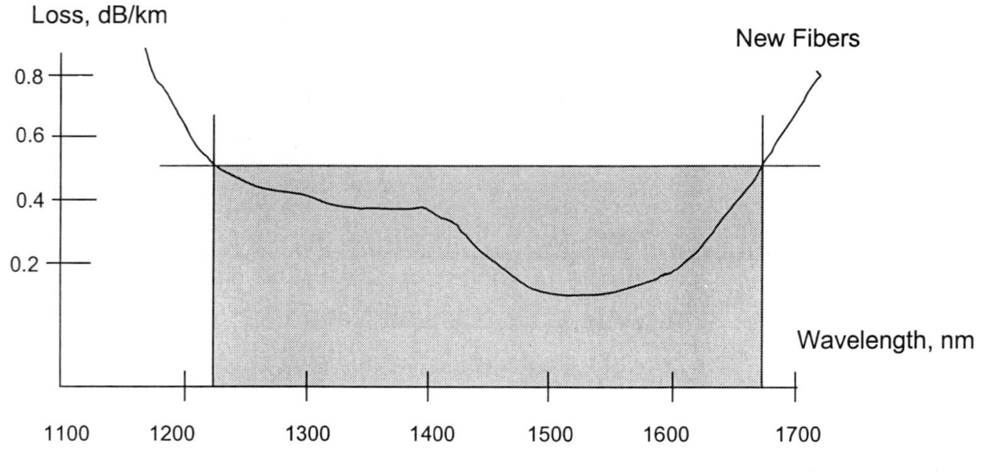

Theory:

Total Bandwidth = 400 nm ⇐ Criteria: loss < 0.5 dB
Used Bandwidth = 400/4 = 100 nm = 12.5 THz ⇐ Reduction due to channel spacing and nonlinear effects
Total Capacity = 12.5 x 4 = **50 Tbps** ⇐ Multilevel transmission; maximum four levels per bit

Practice:
A realistic expectation: 5 Tbps in 5 years, 20 Tbps total capacity in ~ 10 years

Figure 3: Optical Fiber Bandwidth

What are the future expectations? How about 5 Tbps in 5 years, and 20 Tbps in 10 years? *Figure 4* shows where the wavelength division multiplexing (WDM) channels lie, the number of those channels, and the bit rate per channel.

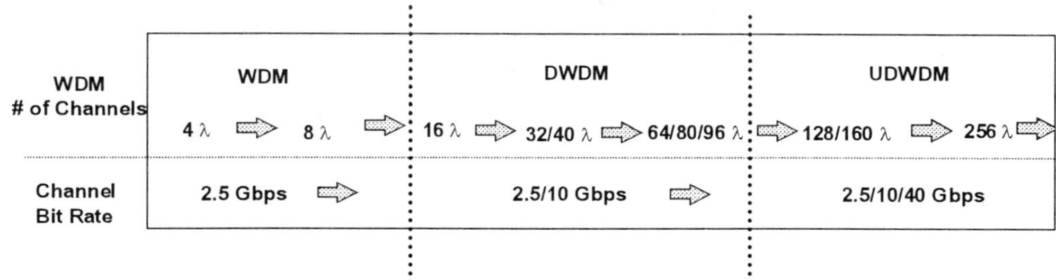

Figure 4: Optical Transport Technology

It is unnecessary to compare time division multiplexing (TDM) and WDM for capacity growth. Capacity consists of the number of channels multiplied by each channel's bit rate. In addition to 2.5 Gbps and 10 Gbps, 40 Gbps systems are becoming very important, and this technology is emerging sooner than the industry expected. From a technology standpoint 2.5 Gbps is simple, but 10 or 40 Gbps are more complex because they approach the very limits of technology. Whenconsidering speeds at four times the bit rate, the complexity is more than fourfold. An open-solution WDM system is set to deal with these complexities.

Dense Wavelength Division Multiplexing (DWDM)

Open solutions are preferred by industry. Clients are connected to terminals with some number of transponders. The clients can be synchronous optical network (SONET) elements, Internet protocol (IP) routers, or asynchronous transfer mode (ATM) switches. In terms of input, as shown in *Figure 5*, the modulation transfer from electrical or from one optical to another optical level occurs at the terminal. Performance—i.e., achieving good signal-to-noise ratio (SNR) and bit-error rate (BER)—is necessary for high-speed transmission.

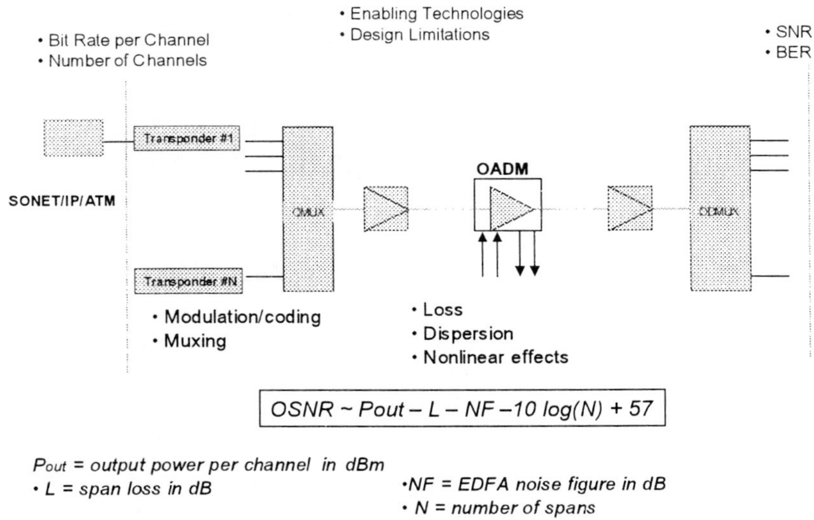

Figure 5: DWDM Systems

Presented in *Figure 6* is a simple formula for calculating optical SNR, which takes into account optical power per channel, optical loss, optical fiber amplifier noise figure, and the number amplifiers.

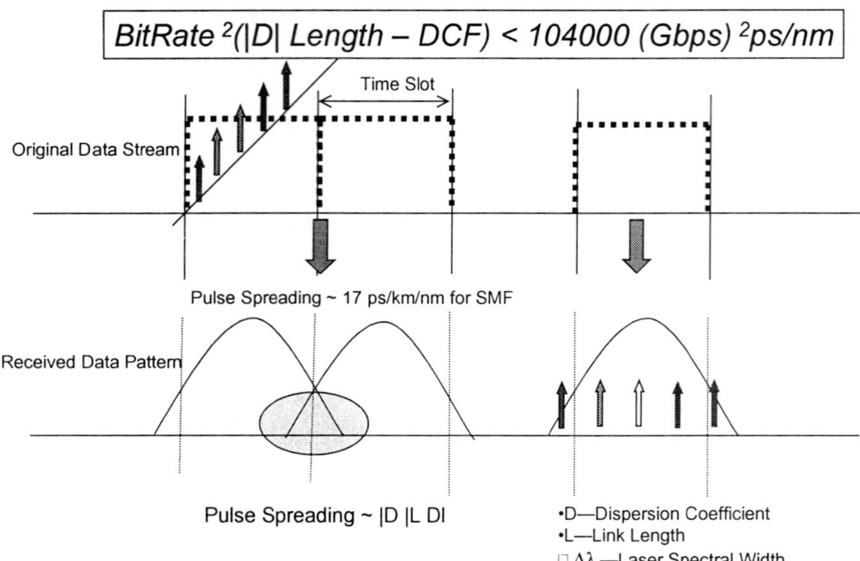

Figure 6: *Chromatic Dispersion as a Limiting Factor*

Chromatic dispersion occurs when neighboring pulses cross time slot borders. The reason for this is a laser spectral width: there is no ideal laser and no ideal source with only one wavelength. It would be nice to have this one wavelength, but there are varying wavelengths within the laser spectrum. The problem is that wavelengths travel at different velocities, so some reach the end faster. Consequently, part of the energy resulting from the disparity in this time slot will be exhibited as an interference noise in the neighboring time slots. The pulse spreading is proportionate to dispersion. Optical fiber dispersion is about 17 picoseconds (ps) at 1550 nm, for standard single-mode fibers (SSMFs) (or around 4 ps for new dispersion-shifted fibers). Thus, when engineering an optical transmission line, dispersion must be considered carefully.

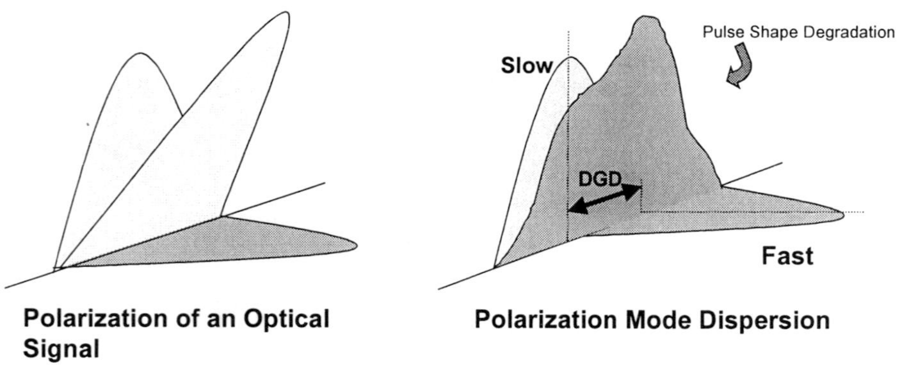

PMD is time-averaged differential group delay (DGD). **[ps/sqrt(km)]**
• Stochastic rotation of the polarization states
• Stochastic distribution of energy between the principal states (50/50 the worst case)

should be < 0.2 [ps/sprt(km)]

Figure 7: *Polarization Mode Dispersion*

For high bit rates, polarization mode dispersion (PMD) is becoming an issue because light in optical fiber can be considered a superposition of two components (see *Figure 7)*. It is good if they travel at the same speed, but if they travel at different speeds, the faster components and slower components will result in pulse spreading. The problem is that PMD is a stochastic process. Mechanical stress, adverse temperature effects, and the degradation of the optical fiber complicate the issue.

Then there are the nonlinear effects or conditions that are generally bad for transmission, but in some cases they may be used to improve transmission characteristics. The nonlinear effects can be divided into two categories: the effects of light scattering and the effects of changes in the fiber refractive index.

Stimulated Raman scattering (SRS) causes crosstalking between channels. Stimulated Brillouin scattering (SBS) occurs when very high power reflects from glass molecules in fiber. It indicates that power should be limited to suppress these nonlinear effects. A second group of nonlinear artifacts includes four-wave mixing, which causes crosstalk (XT) noise in channels due to other channels' influence. The self-phase modulation (SPM) causes pulse spreading of an individual channel. Cross-phase modulation (XPM), which has the same nature as SPM, results from interaction between multiple channels. Both SPM and XPM occur in combination with chromatic dispersion.

A Bit about Bits

Optical carrier (OC)–48 WDM is mature technology and involves no chromatic dispersion. PMD is not critical either. Transponder-based, 3R functionality exists currently for transport systems, although 2R functionality is acceptable for other applications, such as metro. A standard optical amplifier is needed, such as an erbium-doped fiber amplifier (EDFA), to serve as a booster amplifier, inline amplifier, and preamplifier.

OC–192 technology, however, is more complex. The SNR should be 6 dB higher than that for OC–48 bit rate. To achieve higher SNR, precise engineering is needed, and some trade-offs should be expected. The nonlinear effects are becoming more important and event-dominant as pertains to optical fiber. Different fiber types will cause different system behaviors. One engineering approach will be required to deal with SNR, and a second engineering approach is needed for dispersion issues.

Finally, there are the OC–768, 40 Gbps–based systems. For future systems the idea is to mix and match all bit rates (2.5 Gbps, 10 Gbps, and 40 Gbps). However, it is not a simple task. For example a 400 km span is sufficient for 2.5 Gbps and eventually for 10 Gbps, but for 40 Gbps, special methods are necessary. Such methods include an initial spectrum shaping to sort the fastest and slowest components, an advanced coding to reduce the dispersion and nonlinear effects, and an advanced compensation of chromatic dispersion by self-phase modulation (solitons).

Transport

Data-centric networks—IP, ATM, and the like—are the reality and require an efficient networking solution. It is not a SONET world anymore. Wavelength is becoming a networking unit. There should be a solution to enable an efficient, protected architecture that can deliver the speed and reliability comparable to today's SONET. To achieve that, a paradigm shift must take place.

The classical approach was to have Layer 3 (IP) over Layer 2 (ATM) over SONET, and on down to WDM, as shown in *Figure 8*. It was a multilayered approach in which some services were duplicated, and carriers found the arrangement less than economical. *Figure 9* illustrates the next-generation optical network, where multiservice interworking is essential.

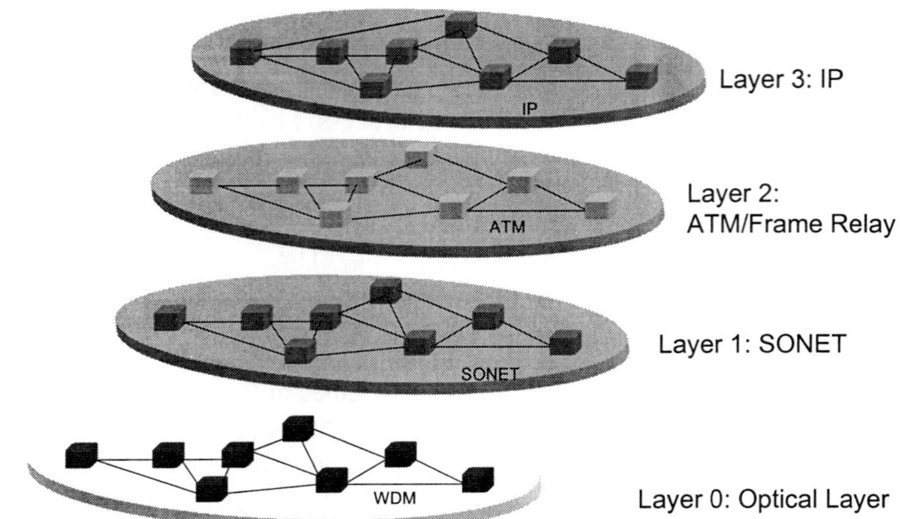

Figure 8: *Optical Network Role—Traditional Paradigm*

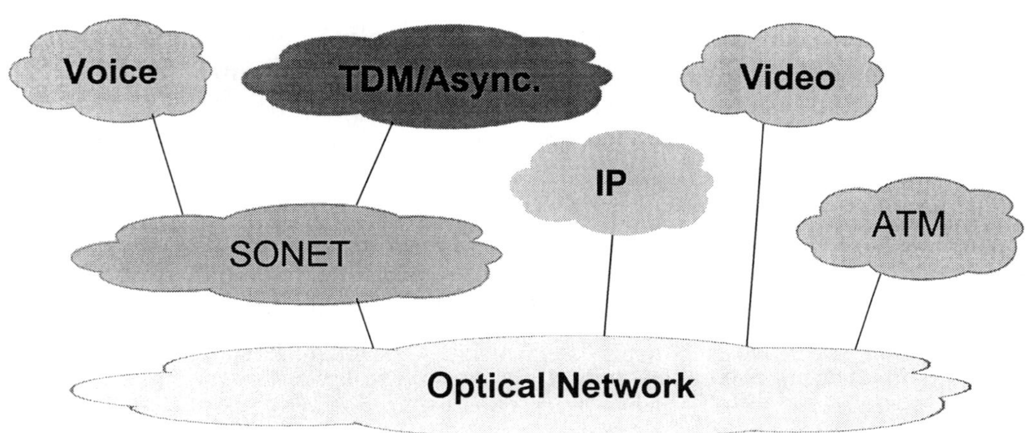

Figure 9: *Next-Generation Optical Network—Multiservice Interworking*

New approaches, as depicted in *Figure 10*, have focused on having an optical network as a core, with everything else as a client to the optical network. SONET, IP, and video should all be transferred to the optical network.

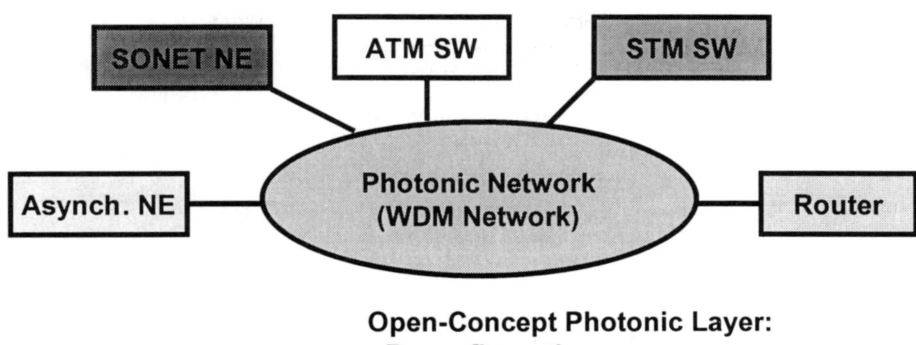

Figure 10: *Photonic Network Evolution Path*

The evolution to next-generation optical network will occur in several steps. Point-to-point is the first step; the next step is to introduce optical rings (as opposed to SONET) to afford protection. Optical add/drop multiplexers (OADMs) are used to bypass the node's electronic layer when traffic becomes overcrowded (see *Figure 11*). The final step is to secure a fully flexible ring and mesh with optical cross-connects (OXCs). An open-concept photonic layer should provide reconfiguration, routing, and protection.

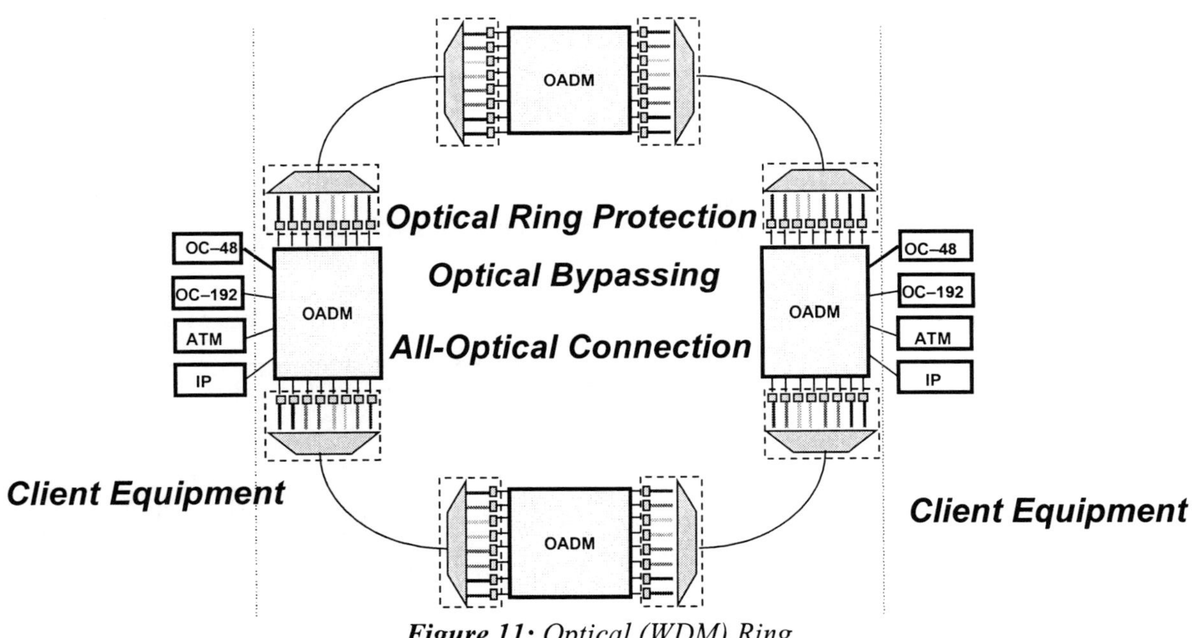

Figure 11: *Optical (WDM) Ring*

Optical ring is similar to SONET, with clients poised at the rim of this optical network. With some very efficient engineering, this arrangement can yield big returns, as shown in *Figure 12*. For example, if one has a traffic request (eight OC–3s) between each pair of nodes, it can be accommodated as usual or by applying some optimization.

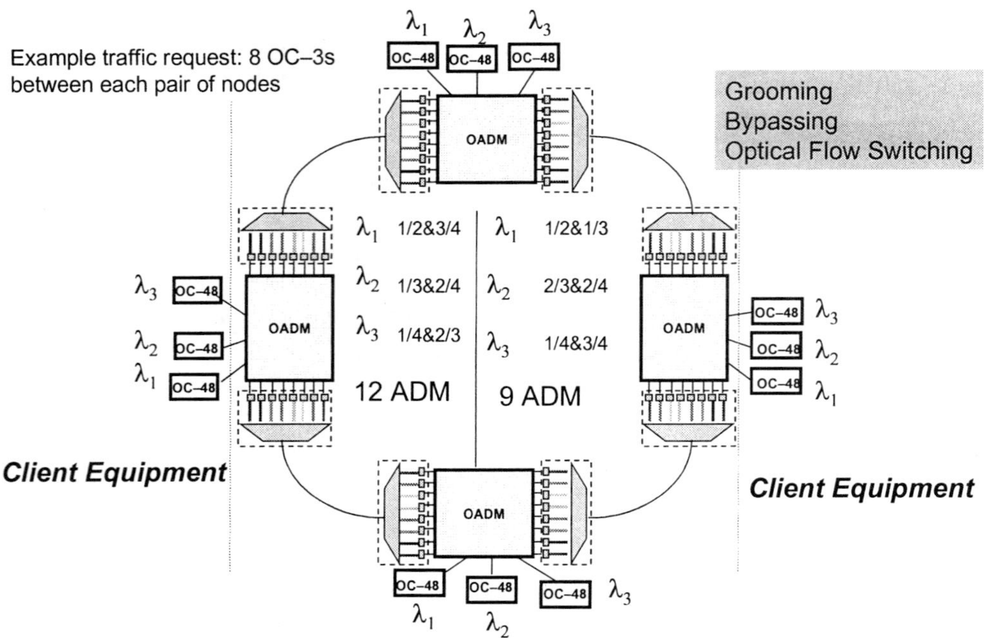

Figure 12: *Optical (WDM) Ring (Continued)*

As result, instead of having 12 add/drop multiplexers (ADMs) to accommodate all traffic, 9 ADMs are needed. This is a simplified example, but it illustrates that, with proper attention, considerable savings are possible.

The Evolution to Optical Networking

The evolution to real optical networking is the goal of the industry. The next generation integrates different services and maps them directly into an "optical frame" (see *Figure 13*).

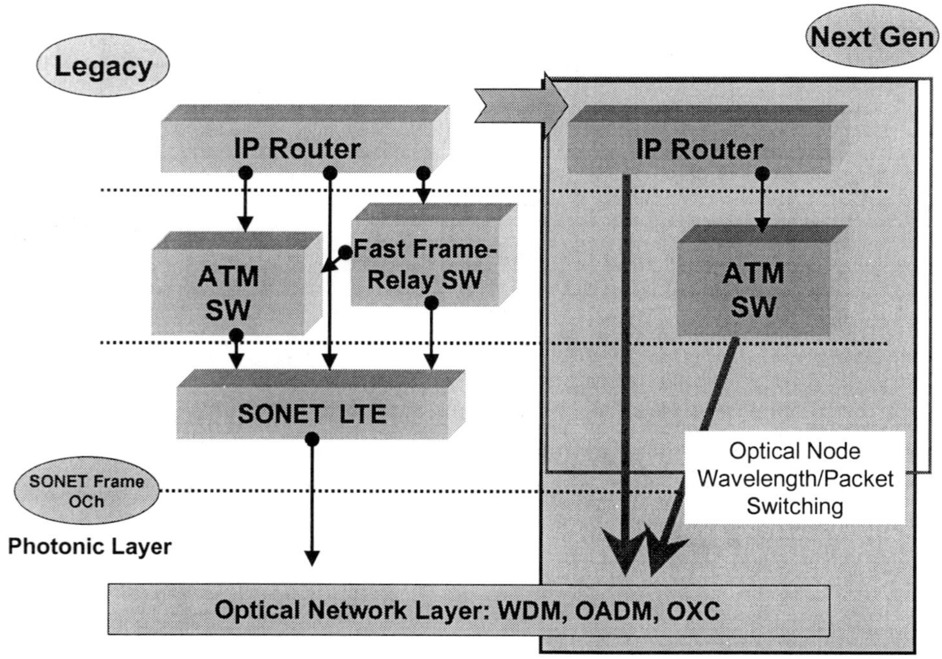

Figure 13: *Evolution to Optical Networking*

Figure 14 presents a rough architecture of such a core optical network. At the circumference are the edge nodes, which surround the core nodes. The legacy equipment, such as the routers, is incorporated at the edge, and signals are put into the optical channel (OCh) format.

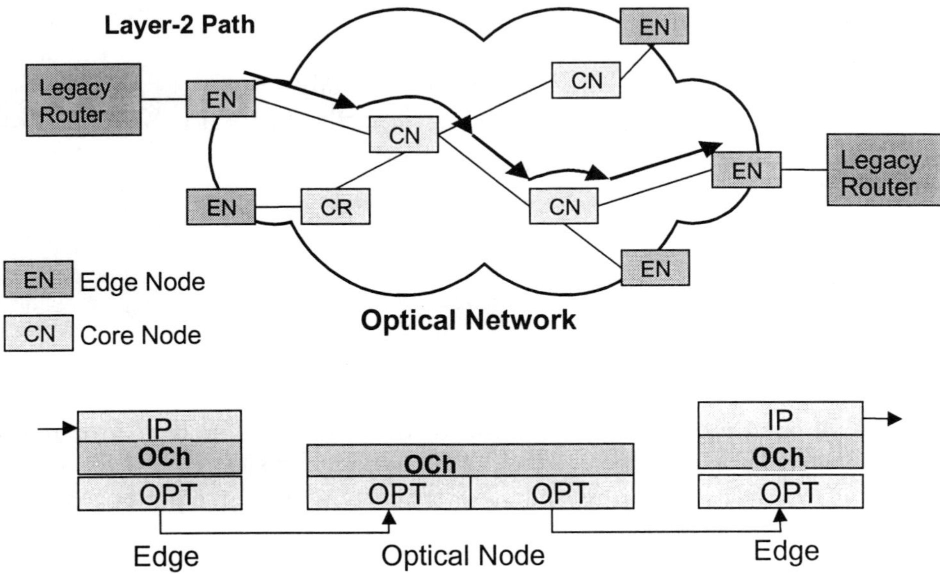

Figure 14: Core Optical Network

After that, the transmission is conveyed within the optical network with WDM technology. Thus, there is only one real layer, and whatever comes to the network should be imported in that optical channel frame. This is what the International Telecommunications Union (ITU) proposed, and it should be standardized sometime next year. Again, instead of having a few vertical layers, such as ATM or SONET, there would be only one layer that would accommodate everything (see *Figure 15*).

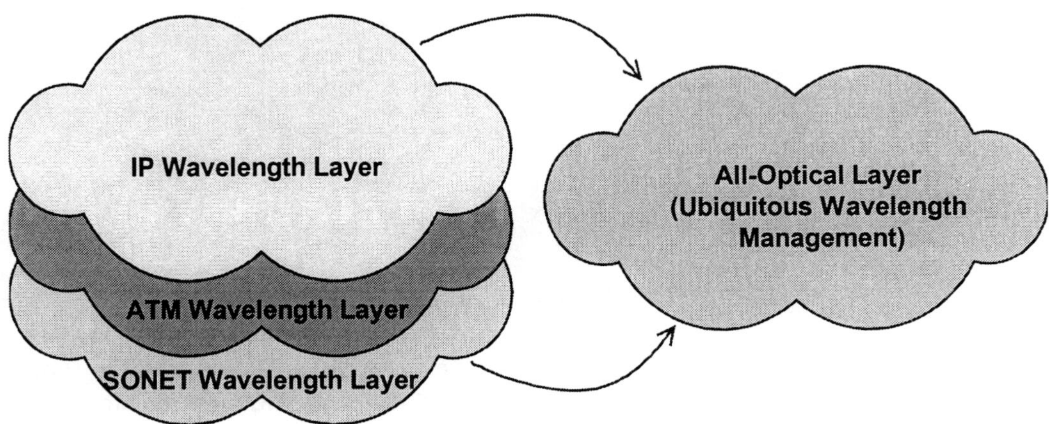

Figure 15: Deployment Strategies

Eventually, the quality of service (QoS) approach with the differentiated services concept within optical channels can be adopted (see *Figure 16*). Individual optical channels, or λs, would carry specified QoS requirements (e.g., a platinum service quality for channel x, a gold service quality for channels y and z, etc.).

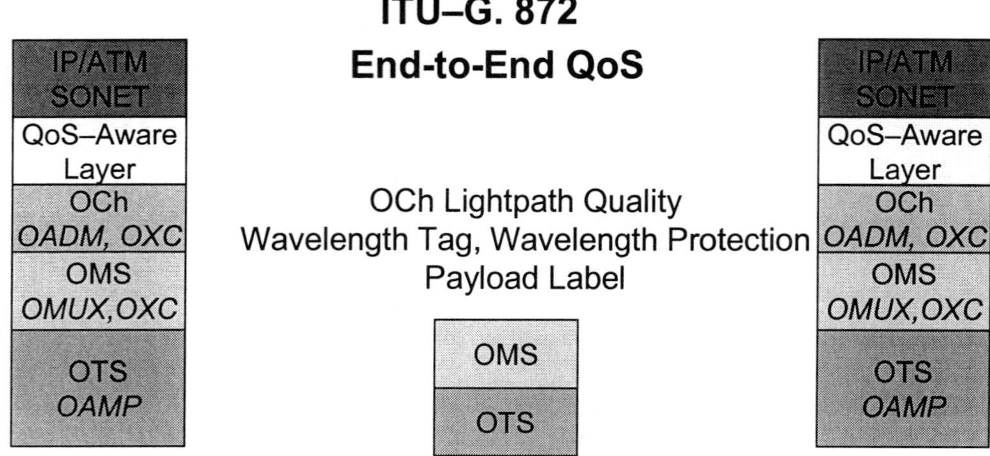

Figure 16: Core Optical Network

Optical networking will be introduced gradually. Meanwhile, there will be IP, ATM, and SONET in the optical network. However, the most efficient protection strategy for such a network remains unclear, as illustrated in *Figure 17*. Point-to-point, ring, and mesh connections are some protection options.

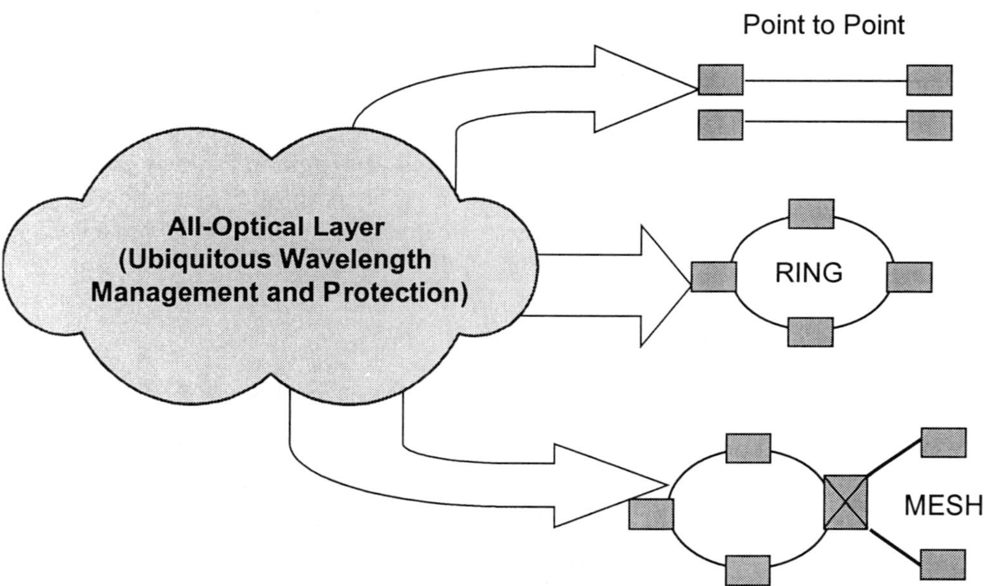

Figure 17: Optical Network Protection Strategies

The most important parameters should be monitored, and everything should be covered (point-to-point, ring, and mesh). To achieve this, different layers in transition stages should work together. If an IP layer is over an ATM layer, which is over a SONET layer that is over a WDM layer, protection can be done either at the IP layer or at the supporting lower layers, as shown in *Figure 18*. The major problem is how to bring the layers to work together, and what would be the most efficient solution. In truth, there is no universal solution, but some guidelines exist on how to approach this.

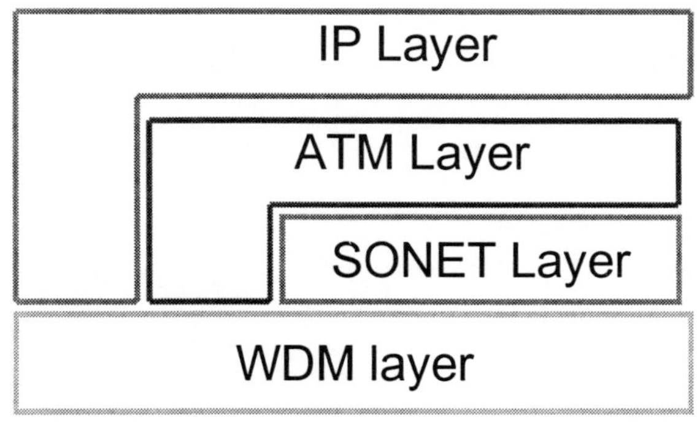

Figure 18: *Protection in a Multilayer Network*

Analysis by *IEEE Comm Magazine* concludes that protection at the lowest layer (WDM layer) is the cheapest route to follow. Recovery at the highest level (by IP routers) provides multiple reliability grades and finer switching granularity. However, at the same time, it is slower than protection at the WDM layer. The ultimate approach is to bring the layers to work together. Spare capacity in the upper layer should be treated as an extra traffic down to the server (lower) layer. This might form the criteria for future interworking.

Network Management

What, then, is important? There will be different elements. There will be different requirements for QoS, and there will be operations, administration, maintenance, and provisioning (OAM&P) function requirements at the optical level. The first step is to help with performance monitoring (Q-factor can be measured instead of SNR). In conclusion, building the optical network entails a complex set of issues that should be addressed on the engineering level.

The All-Optical Network

Glenn Falcao
Executive Vice President
Corvis Corporation

Introduction

It is instructive to note some points that have been made by a variety of industry observers. First, there will be a thousandfold growth in bandwidth because less than one percent of households have access to high bandwidth today. In light of that impending growth, the bandwidth explosion in the core backbone has not even begun to occur. Another important point is that Internet protocol (IP) networks consume all of the bandwidth in their path, and the impact of the growth of the Internet on the all-optical network (AON) will drive additional demand. Everything will scale incredibly quickly, and, as mentioned, the real growth in bandwidth traffic has not even begun. As the network evolves, traffic on the network will be redistributed, and there will be a move from pure bandwidth to services and a dramatic reduction in network cost.

This paper addresses the growth of service deployment in the AON. A rapid move to service deployment will help alleviate the problems that the impending growth in bandwidth consumption will create, and it will also significantly push down the cost of the network. A one percent decrease in costs causes a three percent increase in demand, so everything that will be discussed will also serve to drive top-line and customer-base growth. When an operator looks at new technology, it is important to scale while improving performance and margin. Insurgent technologies must increase service velocity and reduce the total cost of ownership. That is extremely important, because the life cycle costs of these insurgent technologies must be evaluated as well as the initial capital costs.

The AON

The Internet affects everything, and a wide variety of new e-commerce activities have emerged. The business and economic models have changed forever. Several results of this change are a significant growth in Internet subscribers, new applications enabling greater IP access, and an opportunity for a new Internet infrastructure based on an optical network. These changes drive new applications and services. This paper will discuss the impact of the new all-optical Internet infrastructure.

New Network Requirements

The worldwide opportunity in the dense wavelength division multiplexing (DWDM) market will approach 20 billion dollars by 2003. What are the requirements of this new network? The first two requirements are service provisioning and velocity, which concern top-line growth, speed of service delivery, how competitive a player can be in the market, and how to capture new customers and retain existing customers. The second two requirements are high network capacity and scalability. As a network puts in new technologies, it must scale in capacity, margin, and performance. Third, this new network must also be extremely reliable. It is a backbone network, and this is the telecommunications business—where 99.999 percent reliability is imperative. This is not best-effort Internet traffic, but backbone mission-critical traffic, and reliability must be delivered. Fourth, the network must transition from rings to mesh. Service providers have revenue-generating customers, so they will not all jump together to a new infrastructure. They must transition to this new infrastructure, and the AON allows that. A fifth requirement of the new network is end-to-end bandwidth management. Sixth, the separation of hardware and software does not make sense in relation to the delivery of telecom networks. The ability to deliver hardware and the software to install, maintain, provision, and keep networks running is fundamental to all networks. A seventh requirement is a competitive cost structure. Costs must be reduced by an order of magnitude.

Limitations of the Current Network

Today's network is based predominantly on synchronous optical network (SONET) rings, and those rings are becoming a bottleneck as the network scales. Also, there exist too many boxes in the network. All unnecessary optical-to-electrical (O–E) conversions should be eliminated. Every O–E conversion costs money in both the short and long term and also slows performance and the ability to deliver services. All-optical switching is desirable because the laws governing optical switching outstrip even Moore's law. The optical world is simply moving faster than the electronic world, so to scale in this world, operators must move to a true optical-switching platform, eliminating O–E interfaces and conversions. Network management is fundamental to implementation of any network, and rapid service provisioning ensures competitive velocity. A service provider that cannot be competitive will perish.

The All-Optical Solution

What will the AON look like? On the edges of this network is client access. The first function is an edge bandwidth aggregation that allows the network to take traffic from the client, do some grooming and aggregating, and then put it on the backbone. The backbone should be all-optical. There will be large pipes and big traffic flows. Electrical conversion must be minimized in the backbone. As traffic moves back out, it goes into the access and then back to the client layer.

Attributes of the AON

The AON essentially provides an all-optical core that stays optical and manages bandwidth at the optical level. That optical core moves the management of services to the top level with the IP routers and the asynchronous transfer mode (ATM) and SONET layers. It significantly simplifies today's network because the old SONET layer and the electrical layer between the transport and service layer have been eliminated.

What do the layers look like? The very base of the optical network is long-haul transport, while the IP environment is distance insensitive. A visitor to a Web site does not know where the traffic comes from or what server farm is hit, so the network must be able to transmit long distances without regeneration. Provisioning and restoration are performed in the optical-switching layer. As the backbone is under discussion, the switching must be all-optical. The next layer up is network management. It is fundamental to AONs to have software that allows provisioning and network maintenance as the network scales. The final layer is that of services and applications. The layers have moved from pure bandwidth to applications and services, which require more bandwidth and more real-time ability to be handled on the network.

Infrastructure

The optical network infrastructure includes fiber and optical terminals that allow the user to access traffic. Further elements are optical amplifiers, optical switches, and optical add/drop multiplexers (OADMs). Fundamentally, the AON has a very simple architecture.

How does this infrastructure work? Switching optically allows the operator not only to provision services through software rather than hardware, but also to accomplish things such as restoration in the optical domain. These are the factors that make this type of network fundamentally less complex than the existing DWDM network.

Migration

Today's legacy network is a ring network, and the industry is migrating from rings to a mesh network. The first point to consider here is the desirability of moving express traffic to a mesh layer so that the optical network discussed above operates entirely in the optical domain. As much express traffic as possible should be moved to the mesh layer in the optical layer. The rings should be used as feeders and for grooming and aggregation, but the big pipes and the majority of the network should be kept in the optical domain on the mesh layer. That will facilitate the ultimate transition to the all-optical mesh network.

The reason this transition works economically and technically is that 70 percent of the traffic through the nodes is pass-through traffic. That percentage of pass-through traffic is even larger in the IP world, which is beginning to dominate network traffic. There is no reason to access pass-through traffic. There is no reason to interrupt it by converting it into the electrical domain to manage it. All operators want to do is pass that traffic through at the optical layer, and only the traffic that will be accessed will need electrical conversion.

In this economic model staying in the optical layer is cost-effective, because all the electrical interfaces can be removed. The ideal solution in the transition is to use the rings as feeders and collectors. One must use the bottom layer of the optical network for the 70 percent of traffic that is express. That traffic belongs in the mesh layer where the big pipes can handle large flows of traffic and where the network can be rapidly scaled. Once active elements are removed from the network and replaced by an optical backbone with no regeneration and O–E interfaces are eliminated, capital costs will be dramatically reduced.

Reduction in operating costs is even more significant. The majority of the life cycle costs of a network are the ongoing operating costs—the maintenance and provisioning and so forth. Removing active network elements from the network also eliminates manpower costs to maintain those elements. Studies suggest that operating costs can be reduced by 70 to 80 percent over the life of the network. As mentioned earlier, a one percent decrease in costs causes a three percent increase in demand.

Summary

The all-optical mesh network powered by insurgent technology drives top-line growth and competitive velocity with its rapid service provisioning. The key attributes of the AON are revenue velocity, scalability, simplified traffic management, simplified network architecture, reduced life cycle costs, and telecom-grade reliability and performance.

Next-Generation Optical Networking

W. Brooke Frischemeier
Optical Transport Product Manager
Cisco Systems

This paper takes an end-to-end approach to the optical network. It examines how services can be incorporated and looks at the types of network elements (NEs) that will facilitate a realistic transition to next-generation optical networking.

Which Technology?

With what technologies and architectures will next-generation transport networks be built? Many solutions are available in the transport and core spaces. They include synchronous optical network (SONET) with time division multiplexing (TDM), asynchronous transfer mode (ATM), Internet protocol (IP) over multiprotocol label switching (MPLS) and dense wavelength division multiplexing (DWDM) (see *Figure 1*).

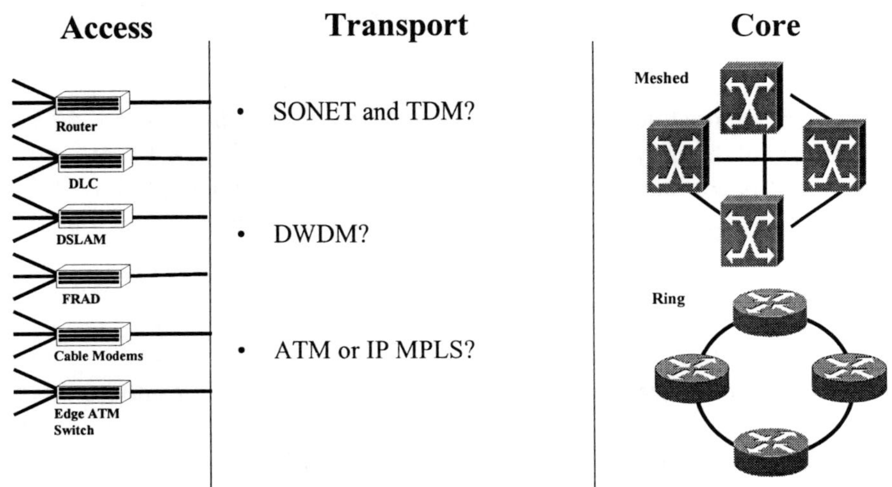

Figure 1: *Which Technology?*

Questions are associated with each of these technologies. Are SONET and TDM, the incumbent technologies, inefficient or inflexible? Can a DWDM architecture with transparent optical switching drop its cost and scale in density to become a viable replacement for low-rate multiplexing for metro and local applications? Can packet- and cell-based technologies such as IP and ATM prove cost-efficient, robust, and voice-capable so that the advantages of statistical multiplexing can be achieved? How can these technologies be used to provide services, and how can they be put effectively into an all-optical network (AON)?

The SONET Bandwidth Bottleneck

SONET framing is nearly universal. There is packet over SONET (POS), ATM over SONET, Ethernet over SONET, and abundant performance monitoring. All of these points are in SONET's favor, but unfortunately, traditional SONET NEs have no data efficiencies. With SONET, circuits are nailed up, and all of that bandwidth is stranded. To protect it, it is necessary to dedicate more bandwidth. Last, SONET rings do not scale (see *Figure 2*).

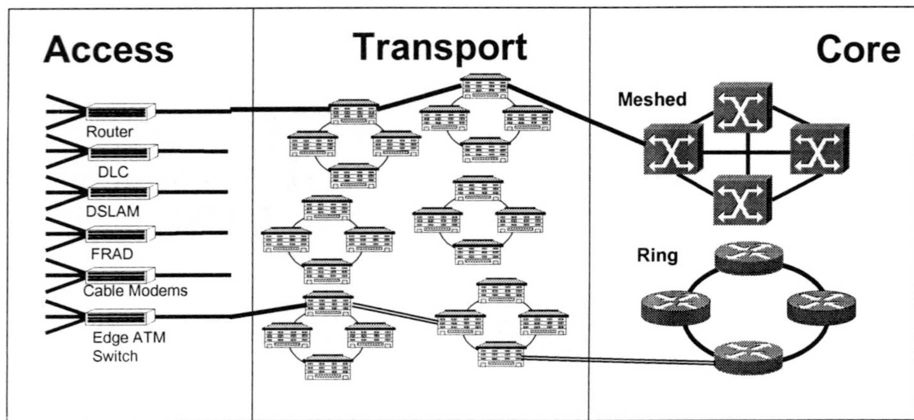

Figure 2: *The Bandwidth Bottleneck*

How much bandwidth is it possible to get through the bottleneck in *Figure 2*? How will it be possible to put a number of rings together to get to another layer in the hierarchy without putting a bigger pipe into the next layer? A more intelligent way to do it than that must exist. People have worked for years on switches, and their backplanes do not resemble rings.

Today, there is an unprecedented demand for high-speed access. Terabit routers and switches combined with DWDM systems enable massive bandwidth in the core. The SONET–synchronous digital hierarchy (SDH) equipment market is dominated by traditional vendors and characterized by little change. Today's SONET–SDH transport systems are optimized for traditional TDM services only. Next-generation SONET must be more scalable and must promote data efficiencies.

WDM Transport

On the other hand, with a pure wavelength division multiplexing (WDM) solution, there are thousands of low-rate access lines per office. Low-rate (digital signal [DS]–1 and DS–3) access lines still dominate end-office and interoffice aggregation needs. Customers still have T1 lines coming into their offices for their virtual private networks (VPNs); they still have cable modems (CMs) transmitting over Ethernet; and they still have digital subscriber line access multiplexers (DSLAMs) transmitting over ATM. Not only is the amount of traffic multiplying, but the number of different traffic types is increasing as well. For instance, not until quite recently would people have considered transmitting Ethernet over a wide-area network (WAN) or wireless intelligent network (WIN). Ethernet over a WIN and over an optical network exist because customers understand and feel comfortable deploying Ethernet solutions. Why should one be forced to convert local-area network (LAN) traffic to frame relay, ATM, or POS? Next-generation networks must be able to transport all traffic types.

Wavelengths remain a limited resource; 1,000+ channel systems are not yet available. Wavelengths are also not necessarily the most efficient avenue for all types of traffic. Consider how a large, multisite VPN or an IP multicast network might be scaled. These traffic patterns require meshes and trees. Unfortunately, scaling either of those services over DWDM involves the connection of pipes rather than packet-based services. Turning pipe networks into meshes or multicast trees requires much up-front provisioning and reprovisioning for every new service termination.

Another issue is that customers may want to buy pipes, services, or both. DWDM transport, however, is still aggregated onto λs (pipes). In a pure end-to-end optical λ solution without any real service level switching, there would be no way to consolidate like services onto one λ. DWDM as a ubiquitous solution, then, has not yet matured.

In addition, no one can afford to waste an entire wavelength on a DS–1, DS–3 or 10 Mbps Ethernet link. Electrical-to-optical (E–O) aggregation by SONET multiplexers will still dominate the transport infrastructure. When will a wavelength cost $100 per port like a DS–1 or Ethernet? (see *Figure 3*)

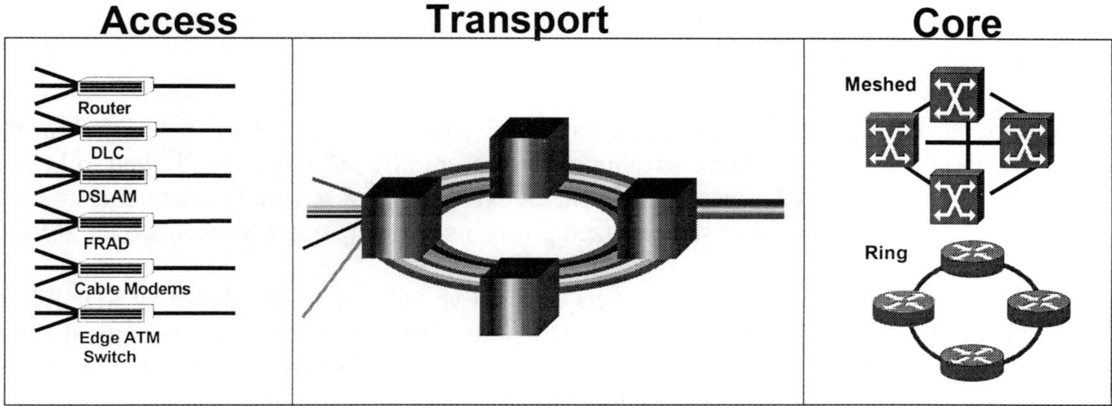

Figure 3: DWDM Transport

ATM and IP Transport?

Other solutions to consider are IP or ATM transport. Might these statistical multiplexing technologies bring cheaper and simpler solutions? An issue that must be addressed is that the number of existing TDM voice circuits is not negligible. When it works, voice over IP (VoIP) or ATM circuit emulation services (CESs) are exceptional, but these solutions must be properly engineered to assure end-to-end quality of service (QoS). Another concern is that voice customers may wish to keep their old private branch exchange (PBX). For voice, IP and ATM offer more complex solutions than TDM.

Statistical multiplexing does not inherently guarantee throughput and delay performance. Packet and cell flows from individual customers (or circuits) must be identified by the transport equipment to be switched and groomed to provide the required performance. ATM does this inherently, while techniques are being developed to fulfill this purpose for IP. These techniques include designing non-overbooked networks, differentiated services (DiffServ), and MPLS.

Other considerations must include the fact that data switch ports cost two to three times what a TDM port costs, and a wholesale retrofit is not always feasible. It is impossible to avoid the conclusion that statistical multiplexing techniques are either not yet widely available (in the case of IP), or are not yet cost-effective for replacing TDM in today's voice and private-line traffic (see *Figure 4*).

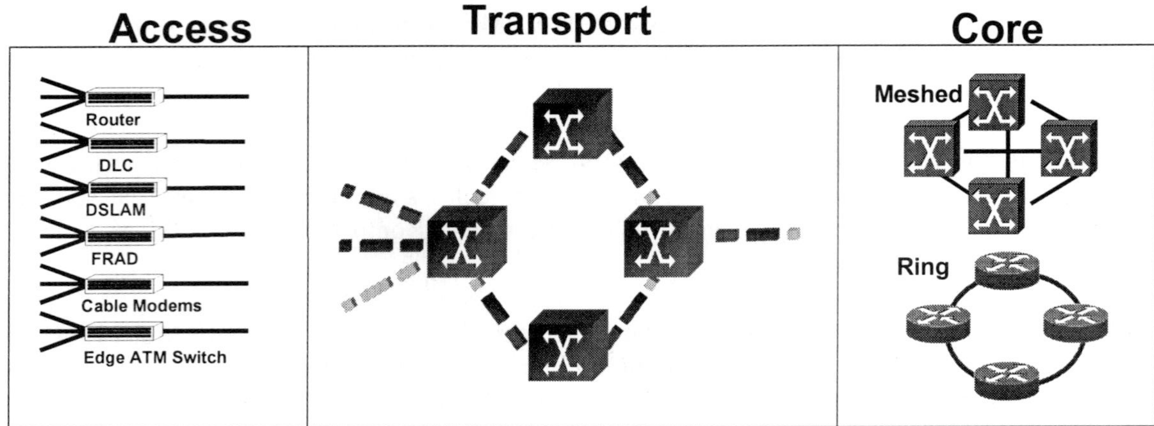

Figure 4: *ATM and IP Transport*

Hybrid Optical/Layer 2/Layer 3

Another scenario to consider is integrating Layer 2 and Layer 3 into the optical NE. While it is also possible to talk about Layer 4 and above switching, the present discussion will be limited to Layers 2 and 3. Wavelengths can carry SONET pipes because the SONET protocol is widely acceptable and promotes high-speed links. Also, there are many standardized protocols that map other protocols over SONET.

SONET pipes carry channels of TDM, IP packets, and ATM cells. Wavelengths carry SONET pipes or any exotic 2.5/10 Gbps signal. In any case, next-generation transport must combine optical networking and service delivery so that network operators can gain the use of DWDM's bandwidth potential and the efficiencies of data network protocols (see *Figure 5*).

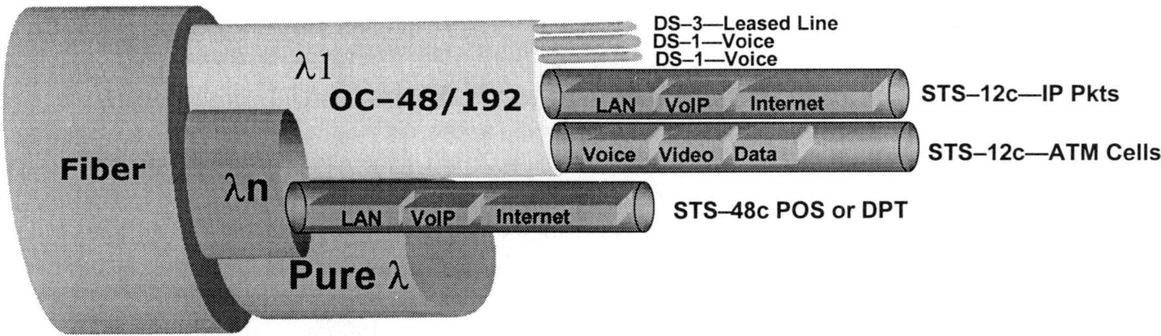

Figure 5: *Hybrid Optical/Layer Two/Layer Three*

Make the Anything-over-Glass Migration Easy

The ultimate objective is to put optical switching, IP over MPLS, and ATM into a universal NE economically. One requirement is to not impose SONET, DWDM hardware, or any proprietary protocols. With traditional NEs, going from one protocol to another may not necessarily be a software load, but may involve significant hardware upgrades or an entirely new NE. But what can be done with the current protocols to improve multispeed SONET? SONET–mapping functions and performance monitoring are industry standards, and SONET remains a growing market with which customers are comfortable. However, SONET rings are not what next-generation networks need because the only way to make them scale is to make them bigger and faster.

SONET must also embrace multiple-λ with its potential in the emerging, ultra-high-bandwidth market. Next, full multiservice switching must be integrated into the optical network device where appropriate. This does not mean that line rate–IP forwarding is needed for every λ or SONET path, but the ability to aggregate packets from different interfaces onto a single synchronous transport signal (STS) or λ is highly desirable. This aggregation and λ awareness can in turn be presented to a unified optical control plane that will facilitate end-to-end configuring and end-to-end reroute.

Path-Protected Mesh Networks

There are many exotic methods for improving SONET network restoration. A very simple and effective way is based on unidirectional path-switched rings (UPSRs). UPSRs take a signal, bridge it at one end, and select the best signal at the receiving end. With UPSR, the intelligence is in the nodes where the signal enters and exits the ring. The intermediate nodes do not have anything to do with the path recovery (see *Figure 6*).

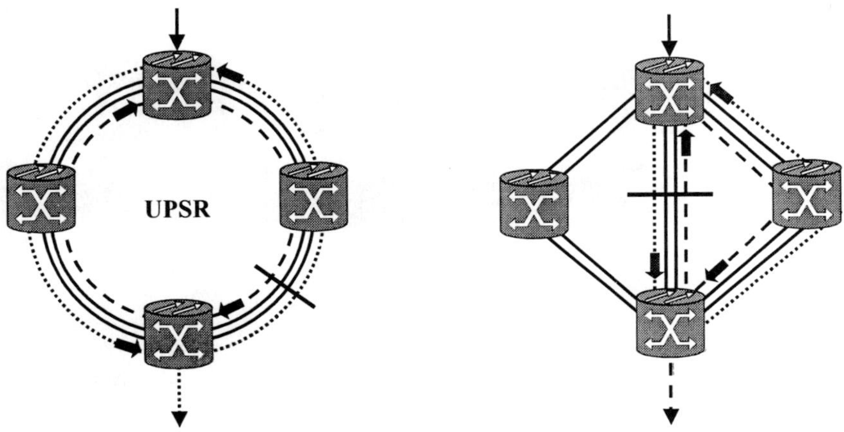

Figure 6: *Path-Protected Mesh Networks*

It should be possible to apply UPSR to a mesh and achieve a one-plus-one path protection. In a mesh network, this would allow the operator to choose the spans over which to send traffic, and this in turn would allow large meshes or switch matrixes to be built. With a ringed architecture, the only way to scale networks is to increase the interface speeds. Of course, the ring interfaces can function only so fast.

Another option is to build intelligent mesh architectures thatallow traffic flows to change as needed. This intelligent mesh architecture must be able to interoperate with current SONET recovery mechanisms, such as UPSR and bidirectional line-switched ring (BLSR) protection (see *Figure 7*).

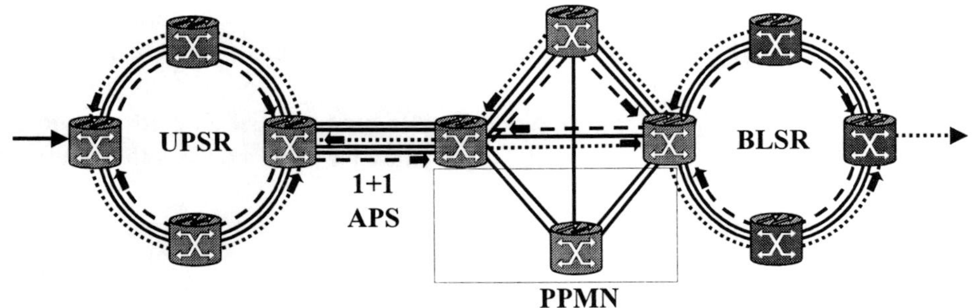

Figure 7: *Integration: Span and Line Protection*

Protection Options

This brings up the topic of multiservice switching. Different layers in the network have different failover speeds. Speeds of less than 50 microseconds are typically associated with Layer-1 solutions, while Layers 2 and 3 typically require more than 50 microseconds to recover. While there are fast-reroute IP functionalities and ATM virtual path (VP) groups that can reach those speeds, Layer 1 is, by nature, faster and easier to manage. Mesh protection also facilitates one-for-many protection, rather than one-for-one protection. (see *Figure 8*).

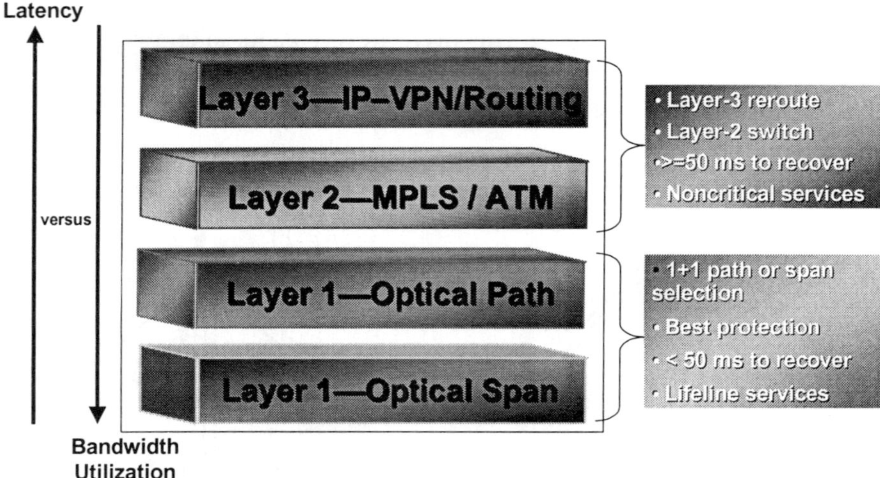

Figure 8: Protection Options

These protection options provide freedom of architectural religion. There are many smart ways to design a network. If this were not true, there would be only IP, only ATM, or only SONET networks. However, because it is true, people want to protect things at different layers. A customer may need both ATM and span protection. In short, every network is different and may require different engineering and protection guidelines. Next-generation transport must provide this flexibility.

Advantages of Service Switching over Multiple-λ

What are the advantages of multiple-λ and multilayer switching? First, λs alone do not provide scalable services. How many λs are enough? If one looks at the current rate of traffic growth, there will never be enough λs. How many λs may be driven on one line before it is necessary to switch out hardware? Multiple-λ switching may be scalable in terms of transport, but it will not always be scalable in terms of service delivery. In a pure DWDM network, things can become difficult with multipoint VPNs or IP multicast, for instance. Integrated Layer-2 and Layer-3 functionality will promote the sharing of λs. If the aggregate traffic of multiple customers is sub-line rate, they should all be multiplexed on the same λ because, no matter how many λs are in a network, none of them are free. Operators can run out of bandwidth for a variety of reasons. A multiple-λ solution with multilayer switching can have a long life before it is necessary to buy more cards or a new network or bigger NEs. One advantage of multiservice NEs is that when providers wish to add new service types, they can add a card rather than a new, overlaid service network.

With multiservice NEs, satisfying a customer may simply be a matter of adding a card rather than a network. That specific switching engine will be smaller and on a particular card, not on a big routing blade or a big λ switch. (see *Figure 9*).

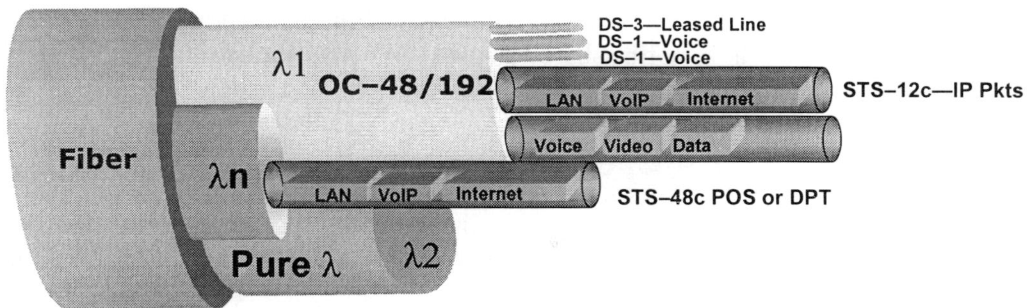

Figure 9: Advantages of Switching over Multiple-λ

Integrated Solution

Integrated service switching allows the flexibility to combine transmission and data networks. However, some providers may still require a more layered, traditional approach. In many companies the data and telecommunications technicians are two separate organizations. Thus, if this integrated solution is to be implemented, it must somehow be manageable by both of these groups. The next-generation NE must provide the flexibility to allow multiple organizational entities to coexist on a single management platform (see *Figure 10*).

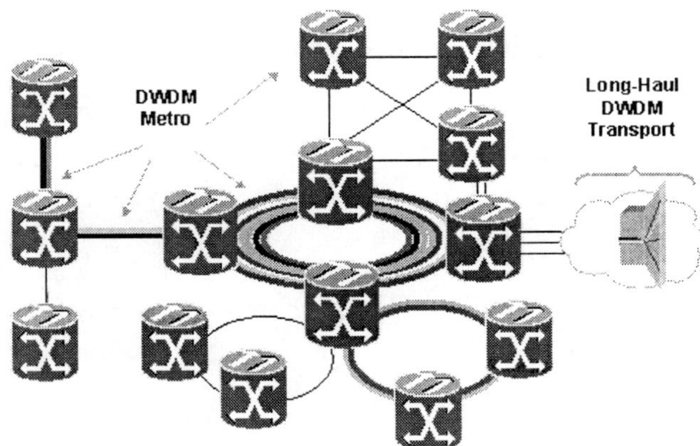

Figure 10: Integrated Solution

Integrated service switching also allows for flexible, multilayer service protection options. Migration between services and peaceful coexistence among them is made easy by add-a-card technology. That technology makes it much easier for customers to go from SONET to ATM or from ATM to router-based networks.

The true multiservice devices of the integrated solution promote freedom of technological religion. Whatever the customer wants the customer can have, and the revenue is won. The provider need not impose an architecture on the customer.

Multiservice Optical Control Plane

Finally, multilayer switching in the optical network sets things up to work well with next-generation optical control plane mechanisms. This envisions a true end-to-end provisioning and recovery solution. Protocols such as resource reservation protocol (RSVP) and MPLS, along with the routing protocols, would be used to set up λ and IP service paths. Many vendors are eager to promote an AON, and a multiservice optical control plane can make those benefits even more useful. The optical control plane can examine things such as whether TDM packet-switching capabilities exist at each node. Are purely separate λs being switched, or is it a fiber-based switch? What is the media type, or what are the encapsulations? Does it switch SONET? Is it coming in as Ethernet, or as some other exotic, unsupported encapsulation that must be passed through? Are IP switching or MPLS available?

What does all this make possible? For instance, if there is a large amount of IP traffic entering a node where it is possible to share or be multiplexed onto a λ, the network is used more efficiently. Therefore it is very important that the optical control plane understands the service layer and its encapsulations.

Finally, service devices must be able to operate over different kinds of NEs and be able to operate multigroup domains. That is, no single element needs to see the entire network; subnets or peer grouping should be available. Optical switches should be able to communicate to routers with or without multiple λs, ATM switches, or any kind of hybrid device.

In short, a multiservice optical control plane will enable a constraint-based, optical-connection setup and integrate the service providing egress points. It will identify the link type—TDM, packet, λ, or fiber—and identify the media—SONET, Gigabit Ethernet, or another encapsulation.

Summary

It will be necessary to support new-world SONET and DWDM simultaneously because they both fulfill many current needs. People continue to buy SONET, and many of those people are considering DWDM for the future.

Network operators must not waste their λs. They must integrate multiservice switching to avoid a bandwidth bottleneck. They will be able to shift customers around on their λs and IP circuits.

Providers must integrate the optical control plane with the service plane. Once that is finalized and functional, it will be a true end-to-end solution that will comprehend the entire network. Such a solution will certainly reduce management time.

Providers should also forget the forklift and find an architecture that supports multiple switching technologies, whether TDM, IP, or multiple-λ. Then it will be possible to migrate from one to another by changing cards rather than installing new NEs, configuring all of its addresses, turning it up, and so on.

Finally, network providers should remember not to outsmart the customer. A provider can always offer the most advanced solution on the market, but if the customer, or the customer's customer, does not understand the solution or does not think he can implement the solution easily, he will not implement that technology.

The Optical Networking Industry

James Kedersha

Managing Director, Telecom Equipment
SG Cowen

This paper outlines the way in which Wall Street and the institutional investment community view the optical industry. It endeavors to provide a sense of why outlandish valuations were paid in early 2000, why they will be again, and what factors are behind such reasoning.

Figure 1 is an illustration of equally weighted price chart indices, and it is clear that the returns have been excellent. The chart indexes everything back to the beginning of 1996, and a group of the optical component names comes out at the top of the list, appreciating a little over 100 percent per year. That group includes names such as JDS Uniphase (JDSU), E-tek (now owned by JDSU), and SDLI (to be acquired by JDSU). What are called the next-generation optical systems names are not far behind—just over 90 percent per year since the beginning of 1996. This group includes Nortel Networks, Sycamore, CIENA, and Harmonic, which is optical for the cable industry right now. Lagging behind that group— and this is the important point—is what is called the older-generation technology or telecom technology companies. They are solid companies, but they are still not perceived as next-generation. Members of this group are the ADCs and Tellabs of the world. One of the key points here is that when the stock market perceives a company to be an Internet infrastructure/broadband communications supplier of some type, the company achieves premium valuation.

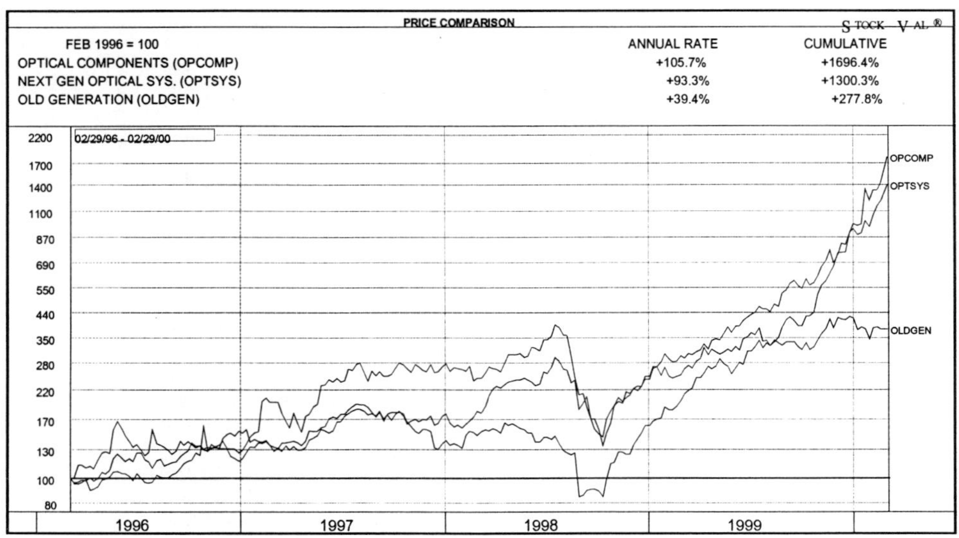

Figure 1: *What Is the Market Thinking?*

The market's view starts with something of which everyone is very well aware, and that is growth of network traffic (see *Figure 2*). Telecom equipment has altered from what was historically a cyclical, moderate-growth industry—something that was probably more based on regional economic factors—into an industry of much higher, secular growth. Of course, a great deal of this is based on the Internet and the growth of network traffic. It is also based on deregulation in the carrier world. The combination of data growth and deregulation has spawned not only network upgrades, but also many new networks around the world. The financial analyst's perception of this situation is that growth is far from over, as broadband access upgrades are just getting started. The hope is that these upgrades will place much more traffic onto the network. Another key underlying point is that this situation applies not only to technology but to the competitive local-exchange carrier (CLEC) community as well. Funding has been available, whether in private equity or public equity markets, for carriers and their technology deployments.

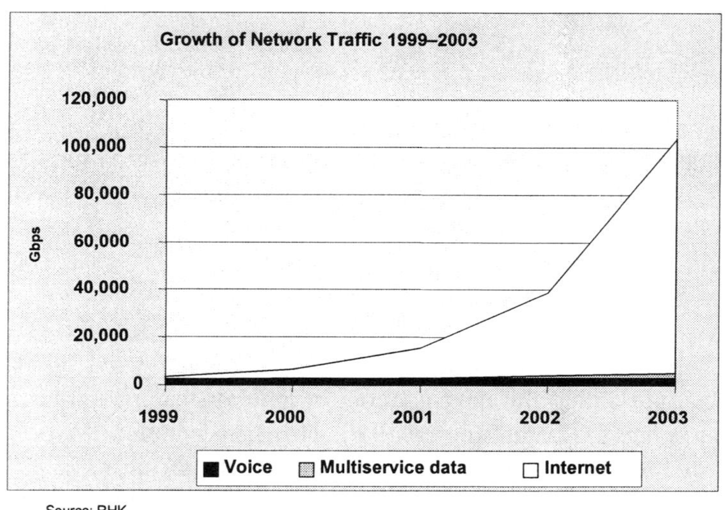

Figure 2: *Secular Growth in Telecom Equipment*

Figure 3 shows that traffic trends, particularly on the data traffic side, suggest an upside to spending. This is not only because more traffic exists or because broadband access is coming, but because the nature of data is different than that of voice. Data's nature is much less predictable and more bursty; therefore, the amount of capacity that is engineered into the network becomes much greater.

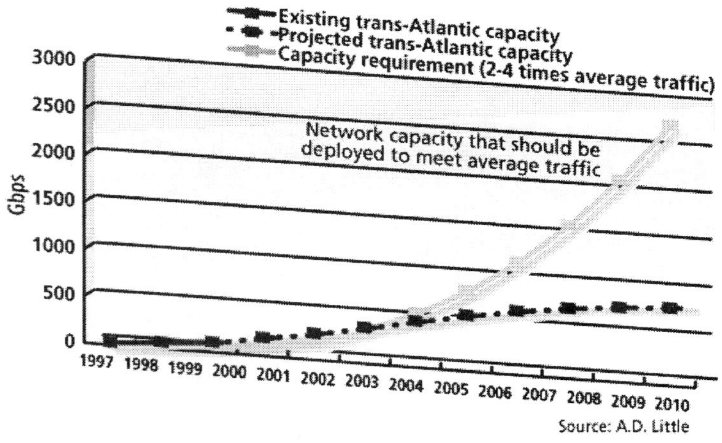

Figure 3: *Traffic Trends Suggest Upside to Spending*

The market then considers optical networking and components. Although the different types of technologies can be debated, all of the networks have one thing in common—i.e., they need real-life services. This factor makes optical a common denominator. It is interesting to note that the perception in 1998 was that optical—dense wavelength division multiplexing (DWDM) in particular—was going to become a commodity. This correlates to the experience of seeing a stock price when it was low as a commodity. One thing to realize about the stock market is that times are never as good as they seem when they are good, and times are never as bad as they seem when they are bad. The reality is always somewhere in the middle. So optical was not quite a commodity back then, but a number of things have emerged since then. Switching and metro markets are on the horizon, and the idea of full-optical systems that could actually generate revenue for a carrier has been born. Much more is involved than simply providing a few more optical pipes between city A and city B. This helps to change the perception that there will be waves of growth in a large market for a long time. Valuation in the stock market is its drive from growth, market size, and market profitability over time.

The stock market generates many key questions, but the commodity question arises more often than not. Today, the more important question, which is asked constantly, seems to be: Will there be a glut of bandwidth? Moreover, what are the key challenges to success from both a system player's and an optical component player's point of view? A consolidation and concentration within the public markets is occurring, while many start-ups on the private side ask what it takes to be successful in the long term.

Optical and communications in general may be considered in terms of an industry food chain (see *Figure 4*). The chain begins with the physical infrastructure—the actual fiber, for example. It moves up into what is a fairly sizable (perhaps five billion–dollar) market for components and subsystems, which rises to $30 billion or so for optical systems. Then the food chain moves into the bandwidth providers, which represent an $800 billion plus market per year, and then to the Internet and e-services providers, which represent a $100 billion market that is growing at a high rate. Of course, there are also end-user bandwidth demands. This is helpful for the market because different levels of the food chain are all growing at high rates. *Figure 5* suggests that what is being seen in equipment and technology growth is sustainable.

Industry Layer	Major Players	Size	Annual Growth
End-User Bandwidth Demand	Consumer, Commercial Users	----	350%
Value-Added Services/Internet Commerce	Amazon.com, E Trade, Ebay, etc.	$102B	174%
Network/Bandwidth Providers	AT&T, MCI WorldCom, RBOCs, CLECs, PTTs	$800B	7% for voice, 40+% for data services
Optical Systems Vendors	Lucent Technologies, Nortel Networks, Alcatel, Ciena, Pirelli, Tellabs, GEC, Sycamore	$30B, including WDM of $3B	50–100%
Optical Component and Subsystem/Module Vendors	JDS Uniphase, Oak (Lasertron), SDL, Lucent, Alcatel, Nortel	$5B, including WDM of $1.4B	50–100%
Fiber	Corning, Lucent	27M klm base-LH single mode	35% in volumes

Figure 4: *The Optical Communications Food Chain*

The Systems Market

When the systems market is examined more carefully, it is recognized as a large market (see *Figure 5*). Everyone tends to use Ryan, Hankin, Kent (RHK) numbers, and the $30 billion includes more than just wavelength division multiplexing (WDM)—it includes the synchronous optical network (SONET) and synchronous digital hierarchy (SDH) markets. The systems market has much potential. Also, switching and metro markets and even ultra-long-haul markets are about to emerge. All of these combined will encourage this growth to the $90 billion rate by 2003.

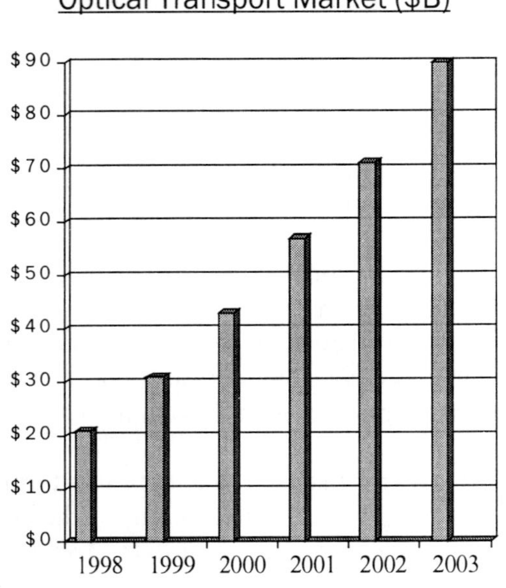

Figure 5: *Systems Market—Large and Growing*

A "club sandwich" debate is underway with regard to what technologies are required or what is the right mix to build the backbone of the network (see *Figure 6*). The world is such that one size does not fit all. A new carrier building a new greenfield facility might choose something far different from what an incumbent regional Bell operating company (RBOC) might choose. The common denominator, however, is that optical technologies, particularly DWDM, will always be there.

The Backbone "Club Sandwich"

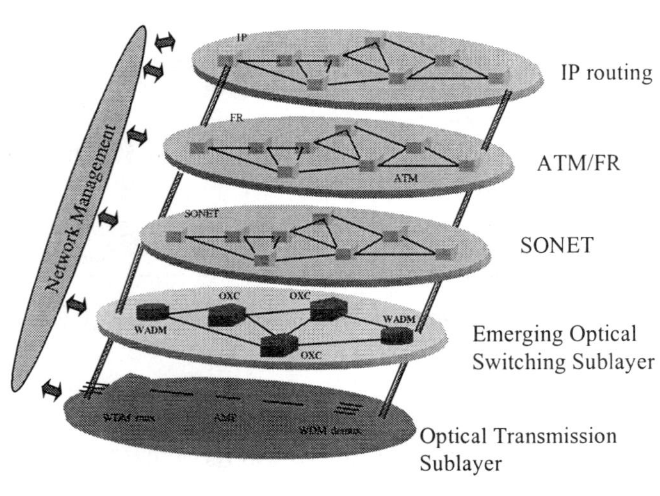

Figure 6: *Backbone Debate—One Size Does Not Fit All, but Optical Is Common Denominator*

Figure 7 depicts the multiple waves of emerging markets. They are not specific only to long-haul WDM undersea markets and cable television (CATV) markets. The metro market will be interesting to see play out. Numerous solutions to the metro market have been proposed. One potentially huge area ranges from metro WDM to a next-generation, data-optimized SONET device. There are even some solutions in the optical frequency division multiplexing (OFDM) arena that may be successful. The prediction is for more varied requirements in metro and, therefore, for a larger variety of solutions that play into it. Of course, one of the most important markets is the switching market. Point-to-point deployments of WDM must be turned into networks, and investors must recognize that once the move is made from point-to-point to networking, a long cycle is begun. Such is the thinking that underlies the general belief that the market may be correcting right now. The feeling is that the market will return, as it is not the seventh or eighth but rather the second or third inning of the game.

DWDM Segment	Long-Haul	Undersea	CATV	Metro	Switching
Size	$3B	$1B	$100M	Nascent	Nascent
Growth	50%+	30%	High, small base	Potentially high	Potentially high
Deployment Cycle	1996–present	1996–present	Early 1999–present	Starts late 1999	Starts 2000/2001
Leading Public Players	NT, CIEN, LU, SCMR	ALA, Tyco, KDD	HLIT, GIC, SFA, ANTC	NT, CIEN	CIEN, SCMR, LU, NT, CSCO
New Entrants	Corvis, Qtera	Few	Few	Many—see 1/00 SOL	Corvis, Qtera, Monterey, Cyras
Rationale/ Requirements	Alleviate fiber "exhaust"	Alleviate fiber "exhaust"	Network simplification / Return path capacity	Fiber exhaust / Data services	Network simplification / Capital costs
Major Carriers	Most IXCs, PTTs	Global Cross., Oxygen	AT&T/TCI	C&W USA, Bell Canada	Williams, Iaxis, and others
Issues	Few now	Maintenance	Adoption by other MSOs	Mkt. size / Timing	Maturity of technology

Source: SG Cowen

Figure 7: *Multiple Waves or Markets Emerging*

Evolution toward Optical Networking

The industry is evolving toward optical networking, although the term *all-optical networking* is a bit too strong to apply yet. However, a movement toward doing away with more and more of the electrical part of the optical-to-electrical-to-optical (O–E–O) sequence is underway, and that requires the deployment of more technology. Software will become very significant in this area. The appropriate technology or the right business mix (i.e., the right channels to the carriers) will not alone make the system vendor of the future successful but, increasingly, software and network architectures will focus on where the value-add is perceived. Vendors have historically perceived their value-add to lie in the vertical integration of components, manufacturing, and systems. However, many vendors have moved toward focusing and adding value with the network at the systems and then at the architectural level. At present, the software level contains the most value. The trend seems to be heading for outsourcing increasingly more of the optical systems, much in the same way that electronic systems are outsourced.

Network simplification is becoming the norm in the industry (see *Figures 8* and *9*). Although the network is becoming complicated and expensive, as the individual switches, the add/drop multiplexers (ADMs), and other elements are consolidated into single devices, the network will become simplified. It is definitely something to consider; it is not simply data-based; it is deregulation-based, which forces carriers to be competitive and cost-competitive if they want to keep up with the new competitors.

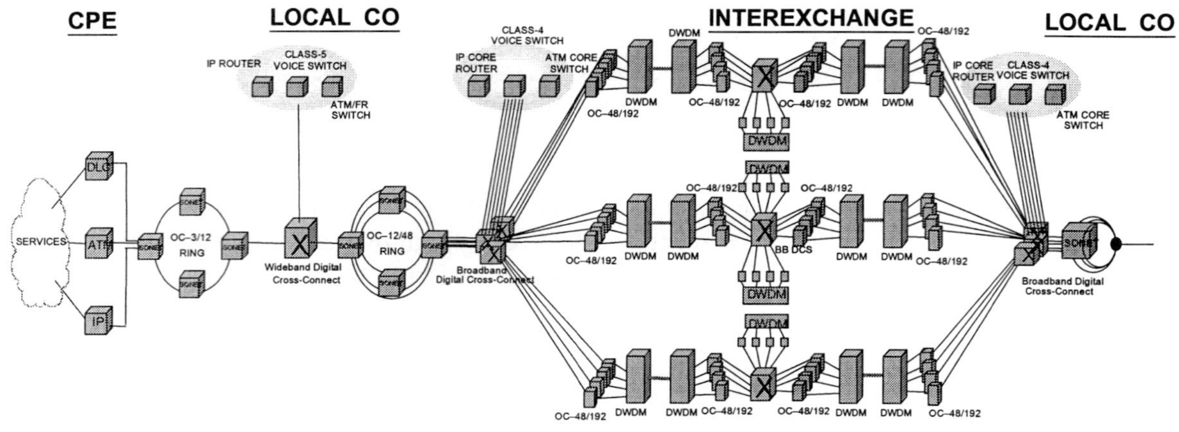

Figure 8: *Legacy Network Evolving...*

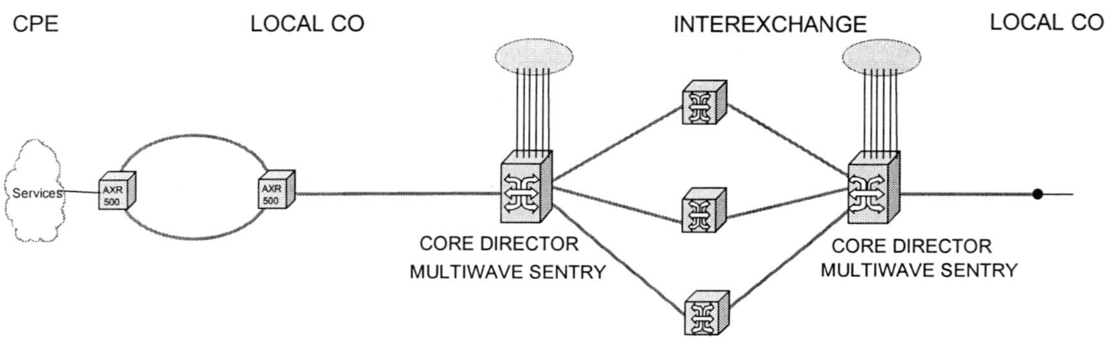

Figure 9: *...To a Simplified, All-Optical Network*

Figure 10 shows the numbers of valuation at the end of February 2000. The market has been paying extraordinary valuations for a number of names—both systems and components. Part of the reasoning for that is the simple view that the revenue and earnings estimates are way too low, so they will be exceeded by huge amounts. The estimates that the institutional investors are discounting are much, much higher than would normally be shown publicly. Earnings estimates are always exceeded, so the estimates are moving up, providing them with momentum. Also, some of these valuations, particularly on the newer companies that have recently made initial public offerings (IPOs), reflect the supply-demand imbalance. There is abundant demand for relatively few names both in systems and components, and there simply are not that many shares when these companies initially go public. Thus, the appetite for them forces the valuations up quite high. There is also an overlay of consolidation as companies are acquired by others. The telecommunications industry has been consolidating for many years, and in the past few years, that consolidation has spread into the optical area. Consolidation, of course, makes investors hopeful that their favorite name will be bought out at a nice premium as well. It is not uncommon to see an existing, successful company trading at 30 or 40 times its revenue. Actually, the metric has gone from price-to-earnings to price-to-revenue; indeed, it is often quoted in price-to–addressable market.

Valuations

	Stock Appreciation		Cal 2000 Val	
	Feb	YTD	P/S	P/E
Optical Systems				
Sycamore	31%	35%	243	-13900
Harmonic	32%	34%	16	134
Ciena	129%	160%	35	269
Tellabs	-11%	-25%	6	29
Nortel Networks	18%	12%	6	90
Optical Components				
JDS Uniphase	26%	59%	56	378
SDL Inc.	52%	80%	49	273
E-Tek Dynamics	42%	92%	45	320
Corning	17%	40%	7	76
MRV	82%	111%	N/A	493

Figure 10: Wall Street View and Valuations

The interesting aspect about the components market is that several years ago, deregulation spawned a war among service providers. Investors want to play the arms merchants and equipment suppliers that sell into them. That has been taken to a new level in recent months; the war is perceived to be spreading into the systems market because of a number of new entrants into that market, particularly on the private side. Therefore, the arms merchants may actually be the component suppliers. Instead of trying to pick the right horse in the optical space, why not buy a company such as JDSU, which would supply components to all, therefore diversifying and spreading out the vendor-specific risk? Of course, lowering the risk usually means increasing valuation.

The components business is seen by the stock market as fairly sizable—i.e., about $5 billion (see *Figure 11*). That includes about $1.4 billion last year for WDM as well as components for SONET, SDH, CATV, and even the submarine markets. It is a market that has very few names in which to invest because JDSU keeps buying them all. The market is really narrowed down to JDSU, SDLI, Corning, and maybe one or two others. One thing many people forget is that the market is much broader than that. There are many large, captive suppliers at Lucent Technologies, Nortel Networks, and Alcatel and in Japan that are not seen as often, but that are still certainly out there.

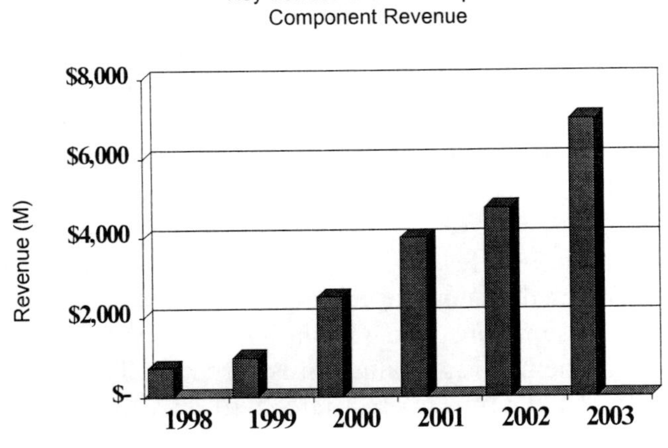

Figure 11: Components Business—Large and Growing

The key components are WDM systems—certainly lasers, modulators, wave lockers, the erbium-doped fiber amplifiers (EDFAs), and all of the first-generation elements needed to get these systems going (see *Figure 12*). Although this technology is moving quickly, visions of the evolution of the all-optical or increasingly optical network and dynamic routing of many wavelengths and provisioning services through a network require that the technology advance at the systems level as well as at the components level. The industry needs components that can provide more wavelengths, send signals farther and at higher rates, and change or tune different wavelengths. If the time-to-market aspect is placed over this concept, it, of course, pushes the envelope on development, therefore enhancing or accelerating acquisitions in the marketplace. JDS recently bought Cronos to get into the optical-switching space. Some key components that will enable the network visions of the future will be in the all-optical, cross-connected, optical-switching space; technologies such as micro-electromechanical systems (MEMS) will become increasingly important.

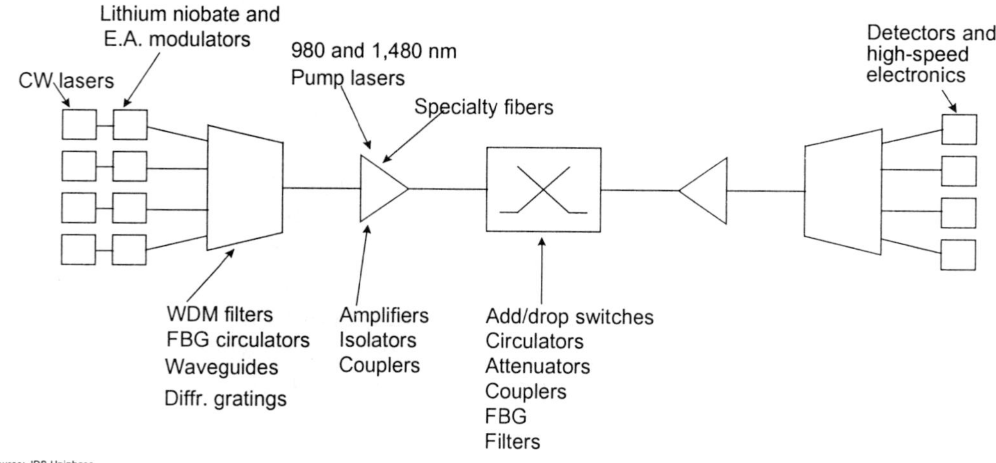

Figure 12: Key Components in a DWDM Network

Consolidation will probably not abate any time soon because IPOs will continue to occur. At a systems level, critical mass is very important. The ability to employ more than simply a single product line and to maintain other aspects desired by service providers (such as financing, support, or global presence) becomes very important. Therefore, the industry will push the systems players to continue to consolidate into larger and larger hands. The Nortels of the world will certainly get bigger, as will the Lucents. It is interesting to watch what happens to the mid-tier companies. A company such as CIENA acquires and fills its story from access to switching to transmission. The big question is: Can such a company stay all-optical, or will some integration with Internet protocol (IP) or asynchronous transfer mode (ATM) or other technologies occur? At a components level, the trend toward consolidation will continue. Two of the key challenges at the components level are scaling manufacturing capacity and getting qualified. If a company emerges as a component player with a device that is smaller, shinier, faster, and better-performing but that takes six months to get qualified, it is very easy to miss numbers. In that game, diversity of product, customer base, and applications will keep things going.

In the second half of 1999 and the first few months of 2000, an enormous run-up in the names on the market occurred. There is not much doubt on the earnings and fundamentals of what is happening in the market. No matter where discussions are held—in the Optical Fiber Conference (OFC), in analyst meetings, or within the carriers themselves—valuation is essential. The market will periodically panic about valuation; in fact, a slight tick-up in interest rates will lead to disquiet. Usually the thinking is that, as interest rates tick up, more focus is paid to the junk bond market. That has slowed recently, and competitive providers currently cannot get their financing. That is the incremental piece of spending; it will slow in a few months and stockholders will sell. This has been seen in the market before, and as long as interest rates do not rise too high (and analysts do not think they will) this phenomenon will fade away. The fall of 1998 was considered the worst period of doubt in recent times. There was no junk bond financing, so the CLECs were all dead; the capital spending survey indicated that no spending would occur in 1999. Carriers will not confess to a sharp increase in capital spending until they are forced to the next year.

It is important to observe trends in deregulation. Even with the year 2000 (Y2K) concerns over the last few years, the prediction is that the fundamentals are there, that the interest rates are not running away on us, and that once the market is through this period (which, hopefully, will not last much longer), the stock market will get back on track. The industry is in the very early stages of network upgrades, and the trends observed on the data side, the Internet side, and the deregulation side are not yet definitive.

Convergence of Optics, ATM, and IP

Vijay Lewis
Chief Technology Officer
Valiant Networks

This paper will outline the convergence of optics, asychronous transfer mode (ATM), and Internet protocol (IP) on a hypothetical full-service provider. It is difficult to address the entire spectrum of services and products a provider can have. A provider can be switchless or have full service with equipment of its own. It could offer only IP services, only voice services, or could offer a multitude of other services such as private line in addition to IP and voice. To help explain the concept, this paper will use a generic model of what a full-service provider must tackle to get into this market. It will also cover some of the technical and business realities faced by a full-service provider. It will outline network topography and design choices (traditional versus convergence design), provide a quick summary of the steps that occur in convergence and service offerings, and describe different implementation hurdles that carriers can face.

Technical Realities

What are the technical realities? Today, data and multimedia basically drive demand for bandwidth. There is also abundant voice traffic, but that traffic is slowly moving over to the data side. Advances in integrated circuit technology are helping emerging telecommunications equipment to achieve higher equipment densities. Dense wavelength division multiplexing (DWDM) has changed the face of what can and cannot be done today. In 1990, companies could get 24 digital signal (DS)–3s on a fiber; today it is possible to get 20 or more optical carrier (OC)–48s on that same fiber. This represents a nearly 50 percent increase in speed and capacity on that same fiber. In addition, ATM switches and IP routers are shipping with OC–48 interfaces, and some routers have OC–192 already in production quantities. IP seems to be ahead of ATM in terms of the 192 mainly because it is easier to build. There is also much hype about what IP can do, and therefore, the demand for it has been tremendous. ATM has existed for some time, but for some of the traditional carriers, putting ATM into the network and using it as the ubiquitous platform to offer services is something new. Multiprotocol label switching (MPLS) is the new introduction and promises to be the agent that brings together the features of ATM and IP. Finally, there is a whole new voice-over-packet paradigm that is making its way into the traditional voice carriers and is being aggressively deployed by newer carriers.

From various industry studies and analyses, it is clear that carriers are installing more fiber and building networks more quickly. This, coupled with the influence of new technology and protocols that exploit the fiber more efficiently on those networks, is leading to an explosion of bandwidth, which in turn is dramatically driving down the cost-per-information-bit transmitted (see *Figure 1*).

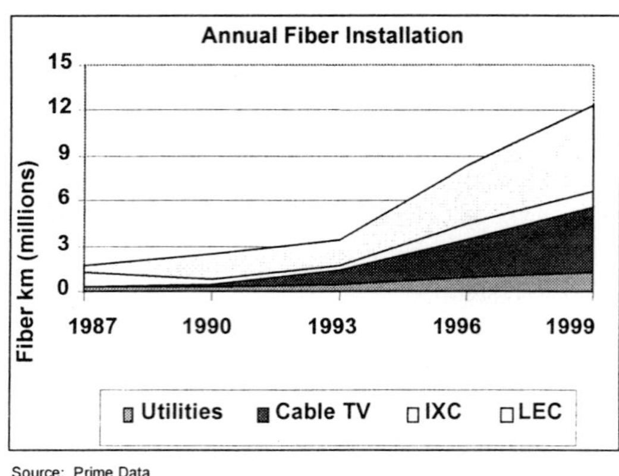

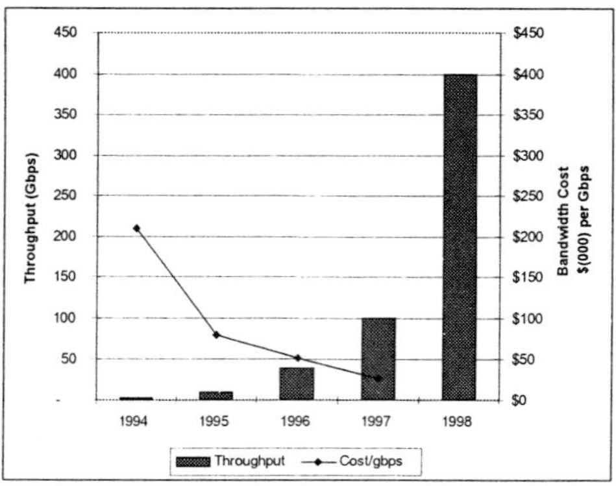

Source: Prime Data

Figure 1: *Technical Realities*

Business Realities

Even though the bandwidth demand on a carrier comes from data and multimedia traffic, the revenues derived from traditional voice and private-line traffic dwarf the data/IP revenues. In some large carriers, the IP business units contribute less than 10 percent of the total revenues. A company that plans to get into this market must realize that it must be reactive to voice users. The term *voice users* encompasses all of the T1 private lines and some of the other optical backbone involved. There is also a market for private lines. For instance, some Internet service providers (ISPs) that receive primary rate interfaces (PRIs) from the incumbent local-exchange carrier (ILEC) in a city use another carrier to back haul these to dial-up equipment installed in the co-location space of the carrier in another city. Large businesses run T1 private lines as tie lines between their private branch exchanges (PBXs) or routers in various branch locations scattered across the country. Whether it is voice or IP, the circuit on the intermediate carrier still uses time division multiplexing (TDM) and must be transported. Therefore, a full-service provider must continue to deal with this legacy service (see *Figure 2*).

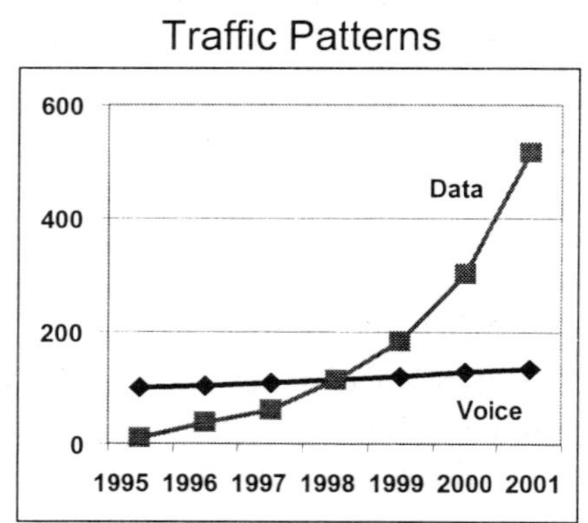

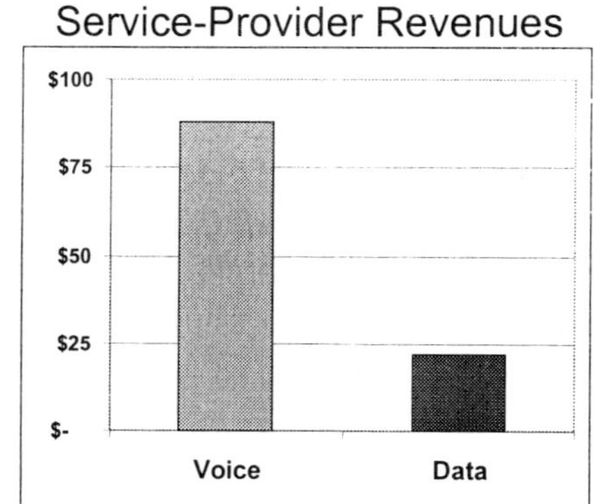

Source: Company reports and FCC statistics

Figure 2: *Business Realities*

Businesses are data/IP savvy. However, most of them have local-area network (LAN) administrators who run the routers and a separate department that runs the PBX. Because of the separation of their budgets, the experience levels of the personnel, and company politics, it is extremely difficult to transition into a single paradigm where one group runs both the voice and data networks of the company. The interoffice voice network personnel either do not have the experience level or simply do not trust a complete data network to transport all their business-critical voice. This forces many businesses to have separate voice and data networks, albeit through the same carrier. The carrier therefore has to cater to this need by supporting traditional as well as data services. Thus, it becomes incumbent on the carrier to evolve ahead of businesses in providing these services reliably across the same access wire. The personnel of these companies must be educated by the carriers on the advantages of an integrated solution that actually scales reliably from today's model to a converged model of the future.

Traditional telcos and ILECs have a very TDM–centric mindset. They understand ATM and are moving in that direction, but they have an existing network of TDM circuits with which they must deal. When they want to get between major cities, they must depend on new carriers to provide these circuits, thereby forcing the latter to build a network that can also support legacy private-line services. In addition, upstart and niche-application ISPs, data local-exchange carriers (DLECs), and application service providers (ASPs) are driving a demand for OC–12 and OC–48 lines, and "waves" between their equipment in various data centers, co-located space, and points of presence (POPs) around the country. They want their own wide-area network (WAN) infrastructure that is under their own control. New carriers are now faced with ensuring that their networks can support such demands.

Most analysts show the same graph, where there is an exponential growth in data bandwidth over voice/private line from now until about 2005. However, the revenues are exactly the opposite. Data is increasing primarily because of the Internet. Over time, as solutions in voice over packet mature, integrated solutions will be developed; voice traffic will move across the data network, and revenues will increase as if they were contributed by data rather than pure voice. The bottom line is that existing and emerging next-generation carriers are forced by the above business drivers to build networks that address the exponential growth of Internet and data traffic while preserving the integrity and revenues from voice and private-line services.

Challenges Faced by a Full-Service Provider

The first point is to define full-service providers' service portfolios. Will they offer frame relay (FR), ATM, or Ethernet local loops, private line, voice, video, etc? Will they target the business segment or the residential segment? Will they be metropolitan players or nationwide players, or will they start as WAN providers, and then drive their long-distance, intercity network into the metro?

Another issue concerns the multiservice nature of the network. Does the provider optimize the network for data services and leave the legacy TDM? Can it provision its service portfolio easily as these new services and markets are introduced? How does it leverage technology innovation? At this point, the number of so-called multiservice aggregating boxes is around 42 vendors. Those 42 vendors target seven mean carriers. Engineers constantly face problems trying to determine the best solutions. When they find a solution, the next problem is how to integrate it. A network must be built that faces the problem of multiservice aggregators. When a new box comes in, how will it be put onto a network? The network must be sufficiently flexible to absorb these new technologies without disrupting what is already in place. Future-proofing the network to absorb new technology and optimizing the use of its network assets (fiber and switches) are challenges faced while designing that network.

Service differentiation is important for a provider to understand. The provider must know what its costs are versus its competitors, and it must force these costs to be lower. In the case of provisioning and service delivery, some wish that an OC–3 could be provisioned within 15 minutes; others want it done within 5 minutes. There is a difference between provisioning a service and actually activating it. For example, both can be achieved in 15 minutes if the carrier owns the local loop; if another carrier is involved, it can take 45 days before circuit activation. The reality is that the provider should able to provision, test, and deliver a service faster than the competition. In addition, it needs to accomplish this with equal or higher reliability than traditional networks. To guarantee reliability, right-of-way (ROW) issues must be resolved to ensure redundancy in the fiber paths between the central offices (COs)/POPs.

In a next-generation carrier or even in the traditional networks, synchronous optical network (SONET) (as a ring implementation or as a mesh) plays an important role in achieving this redundancy and reliability. There is also a need to decrease staff count, because the higher the staff count, the higher the operating cost. Companies want to conserve on the POP and to make their CO space smaller than that of the competition. If the carrier next door has three racks of equipment to do business, the provider hopes to have half a rack. All of these affect the efficiency and clarity of direction of the carrier in the service marketplace.

Given the spectrum of technologies and services, carriers must decide how to connect optical devices and network them. Should SONET be used at all? Can a carrier reliably offer FR services across an IP backbone? Must Ethernet connect to FR? Will the carrier provide voice services, and will voice be provided traditionally or through the use of IP? Will video be broadcast via point-to-multipoint or more like Internet streaming video? What kind of network management system (NMS) must be acquired or built that will integrate these technologies and systems now and in the future? These questions are some that will be faced by the carriers' network architects, engineers, and operations personnel as they begin to offer services and as the network evolves.

Network Design Alternatives

Network architects face a dilemma when they try to enhance their existing network or build a brand new one from the ground up due to the variety of service types that it must support. Should they build a traditional network with cross-connects and SONET rings and overlay ATM and IP on top of it? If it can be managed, and can be scaled limitlessly, then it could work. A second alternative would be to build an all IP–optimized network consisting of large IP routers interconnected by a mesh of packet over SONET (POS) trunks that light the fiber directly. Some have argued for this case and have justified its applicability to capture the private-line market using a proprietary scheme of circuit emulation over IP, which has failed temporarily due to the functional shortcomings of current routers in the marketplace. For a full-service network, an IP–optimized backbone is not the most desirable until the technology to support all the services is actually implemented in the routers. The third, more innovative, approach involves starting with technology that is available today, such as first-generation optical switches (actually optical-to-electrical-to-optical [O–E–O] or SONET switches) and DWDM transport equipment. A SONET switch can switch from one port to any other port at the synchronous transport signal (STS)–1 level. Plus, it has the routing and signaling intelligence of traditional ATM networks. It can route from one switch to another switch over the wavelengths ("waves") provided by the DWDM equipment and can ensure different levels of service quality and OC–n private-line services. Overlay this optical backbone with either a pure ATM or pure MPLS or a hybrid ATM–MPLS core. For instance, one could begin with ATM to support traditional services as well as IP and then migrate to MPLS for network convergence. If a provider runs only IP and then moves into this traditional space, it is better to start with an MPLS core, so as to merge IP traffic to different points in its network, and then gradually support private line and voice once the technology has stabilized. As MPLS and pure optical switches come to market, the evolution to an MP–λ-switched network can begin. In terms of achieving revenues, however, none of the MPLS switches available today can actually support the private-line process. Thus, ATM may be a good consideration for that purpose; a gradual migration to MPLS would be the logical follow-up.

In essence, any of these convergence designs boil down to some form of entrance facility, some way of getting into the POP, CO, or co-location. Once in, traffic would go into some kind of service aggregation device; some call this a multiservice aggregation switch. This switch can support all traditional services such as private line, FR, and edge ATM. In addition, some of the newer devices also support most of the edge IP functions. All aggregated traffic would be fed into some kind of core device, either ATM or MPLS or a good hybrid system. Voice service is a totally different situation. When dealing with the current public switched telephone network (PSTN), there are bunches of information, services, and features that must be done on the platform. Trying to push all this into the multiservice aggregator may be pushing the technology edge more than it can practically achieve today or in a couple of years. It would be more prudent to support voice over packet on a separate platform; incoming voice traffic is guided from the entrance facility via the multiservice aggregator to the voice-over-packet switch. The latter then trunks across the core ATM–MPLS device in a gradual migration from the PSTN to the all–IP network.

Finally, the core ATM–MPLS switches connect into the optical backbone, which would consist of SONET switches that have built-in DWDM features to connect directly to inter-CO–POP fiber. Today, it is probably best to couple a SONET switch with DWDM transponders, essentially building a network of SONET trunks between SONET switches using the "waves" provided by the DWDM transport equipment. Going forward, numerous companies are interested in integrating the DWDM and transponder as part of their SONET switch. The goal is to have fewer boxes to manage (see *Figure 3*). The goal of any new provider entering the market is to reduce the total amount of equipment so that the provider can hire a smaller staff to manage it, making its margins much bigger. The convergence design so rendered becomes a fusion model wherein one service layer implicitly merges into the one below it (see *Figure 4*).

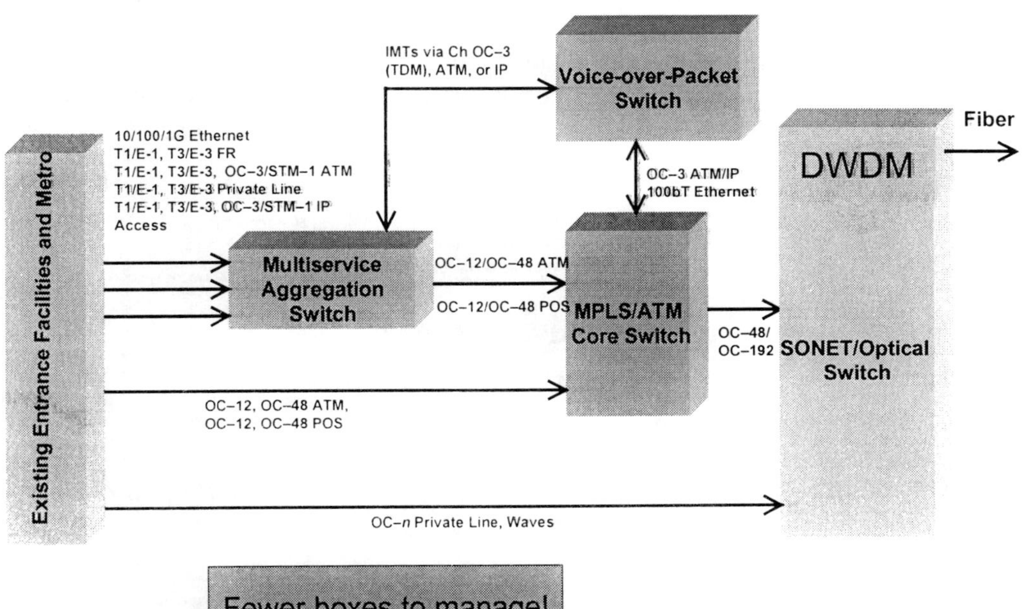

Figure 3: Optics, Voice, ATM, IP Convergence Design

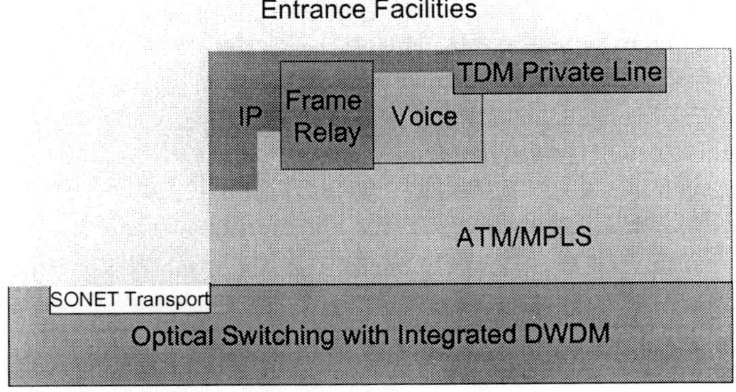

Figure 4: Results of Convergence Design

Steps in Achieving Convergence

Building a converged network involves three fundamental steps. The first step is to build a resilient and restorable intelligent optical network (see *Figure 5*). It involves lighting up the fiber using transponder-based DWDM transport equipment. This allows the fiber to be divided into channels or "waves." Most of these can be resold as "dim-fiber" or "wave" private lines allowing other carriers/ISPs to interconnect their network islands. A few can be used to overlay a SONET switching network. Once in place, SONET switch technology may be leveraged to build a meshed or ringed optical core. The decision to go with mesh or with rings or to have a hybrid depends on where the network is being built, the costs involved, the distribution of COs/POPs in a city or across the country, and on the mindset and comfort level of the carriers' staff with regard to SONET switching technology. SONET switches are a good, intermediate solution until intelligent optical switches are available. They are scalable, have higher port densities, with a footprint smaller by one order of magnitude. Not only do they provide intelligent automatic routing with flexibility of restoration options—priority reroute, 50 ms, 2 seconds, or no restoration, which can be leveraged to provide classes of service for private line—they perform all functions of existing SONET terminal/mux systems in addition to switching. They allow the creation of virtual rings and support all OC–n/synchronous transfer mode (STM)–n private-line services up to OC–48. Point-and-click provisioning makes it easier to deliver services quickly. One way of using SONET switches is to locate them at long-distance, high-traffic volume POPs interconnected with a "mesh" of OC–48/OC–192 trunks using DWDM transponders on the WAN side. Feeding into them would be OC–48 "wave rings" on the metro side using optical ADMs (OADMs) integrated with on-board SONET muxes.

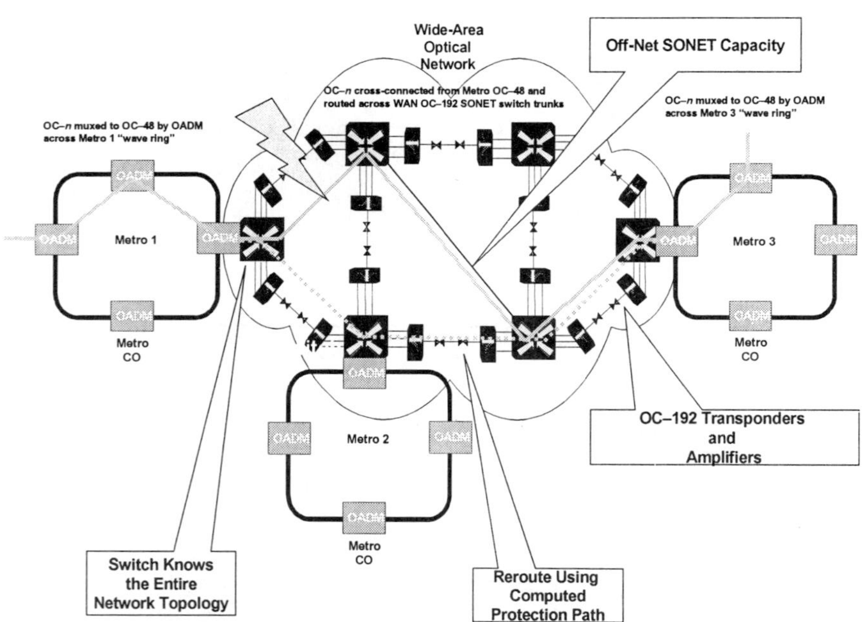

Figure 5: Fault-Tolerant Optical Core

Step two is to build an IP–aware ATM–MPLS core that is trunked using OC–48 or OC–192 across the optical backbone (see *Figure 6*). OC–12 ATM interfaces provide standard ATM and IP access and aggregation services in addition to trunk connectivity to multiservice aggregators at city POPs as well as metro COs. OC–12 POS and Ethernet interfaces provide pure IP access and aggregation as well as peering and transit functions and connect directly to private peers and network access points (NAPs). This core also provides border gateway protocol (BGP)–4 functionality on all peering and transit access ports with BGP/label distribution protocol (LDP) and private network-to-network interface (PNNI) functionality on all trunks. On all high-speed access interfaces that cannot be accommodated on the multiservice aggregators, the switch provides standard ATM services and signaling to build point-to-point permanent virtual circuits (PVCs), and to provide quality of service (QoS), policing, etc. In addition, these same interfaces can be provisioned such that they provide IP edge router functions such as IP flow classification, differentiated services (DiffServ)/IP–classes of service (CoSs), IP–virtual private networks (VPNs), etc. The switches are interconnected using OC–48 trunks; trunks can be POS–MPLS, or ATM–MPLS. The ATM–MPLS switches do virtual circuit (VC) merging whenever required and provide the mapping between ATM and MPLS/IP–CoSs. Finally, this core provides a second level of restoration and fault tolerance above the optical core.

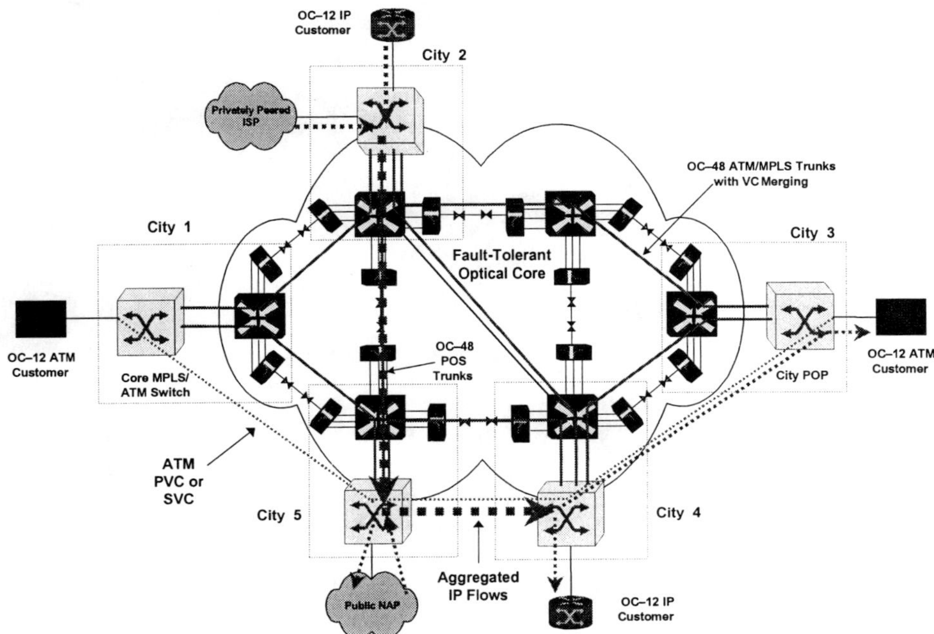

Figure 6: *IP–Aware ATM Backbone*

The third step in completing a converged network is to surround the ATM–MPLS core with a set of service devices (see *Figure 7*). A multiservice aggregation switch essentially performs all aggregations for low-speed customer access and cross-connects, the voice-over-packet switch provides all the voice services, and a set of ancillary devices known as broadband–remote access servers (B–RASs) terminate dial-up, xDSL, and cable-access customers.

Multiservice aggregators typically provide access cross-connect and grooming of incoming circuits to service modules. Access interfaces include OC–3/STM–1, DS–3/E-3, DS–1/E-1, fractional (F)DS–1/FE-1, and 10/100/1,000Base Ethernet. Trunk interfaces that typically connect it to a core ATM–MPLS switch would be OC–3/12 ATM/POS or Gigabit Ethernet. The service modules support LDP, PNNI and VC merging on the uplinks to the ATM–MPLS core switch, private-line services using circuit emulation over ATM and/or MPLS, IP aggregation and IP edge router functions, IP flow classification, DiffServ/IP–CoS, IP–VPN, native ATM and FR services, FR–ATM interworking, FR–IP interworking, ATM–IP interworking, and rudimentary test, turn-up, and troubleshooting functions. These devices come in different sizes and densities so that a carrier can match them to their location needs—at a metro CO versus a large POP versus one that sits at the customer premises or in the basement of a building. The devices would also serve as the entry point for the integrated access device (IAD)–customer-premises equipment (CPE) that a customer uses to get unified access into a carrier.

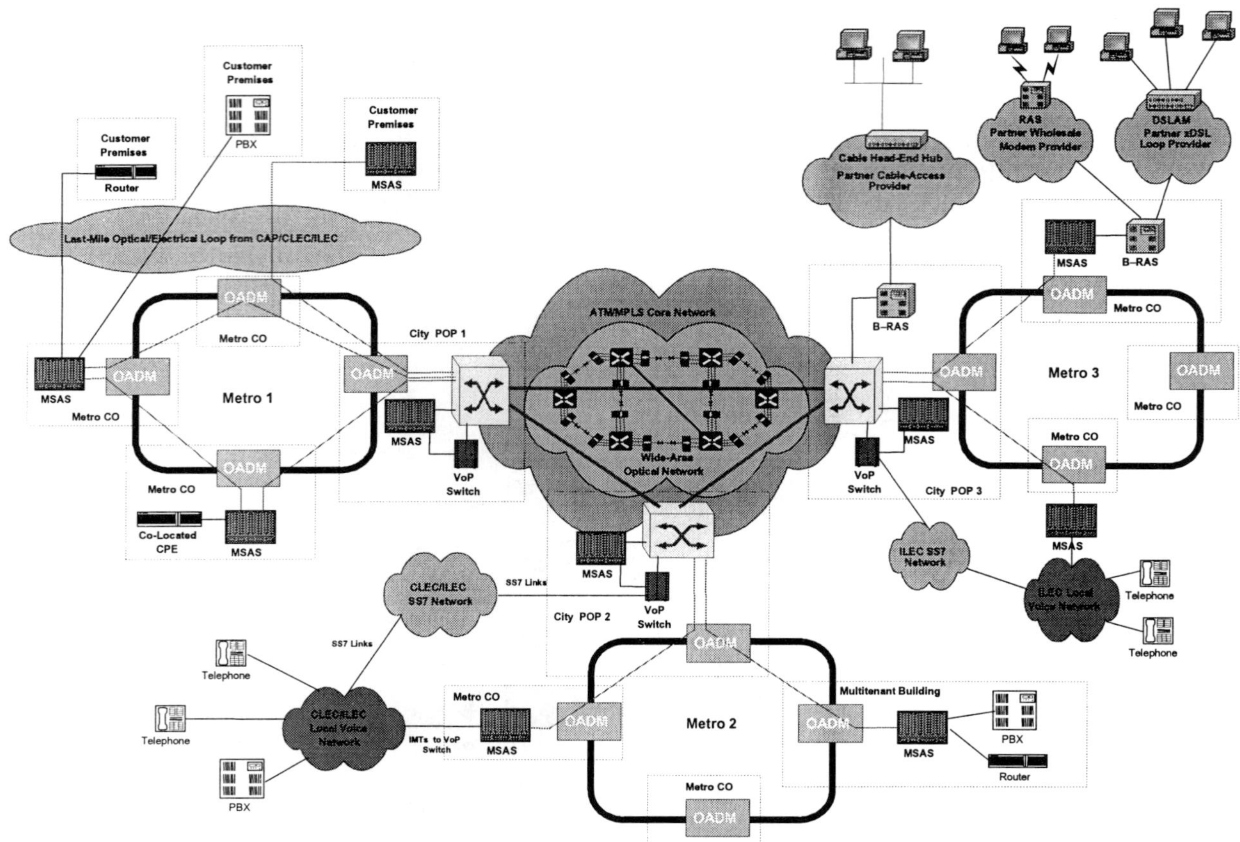

Figure 7: *The Big Picture*

Voice-over-packet switches get their access circuits from the multiservice aggregator and connect on the uplink trunk side using OC–3 ATM–MPLS–Ethernet to the ATM–MPLS core. These switches include a gateway function that provides trunk connectivity with the traditional PSTN. In addition, it includes a signaling gateway that works with signaling system 7 (SS7) network of the PSTN. It also supports the use of a softswitch that dictates how it handles voice over IP (VoIP) calls that originate or terminate directly at customer premises that are equipped with IADs. Functions supported by the voice-over-packet switch include local and toll voice services, SS7 for legacy voice handoffs from ILECS and competitive local-exchange carriers (CLECs), media gateway control protocol (MGCP)/H.323/session initiation protocol (SIP)–enabled gateway for VoIP traffic to and from the IP network, MGCP–integrated services digital network user part (ISUP), MGCP–Q.931 and Q.931–Q.2931 interworking, MGCP–SIP–ISUP+ signaling between switches with tandem-switching functions, compression, silence suppression, and echo cancellation.

B–RAS devices terminate IP dial-up, xDSL, and cable traffic from DLEC/CLEC/ILEC partner networks. Depending on their size and density of termination, they connect to the multiservice aggregator and/or the ATM–MPLS core switches.

It must be noted that in every step the intelligence is given to the network to make some decisions, with some of it controlled by the provider. By doing this, much of the provisioning complexity can be reduced. The complexity comes not from next-generation networks but from traditional legacy equipment with which they must interconnect.

Possible Service Offerings from a Converged Network

Given the functionality of the above devices, a plethora of services can be evolved and offered. The first involves the leverage of a carrier's construction and CO/POP ownership—space and power and outsourced management of equipment in co-located space. In addition, over its ATM–MPLS core, it could offer long-distance and local voice services via TDM, ATM or IP access, T1/FT1/E-1 and T3/E-3 private line, native FR VPNs, (DS–0/FT1/T1/T3/E-1/E-3), native ATM VPNs, integrated access services, IP services—transit, access, dial-up, xDSL VPNs, CoS, etc.—and Ethernet local loops for IP and integrated access.Over its optical core, the carrier could provide OC–n private-line services and "wave services" and could leverage its ownership in right-of-way to sell dark fiber.

Implementation Hurdles

When a network is designed and built out, a number of implementation issues arise. These cause delays in service rollout and in some cases may even force the carrier to redesign a section of the network.

Delays could be caused by ROW procurement, fiber availability, construction licenses, procurement and leasing of real estate, and general network construction. If some of the equipment to be used is from start-up vendors, or if the technology being used in a piece of equipment is new, one must accommodate network design cycles and deal with testing, interoperability, stability, and operational reliability issues. Network management and back-office systems integration issues and problems are also a source of delay leading to extra time spent on wringing temporary work-arounds.

Network designs and rollouts are based on customer demands and marketing forecasts. However, they can change abruptly, leading to prioritization and rollout dilemmas and delays for services that may be offered. Budget allocations and reassignment of capital for unforeseen increases in demand of one service or prior miscalculations on cost for one kind of equipment could push another piece of equipment or functionality or service to the back burner; this could well delay a service or totally move it from the carrier's current portfolio of services to a future offering.

Customers' (dis)comfort with a technology and its deployment also plays an important part in the final evolution of a network. For instance, many customers might ask a carrier how it offers low-speed private-line services. If the carrier uses circuit emulation to provide these services, it is possible that the customer does not trust the technology. The possibility of losing this customer, especially if it is a large one, may push the carrier to redesign its network using a TDM model, or it may require the carrier to spend a lot of time and effort going back and forth trying to convince the customer that the offering is as good as a legacy network. During this time of flux, the network and its design come to a standstill until one or more customers decide which direction the network should go.

There exists a lack of available, innovative engineering talent and operational personnel; the talent pool in the industry is finite, and a carrier needs to move quickly to hire the best talent. Delays may be caused by a lack of critical mass of experienced yet innovative engineering talent within the company. Experience also shows that once people understand a certain technology, it is difficult to get them to learn and accept another quickly. A company's personnel training may need to be adjusted to cope with the new technologies. If a new technology is placed in a network, it will work only with cooperation of the personnel involved in implementing and managing that technology. If new technology is involved, operations people must be involved in the technology decision. Even if they ask the most stupid questions, they must be tolerated and accepted. They must be told what the technology can do and shown what wonders can be worked with it. It is important to involve personnel in the decision-making process to eliminate problems when the new technology is implemented.

Final Thoughts

It has been shown that to become a full-service carrier or even a niche provider, one must make many decisions. While balancing the technical innovations of the past decade with the business demands of the new Internet, one is compelled to design a network that achieves convergence in a way that is practical and revenue generating. The best approach seems to be the fusion model wherein one layer of the network fuses with the one below. The optical core of DWDM transponders and SONET switches is wrapped within an ATM–MPLS core. Feeding into it are multiservice aggregating switches and voice-over-packet switches. The portfolio of services offered by a carrier is a direct reflection of the mixing of feature sets available on these edge and core devices. The completion of the network and the availability of services directly hinge on a strong network management system that can scale and accommodate new technologies as the network evolves. Finally, network designers must remember that a converged network can be guaranteed only if issues such as customer perception, network construction, budgeting, and involvement of operations personnel are recognized and resolved at the beginning of the design process.

Beyond Data-Aware SONET

Alnoor Shivji
President and Chief Executive Officer
Cyras Systems

Today's networks are based on synchronous optical network (SONET)–synchronous digital hierarchy (SDH) in the metro. In the core, SONET SDH runs over wavelength division multiplexing (WDM), and essentially these are networks that were architected for voice lines as well as point-to-point circuits. Looking at the traffic today, 80 percent is based on voice and point-to-point circuits such as digital signal (DS)–1 and DS–3 circuits, and 20 percent is based on packets of frame relay (FR), asynchronous transfer mode (ATM), and native Internet protocol (IP). By about 2005, that trend should reverse so that the majority of the traffic is the data-oriented packet, while 20 percent is voice and point-to-point.

The trends are very clear. The industry is shifting to data, but today's networks are geared toward voice and point-to-point. The infrastructure must be changed to accommodate these trends. In the metro area, the trend is tremendous growth in bandwidth. The growth of optical carrier (OC)–192 from 2003 to 2008 is projected to represent a fourfold increase. That is a tremendous increase, which indicates that there is a real need for high bandwidth in the metro space. People used to have OC–3 lines on the access side. Now those are moving to OC–48, and global networks are moving to OC–192. There is tremendous growth, mostly in the OC–48 and OC–192 range. The trends are continued growth at SONET–SDH deployment with a massive amount of bandwidth.

Core Network

Some carriers are doing innovative things in the core network—i.e., in the long-haul network. In 1998 they were deploying OC–192 with perhaps eight or 16 λs, and in 1999, they moved to 16, 32, and even higher. Again, the same trend is clear even in the long-haul that there is a tremendous need for very high bandwidth. The need for 10 Gbps solutions and beyond is becoming more and more important. The 2.5 Gbps solutions are becoming less important in the long-haul networks, with more λs becoming the norm. One thing that is very clear is protection strategies. These continue to use SONET SDH, so even though WDM technology is used, these protection strategies run over bidirectional line-switched rings (BSLRs) for most of the carriers, with only one carrier using a proprietary scheme. The majority continues with the traditional protection schemes for which SONET is known.

SONET–SDH rings remain prevalent for many reasons. First, they have the ideal kind of fault tolerance. They represent the industry standard, have bandwidth flexibility, and are geared toward network reliability. The reliability of SONET exceeds several ninths, which means it is ideally suited for delay-sensitive traffic. In addition, it is architected to support several customers and services. However, there exist many problems with today's networks, primarily based not on the SONET protocol but on the SONET equipment. The equipment requires abundant rack space, and rack space is becoming a new challenge with core location for central offices (COs). The first generation of SONET equipment was very costly and the integration level was minimal, so racks and racks of space were needed to support very limited bandwidth. The cost was prohibitive, the space was an issue, and only traditional telco interfaces were supported. Because this was geared toward voice networks, there were point-to-point interfaces such as DS–1, DS–3, and OC–3. There was no support or optimization of data. This was also strictly a Layer-1 product, so there was no integration that would provide the efficiencies wanted. To achieve the efficiencies for today's networks, it would be necessary to reuse the level of processing that occurs in any of the layers. One would have to move down the layers in the chain, but by having a Layer-1 kind of device that would integrate with other layer devices, loss of optimization would occur.

First-Generation Networks

In the evolution of networks, there were the first-generation add/drop multiplexers (ADMs) that were like specific hierarchies, such as OC–3, OC–12, and OC–48, as shown in *Figure 1*.

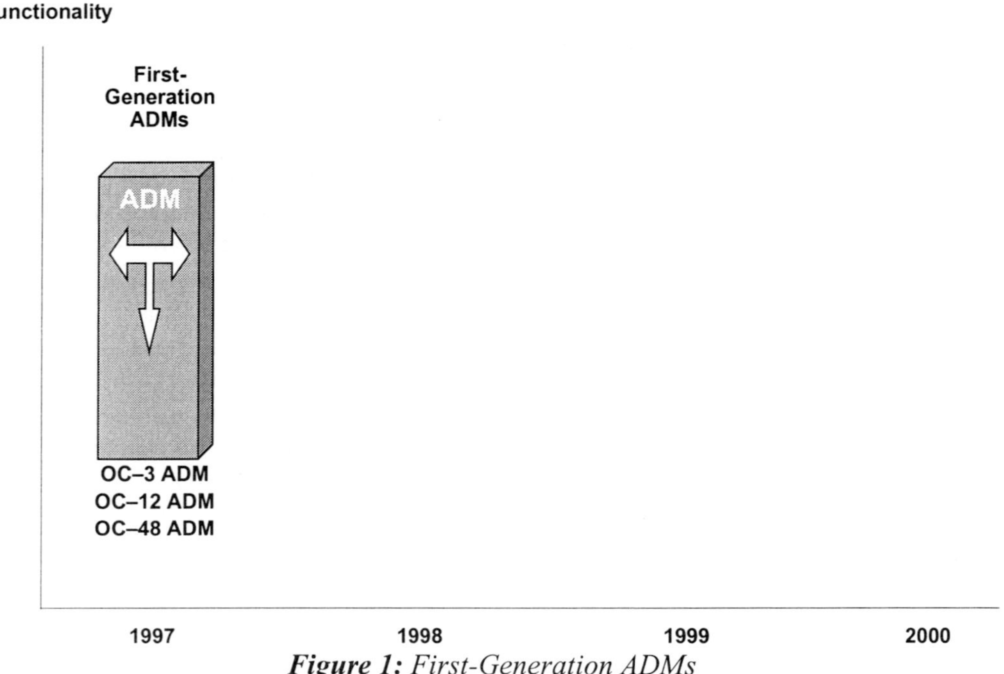

Figure 1: *First-Generation ADMs*

These were discrete systems that were geared only to voice. Then there were the next-generation ADMs that essentially integrated some of those hierarchies, as shown in *Figure 2*.

OC–12 and OC–48 were integrated in a single box, and the footprint was reduced. That resulted in cost savings as well as bandwidth optimization. The next-generation ADMs, however, did not have any kind of data optimization. Thus, it was necessary first to have all the data devices, such as ATM boxes, IP switches, and Ethernet routers, feed into this device and then to carry that traffic through the metro area onto the long-haul network. The efficiency was not there for data, and this was not architected with data in mind. The data-aware SONET devices that were specifically architected to run data came next, as shown in *Figure 3*.

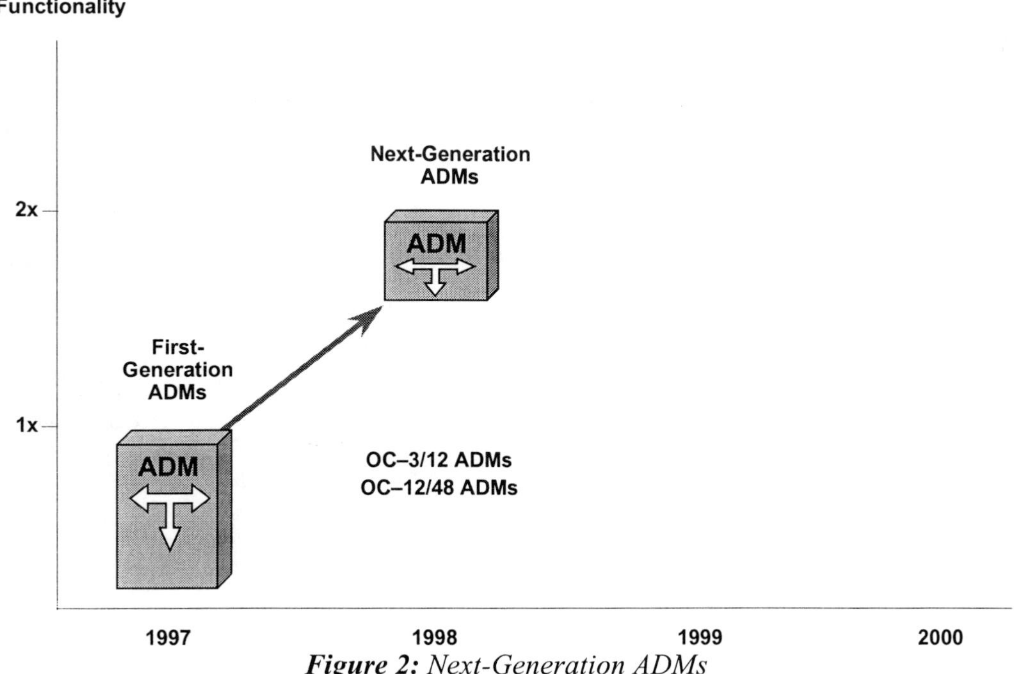

Figure 2: *Next-Generation ADMs*

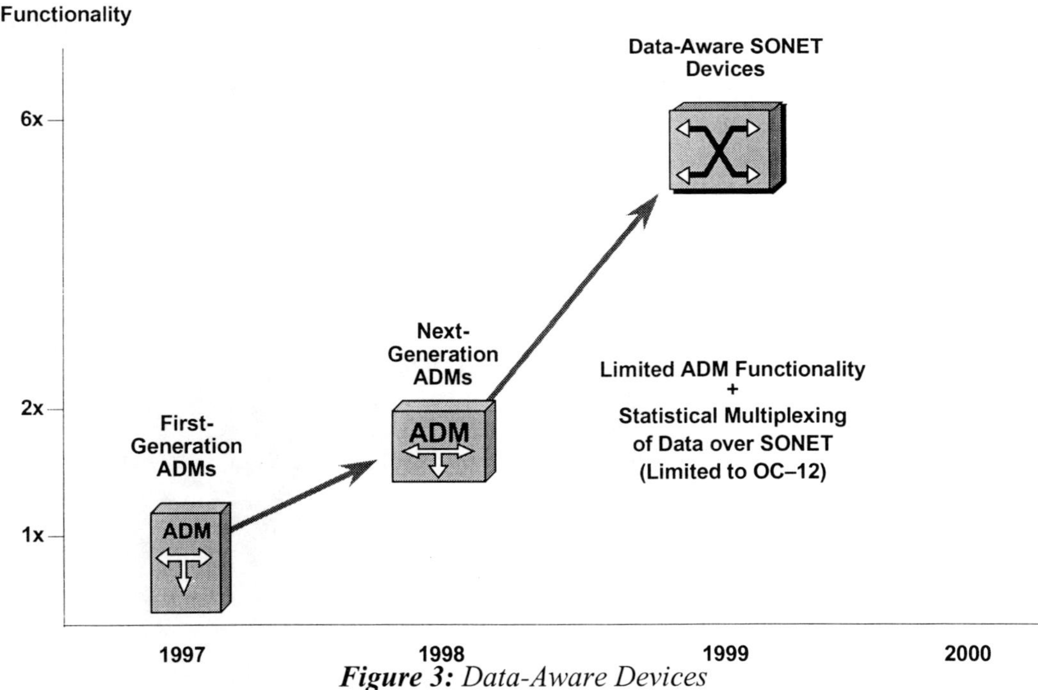

Figure 3: *Data-Aware Devices*

Essentially, these boxes converted all the traffic into a common form, in this case by using ATM and circuit emulation for voice. However, there is a problem if most of the traffic is FR. The ideal solution, if massive scalability is desired, is to address the networks of today to have minimal processing to take traffic as natively as possible, as outlined in *Figure 4*.

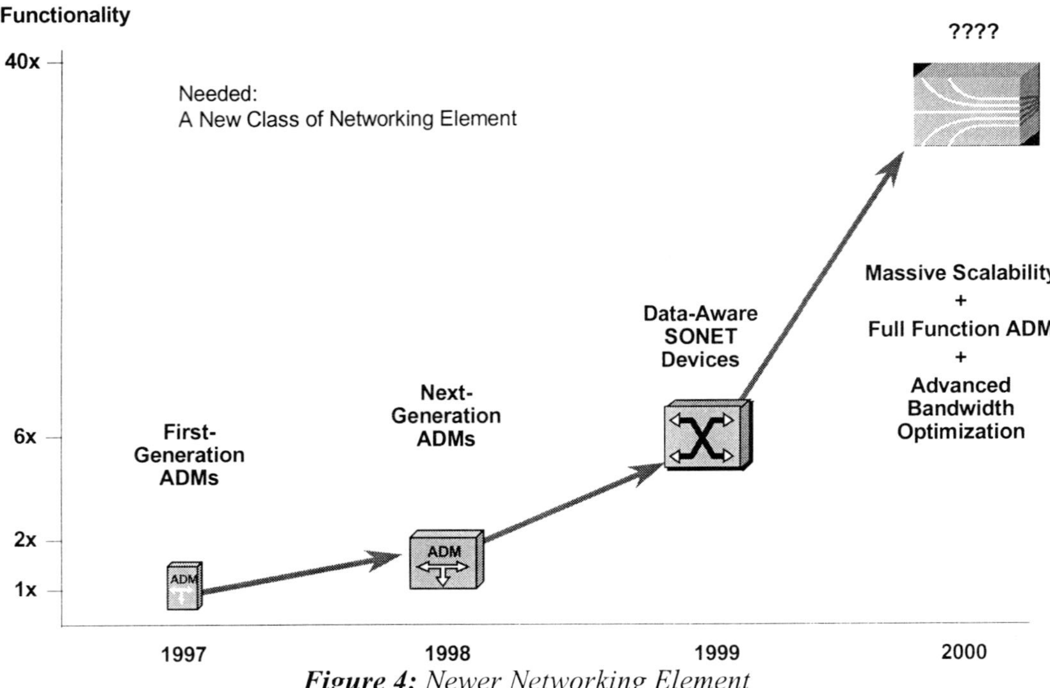

Figure 4: *Newer Networking Element*

The Internet, ATM, and IP should be taken as natively as possible and used over a good transport mechanism such as SONET, which provides ring protection and bandwidth management. This appears to be the evolution. It is a new class of product that includes data and advanced bandwidth optimization features that go over the SONET hierarchies and overcome some of those hierarchies. It provides a carrier with full ADM capabilities, as ADM is needed in such equipment with massive scalability.

In the hierarchy of the metropolitan networks, much equipment is involved. Much of the digital subscriber line (DSL), Ethernet, FR, and other equipment feeds into service access points or ATM multiplexers that feed into ADMs, as shown in *Figure 5*.

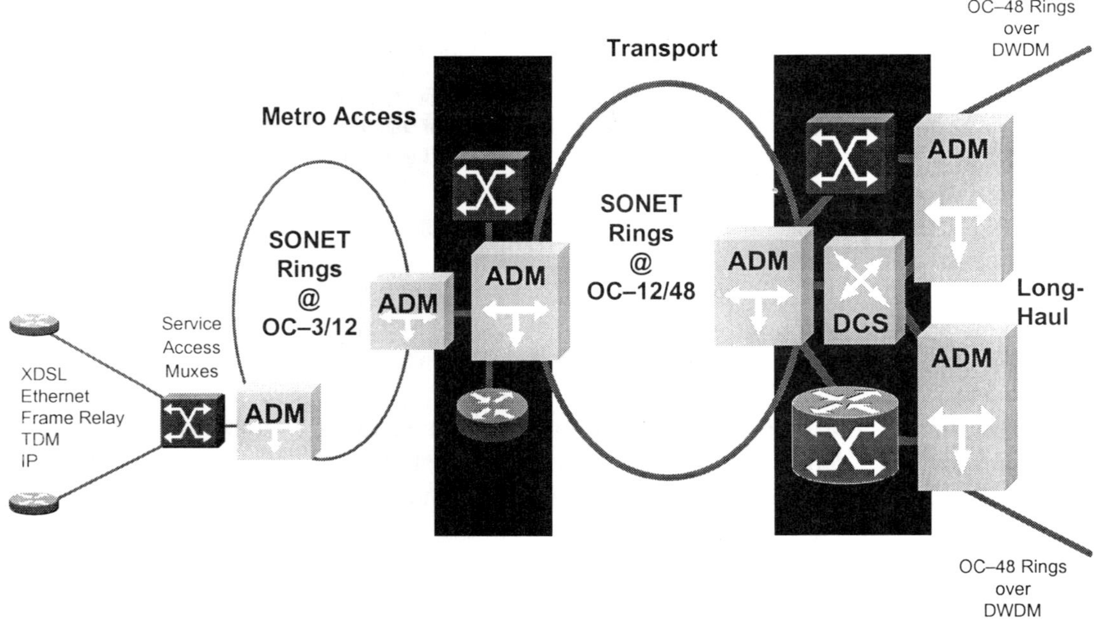

Figure 5: *Hierarchy*

These are found at access rings running at OC–3 or OC–12 that feed into other ADMs and add switches or aggregation routers into interoffice rings. The interoffice rings are primarily OC–12 or OC–48 and feed into various types of core units such as the very-high-end Terabit ATM switches, the very-large digital cross-connects, or the routers, which then pass this on to the long-haul network. Many ADMs participate in this. To provide an end-to-end solution, it is necessary to go through several boxes, which becomes very inefficient. In terms of management, it is a real challenge, because now there are IP switches, ATM boxes, and ADMs. This is why many carriers have segregated the data wall from the voice wall. Management systems used to be very different, and all this gear was managed separately.

Massive Switch

This process can be simplified by having a massive switch that operates in the trans-metro area, as shown in *Figure 6*.

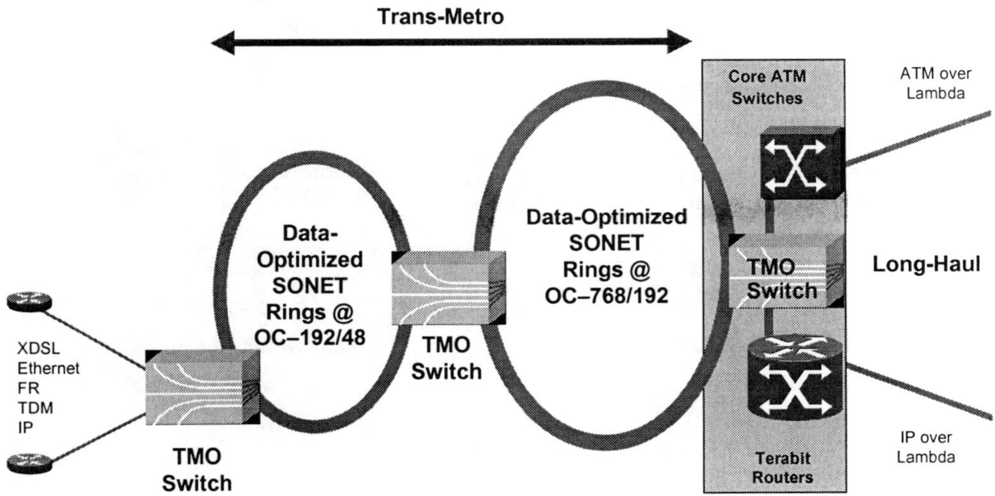

Figure 6: Trans-Metro Optical Switch

The trans-metro optical (TMO) switch is an optical switch that can handle Internet, FR, time division multiplex (TDM), IP, DSL, any kind of data traffic, voice traffic, and point-to-point traffic efficiently. Basically, this equipment is optimized to carry data efficiently to perform stack boxing found in ATM or IP switches, or to do stack boxing and feed it onto large rings. The management of this also becomes simplified, as it is possible to have a single management system and perform end-to-end service provisioning. In today's network, end-to-end service provisioning is very important. The key for carriers is to be able to provide services rapidly.

In terms of technology evolution, first there was SONET, which had some issues in terms of its hierarchies and what it was geared toward (primarily voice traffic). Then there was data-optimized SONET, which focused on handling voice traffic and point-to-point traffic. Voice and point-to-point traffic is growing, but not as rapidly as data traffic. Thus, it is insufficient to replace everything with data-oriented equipment, as there remains a large need to handle voice traffic and point-to-point circuitry. This equipment would allow a carrier to handle special traffic efficiently.

Migrate Seamlessly

The key is being able to migrate seamlessly in today's networks. It is essential to allow carriers and customers to migrate slowly and easily into more efficient equipment that will be able to handle the massive amounts of data of tomorrow's networks. The industry is looking at a switch that replaces digital cross-connect and provides both additional cross-connect functionality and full SONET–ADM functionality. Such functionality would enable time slot into change, drop-in continue, hairpinning, and all of those functions that are essential in today's networks. It would also replace ATM service-access boxes, IP switches, multiservice switches, FR access switches, DSL access multiplexers (DSLAMs), and multiprotocol label switching (MPLS) switches. To build a global network, such equipment is crucial.

The equipment that combines those functions into a single box creates a tremendous increase in efficiency. The network can carry much more traffic on the rings. A carrier may thus carry a tremendous number of customers in that ring. This represents a 10- to 40-fold increase, depending on whether the network contained primarily voice or primarily data traffic (see *Figure 7*). If the traffic were mostly voice, the increase would not be as significant as it would be for a data traffic.

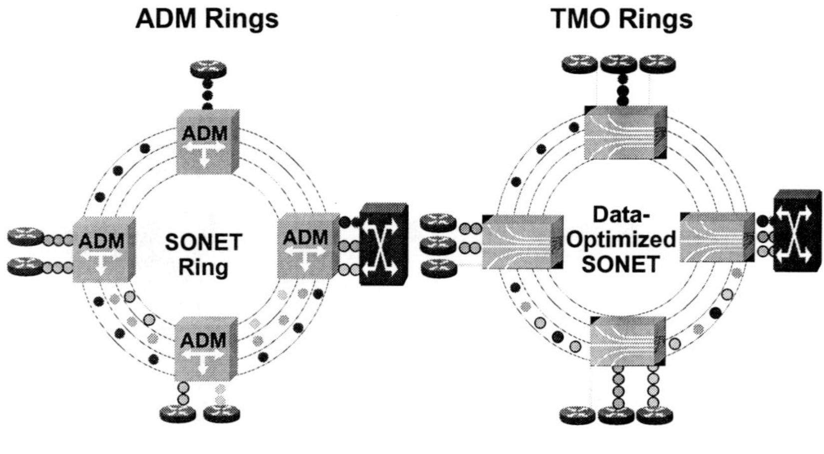

Figure 7: Unsurpassed Price–Bandwidth Efficiency

In terms of price performance, there is a huge difference between traditional ADM, which was large and low-key, had very little integration, and required other equipment to complete the function, and a TMO switch, as outlined in *Figure 8*.

$ Price per *usable* STS–1 (Bandwidth-Optimized)

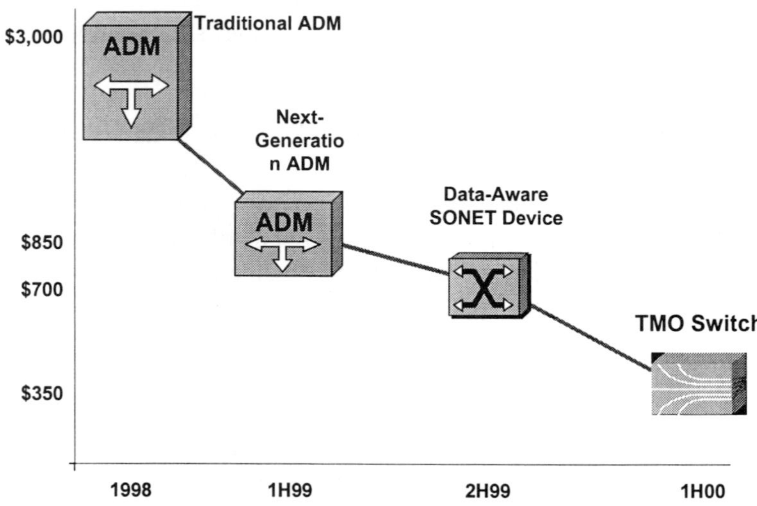

Figure 8: Price and Performance

With regard to the traditional ADM versus the next-generation ADM, there is an increasingly greater degree of integration, optimization, and scaling. Scaling becomes a very important factor because today's carrier must be able to handle any kind of customer. This means that requirements may range from a DS–1 circuit all the way to OC–192, so scaling becomes very important.

In terms of building networks, this box can take any kind of local-area network (LAN) or customer-access traffic such as ATM, FR, or voice traffic and feed it onto the long-haul network through the access loops and the metro rings, as shown in *Figure 9*.

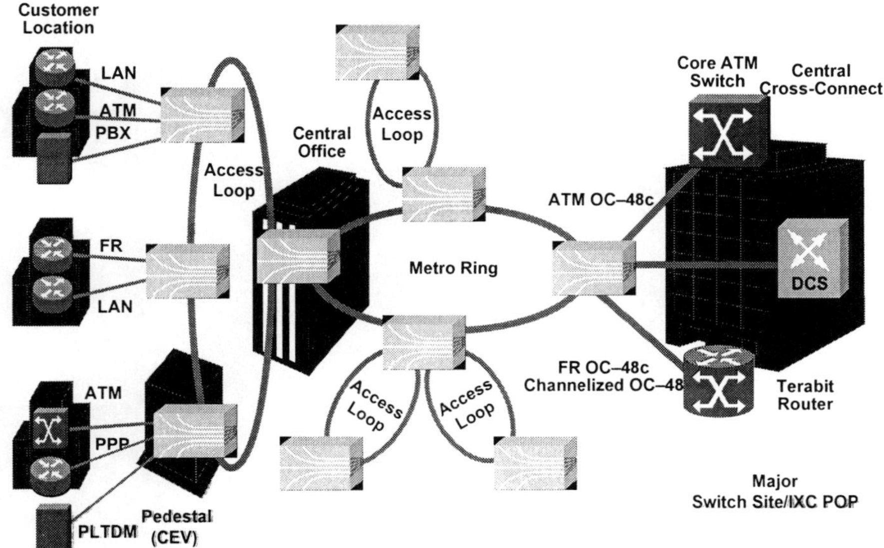

Figure 9: *TMO Switch Traffic Aggregation*

Today, it is possible to deploy global networks using equipment that has carrier-class reliability, which is fundamental. In fact, it is given that the equipment must be carrier-class and that a carrier cannot base its networks on enterprise equipment. It must have the carrier-class reliability of the traditional telco equipment. It also must have telco and LAN interfaces, ATM, Internet, IP interfaces, and packet-to-door SONET. Interoperability cannot be ignored. It is very important, because a carrier fundamentally wants to be able to have equipment that can go easily into today's networks. It is critical to be able to support what is out there and to have interoperability. The carrier should be able to manage the equipment and be able to interface into the operations support system (OSS) and into the vendors' managers.

Fault Protection

Fault protection is another important point. Carriers are unwilling to relinquish SONET protection; they will not give up the kind of reliability with which they have become accustomed; and they are not willing to pay the price for having data equipment that fails and does not restore traffic within 50 milliseconds. Fifty milliseconds has become the norm, so anything more than 50 milliseconds becomes unacceptable. Density and scalability and the need for massive bandwidth are becoming the norm. Because Internet service providers (ISPs), application service providers (ASPs), and enterprises that are beginning to replace their infrastructures all require more bandwidth, the scalability of networks is becoming more important. Likewise, densities become important for carriers that hope to service those customers who use more bandwidth. Another fundamental concern is the ability to change the SONET–SDH networks in such a way that they are optimized for data, unlike the traditional telco ADMs that were optimized for voice. Today's equipment must be optimized for SONET SDH.

Next-Generation Optical Transport Networks: Terabit Optical Link Architecture Based on 40 Gbps Transmission

Albert Vanthilt
Director and Principal Transport Engineer
Qwest Communications

Niloufar Tayebi
Advisor, Optical Networks Planning and Architecture
Nortel Networks

Denis Fluet
Senior Brand Manager, OPTera Networks Solutions
Nortel Networks

Introduction

Capacity on the long-haul backbone optical networks is key to successful realization of the data communications networks with ever-increasing demand for bandwidth. Total capacity deployed on the backbone routes will reach to multiple hundred gigabits per second by the end of 2000. Long-haul carriers now plan to deploy systems that scale to terabit capacities. The goal of achieving Terabit network capabilities is market and technology driven. Market drivers are the emergence of new applications and services and the substantial growth of the traffic. Technology drivers are higher spectral efficiency and wider optical bandwidth that enable higher capacities on the links. The market and technology trend continues to decrease the cost per bit per km and to increase the demand for more bandwidth.

Higher spectral efficiency is an enabling factor to achieve multiple-terabit capacity. Optical fiber is an important element in design and implementation of terabit links. Fiber dispersion impairments demand more precise dispersion management at higher transmission rates. At 40 Gbps transmission, dispersion slope compensation may be required in addition to high-precision dispersion compensation. In terms of polarization dispersion compensation, post-1994 fiber is usable to about 40 Gbps line rate, while at above 40 Gbps line rates, improved fiber performance is required in terms of polarization mode dispersion (PMD) [1].

Understanding network attributes and requirements plays a key role in defining the target terabit link architecture. In this paper, the backbone and regional transport architecture of a long-haul, high-capacity network is characterized in terms of physical infrastructure and in terms of traffic. This characterization aids in identifying network requirements in each segment. Raman amplification and new forward error correction (FEC) schemes increase the optical reach while keeping the amplifiers spaced at currently installed huts. Matching between the technology and the infrastructure enables carriers to plan for deployment of terabit links without substantial new investments in physical and fiber infrastructure.

This paper defines a target fiber cross-section capacity in the order of multiple Tbps on the express backbone links and examines the available options in reaching this target. The target terabit link architecture is based on 40 Gbps transmission with spectral efficiency of 0.4 bit/sec/Hz. The model in this paper uses up to 80 nm optical bandwidth. The paper also considers the attributes of a real backbone network in terms of traffic and physical infrastructure. In this model, 2.5 Gbps or 10 Gbps interfaces provide the connection of core service platforms to the backbone optical transport. *Figure 1* shows a schematic of the link architecture.

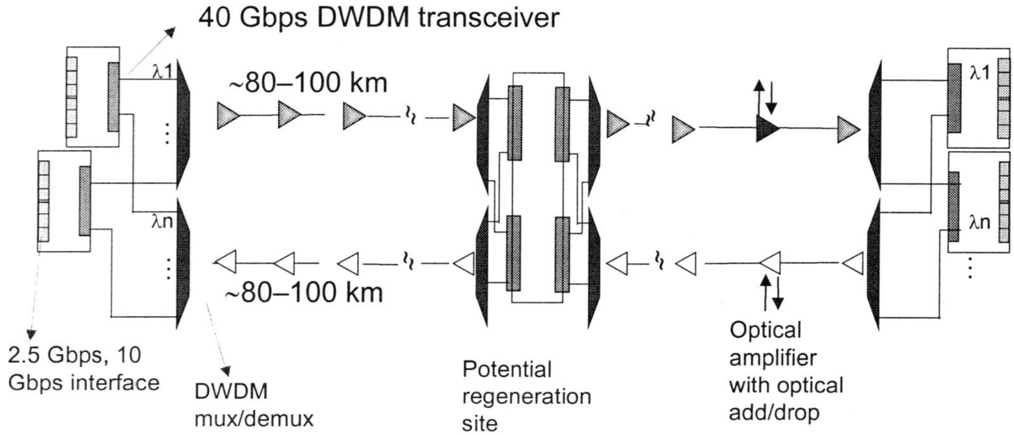

Figure 1: Terabit Link Architecture

Three types of criteria are used to assess the performance of the terabit link architecture. The first criterion is cost per bit per km as a common metric used in assessing the total network cost. The other two criteria become increasingly important in terabit networks due to the magnitude of the capacity that must be provided in the terminal sites. These are floor space and power consumption, which are discussed in terms of the number of network elements (NEs).

In conclusion, the paper demonstrates that the network architecture, using 40 Gbps transmission, achieves lower network cost per bit per km compared to alternative architectures and to the present mode of operation. In addition to cost savings, the 40 Gbps–based architecture requires fewer NEs and less floor space, and it is more efficient in power consumption. As a result, 40 Gbps–based architecture enables faster time to market and faster service velocity.

Network Model and Characterization

Figure 2 shows the physical infrastructure map of an example long-haul, high-capacity backbone network. Fiber links connect backbone hub or junction offices [2]. Increasing growth of traffic between wavelength-managed services, core Internet protocol (IP), and asynchronous transfer mode (ATM) service platforms yields to higher demand for capacity on the backbone fiber-optic links. The optical link architecture and the optical transmission equipment in the backbone offices must be able to keep pace with the accelerating demand for capacity. Capacity on the backbone fiber-optic links can reach to several Tbps in two to three years (see *Figure 3*). Based on this assumption, the number of backbone links requiring more than 1 Tbps capacity would be considerable in the network.

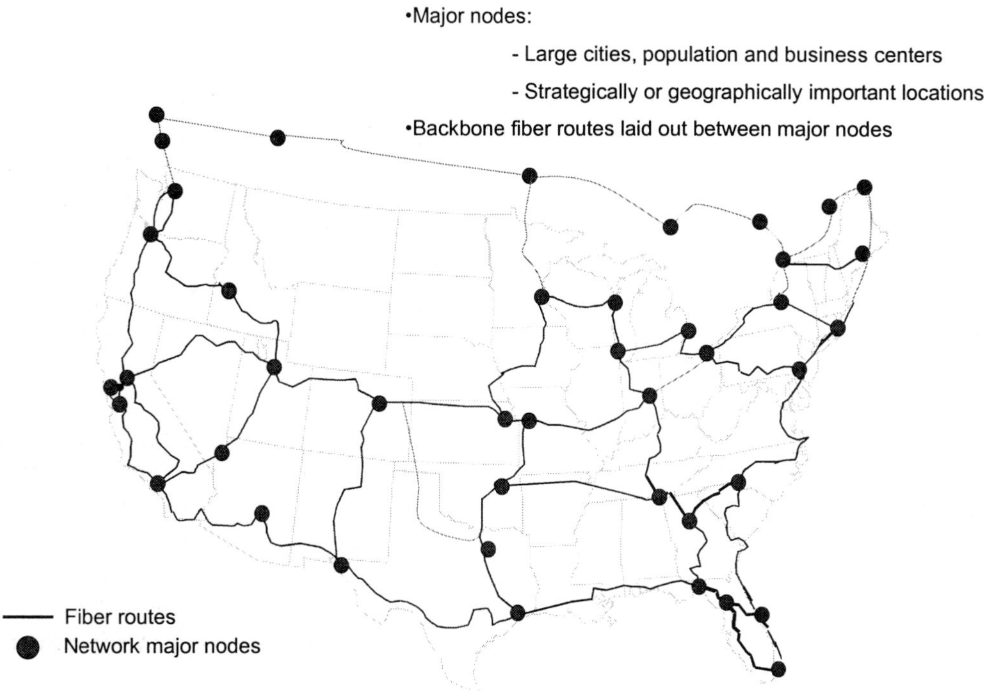

Figure 2: *Generic Backbone Network Model*

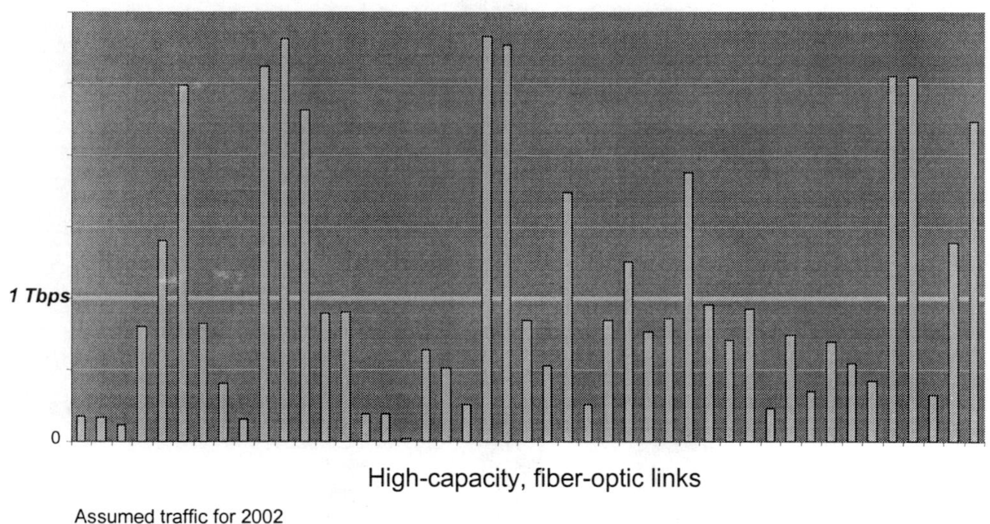

Figure 3: *Capacity Cross Section*

Traffic on the regional collector links would require less capacity; spacing between amplifier huts or span lengths are similar to backbone links, and there is potentially less optical reach required. Hence the technology applied to the backbone network is not necessarily optimized for the regional collector network. While backbone links would benefit from optical link architectures that provide terabit capacities, applications of the 10 Gbps transmission with various optical reaches will remain in the regional collector network.

Link Capacity and Cost Modeling

For cost modeling of the optical transmission link, the paper considers a linear model that represents the backbone links between the major nodes in the network. The cost performance of the 10 Gbps–based and 40 Gbps–based architectures is evaluated for various capacity cross sections in the network. The link architecture has been shown in *Figure 1*. Wavelength-managed services, core ATM switches, and IP routers connect to the optical transmission equipment with 2.5 Gbps and 10 Gbps short-reach optical interfaces. The key difference between the selected optical link models is in their spectral efficiency. Two varieties of optical reach are used for the 10 Gbps–based architecture. One uses shorter optical reach but can achieve higher total capacity. The other variety uses longer optical reach. The two models of the 10 Gbps–based architecture may represent the state-of-the-art technology in optical transmission at transmission rates slower than 40 Gbps.

A transmission rate of 10 Gbps has been successfully deployed on the long-haul backbone fiber-optic links and has proven to provide lowest cost per bit per km for optical transmission. A transmission rate of 40 Gbps reduces the cost per bit per km even further under the specific network traffic profile. *Figure 4* shows the network cost performance. The 10 Gbps–based architectures in the model have similar (bandwidth x distance) products and can represent the technology deployed in the networks to date. The 40 Gbps–based architecture shows strong network cost savings. For terabit links, the 10 Gbps transmission requires overlays of amplifiers, while the 40 Gbps transmission can consistently reduce the cost per bit as capacity on the link grows.

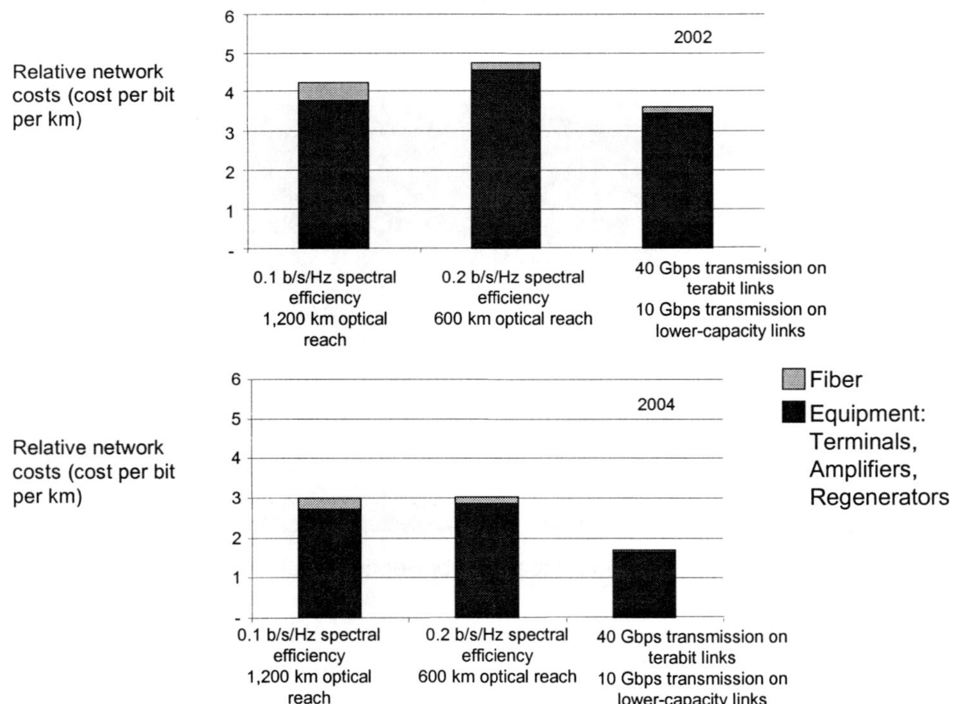

***Figure 4:** Relative Network Costs for 10 Gbps and 40 Gbps Transmission*

As the network demands higher-capacity optical transport, more links will use 40 Gbps transmission, and hence the network cost savings will increase (see *Figure 4)*. Architectures designed for terabit capacity have a clear economic and operational advantage.

Terabit Offices

Corresponding to the capacity on the backbone links, major hub offices and junction sites would require platforms optimized for terabit capacity. *Figure 5* may represent an assumption of the capacity required in the backbone offices in two to three years. Terabit offices would dominate the backbone network.

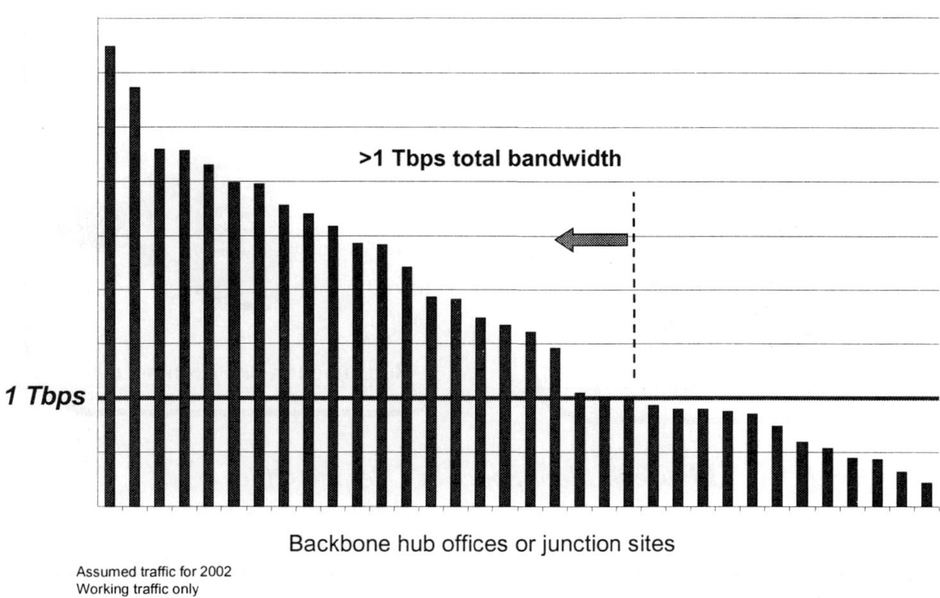

Figure 5: *Total Bandwidth in Backbone Nodes*

High-capacity platforms need enormous amounts of power and space in terabit offices. The power and space requirement with the currently deployed technology may not be pragmatic in future terabit applications.

An example office on the junction of three fiber links will demonstrate the power and floor space required for a terabit office. This office may also represent a major hub office because many of the backbone hub offices have more than two fiber links connected to them.

Figure 6 demonstrates the floor space improvement in the example office using 40 Gbps transmission and terabit terminal platforms. For terabit capacities in the backbone offices, the typical high-capacity platform with 300 Gbps per standard bay will need several bays. Terminal technologies based on 40 Gbps transmission can provide terabit capacity in substantially less floor space. Higher density in less floor space enables the carriers and service providers to free up the office floor for more revenue-generating equipment, instead of for sole transmission. In addition, higher efficiency in floor space and power consumption aids in reducing the network cost.

Power consumption of the platforms developed and deployed to date may not be pragmatic for terabit office applications [3]. With accelerating growth of traffic in the backbone networks, the optical transmission equipment would face challenges in keeping the power consumption and heat dissipation at low levels and within practical limits. *Figure 7* shows the power consumption of a typical high-capacity platform compared to a terabit platform in a backbone office. Terabit platforms enable the carriers and service providers to increase efficiency of transmission for the consumed power in the backbone office more than 100 percent.

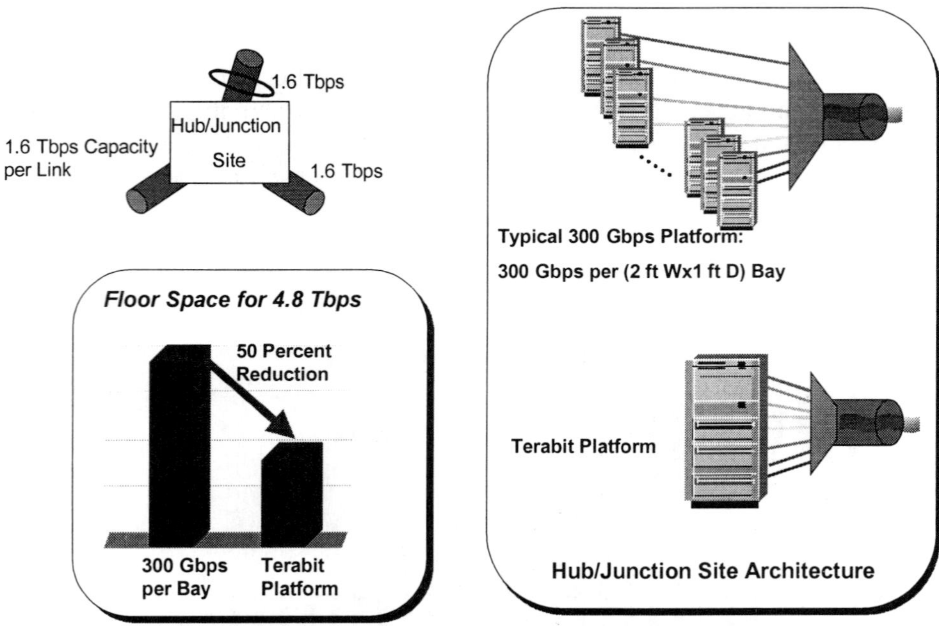

Figure 6: *Floor Space in a Terabit Office*

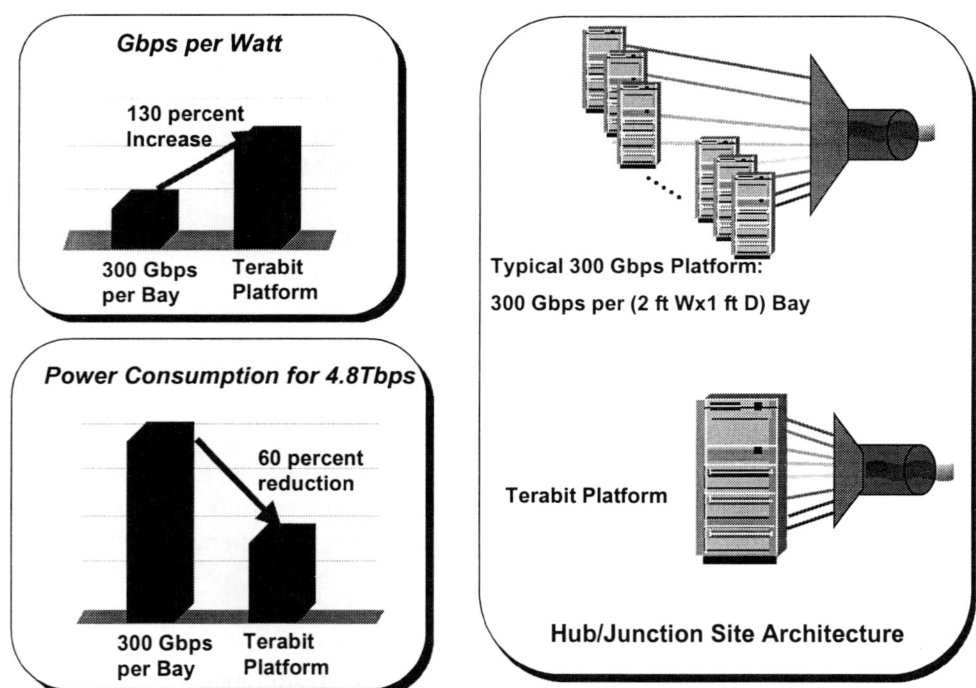

Figure 7: *Power in a Terabit Office*

NEs and Fiber

Operational benefits of 40 Gbps transmission can be expressed in terms of deployed fiber and NEs. The number of fibers connected and managed in the backbone terabit offices is reduced by multiple factors with 40 Gbps transmission. Operations and maintenance of fibers in the network become simpler, resulting in savings in fiber management and operational costs. With 40 Gbps transmission and high-density terabit platforms there will be fewer NEs to operate, monitor, and maintain in the network. Combined with the ability to manage and configure network resources and services intelligently, terabit platforms will result in simple network operation and problem resolution. Improving NE density and spectral efficiency yields faster time to market and faster service velocity.

Backbone Network Evolution—Bandwidth Management at the Terabit Office

While optical links and backbone hubs are evolving toward terabit capacities, there will be requirements to manage the pass-through and add/drop traffic in the terabit hubs. Bandwidth management is needed at high-granularity, wavelength-level traffic, as is management at lower service rates. Terabit hubs must consolidate and groom the lower service rates prior to optical transmission on the backbone links for cost efficiency. Simple service provisioning and problem resolution techniques are required to drive down the costs of operation and management.

Conclusions

With the accelerating growth of traffic in the long-haul backbone networks, technologies that enable terabit capacities will be in high demand. A transmission rate of 40 Gbps has been evaluated in this paper and recognized as a building block for high-density and high-efficiency terabit platforms. Terabit fiber-optic links with 40 Gbps transmission demonstrate strong network economic performance in terms of cost per bit per km. Transmission at 40 Gbps reduces the network cost significantly compared to the presently deployed technologies.

References

1. Cowper, R., et al. 1999. A view of next-generation optical communication systems—possible future high-capacity transport implementations. *Conference of Optical Fiber Communications, OFC99.*
2. www.qwest.com, Network Map
3. Vukovic, A., et al. 2000. Power-density challenges of next-generation IP networks. *World Telecommunications Congress/International Switching Symposium, WTC/ISS2000.*

Meeting Carrier and End-User Needs with a Dynamic Optical Services Network

Jonathan Andresen
Global Marketing Manager, OPTera Connect and ASON, Optical Internet
Nortel Networks

Mark Ndesandjo
Senior Manager, Services Marketing, Optical Internet
Nortel Networks

Introduction

The demand from end users for new features and for greater speed, quality, service, and reliability is increasing exponentially. New, innovative services and applications are driving the evolution toward an intelligent optical services network (OSN). G.ASON, an emerging global network standard that defines dynamic optical connectivity, is a key component of the OSN. It is a dynamic, edge-to-edge services creation platform powered by an automatic provisioning and signaling architecture. Dynamic optical connectivity enables service providers to support emerging service and applications demand by taking intelligence from the backbone to the edge, from the network management to the optical layer, and with true granularity, quality of service (QoS), and location independence.

This paper outlines the key revenue-generating applications and services powered by a G.ASON–enabled OSN and explores the network capabilities required to deliver these applications and services.

Definitions

Service is defined as the solution a service provider sells to an end user. Virtual private network (VPN) is an example of a service.

Service attributes are end-user measures of the service. These include quality, reliability, protection, and time to provision and are determined by network capabilities and service agreements.

Applications refer to the specific and tailored use of a service by the end user. For example, a hospital uses wireless VPN for physician-to-physician communication within the campus.

The *OSN* refers to the complete set of products and capabilities needed to meet end-user demand for next-generation profitable and customizable optical applications, edge to edge.

The Dynamic OSN

A global carrier delivering a Web cast of a World Cup match between distant cities may have a usage spike between locations that threatens to overload the network. Using dynamic optical connectivity, the carrier's network requests a connection from another provider's network at specified service-level agreements (SLAs). The network element (NE) in the provider network checks to see if bandwidth and SLA guarantees can be met. They can, and the bandwidth is instantly delivered at an agreed-upon price. The whole process takes place in seconds. End users continue to watch the Web-cast without interruption. The global carrier is billed automatically for the service.

Faced with an extremely competitive marketplace, service providers are attempting to meet growing end-user demand for innovative applications with true, international, end-to-end connectivity at lower costs, characterized by faster time to service and differentiated grades of services.

To do this, service providers require differentiated levels of service protection that reflect the priority and sensitivity of the underlying traffic. They also require a network that provides a range of service-level guarantees (classes of services [CoSs]) that can be provisioned on a per-connection basis and over the lowest-cost route. Meeting these challenges requires services providers to move toward a fully optimized network geared toward efficient services creation. In other words, it requires the adoption of an OSN model.

The OSN is a service-oriented network that consists of a complete set of network products and capabilities needed to meet end-user demand for a new wave of profitable, innovative, and customizable optical applications, edge to edge. Underlying the OSN is a full suite of optical intelligence capabilities throughout the network that is necessary to realize fully enhanced services differentiation through efficient usage of optical bandwidth.

Dynamic connectivity is the automatic routing and switching of optical bandwidth and is defined by the emerging G. ASON global architecture standard for an intelligent optical switch control plane that automatically provisions optical bandwidth according to a set of customizable and dynamic features.[1] Bringing intelligence, such as connectivity, to the optical layer transforms today's static optical transport network into a dynamic, service-oriented platform and converts the optical core into a more robust network that delivers unprecedented revenue-generating potential to service providers.

G.ASON extends proven synchronous optical network (SONET) and dense wavelength division multiplexing (DWDM) capabilities to create a truly intelligent network, enabling rapid provisioning and reconfiguration of diverse end-to-end services across multiple network domains with either end-user or network-operator control.

Sustaining a simpler and less resource-intensive network with automation, the dynamic connectivity leads to significant reduction in capital, operational, and maintenance costs for both service providers and end users. Greater network optimization offers the following benefits:

- Automatic signaling, routing, provisioning, and restoration of optical services based on intelligent routing and signaling algorithms
- Auto-discovery of network topology to simplify the provisioning of new switches and connections, resulting in fast and reliable setup and teardown of connections

---

[1] G.ASON is presently under review in the International Telecommunications Union (ITU), while signaling protocol definition is underway in the Internet Engineering Task Force (IETF), Optical Internetworking Forum (OIF), and T1X1.

- Automatic prioritization of traffic (based on different CoSs [e.g., mission-critical data])
- Support of a mesh (or ring/mesh hybrid) architecture for greater route diversity, bandwidth efficiency, and upgrade flexibility, as well as faster network deployment
- Point-and-click connection provisioning based on CoS priorities and flexible restoration
- Full protection/restoration for enhanced services
- Client signaling for end-to-end dial-up connectivity supporting personalized end-user services and applications
- Consolidation of CoS and restoration points

The customizable applications and services supported by dynamic connectivity include managed wavelength services, end-to-end optical Ethernet services, storage networking services, managed fiber services, and bandwidth brokerage services, as well as others defined later in this paper.

Drivers for Dynamic Connectivity

Several factors contribute to the demand for smart connectivity and to service providers' desire for greater levels of network performance. Chief among these include the following:

- *Increasing customer demand:* End users, including enterprises and carriers, support innovative applications that require high-capacity transport and fast time to service. In addition, the proliferation of service-provider entrants and new retail and wholesale market opportunities is increasing the mix of applications and services. For service providers that seek to retain existing customers while attracting new customers, delivering services that support these applications flexibly and rapidly will determine first-mover or fast-follower advantages.
- *Service rates are approaching optical line rates.* Asynchronous transfer mode (ATM) and packet over SONET (POS) interfaces on switches and routers are approaching line rates of 10 Gbps, comparable to the fastest optical line rates available today. Customers with existing infrastructure investments in SONET and ATM are realizing the need to interface the network on a wavelength basis with technologies such as DWDM.
- *Cost-effective enforcement of SLAs:* The development of dynamic connectivity in optical networks enables managed optical services that can be provided according to flexible criteria, such as route diversity, and at lower cost. For example, mission-critical and delay-sensitive traffic can be carried on protected routes, while best-effort traffic is carried on unprotected routes.

A New Breed of Applications

Below are some applications consistent with an OSN and enabled by dynamic connectivity:

- *Bandwidth trading:* Real-time transactions of bandwidth products between buyers and sellers negotiated through brokers. Contract initiation and provisioning are enabled by Web-based and dynamic NEs.
- *Service farm applications:* Service farms provide Internet service providers (ISPs) and application service providers (ASPs) with additional bandwidth at specified times according to a predetermined demand profile. Network grooming and automatic provisioning enable efficient delivery of bandwidth according to various CoSs.
- *Optical Internet data centers (OIDCs)*: An OIDC extends high-performance optics into and between data centers to deliver storage networking solutions, addressing performance and throughput requirements, intelligent traffic management Jond security, and QoS and SLA commitments.

- *Remote data storage*: Storage networking service that provides bandwidth and storage capacity to small and medium-sized businesses when needed for database backup and restoral applications. For storage wide-area networks (SWANs) the dynamic OSN automatic provisioning provides customized, on-demand connections for more efficient delivery of mission-critical data.
- *High-bandwidth content delivery*: For more efficient delivery of content and services, such as video on demand (VOD) and digital movie distribution
- *Local-area network (LAN) interconnection*: High-speed and lower-cost networking to all segments of the network, from campus LANs to metropolitan-area networks (MANs) to points of presence (POPs) to long-haul networks
- *Transactional mirroring*: The ability to mirror critical, high-value transactions, e.g., financial institutions
- *Video-area network*: Enables customized connections for more efficient delivery of high-bandwidth content for services such as VOD and digital movie distribution and for the move to uncompressed high-definition television (HDTV) signals (1.5 Gbps)

Additional services and applications are listed in *Table 1*.

Application	Description	Value
Storage on Demand	Leased storage space for businesses looking for economies of scale, security, S/W capability, and protection that are enabled through high-bandwidth, dialable optical pipes to storage solutions providers	Minimizes investment risk in implementing storage solutions (forecasting, capital cost, operations cost) Enables storage solutions provider business
Optical VPN (This can also be a service.)	Access to dedicated, secure, high-bandwidth capacity and connectivity (e.g., lambda) over shared facility	Enables departmental connectivity and minimizes investment associated with corporate backbone communications networks
Intranet Connectivity	Connection of corporate intranets with each other, using native protocols over high-speed optical pipes	Extends the reach of transparent data communications within a company
Bandwidth Trading	Negotiation for, and execution of, contracts, either scheduled or on the "spot market" to secure (or sell) bandwidth specifying endpoints, capacity, service level, schedule, etc.	Optimal use of network (either enterprise or carrier) capacity—sell-off of unused capacity, acquiring of idle capacity to fulfill peak demands
HDTV Live-Event Content Creation	Cameras networked to production studio sending uncompressed HDTV signals	Consolidation of production facilities. Reduced expenses through elimination of 18-wheeled studio
VideoMAN for Movie Production	The VideoMAN approach involves the integration of modern technologies to enable digital video/audio storage, retrieval, and file sharing capability among several different locations within a metropolitan area	Economical, shorter time to market for video productions, and complements existing LAN/WAN networks. Cost optimized for healthcare, education, financial, and entertainment industries
Instant Network (Rent to Own)	Customer initially leases wavelengths to provide connectivity; as demand on select routes increases it has the option (discounts, etc.) to purchase entire fiber	Manageable financials, pay as you grow

	(other customers are transparently switched to other fibers); in time an entire network is built out	
Digital Cinema Content Delivery	On-demand delivery of digital cinema content to theaters	Enables expanded use of theaters and theater "experience"—film festivals, foreign language screenings, business conferences, virtual presence at live events
Digital Video Disc (DVD) Content Delivery to Retailer	"On-demand" DVD download for burning in local video rental store	Eliminates need for in-store DVD inventory Consumer access to all (6,500+) DVD titles
"Napster" for DVDs	Search, purchase, download, and burn DVD movies in-home	Eliminates the "trip to the video store." In-home access to all (6,500+) DVD titles
Tommy's Terabyte	EMC term for networked digital storage locker for personalized content accessible via high-speed connection	Networked access to personal content; Moving Pictures Experts Group–Layer n (MPn) files, compact discs (CDs), home video, etc.

Table 1: Additional Services and Applications

Business Value of the Dynamic Optical Network

Based on customer and industry perspectives, the following service attributes have been identified as critical components of any optical service:

- *Return on investment (ROI)*: For a given use or investment of money in an enterprise, the ROI is how much "return," usually tracked back to profit or cost saving, results from that use. The dynamic optical network allows service providers to offer highly differentiated, revenue-generating managed optical services.
- *Time to revenue*: Refers to the speed at which the initial investment shows a return. Dynamic connectivity simplifies network operations and facilitates network management functions, freeing service providers from manually configuring service-provisioning requests. Managed optical services can be provided faster due to automatic connectivity and signaling at the optical layer, which reduces provisioning time to hours instead of weeks
- *Total cost of ownership (TCO)*: TCO is the cost of operating the solution. Operating costs are a factor of the simplicity of the solution—the ability to use fewer people, less space, less equipment, less technology—to initiate the service. TCO is also dependent on the ongoing maintenance and support that the solution requires. Dynamic connectivity reduces the cost of deploying, operating, and scaling the optical network, whether through ring or mesh topology, by automating traffic engineering underpinned by an unprecedented capacity to achieve low cost per managed bit over any distance.
- *SLA*: SLAs are contracts that enable service providers to tailor their service offerings to specific markets. SLAs are realized by extensive service reporting, service guarantees, and policy management. Using a client-drive network interface, SLAs and can be customized to greater specificity. With dynamic connectivity, service providers can now easily provide differentiated levels of bandwidth services across the entire network based on various CoS demands, flexible restoration priorities, and full protection capabilities.

- *Time to service*: The ability to satisfy customer demand quickly, which includes the ability to check policy and SLA and then to allocate bandwidth. Automatic provisioning based on CoS priority reduces the time needed to set up new services from days or weeks to hours or seconds. Direct, client-driven signaling has the potential to reduce time to service radically.
- *Time to restore*: The ability to provide working, protection, and restoration paths on the network topology. Protection is optional, graded, specified in the SLA, and enforced by policy management. Time to restore is a key capability of automated connectivity. Dynamic connectivity provides flexible restoration according to different levels of priority, allowing greater control over the time it takes to restore paths and prioritize traffic.
- *Edge flexibility*: Edge flexibility means supporting most network interfaces and protocols and capacity independence where appropriate, and supporting most common and widely used proprietary interfaces. Using standardized network interfaces, dynamic connectivity provides customizable end-user bandwidth provisioning.
- *Investment protection*: The network provides both backward compatibility with current networks as well as a platform (migration path) for future growth. Dynamic connectivity is fully future-proof. End-to-end scalability, from access to metro to the high-capacity optical backbone network, as well as interoperability based on the G.ASON global standard, ensures future investment returns.
- *Vendor independence*: Interoperability with a broad range of systems and equipment from different manufacturers gives the network operator freedom to buy NEs from the vendor with the best technology, price, and service. Interoperability is ensured by the G.ASON open standardization as well as by the standardization of all relevant interfaces.
- *Security*: Capability to support dedicated and exclusive capacity on a single wavelength for a specific customer while meeting authentication requirements. Open interfaces with full transparency are required to pass all traffic unaltered. Dynamic connectivity supports the ability to authenticate transactions.

Dynamic Connectivity—Network Capabilities

To be viable, the OSN requires capabilities that can support the above key service attributes *better* than alternative network architectures available to carriers and users.

These capabilities include the following:

- *Auto-discovery:* Together the network management system and the control plane automatically detect the existence and status of all NEs and their connectivity. Selecting and turning up an end-to-end path for a given service is almost totally automated. New revenue streams can be realized more quickly because manual data entry is not required at the time of commissioning.
- *Service definition:* Once service-enabling elements become visible in the management system's network view, a capacity database can be created and reported for ordering and planning systems. This provides information feedback to prevent overprovisioning of network resources. Both total bandwidth available and current bandwidth usage are reported, plus information to help in the management of service quality. Service paths can be determined from end to end, including routing through various optical components.
- *Service policy:* Policies can be established against key service parameters such as protection, signal quality, bandwidth allocation dynamics, and route diversity requirements. These policies are especially valuable in defining performance characteristics (e.g., hold times) and in determining routing preferences and can provide a vehicle for differentiation among service providers. For example, a carrier could offer an SLA that specifies certain wavelengths that are

available during specified time periods or only on weekends. An intelligent optical network can enforce flexible policies such as these.

- *Service ordering:* The network should offer flexible e-commerce solutions for order entry in a dynamic environment. These solutions link customer-service requests with service activation. Service-enabling elements are automatically allocated to services for turnup during the activation phase. The powerful combination of linked and coordinated service ordering, activation, and policy functions efficiently integrates optical network management into the carrier's core business processes. Linkage to the service ordering function requires flexible, open information interfaces between management and ordering systems.

- *Service activation:* Point-and-click provisioning tools use policy-based criteria to route connections per service-order requirements. The service activation function encompasses both electro-optical and photonic switching. Flexible, open information interfaces are required to support service activation. As service activation becomes an increasingly dynamic process, the next-generation management system must also support accurate wavelength-level service accounting functions

- *Service assurance:* The network must offer fault management architectures that scale easily from small to very large network applications. Service assurance tools should handle the necessary volumes of data for carrier-grade performance reporting. Also, performance data must be compared against policy parameters so that network performance can be reported in the context of customer SLAs.

- *Customer-service management:* Network architecture should allow partitioning of the network so that customers can self-manage their services, including service ordering, service activation, and service-assurance functions. It is highly desirable to combine this partitioning with Internet e-commerce technology that permits access to management functions in a secure information environment.

- *Service granularity:* The light path and associated properties define the physical characteristics of the optical connection. Granularity allows the user to consume optical services at the sub-rates of the light path, with subminute provisioning extensions into the edges of the network. Light paths should support both protocol-independent and rate-independent wavelengths.

- *Directories:* The network supports multiple higher-layer services. Directory services provide address translation between higher-layer service addressing and physical entities.

- *Protocols supported:* Common routing protocols are used within the domain of the intelligent optical networks. Signaling protocols are used over the intercarrier interfaces.

- *Destination:* To deliver the new breed of services and applications, the intelligent optical network supports multiple addressing schemes for both Internet protocol (IP) and non–IP services. Physical entity, IP, and network service access point (NSAP) addressing allow network intelligence to grow toward the edges.

- *Location independence:* Customers require high-bandwidth connections wherever they are. Ubiquity of service is supported through interoperability with multivendor networks and a standardized platform. Pooling points, for example, which are used to enable bandwidth trading, can now be supported with faster provisioning across more POPs than before.

Rollout of the Dynamic Optical Network—Enhancing Services Today and Evolving to Bandwidth-on-Demand Capabilities

A core enabler of the continuum of evolving optical services, dynamic connectivity will offer true bandwidth-on-demand capabilities—anytime, anywhere. This includes enhanced support for current managed optical services as managed wavelength services, end-to-end Ethernet services, storage networking, and managed fiber services:

- *Managed wavelength services:* Otherwise known as leased lambda service, this solution offers customers optional service levels, including protection options via wavelengths.
- *End-to-end optical Ethernet services:* These merge the power of optics with the simplicity of Ethernet to bring connectivity to the home, business, and across communities.
- *Storage networking services:* These services and their interconnection represent the future of enterprise storage and create new business opportunities for service providers. When implemented, interconnected storage-area networks (SANs) provide an unprecedented ability to share, consolidate, manage, and protect information technology (IT) computing resources timely and reliably.
- *Managed fiber services:* Optical private networks are private, high-capacity, high-reliability optical connections between enterprise facilities.

These services are only a few of the revenue-generating solutions providers can offer to enhance and supplement their existing portfolios. Vendors today are deploying products designed to support these and more services as provider networks gracefully evolve to allow dynamic end-to-end, mesh, ring, or hybrid-based dynamic connectivity.

The dynamic OSN represents a strategic paradigm shift toward service differentiation and greater services value to end users. It is a simple and cost-effective solution for migrating toward mesh architectures and for providing profitable next-generation services. Automatic dynamic switching replaces time-consuming manual switching, allowing services and service attributes to be tailored in real time to specified customer requirements. The result is significantly greater customer satisfaction, reduced network costs, more efficient usage of bandwidth, and faster time to market with new revenue-generating services.

As the optical transport network evolves to a real-time switched, on-demand network, a service provider will be able to offer capacity for shorter periods of time, opening new revenue opportunities. These developments will propel a new era in intelligent optical connectivity.

Impact of Optical Networking on Service and Transport Network Planning

Parish Autry
Product Manager
Cisco Systems

The world is steadily moving toward an all-optical network (AON), and this impacts network and service planning and creates new challenges. There were expected to be 3,500 new telcos in 2000, and data traffic has surpassed voice in most backbone networks. As shown in *Figure 1*, the growth in voice traffic has been linear, while the growth in data traffic has been exponential. Driven by the Internet, bandwidth demand is exploding.

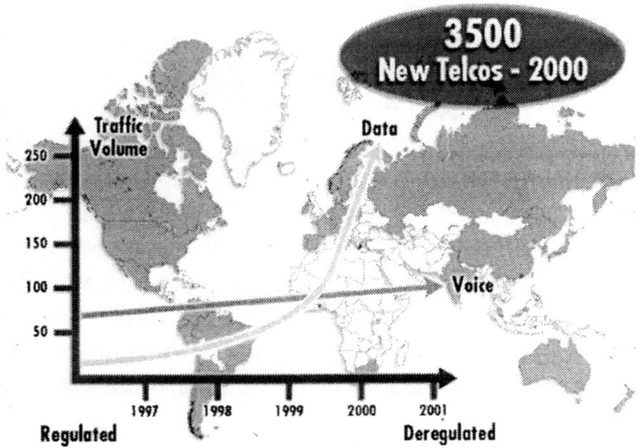

Source: *Fortune*, March 15, 1999

Figure 1: *Challenges in the New World*

New, multiplying service types are coming out with different applications that utilize the Internet protocol (IP) layer. At the same time, the location of these service types and applications has become unpredictable because of the Internet. Traditionally, a carrier knew where to look for source and destination traffic patterns, so it could overbuild and plan for the future. The Internet, however, has changed all of this. The challenge, then, is to plan networks accurately when the growth is driven by services that are unpredictable in size and location.

The Core Network Architecture

Before getting into the planning challenge, first consider the core network architecture today, as shown in *Figure 2*. This resembles the "club sandwich" network architecture that has an IP layer, an asynchronous transfer mode (ATM) mesh over synchronous optical network (SONET) rings, and point-to-point dense wavelength division multiplexing (DWDM).

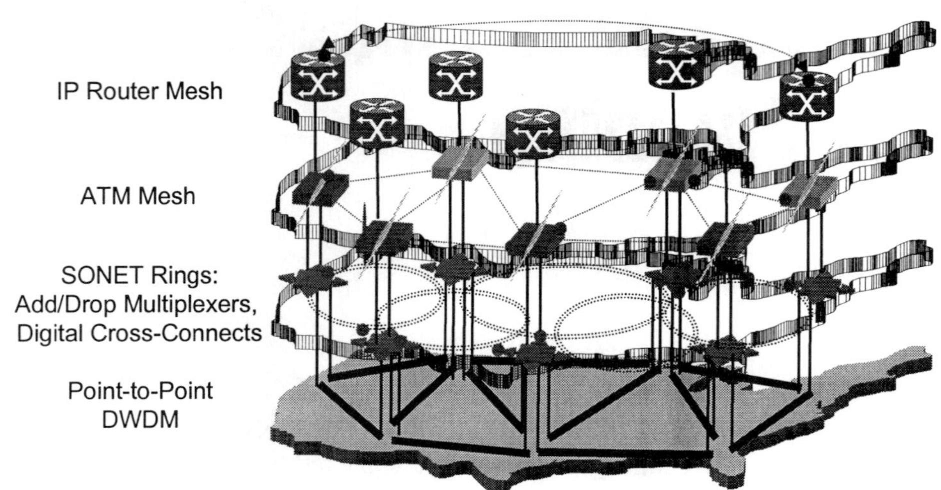

Figure 2: Core Network Architecture Today

To get an IP service from source to destination, it must be provisioned at the IP router mesh. An ATM mesh is typically the path for traffic engineering, while transport is accomplished across SONET rings for survivability. Then data passes across point-to-point DWDM, which is typically deployed for cross-country fiber exhaust. Looking at these different layers from a network-planning and service-velocity perspective, this kind of a network is slow to scale. There is some functional overlap; some layers currently perform the same functions as other layers, with respect to traffic engineering and restoration, and retain some outdated functions.

It is often said that SONET is not necessarily the most efficient way to transport IP traffic because of the amount of overhead that is required. SONET was developed (and very well) for carrying voice traffic and circuit-switched networks, but it is not as efficient for an IP layer. DWDM was developed in the late 1990s and was needed in the backbone networks because the winning business strategies in the carrier market were, and still are, dependent on service velocity. A carrier must get a service deployed and turned up quickly in order to get revenue from it. Profits are inherent in the paradigm shift from a deployment time of a couple of months to days or even minutes. DWDM allows carriers to multiply the fiber capacity, and typically the DWDM deployments are faster than pulling in new fiber, especially in a long-haul environment where a DWDM system that is capable of multiple channels is deployed. The core structure is there with amplifiers, and it is simply a matter of adding channels and transponders instead of pulling fiber every time capacity runs short.

DWDM creates challenges from a network-planning perspective. The optical span designs now become critical when there are high channel counts and high-power optical amplifiers. The span designs must also look slightly different with respect to whether the core is 10 Gbps or 2.5 Gbps per wavelength, and the outside fiber plan also weighs heavily into this. Thus, when planning span design (the number of points required for regeneration from line termination to line termination, or for every time the network has to go electrical), keep in mind the pending changes with ultra–long-haul DWDM. The real message is that much planning is involved in network design; the carrier must account for the final architecture from the onset and then plan the upgrade path.

The Effect on Service Planning

From the surface, it would appear that DWDM creates abundant raw bandwidth. It can take one fiber, put a 160-channel DWDM system on it, and raw bandwidth is created. From that perspective, DWDM is simply a dumb fiber multiplier, and it is capable of supporting the exponential demand that the Internet is driving. Without intelligence in the core, however, it is difficult to harness that bandwidth for service, and carriers need service planning. Carriers are competing for services, and being able to get to those services more quickly will make the difference.

Difficulties in harnessing these services are a result of cumbersome core network architecture, as was shown in *Figure 2*. The network planning function is tedious and lacks intelligence in the optical core. Granted, vendors have envisioned different intelligent optical devices for the core to make service planning and service velocity quicker. However, until these are fully realized, there will be a bandwidth scarcity from a service perspective, despite having bandwidth abundance with high–channel count DWDM.

Service and Transport

Currently, restoration and provisioning for the optical core are primarily performed with SONET rings. New services from source to destination would require growing a SONET ring or adding a new SONET ring if the capacity were not already present. The time it takes to deploy and implement SONET rings is not fast enough to keep pace with service demands, especially when traversing multiple rings to get from source to destination. When all the needed capacity is available except for one link, an entire ring may have to be overbuilt, or channel counts and amplifiers must be added simply to get that service from source to destination. In addition, the provisioning of new services today is a slow, manual process. Planning is involved from the onset; to provision across the rings, the selection of the nodes, ports, and wavelengths must be known. Large carriers have departments full of people who do nothing but write documents on the ports needed to provision each of the elements and boxes and on how the rings may be physically connected. Sometimes it takes a month or two simply to get an optical carrier (OC)–48 or higher service from source to destination if the network is cross-country. At that slow implementation speed, it cannot meet demand.

Most of the tools presently used to model the network are off-line tools, and much manual interaction occurs in entering network data and traffic pair demand. The tools are often static and based on outdated traffic patterns. The carrier is dependent on that person whose planning function is to enter the data from the network end to make sure it truly represents the state of the network so that the locations for needed capacity can be planned. This is often also tied to equipment delivery cycles. Thus, planning tools as well as the network elements must become intelligent. They must be able to take advantage of new optical-networking solutions, such as the optical add/drop multiplexers (OADMs), optical cross-connects (OXCs), and optical routers.

The Evolution of the Optical Core

If the network is considered to be two planes (layers) when focusing on an IP service, new functions can conceivably be added to the mesh optical core, as illustrated in *Figure 3*.

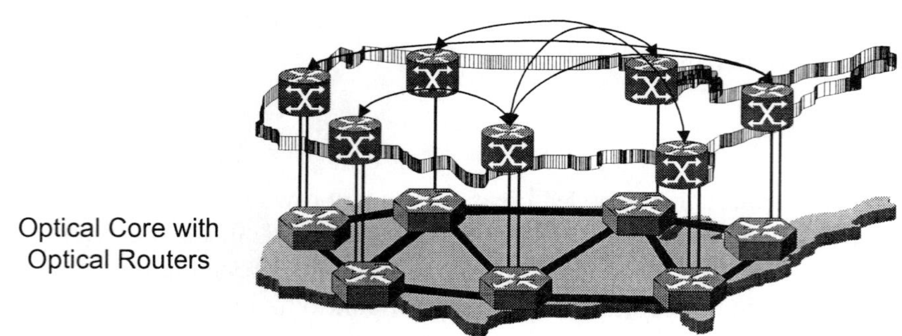

Figure 3: *Optical Core Evolution*

The additions shown here, at the core, are the optical (wavelength) routers. Other vendors have products with similar names, but the key aspect of the optical core where DWDM is deployed is that, in most networks, this is already physically set up. With this connectivity, companies want to be able to take advantage of the different connection points between the network's major cities or major junction points. The topology shifts from ring to mesh. Providers hope to take advantage of the connectivity in a mesh, while meeting the SONET–like ring speeds for restoration and switching, and to find many more ways to get from source to destination without as much concern over planning ring by ring. This would allow the shift of the provisioning and restoration out of the SONET layer and into the optical layer—that is, once the elements become intelligent. The layers become streamlined, and some of that functional overlap is removed. Network planning also becomes easier. The goal is to be able to deliver the optical transport and the traffic engineering at a wavelength level while complementing the IP and ATM layers.

Provisioning and Restoration Planning

Provisioning and restoration planning is moving toward an automation that will dramatically reduce the operational planning costs by letting the network elements perform the provisioning functions themselves. The network elements will have common protocols within and between the layers so that when bandwidth is needed, the carrier can focus on the IP service and the application and, knowing the source and destination demand, let the network elements communicate with each other and request the bandwidth they need between the layers. Many standards, within both companies and the open Optical Internetworking Forum (OIF), are attempting to implement a common protocol so that the layers can communicate and be interoperable, thus enabling more intelligent network (IN) elements. Once the network elements are intelligent enough to do the provisioning, the need for the planning departments (concerned with the nodes, individual ports, and links) is eliminated. Carriers can then turn their focus to the network on a link-by-link basis, adding in the bandwidth only where it is needed. They are free to focus on the choke points or the bottlenecks, rather then muddling over building rings or adding spare capacity that may not be needed. They simply overbuild certain links on a span-by-span basis to ensure that the network remains covered for new services, and restoration requirements will be met. Ease of bandwidth delivery is the real goal of one company's optical networking focus, as it tries to get faster service capacity and to allow the carriers to offer new services more quickly as they deploy new products and build their intelligent optical cores. Smarter network elements, which enable flexible traffic engineering, are preferable to the manual planning of paths that was typically done with SONET or unintelligent optical devices.

In addition to network elements becoming more intelligent, restoration may be achieved by making network-planning tools more intelligent. The tools themselves must model the bandwidth capacity, apply the proper resources on a link-by-link basis, and add the new nodes and links to the network where needed, but they can also model the effects of growth as it relates to restoration. SONET protection switching speeds still must be met; intelligent modeling tools—unlike rings—enable the amount of restoration bandwidth required in a mesh environment to be calculated while maintaining those speeds. Another way to accomplish this is to take the intelligence that is in the network elements, the protocols, and the algorithms and duplicate it into the planning tools so that the network operations can be simulated. The last step to making the tools intelligent would be to connect them at either the element or network management layer or to the network element itself to be able to get up-to-date topology information. The planning tools are run through an off-line mode of "what if" scenarios or are connected into the network so they get real-time updates of all topology changes. These intelligent tools can run some of the same routing protocols to do the same real-network modeling. This will allow the carrier not only to plan new capacity, but also to simulate cuts in different points of the network. For example, if a major artery is taken down, is the network still restorable? How many failures can be handled? Where is the network most vulnerable? Planning is possible on a link-by-link basis, rather than by overbuilding rings, as in the SONET layer.

Finally, IN planning tools would provide instant feedback to network changes, so in the face of a real event, with a live planning tool feedback is returned on that new service, despite major work occurring in a certain area of the network. The planning tool is updated the same way that all network elements in the optical core are updated; it gives the company feedback to accurately model where that service should be deployed.

Tomorrow's Network

Where is all of this headed? *Figure 4* presents a simplified version of the next-generation network in terms of application layers as an "IP cloud" and an "optical cloud."

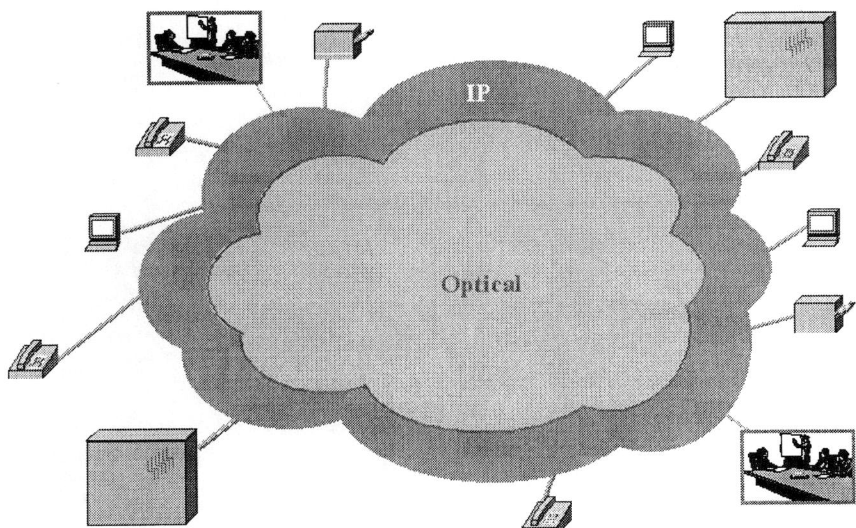

Figure 4: Tomorrow's Network

There will be much discussion about IP over glass and the vision of the IP service directly over the AON. There is clearly a migration toward the AON, but steps must be taken to make the IP layer and optical layer intelligent enough to communicate with each other. To the end user and the service provider, it should really appear as IP over glass, because although they focus on planning and provisioning the IP service, the layers communicate with each other. As the world migrates toward the AON, it may truly become IP over glass, but in the short term, appearances are sufficient, and making the network layers more intelligent enables transparency.

Summary

The optical network in the future will have a distributed intelligence with a common optical control plane between the equipment and the layers for the seamless provisioning of services. This is the end-to-end provisioning that drives the service velocity needs. In addition, scalability will meet the growth capacity evident in the data explosion that is occurring along with the demands of Internet-driven traffic. However, networks must remain restorable. Carriers cannot step backward into SONET speeds. Restoration must be possible with more than single failures across the network and, at the same time, must meet the infrastructure class reliability requirements. If a cross section of the optical core is taken, and the amount of bandwidth passing through is considered, the carrier cannot afford to overlook equipment redundancy, diverse routing of power, and optical signals within the elements themselves. In all of this, the infrastructure class reliability requirements to which the carriers in the core are accustomed must be maintained. What, then, is the answer? It is simple: Carriers must provision intelligently at the services layer and let the optical network take care of itself.

The Digitization of Radiology within the Army Medical Department[*]

Anna K. Chacko
Special Assistant to the Commander
United States Army Medical Research and Materiel Command
Fort Detrick, Maryland

The medical arena has been slow to accept new change, not only in the United States but globally, as well. Part of this reluctance is a result of the turf battles in which doctors engage; they have become used to charging for what they consider medicine-as-an-art rather than medicine-as-a-science. And they are free to do so. If they can be exclusive they can justify a very large fee. Physicians in the U.S. Army Medical Department or in the U.S. Department of Defense (DOD), however, are paid a flat salary, irrespective of how many people they see. Some of these medical professionals work very hard, and some do not, but this economic equity has liberated military doctors, giving them the ability to accept and implement new changes in medicine—changes that are slow to come in the civilian community. As a case-in-point, 10 years ago at Fort Sam Houston, the Army was the first to establish a complete digital radiology department to serve not only the Army, but the Navy and Air Force, as well.

In 1998, Army administrators lobbied for a pay increase for radiologists, hoping to improve the retention rate for quality physicians. Currently, even the most senior colonel in radiology earns less salary than a starting radiologist in the civilian world. *Investors Business Daily* published a survey on salaries for the different subspecialties in medicine noting that the median income for a radiologist in the United States was \$287,000 annually. The remuneration for a DOD radiologist is significantly lower than that, so frequently, after the radiologists have paid their debts and obligations to the DOD (which defrayed their medical school expenses), they leave for civilian positions with more attractive paychecks. The DOD has 600 billets (places) for radiologists. By the year 2001, it is expected that only approximately 60 percent of the billets will be filled. The DOD is left with the option of hiring contractors to perform the same work as the radiologists in uniform. This brings with it the tremendous morale and retention problems that arise when a radiologist, who makes \$128,000 per year, ends up sitting next to a contractor who makes substantially more per year for doing less work and not being on call.

A Paradigm Shift

This problem is being considered in new ways. One view is that it is simply cheaper to grow a radiologist than it is to grow a line officer. Taxpayers pay a hefty sum to train young line officers, and it is cheaper to buy a radiologist's services than it is to train somebody who will go out and wage war. Thus, if the military can be divested of 600 radiology slots, they would have the option of hiring 600 officers and off-loading the clinical work to people who would perform those tasks remotely. One option would be for the

---

[*] Disclaimer: This article does not purport to represent the view of the United States Army Medical Department.

organization to strip itself of subspecialties such as radiology. If access to the images and to expertise could be provided remotely at civilian institutions, the organization could divest itself of this function.

The Army Medical Department is now digitizing the radiology environment in multiple geographic areas. The manner in which this effort has been deployed in the Great Plains Region represents a new paradigm in picture archiving communication system (PACS) implementation—Virtual Radiology Environment (VRE). *Figure 1* depicts the VRE and its customers, the three stakeholders—the clinicians, soldiers (patients), and commanders. The public frequently believes that the Army deals only with healthy people. This is not true. Soldiers have families that they leave behind, including elderly and aging parents. In a typical military hospital, military physicians see just as many varieties of ailments as they would see in a civilian hospital—in addition to what they see in war, which are combat-related injuries and diseases.

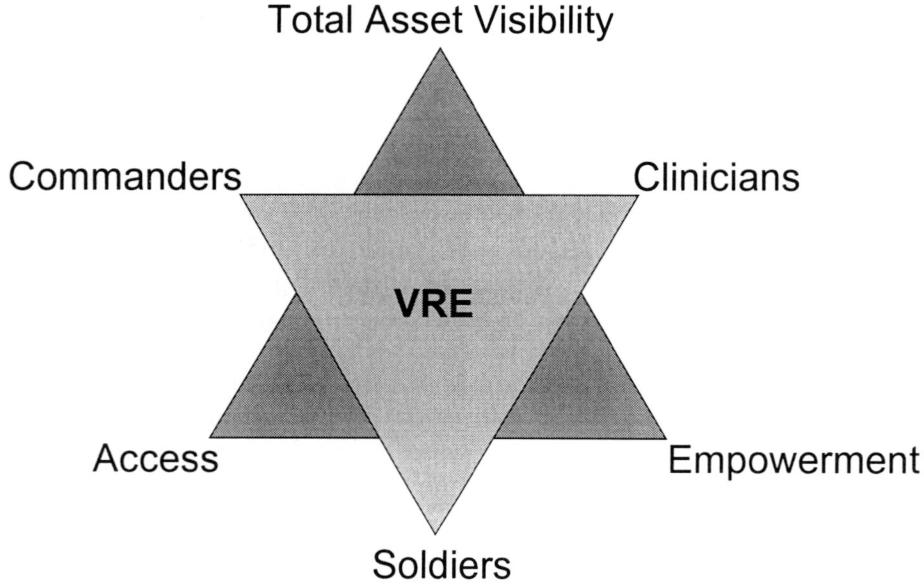

Figure 1: The Virtual Radiology Environment

As represented by the background triangle of *Figure 1* the Army, as the purveyor, has agreed to provide its stakeholders with total asset visibility to empower them and allow them access. The clinicians should be empowered to consult any radiologist they desire. They should be able to access any patient's record and, irrespective of geography or time, be able to achieve total asset visibility with regard to the medical records and the available tools.

What about commanders? People who command hospitals must know exactly how many radiologists, pathologists, dermatologists, and so forth are available for them to access. The commanders, too, should be empowered to use whatever tools they have within their hospitals or within the enterprise. Patients are similarly empowered by access to a virtual environment. They should be able to access any clinician they desire, their records, and the diagnostic capability that is most appropriate.

The Power of Technology

Most people underestimate the enormous power of information technology and dissemination of information to multiple and disparate purveyors. One can go to any part of the world and still access one's bank accounts—yet one cannot do that with medical records or things like X-rays. About 15 years ago, the Army conducted a wide-scale investigation of what was happening in radiology in the United States and the military. They found that, out of 100 patients, only 21 had access to all of their radiology records. What happened to the others? Those patients' reports were closeted in doctors' offices. One hospital had 14 file rooms manned by six file-room technicians. Much time was spent trying to access patient records and patient films. If the films or records do not materialize, the patients have to be reexposed. This creates numerous problems for patients as well as for clinicians. It was this revelation of inefficiencies that motivated the DOD's decision to digitize the entire department of radiology.

Health-Care Costs and Survivability Issues in the Arena of Mammography

Of all the subspecialties of radiology, mammography is the only sector just now becoming digital. All the others have gone digital in the past decade. Mammography is one of the most highly effort-intensive, highly memory-intensive subspecialties of radiology. It also has the dubious honor of being one of the most highly litigious. Consequently, physicians are wary of losing resolution by converting to digital. Going from analog to digital in mammography is not widely accepted at this point because of a possible loss of resolution and increasing risk of nondiagnosis. The arrival of digital mammography will change the face of radiology throughout the world. It is anticipated that 90 million women will be screened every year for breast cancer. Many are not currently screened because of lack of access to mammographers who can read their radiographs, lack of access to places where they can get mammograms, and lack of resources. A mammogram costs $37 per screening alone. If an abnormality is found, it costs $140 for a diagnostic mammogram. Few people can afford this. However, when digital mammography is allowed in and made available to anybody in the world, the price of mammograms will drop and the service, as well as the expertise needed to evaluate that service, will become far more accessible.

By 2004, the American Cancer Society and the National Cancer Institute are determined to ensure that every woman 40 and older receives a screening mammogram. Should this effort be completely successful, the question then becomes: Who will read the more than 90 million mammograms that will be performed every year? There is insufficient expertise to cope with this task. The U.S. Food and Drug Administration (FDA) mandates that certified radiologists read 400 mammograms per year. This is a major conundrum. If professionals do not read enough mammograms, they are not certified. If they are not certified, they cannot technically read mammograms. A cure for this lies in disseminating these readings to the digital environment—not just in the military, but also on the economy.

Looking at the Points and Perks

The Army Medical Department is comprised of a varied mix of large, medium, and small fixed facilities, clinics, and mobile hospitals. Digitizing radiology and deploying it throughout the military will provide access to images anywhere irrespective of the location of the patient or the radiologist. This therefore, provides a great convenience during conflict or operations other than war. Ubiquity of images and attached information is a great boon in the military because the military is not bound by the licensing constraints that civilians are. In the civilian arena physicians are constrained to practice only in states where they are licensed to do so. In the military, however, if a physician is licensed in any of the 50 states and keeps that license current, he is able to practice in any federal facility. Thus, the deployment of a digital environment or completely virtual environment within the military becomes relatively easy and economically viable instantly.

Aside from digitizing radiology (the image environment) in terms of X-rays and mammography, other specialties that lend themselves to digitization are dermatology, pathology, and cardiology.

The Effects of Digitizing Radiology

The advantages of moving to the digital environment are many. One is the speed that the digital environment makes possible in terms of transport of vital information. When it comes to medical diagnosis, time is of vital importance. In many cases, speed saves lives. Digitization will also reduce the number of necessary radiologists. It will facilitate a cross-leveling effect on physicians' workloads, allowing them to be available for service elsewhere when and if needed. For example, a doctor was sent to a foreign location for nine months. In those nine months, he performed only 30 X-rays. If those 30 X-rays were brought back to the mainland United States, his time would have been more efficiently utilized. The organization would not need to hire a highly paid contractor to replace him. Alternatively, the radiologist could continue to do his stateside work while taking care of the minimal workload he was given while away. Radiology, in particular, lends itself to the digital environment; approximately 90 percent of all imaging exams can be diagnosed remotely. Also, with the established medical network that exists in the United States, the basis already exists for an efficient teleradiology infrastructure to facilitate insourcing and outsourcing.

The Challenges of Digitizing Radiology

Several obstacles stand in the path of digitizing radiology. First, providers of digital expertise must be trained. This technology radically changes the culture of health-care delivery, so it is important to train and prepare the medical community to deal with this new way of practicing medicine. Technologists and clinicians will require training in how to convert analog to digital, how to read those digital images, and how to transport them to the next destination—all while maintaining the same high quality of yesteryear. Consequently, quality control represents another challenge.

There must also be a system devised for failover and recovery. Should the system fail, should there be a broken link in the chain, a bypass must be planned at the onset to maintain an accurate and expeditious reading. Provision has to be made for archiving and transmission. With the move to a digital environment, the face of storage will change—disk drives, optical media, and more innovative electronic storage media are certainly on the horizon.

Finally, the tools must be in place to transmit digital data. This will require an overhaul of the infrastructure to incorporate the networks capable of sending, retrieving, and storing this information. This may be the greatest challenge of all, because network technologies are changing at lightning speed. Yesterday it was electronic networks and communications based on wireline technology; today it is optical networks and wireless global positioning systems (GPSs). Who knows where technology will be tomorrow? It will take a monumental effort to catch up with, run steady with, and anticipate the next moves of this rampant technology.

Figure 2 illustrates some of the equipment that will facilitate this move to a digital environment. At the core of the system is the digital network, which already exists in one incarnation or another in most medical institutions. With the expanding sophistication of network technologies, such as the migration to all-optical systems, engineers have, indeed, taken into account the integration of these new technologies into the legacy (currently established) networks. Their effort is to make the transitions as seamless as possible, even in this mission-critical environment.

Input devices are segregated to the left of *Figure 2* and include equipment such as CR readers and direct, digital-capture, diagnosing, imaging devices. To the right are the output devices, the GPS communications technology that will disseminate this data, and the end-user workstations, which include printers and endpoint application software.

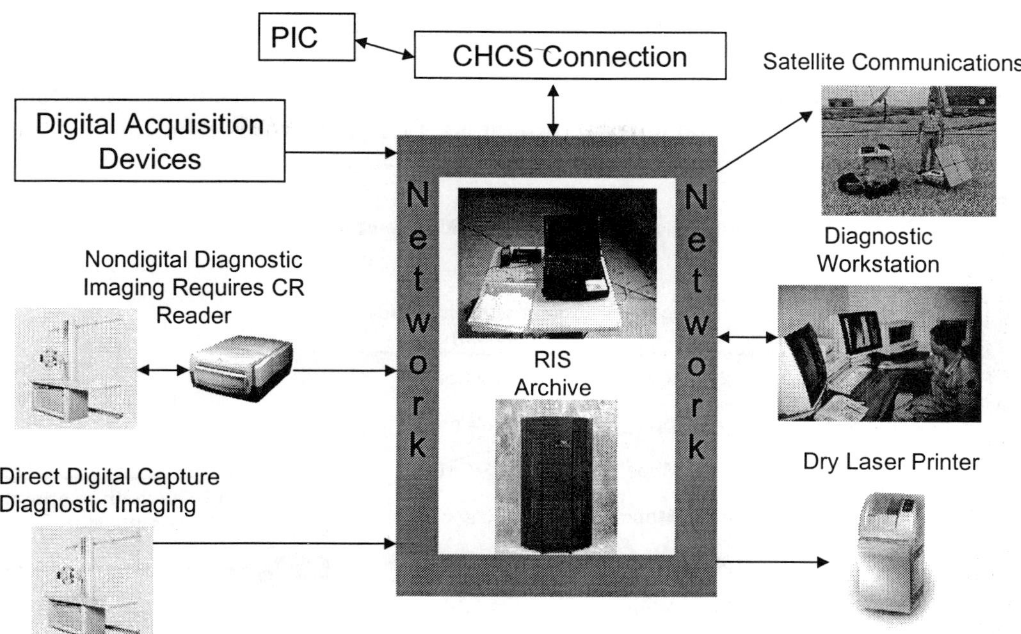

Figure 2: *Equipment Necessary to Digitize Radiology in the Table of Organization and Equipment (TO&E)*

A Strategy for Deployment

How should the migration to digital begin? A prudent move would be to begin with starter-kit systems—i.e., scaled-down versions of digital information network—PACS (DIN–PACS). The systems should be scalable and modular, enabling the network to be tailored to the type of medical unit and facilitating future upgrades as the technologies advance. Such an advantageous beginning would do the following:

- Reduce the number of radiologists needed
- Enable teleradiology, increasing the speed with which tests are transported from place to place
- Improve image quality
- Obviate the logistics burden of darkrooms with their inherent and expensive chemistries
- Eliminate the need for film
- Enable the organized archiving of images and reports on the primary interexchange carrier (PIC)/local.

To facilitate all of this, the composite health care system–PACS (CHCS–PACS) must have a bidirectional interface to allow for data input and diagnosis delivery. The legacy systems must be interfaced to retain the older, necessary structural components while they are in the process of upgrade. Outyear maintenance, as well as sustenance and training, will be necessary. And these next-generation systems must have interconnectivity with other legacy networks, such as wide-area networks (WANs), local-area networks (LANs), and metropolitan-area networks (MANs), and be able to handle optical fiber networking. The medical environment requires handshaking over many miles and many networks. Finally, security is a critical issue, safeguarding not only the delivery of data, but also the operability of the network itself.

Optical and Its Role in the Medical Community

Optical, with its speed of transmission and capability for large-scale data transmission, presents an excellent venue for connecting major medical centers that have large numbers of physicians on staff who treat a large number of patients and who require centralized storage areas for studies and records. These networks could best be serviced by an optical backbone designed to serve several locations. *Figure 3* illustrates some of the advantages of using such a backbone to process a mammography study. The speed afforded by optical is astounding, and the cost (utilizing a T1 bandwidth) is efficient.

Example:

Sending a single mammography study between two medical medical centers

Transmission Type	Speed	Mammography Study (240 Mb)
Modem	28.8 kbps	18 1/2 hours
2-Channel ISDN	128 kbps	4 1/4 hrs
T1/DS–1	1.544 Mbps	20 3/4 min
T3/DS–3	44.736 Mbps	43 sec
OC–3	155.52 Mbps	12 sec
OC–12	622.08 Mbps	3 sec
OC–48	2.488 Gbps	less than 1 sec

T1 bandwidth or above is sufficient

Mammography study: 8 images @ 4 kb x 5 kb x 12 bits = 240 MB, noncompressed file; this file size is adopted from the benchmark established for mammography by the Joint Working Group on Telemammography/Teleradiology and Information Management, sponsored by the U.S. Public Health Service's Office on Women's health (PHS OWH) and the National Cancer Institute (NCI); cosponsored by the American College of Radiology.

Figure 3: *What Can Optical Do for Medical?*

Figure 4 demonstrates the productivity increase caused by stepping the network up to T3. A radiologist could conceivably read the equivalent of 500 mammographic studies per day, thus putting the FDA's conundrum of 400 mammograms per year to shame.

Example:

A typical medical center's workload is receiving the "equivalent" of 500 mammography studies across the network in a single day.

Transmission Type	Speed	Mammography Study (240 Mb)	500 Studies
Modem	28.8 kbps	18 1/2 hours	over 1 year
2-Channel ISDN	128 kbps	4 1/4 hrs	87 days
T1/DS–1	1.544 Mbps	20 3/4 min	7 days
T3/DS–3	44.736 Mbps	43 sec	6 hrs
OC–3	155.52 Mbps	12 sec	1 hr 43 min
OC–12	622.08 Mbps	3 sec	25 3/4 min
OC–48	2.488 Gbps	less than 1 sec	6 1/2 min

T3 minimum bandwidth

Mammography study: 8 images @ 4 kb x 5 kb x 12 bits = 240 MB, noncompressed file; this file size is adopted from the benchmark established for mammography by the Joint Working Group on Telemammography/Teleradiology and Information Management, sponsored by the U.S. Public Health Service's Office on Women's health (PHS OWH) and the National Cancer Institute (NCI); cosponsored by the American College of Radiology.

Figure 4: *What Can Optical Do for Medical? Moving to a Minimum T3 Bandwidth*

In addition to an increase in productivity, a significant modification and improvement in operation and in the quality of the data received is achieved. This is referred to as an interactive operational paradigm change. Finally, *Figure 5* steps the system up to an optical carrier (OC)–12 or OC–48. High-resolution data archiving is possible from 20 separate medical facilities with speeds running from 622 Mbps (OC–12) to 2.5 Gbps (OC–48). These figures assume a typical mammography study consisting of eight images at 240 Mb per image, noncompressed.

Example:
Data archiving center workloads from 20 separate medical facilities assuming all images are high resolution comparable to Mammography.

Transmission Type	Speed	Mammography Study (240 Mb)	500 Studies	20 Sites @ 500 Studies
Modem	28.8 Kbps	18 1/2 hours	over 1 year	21 years
2-Channel ISDN	128 Kbps	4 1/4 hrs	87 days	4 3/4 years
T1/DS–1	1.544 Mbps	20 3/4 min	7 days	144 days
T3/DS–3	44.736 Mbps	43 sec	6 hrs	5 days
OC–3	155.52 Mbps	12 sec	1 hr 43 min	1 day 10 hrs
OC–12	622.08 Mbps	3 sec	25 3/4 min	8 1/2 hrs
OC–48	2.488 Gbps	less than 1 sec	6 1/2 min	2 1/4 hrs

OC–12 or OC–48 Bandwidth Required

Mammography study: 8 images @ 4 kb x 5 kb x 12 bits = 240 MB, noncompressed file; this file size is adopted from the benchmark established for mammography by the Joint Working Group on Telemammography/Teleradiology and Information Management, sponsored by the U.S. Public Health Service's Office on Women's health (PHS OWH) and the National Cancer Institute (NCI); cosponsored by the American College of Radiology.

***Figure 5:** What Can Optical Do for Medical? OC–12 and OC–48*

Conclusion

Converting the radiology department in a medical center to optical network yields many benefits. Immediate diagnosis will result in better, timelier patient care. There will also be faster access to patient studies and more accurate historical comparison of those studies. A centralized storage and archival system can be more easily managed and more quickly accessed. The new-generation networks will allow physicians to perform patient diagnosis remotely, thus economizing the physician's time and freeing the patient from sometimes long, arduous treks to qualified facilities.

Time and work efficiency will be improved. The physician's time and workload can be prescheduled, and a sharp increase in these workloads can be managed by distributing cases across the network. Better image quality will also result from the move to the optical network. Images will have a greater resolution with increased frequency of images (more images per second), and compression of images will no longer be required. Finally the move to the optical network will facilitate new revenue-generating business models that can extend services to medical facilities outside the network for incremental revenue and the betterment of the community at large.

Real-Life Lessons Learned

Fred Harris
Director, Network Planning and Design
Sprint

Splice Characteristics

When a fiber network is built, splices occur where the fiber comes together, and losses are created at those splices. Sprint, which will be used as a case study to explore the fiber network, has developed a system to track every splice in its network. When technicians are sent to work on a cable, they can call up the characteristics of each splice so that when they finish whatever they are going to do to that cable, the characteristics of the fiber remain the same. This means that the design characteristics of the fiber plant are maintained over time. This is important, because if those characteristics change, the technology riding on the fiber will experience difficulties.

Technologies

Many technologies ride on Sprint's fiber network. When Sprint's fiber plant was deployed, asynchronous transfer mode (ATM), synchronous optical network (SONET), and dense wavelength division multiplexing (DWDM) did not exist. The fiber plant had to be built such that those technologies, when they arose, were not excluded. Thus, the selection of the fiber plant was very important. Single-mode fiber (SMF)–28™ was selected by Sprint and has proven itself over time.

No fiber exceeded the capabilities of SMF–28™ until LEAF®, a single-mode, nonzero (NZ)–dispersion-shifted fiber (DSF) by Corning, was developed. That fiber has its own idiosyncrasies. With all of the variations in the type of fiber plant, standard SMF (SSMF) is good for the technologies that have been developed since it was first put in the ground.

Mistakes have been made along the way. Sprint, for example, performed a couple of fiber exchanges with competitors where DSF was used, so the company experienced some problems in those areas. Technology variations have been developed to allow Sprint to overcome these issues. Fortunately, these fiber exchanges have been in corridors where limited capacity can still satisfy overall network demand. Currently, there are two spots in the network where the company uses leased fiber. To the lawyers writing the contracts, fiber is fiber, and when the company leased this fiber, after the contract had been signed, it turned out to be DSF. However, in the 32,000 route miles of Sprint's network, only two small dispersion-shifted pieces exist. The rest is SMF–28™, and the DWDM technology deployed currently operates at 40 windows and will be upgraded to 80 and 96 windows. Those will be fully operational.

Sprint has had DWDM technology in the network since 1995. Thus, the company has a history of using this technology to create the scalable optical cord necessary to compete in today's environment.

Bandwidth

In the early 1980s, no one could predict what was going to happen to bandwidth. The idea that bandwidth would reach the levels that it has today was not even on the most distant horizon. How fast has capacity grown? Up until about 1992, growth in bandwidth was flat as a pancake. With the explosion in the data world and the Internet, bandwidth use began to shoot up in the mid 1990s, and similar growth is projected for 2000 and 2001. Bandwidth has moved from 2.5 Gbps windows to 10 Gbps windows to equal 160 total windows. Clearly, it will climb beyond that (see *Figure 1*).

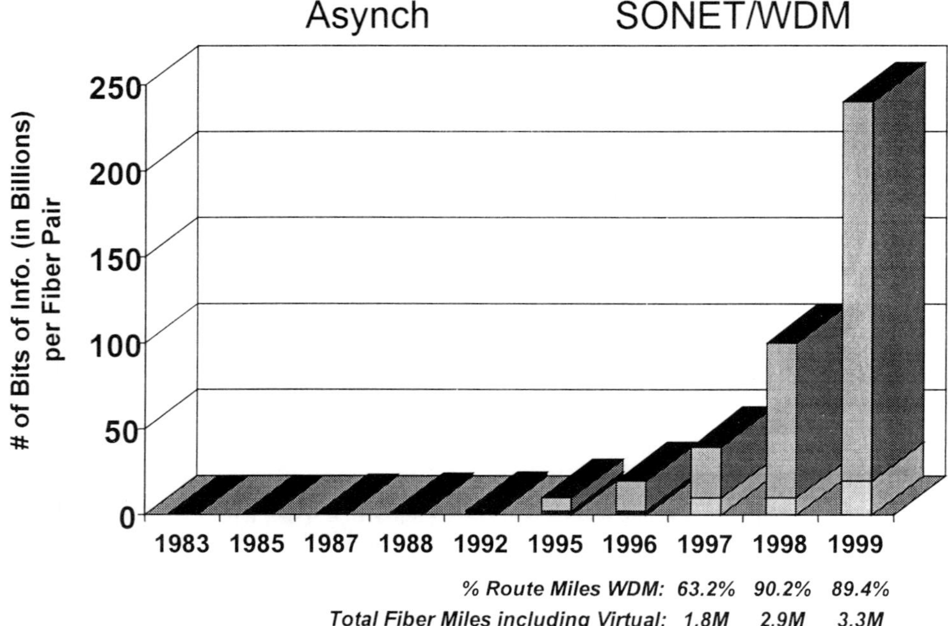

Figure 1: *Capacity of One Fiber Pair*

It seems that a 400 window of 10 Gbps capacity cannot be too far off, and 20 and even 40 Gbps windows should be attainable. Of course, there is good cause to be skeptical of these projections. While 40 Gbps may be possible in the laboratory, the flapping of a butterfly wing at the site will cause distortion in the fiber circuit, given such high capacity. Thus, the ultimate speed of those windows remains uncertain.

Nobody can predict these things. No matter how many lessons the experts believe they have learned, prediction is impossible. Consequently, the network must be designed for the unexpected. It must be able to evolve as the technology changes.

SONET

A different paradigm was introduced with SONET. When Sprint deployed SONET, it deployed a version that had not been previously deployed. That version was four-fiber, bidirectional line switched ring (BLSR). This version changed the paradigm of service restoration in the environment.

Companies must consider how they can use technology to their advantage to change the way customers view specific characteristics of the network. SONET provided Sprint with the opportunity to tell its customers that it could change the entire service-restoration paradigm. With four-fiber rings, if a fiber is cut, traffic can switch around the other side of the ring in 60 to 80 milliseconds. This means that customers would not even notice a fiber cut if it happened in the middle of a call because it would have no impact. With every other restoration technique available at the time, the customer's connection would be cut, and it would be necessary to redial. Sprint's introduction of four-fiber ring SONET was a substantial change in the paradigm (see *Figure 2*).

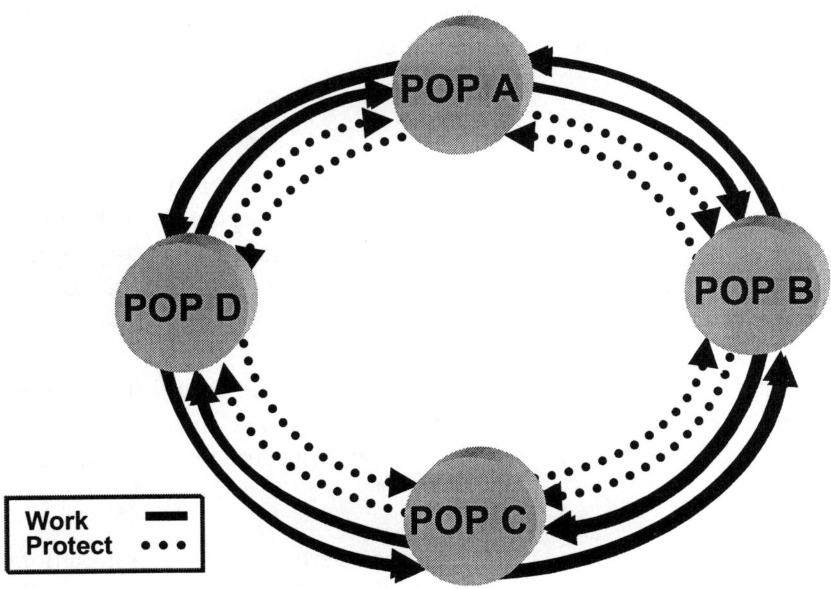

Figure 2: Did It Again

How should networks be built so that the builder can take advantage of changes in technology? The introduction of four-ring SONET shifted the paradigm, and now all companies must employ ring technology in the core of their networks if they want to stay in business.

Of course, in places such as the Internet, that technology is not necessary. In the best-effort environment of the Internet, Layer-3 restoration works reasonably well. Also, routers can be interconnected directly to wavelengths. If, for instance, a network runs all traffic via packet over SONET (POS) at optical carrier (OC)–48 levels, a fiber that can support DWDM is needed so that multiple channels of that router technology can be put in the network. Again, the ability to take advantage of changes in technology is important.

Span Protection

Another important aspect to remember in ring technology is that the carrier itself causes most disruptions in any carrier network. Fiber cuts will cause the ring to switch, as will any one of a hundred maintenance events. However, four-fiber SONET has a capability called span protection. For example, if maintenance needs to be done on one laser, the signal can be switched around that laser on the protect channel. Maintenance can be done to the laser in question. The signal is then switched back to the work channel without any disruption. As maintenance activities are constant in carrier networks, span protection occurs daily throughout the country. Such things as software upgrades on add-drop multiplexers (ADMs) drive span protection. These kinds of activities are continuous. It is not unusual in a large network to have 200 span-protect switches every day. These must be considered when the network is designed, or the deficiencies in the underlying plant will be discovered after deployment.

Protect Systems

It is difficult to overstate the importance of protect systems. While fiber rings work well and the service impacts are significant, the operation and maintenance of the network is by far a larger problem and one that is not always given the attention it warrants. For a carrier, however, maintenance is critical. If the carrier does not consider protection schemes while the network is being built, the not only will the design be faulty, but the price will be paid in terms of customer service.

Performance Management

Performance management is another key concern. One of the great problems with optical networking is the difficulty of discerning what is occurring with a particular signal. A light pulse goes in one node, and a light pulse comes out the other end, but who knows if it is the same light pulse? How can quality of service (QoS) be measured in an AON? Today, there is no parametric expression of QoS in an optical network.

This means that if a carrier makes its core network purely optical, it will have no means of performing service assurance. While new carriers may talk about moving to AONs, rarely will a carrier with many customers do so. This is because the ability to measure and manage QoS is absolutely critical to serving the customer.

Consider a customer like Bank of America, which has numerous financial transactions moving across its network. Suppose they have problems with some of these financial transactions. When they call their carrier about these problems, they expect that carrier to be able to look into the network to find out exactly what is going on (in terms of causes and effects) and to be able to restore the QoS expected by the customer. Thus, performance management service assurance is absolutely essential and must be built into the network. It is not the sort of thing that can be added at a later date.

Risk Management

Another key aspect is risk management. Carriers must understand what they have built and where they are at risk. If there is a common, single point of failure in a network, Murphy's Law says that it will fail. The entire company is at risk because of those kinds of single points of failure.

Thus, an understanding of the risks in the design of the network becomes absolutely critical. It is easy to overlook them, because more than a single point of failure may exist. There may be two points of design in the network that put the company out of business if they fail simultaneously. Murphy's Law guarantees that they will fail.

Third-level and fourth-level engineering in the design of the network is required to ensure that the risk in the network is understood. For example, when one company first built its network, it used a number of railroad rights-of-way. A plain old telephone service (POTS) building was put near the railroad tracks, and the fiber ran along the right-of-way into the POTS building and then back out to the right-of-way. When trains derailed, the railroad sent heavy equipment to put the train back on the tracks. In the process, the ground was churned, cutting both the entrance and the exit cable into the POTS building. That POTS building was therefore isolated from the rest of the network.

Risk management in such a case could have been as simple as ensuring that the entrance and exit cables were at least two car lengths apart, because generally only a couple of cars come off the tracks in derailments.

Another interesting point is that there are certain corridors in the United States in which AT&T, MCI WorldCom, Sprint, Level 3 Communications, Qwest Communications, and others are all on the same right-of-way. This provides an excellent opportunity for Murphy's Law. In several cases, everything that can go wrong has, and every company's cable has been cut at the same time. This situation is especially prevalent in such places as Chicago, where there are limited rights-of-way.

Ring technology offers at least a partial solution to risk management, and as the move continues toward optical networks it is necessary to remember the Chicagos of the world and design the network to survive Murphy's Law. Risk management is fundamental to network development.

Network Management

Network management is the ability to control the network. The granularity of that control must be at least equivalent to the granularity being sold to the customers. If the customer is sold a digital signal (DS)–0 voice circuit, that voice circuit must be visible everywhere in the network. While between two points there may be little that can be done to DS–0, it is essential to be able to isolate it, look at it, sample it at different points, and ensure that degradation is measurable in different parts of the network.

Another important thing to remember is that while it might seem that a laser should either work or not work, sometimes it will not work only on, say, every millionth pulse. Though these issues arise only infrequently, they are exactly the kinds of problems that create havoc in networks.

Summary

Should the network be all-optical? Should it be optical-to-electrical-to-optical (O–E–O)? The economics favor AONs, and they are clearly the future. Networks will certainly be all-optical when a device that can look at an optical stream and provide information on the content of that stream is created. It may take longer than some of the futurists predict, but AONs are inevitable.

Small Operator Takes Network to Marketplace in a Big Way

Eamonn McCusker
Network Design Authority
iaxis

This paper examines what one European operator did in the marketplace to establish a network that spans parts of Europe. Originally, the aim of the company was simply to build a pan-European network that offered everything (right up to wavelengths) to carriers around Europe. For a long time, the industry was comprised of only the Post Telephone and Telegraph Administrations (PTTs) and the link-up between the PTTs such as British Telecom, Deutsche Telecom, France Telecom, and Telecom Italia. However, an enormous number of pan-European carriers have since emerged.

The European operator, if it followed the building strategies of competitors, was basically going to be one among many. As the European operator was not as big as the rest, it could not compete with the larger companies on their terms. Thus, it built a network that covers about 3,000 kilometers across most of Northwestern Europe. The network covered London, Paris, Brussels, Amsterdam, and three cities in Germany. This was done in six months. In Europe, that was considered quite fast. The situation was really quite simple, however, in that the network had to be up and running in six months or the operator would not exist anymore; other companies with bigger networks and more marketing muscle behind them would simply take over.

Interestingly, the operator found that many of the same issues experienced by others did not apply to them. For example, instead of digging its own fiber, the operator leased fiber from others. A smaller operator simply does not have to confront the same issues faced by many larger companies.

Market Window

To build the network, the operator used the market window effectively to get the network up and running in time. Basically, it showed a reduced time to market. The next step was to build extensions to the network, so the company built a Phase Two network in two sections. One was a German national network, and the other was a network that covered Switzerland and Italy, going into Milan and then the south of France. That added another 4,000 kilometers to the network. Thus, the operator produced the second-biggest network in Europe within a year. The aim this year is not only to consolidate the current network, but also to increase what can be offered to customers so that, at the end of this year, the European operator will operate the largest network in Europe.

Offerings include digital signal (DS)–3, synchronous transfer mode (STM)–1, STM–4, and wavelength services (which include STM–16 and STM–64). Another strategy is to offer co-location at the operator's sites in the European cities. The operator also just began to offer services on a backbone Internet protocol (IP) network that offers everything from E-1 to STM–16. The operator has also taken space in many major European cities in the existing carrier hotels and is building its own in two or three streams. It is co-located in existing points of presence (POPs), is building its own buildings, and is linking up with another company to build various buildings throughout Europe.

Out of Date

Much of the network is outdated, and the operator has chosen to expand its realm to Eastern Europe. In the initial network, the operator examined the many different telecom suppliers in existence at the time. It found that companies were offering 16λ systems a year ago. Had the operator taken the 16λ system a year ago, it would have filled it up. Instead, the company wanted to know the third upgrade that the vendor could offer on the equipment. The operator wanted proof that the upgrade worked and that it would accommodate as much bandwidth on the single fiber pair as possible. That is where the operator started.

Design rules have worked and have been consistent. Everything that the company anticipated in the network has pretty much happened to date. When looking at suppliers, the company went for three different areas in which to increase bandwidth. Initially, it looked at 2.5 Gbps, 50 GHz channels using a conventional (C)-band, and that is where the company is today. The company told its equipment providers and dense wavelength division multiplexing (DWDM) providers that it wants to continue to increase—perhaps to 10 Gbps this year. Meanwhile, the company is staying at 50 GHz, even though there are many 10 Gbps platforms that run at 100 GHz, and it is looking at using the long (L)-band amplifier when it is available.

Staying Small

Because the operator is one of the newer network operators in Europe, it found that it did not really have to follow what many other people did. It discovered that, in remaining small, there were many things that it could not really do. For example, it could not afford to have enormous numbers of network planners and enormous numbers of network designers sit around and plan networks and design networks and make sure that this customer is routed this way. The operator simply could not afford any of that. It had to keep everything as simple as possible by reducing layers in the network. The company is now looking at an optical network. Synchronous digital hierarchy (SDH) services factor into this network. It is provisioned around it, and there are basically few layers. There are probably about two layers in that network, such as optical and then a layer above it over which Internet protocol (IP), asynchronous transfer mode (ATM), and SDH exist.

An SDH network is the goal of the company. In such a network, IP and ATM networks feed into an SDH backbone, and the SDH exists over this optical network only to exploit the fiber pair out there. The operator, to keep from going for a second fiber pair, plans to keep rolling out SDH networks.

Currently, the operator is at the IP network that it is deploying, going directly onto the DWDM layer. The M-40 routers are going directly onto the optical layer at STM–16. However, the problem is that it cannot offer voice over IP (VoIP), real-time IP services, or video over IP. It is fine for World Wide Web (WWW) hosting and for WWW services, which demand best effort, but now customers want the IP network to deliver voice and video services over IP. Thus, the optical layer must be managed. The operator is currently trying optical switches and will deploy optical switches this year.

Optical Switch

Everybody seems to be pushing for the optical switch. What the optical switch will do is probably still open to some interpretation, depending on the company, but at first only straight wavelength switching was considered. The company wanted to take IP routers onto the wavelength division multiplexing (WDM) layer. If it was not within 50 milliseconds, but quite close, then it would be possible to tell customers about optical switching and about real-time services at the IP layer. About a year from now, however, it will hopefully be possible to see real all-optical switches. If they emerge and are effective in the network, they may be deployed for the straight instances of wavelength switching. These optical cross-connects (OXCs) can actually add much more functionality to the network.

In the management and application of services in the network, two types of services could be offered to customers. One was the network itself, and the other was the management of the network. In each case, existing companies tended to have many layers and many different technologies around the network, all of which were never truly integrated. An attempt was made on the network-management side to integrate all of these platforms. It did not offer all of the services, but it provided to customers a single alarm for service failure. Any fault in any of the systems would be fed to the central database, which would filter alarms. If anything happened at the optical layer, customers would not pick up three alarms—one of which dealt with the router, another that dealt with the add/drop multiplexer (ADM), and another that dealt with the WDM terminal. They would receive one alarm explaining what happened to their service. The same thing is done in the network to avoid having different layers and functionality across it and so much user involvement. Plans are eventually to get the network pretty much to manage itself.

Provisioning Circuits

In the past, people have tended to say that they provision circuits on the network. What actually happens is that they take all this information off the network, and any time they want to provision something, they look at what is on the machine to see where capacity exists. It is easier to say that the best source for all that information is actually on the network itself. The best place to discover what is on the network is the network. When attempting to make a connection, customers do not necessarily need to know what bandwidth exists on what links; they need only to know what can be provisioned off the network. When more bandwidth must be provisioned, the network can report out. It is possible that the network may eventually be effectively self-managed. Whatever information must be known can be received from the network quickly. Some network systems are not particularly good at being interrogated and passing information. For example, should it be necessary to discover how many trip cards are currently on the network, the present network management cannot provide that information, even though that is precisely the kind of information needed. Is it possible to recover ports or recover trips? What trips are not in use? Are there protected cards available that have never been used or connected? These are precisely the kind of questions for which experts seek definitive answers.

Next, if it is possible to reach the stage at which so much intelligence is on the network layer, someone who came from network design or product design is not needed to provision a circuit across that network. It is only an STM–1. The design rules are that an STM–1 runs at 155 Mbps. There are 16 of them in an STM–16. With changes, customers can be told that the carrier will pre-provision bandwidth on the network, but this is so simple that they can do it themselves. If they want to provision circuits around the network they can log onto the Web site that is used at the moment for their alarms and database applications.

Complications

Currently, on all the management systems now in existence, this process is still extremely complicated. Carriers seek a port in and a port out. For example, a connection must be established between here and a point in Berlin to a point out. It would be necessary to make a connection in Paris, where it would come out of a trip and go into another trip on the Phase One network. It would then get carried over onto Frankfurt, perhaps, as it would actually have the shortest link. Then it would get carried off to Berlin.

There would be three connections and two trip-to-trip physical connections. It would be nice to get to the stage where a customer could log on, view a simple representation of the network, and choose to interconnect through Berlin. Customers could hit a simple button to allow the provisioning.

Carriers, however, wonder if they can do such a thing in the network, if indeed provisioning is that simple. How else can they develop the software on the network elements (NEs) and on the management side to provide other services to customers? The original network used standard SDH systems, which were also working in North America, as well as synchronous optical network (SONET). The European operator offered a service that nobody else in Europe had ever offered before. It switched on the lines on the STM–16 multiplexers so that they would not allow protection. The network could perform a roam protection at Layer 2 or Layer 3.

Unprotected Services

The company was the first pan-European network operator to offer unprotected services. Customers were saying that they had been in discussion with other carriers who claimed that the operator could not offer unprotected services if they were actually protected—that they were simply sold at a less expensive rate. Well, such services really are unprotected. If the connection breaks, the service is lost, but it is considerably cheaper. In the networks being rolled out, the unprotected networks often perform much better than protected networks. That service is offered to the customer. If taking the fully integrated network, the carrier is looking at simple STM–1 and STM–4 applications out in the field, probably where rings meet at the network. Then there is the application in an optical switch at the optical-to-electrical-to-optical (O–E–O) level with full virtual circuit (VC)–4 or STM–1 grooming. If this can be accomplished, services can be offered to customers.

It is possible to begin to offer services based on customer requirements. If a customer wants an unprotected network, so be it—they should receive that level of service. The connection will be up and running, but if it breaks, it is lost. If, however, the customer wants the full, less-than–50-millisecond protection switching, then that is the top service. There are services in between. If customers want a top-level service, but what is received is a service-level agreement (SLA) of 99.95 percent, then, by reducing time spent provisioning the service, the carrier can run through scenarios and risk analysis on that section of the network. The carrier may then, by researching the customer's particular circuit, offer the customer an even better service—one that, for example, could take the SLA to 99.99995 percent.

Network customer updates are offered now and, hopefully, can be offered off this new platform. For example, a Web site is offered that is essentially a means for any customer who has service in the network to log on and find out what is happening live on their circuits. If it fails at any point, and if it does not meet the performance monitoring targets, the Web site announces this. The entire process is automated.

Customers have always been hyped about the future, but sometimes the future takes time. This way, if the circuit goes down, customers are informed about it. The database is updated every 15 seconds, so they know about anything that happens in the network. The more information that can be provided—the more information that is actually available in the system—the more information the customer picks up. It is not hiding it because it is their service, and they have paid for it. Therefore, should the slightest problem occur with their service, they need to know.

Network Elements

NEs out in the field often work particularly well. They do exactly the job that they are supposed to do. In the management of the systems, however, this process breaks down. It is almost as if the builder puts together the hardware, and while the carrier lets the software do most of the work, the management arrives later. The management should be integrated with the NEs and delivered as one package, but often the management puts as much information on the network as needed, almost interrogating it. It should be a version of the World Wide Web for the network. Simply ask it a question, and it will come back with the answer. For example, how much capacity would exist if the builder wanted to make a connection between London and Milan?

No decision has been made on which way to go on this; that will be done this year. But the suppliers are being questioned about all of these things. What are they going to provide? What services can they provide that can be developed and passed on to customers? A business does not invent products and services in the hope that some day somebody somewhere will buy them. Rather, these are things that customers in Europe are actually asking about on a day-to-day basis. They want to know when they can provision their own circuits. They want to know when they can get more information off the company's Web site. They say they like the alarm information, but that they want more information coming through—such as regular reports on what is happening. The business wants to get the equipment and then get feedback directly from customers on what they want to see. This network does not exist at the moment, but much work is being done on it so that it may, hopefully, be deployed within the year.

Guided-Wave Optical Components for DWDM Transmission Systems Using ADMs

H. Arai, S. Takasugi, S. Sato, K. Maru, and K. Ohira
Researchers, Optoelectronic System Lab
Hitachi

Introduction

The recent explosive growth of the Internet has led to the development of dense wavelength division multiplexing (DWDM) systems that increase the capacity and flexibility of optical network systems. The optical signals in the system are generally in the 1.55 μm wavelength region, where erbium-doped fiber amplifiers (EDFAs) are applicable. The densely arranged multiple wavelength signals in this region are multiplexed and coupled into a single fiber, then demultiplexed at the end terminal (see *Figure 1(a)*).

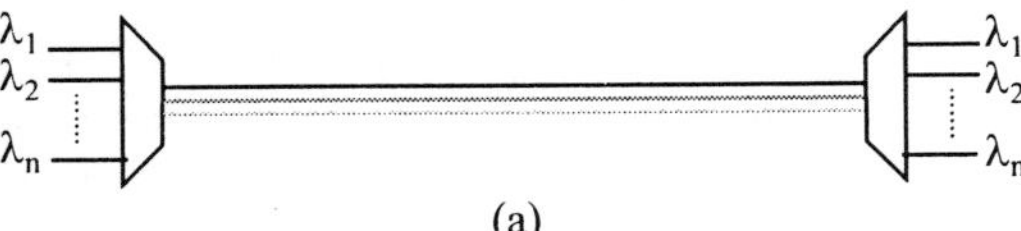

Figure 1(a): *Conventional WDM System*

The next-generation DWDM system will be an add/drop-multiplexer (ADM) system like that in *Figure 1(b)*. Optical signals will be dropped and added in accordance with the traffic demand, thereby making the system more flexible. Key components for this system have been developed: an arrayed waveguide grating (AWG), a fiber Bragg grating (FBG), a planar lightwave circuit (PLC) grating, an optical switch, and a wavelength interleaver. This paper describes the recent advances in these components.

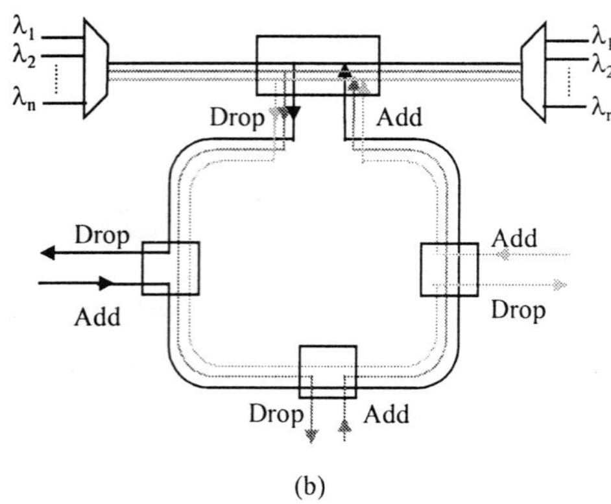

Figure 1(b): *Next-Generation WDM System*

Athermal AWG

AWGs are basic components used for multiplexing and demultiplexing different-wavelength signals in ways similar to how prisms and relief gratings work. They are especially important when the system has many channels or closely spaced channels. AWGs with various numbers of wavelength channels (16–40) and channel spacings (50–200 GHz) have been developed and are being used in DWDM systems. Until now, AWGs have been used in the end terminals of point-to-point DWDM systems. In the next-generation DWDM systems, they will be used in the repeater sites in combination with the switch fabrics to provide the add/drop function (see *Figure 2*). An "athermal" AWG will thus be needed if the optical components are to be user-friendly. Conventional AWGs are used under constant-temperature conditions because their center wavelengths have temperature dependence due to the refractive index change of silica material. *Figure 3(a)* shows the configuration of a proposed AWG consisting of an AWG chip, an input waveguide chip, and fibers. To realize temperature-independent devices, an approach has been introduced in which, to eliminate the temperature dependence, several crescent-shaped trenches are formed in the slab waveguide [1]. Then, material (silicone) with a temperature dependence opposite that of the waveguide material (silica) is inserted into the trenches. The temperature-dependent propagation phase change in the silicone-filled trenches cancels out the phase change in the silica waveguide.

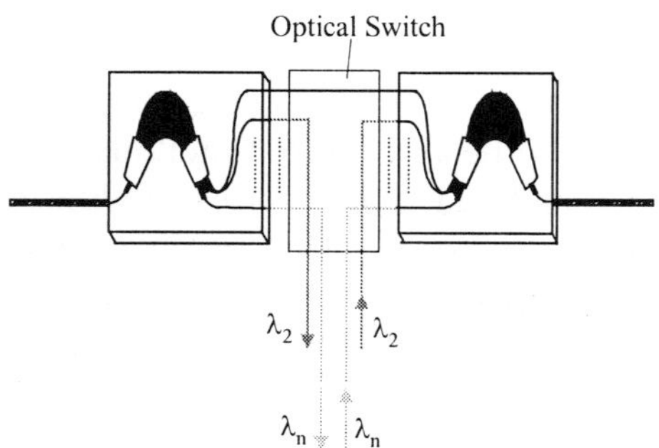

Figure 2: *Optical Add/Drop Multiplexing Using AWGs*

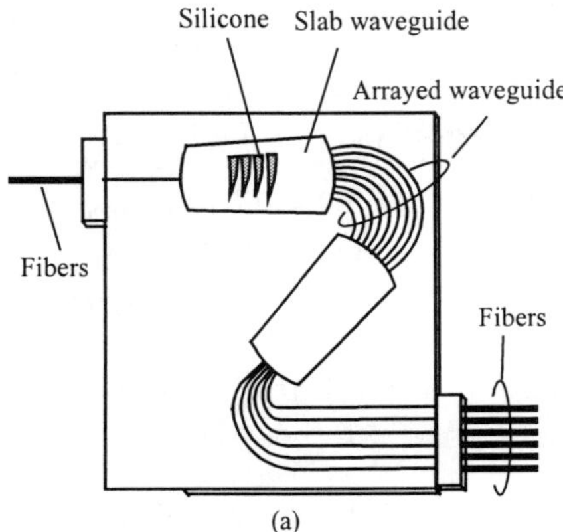

Figure 3(a): *Structure of Athermal AWG Chip*

Figure 3(b) shows the outside appearance of an athermal AWG module. The package is slim because there is no need for a temperature sensor or heater. *Figures 3(c)* and *3(d)* show the spectral response of athermal and conventional AWGs at various temperatures. The temperature dependence of the athermal AWG is –0.0013 nm/°C, an order of magnitude lower than that of the conventional AWG. The athermal AWG thus has characteristics that make it suitable for practical use.

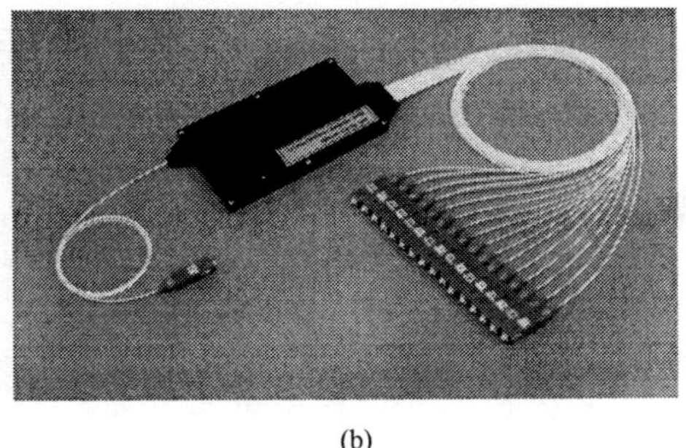

Figure 3(b): *Appearance of Athermal AWG Module*

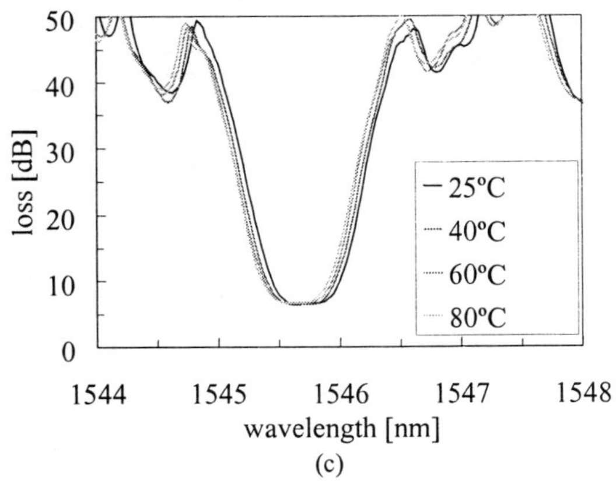

Figure 3(c): *Spectral Response of Athermal AWG*

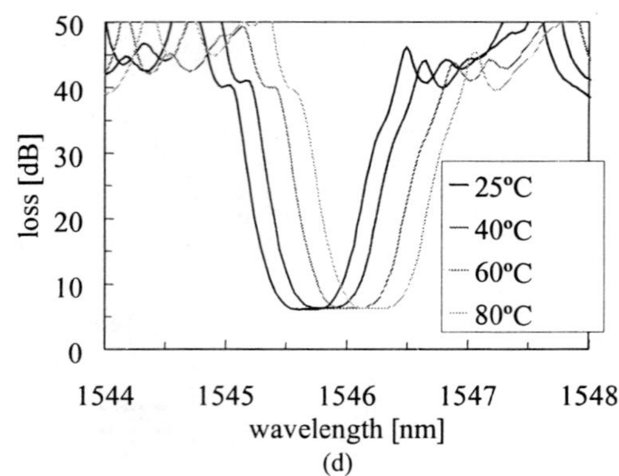

Figure 3(d): *Spectral Response of Conventional AWG*

FBG/PLC Grating

FBG

An FBG is an optical fiber with refractive index modulation formed by utilizing the photosensitivity of germanium (Ge)-doped silica glass [2]. As explained later in this section, it can be used to make an optical ADM (OADM) with two optical circulators. Its use is cost-effective in systems that do not have to add or drop many channels. FBG–type OADMs have been developed for systems with 100 GHz and 50 GHz channel spacings.

The method used to fabricate an FBG is illustrated in *Figure 4(a)*. A KrF excimer laser with a wavelength of 248 nm was used to irradiate the optical fiber through a phase mask, which is a kind of diffraction grating. The interference fringes of ±1 diffraction-order light are formed directly under the phase mask and are written onto the optical fiber.

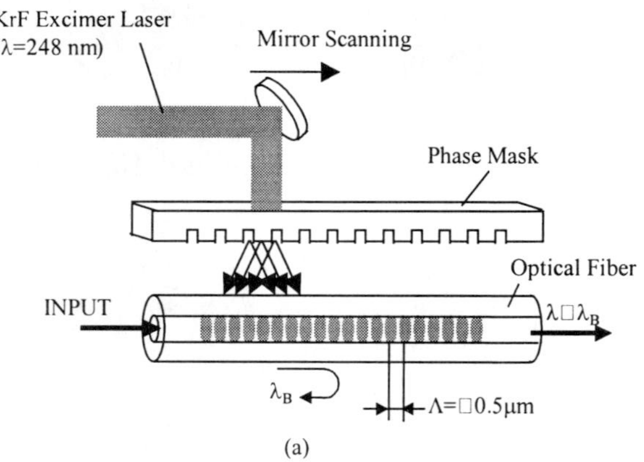

Figure 4(a): *Formation Method of an FBG*

An FBG acts as a wavelength-selective reflection filter. The wavelength λ_B of reflection light is described as follows:

$$\lambda_B = 2N_{eff}\Lambda$$

where N_{eff} is the effective refractive index of the optical fiber, and Λ is the grating period. Other wavelengths are not reflected at the grating; the loss increase at some wavelengths is less than λ_B. To prevent a loss increase, a grating should be imprinted into the cladding by doping Ge because the signal light is not completely confined to the core, but slightly coupled to cladding modes. Using special fiber-

doped Ge and fluorine (see *Figure 4(b)*), the FBG was developed for a system with a channel spacing of 50 GHz for a 50GHz OADM [3]. Its spectral response is shown in *Figure 4(c)*. The crosstalk (XT) was less than –30 dB, and the increased loss at shorter wavelengths was 0.3 dB.

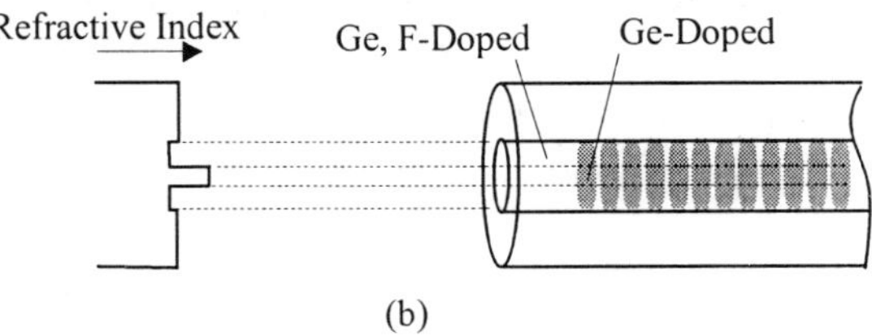

Figure 4(b): *Ge-Doped Cladding Fiber*

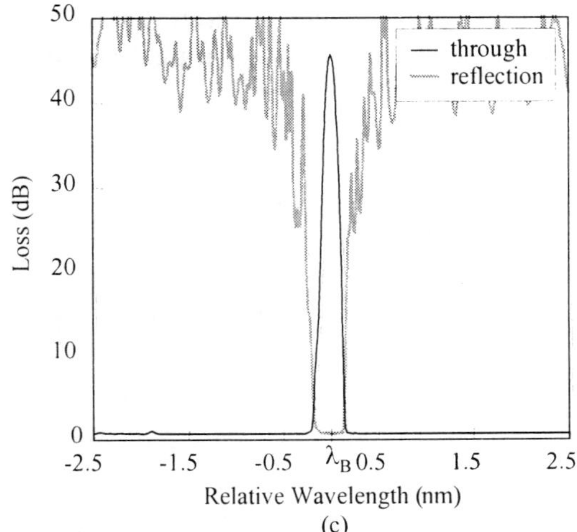

Figure 4(c): *Spectral Response of an FBG for a 50 GHz OADM*

Because an FBG is made of silica glass, its reflected wavelength has a temperature dependence of about 0.01 nm/°C. Therefore, a temperature-compensating package is used (see *Figure 5(a)*). It consists of two metals with different thermal expansion coefficients. This difference in thermal expansion produces strain with a negative temperature dependence. Therefore, the positive temperature dependence of the FBG due to the refractive index is canceled out. *Figure 5(b)* shows the temperature dependence of reflection wavelength λ_B. Because the wavelength shift at the temperatures from –40°C to 80°C is less than ±0.015 nm, this FBG is suitable for practical use.

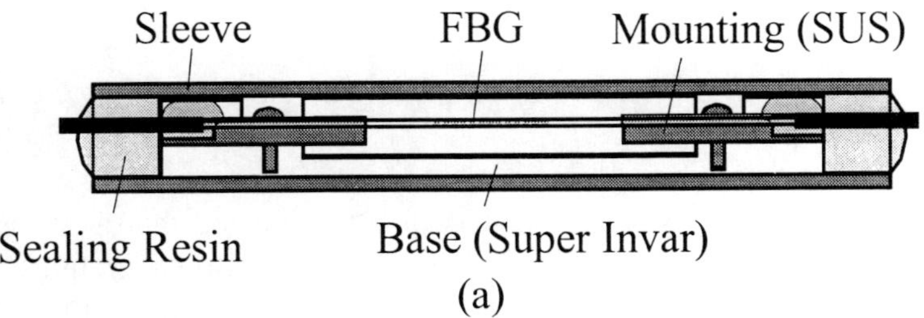

Figure 5(a): *Structure of Temperature-Compensating Package*

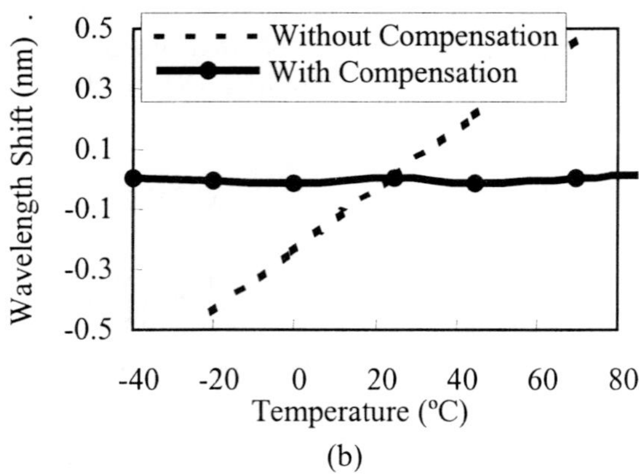

Figure 5(b): Temperature Dependence of Bragg Wavelength

As shown in *Figure 6(a)*, an FBG with this temperature-compensating package and two optical circulators acts as a one-channel OADM. The FBG and circulators are packaged compactly to make the OADM shown in *Figure 6(b)*.

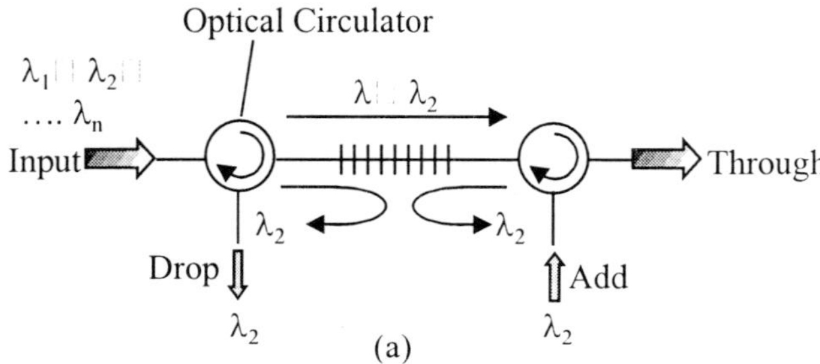

Figure 6(a): Schematic Diagram of OADM Module Using an FBG and Two Circulators

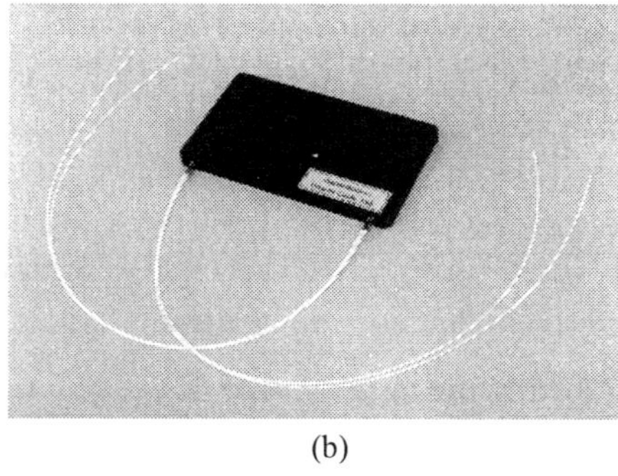

(b)

Figure 6(b): Photograph of OADM Module Using an FBG and Two Circulators

PLC Grating

A Bragg grating can be formed on a Ge-doped, silica-based PLC by using the same method that is used to form an FBG in a silica fiber. A PLC–type OADM that consists of Bragg gratings and a Mach-Zender interferometer (MZI) on a PLC is in development (see *Figure 7(a)*). Bragg gratings are formed on two waveguides between two 3 dB couplers. The couplers act like optical circulators in this configuration, so optical circulators are unnecessary, which should reduce the cost. *Figure 7(b)* shows the spectral response of a PLC–type OADM. The XT was less than –30 dB, and the increased loss at shorter wavelengths was 0.3 dB.

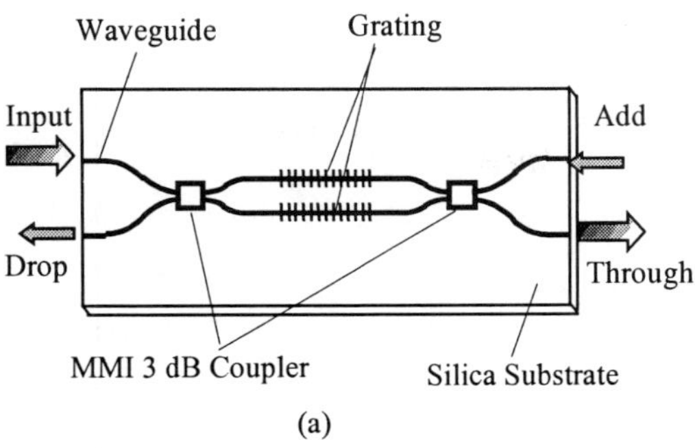

Figure 7(a): PLC Grating for an OADM—Optical Circuit

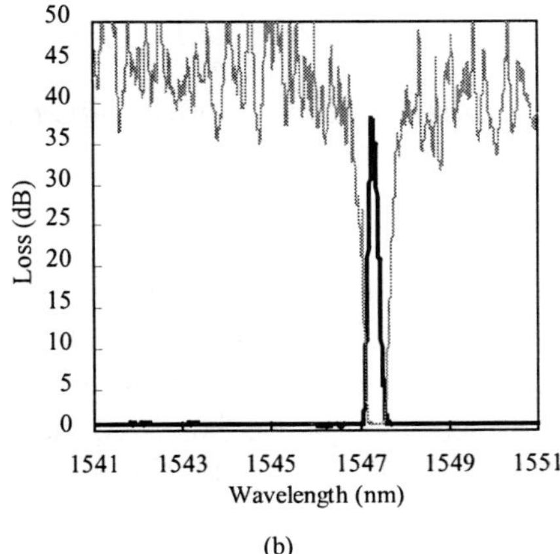

Figure 7(b): PLC Grating for an OADM—Spectral Response

The temperature dependence of a PLC grating, like that of the FBG, must be reduced. Thus, a temperature-compensating structure is being developed using an aluminum plate attached to the bottom of the PLC chip [4]. Because the thermal expansion coefficients of silica glass and aluminum are very different, this structure acts like a bimetal, as shown in *Figure 7(c)*. The temperature dependence is compensated for when the temperature dependence of the refractive index is canceled out by the change in the grating period. *Figure 7(d)* shows the temperature dependence of reflection wavelength for PLC gratings with and without an aluminum plate. Because the wavelength shift at the temperatures from–5°C to 70°C is less than ±0.011 nm, this PLC grating is suitable for practical use.

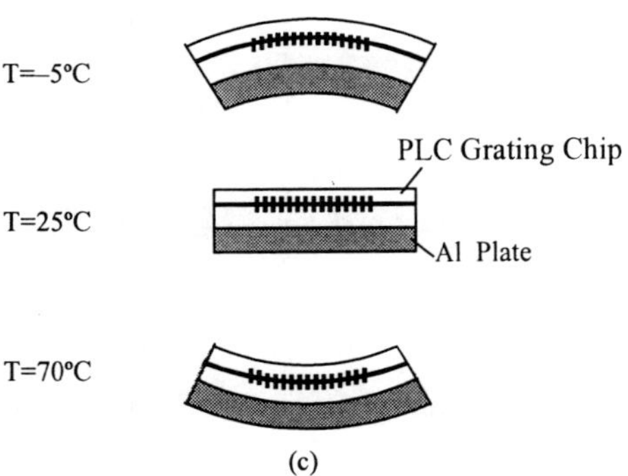

Figure 7(c): PLC Grating for an OADM—Concept of Temperature Compensation

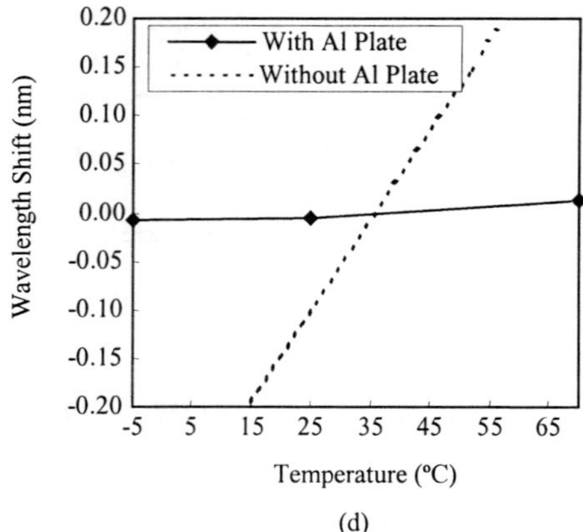

Figure 7(d): *PLC Grating for an OADM—Temperature Dependence of Center Wavelength*

Wavelength Interleaver

A density of channels as well as a bit-rate per channel in DWDM systems has been higher and higher to accommodate exploding traffic demand, which expects overly demanding performance from optical components. To improve such a situation, the interleave filtering technique, which separates the incoming spectrum into two complementary sets of periodic spectra, has been viewed with keen interest. To illustrate, consider the system shown in *Figure 8*. It uses a "50-to-100 GHz interleave filter." Initially, a single AWG, for *n* channels spaced 100 GHz apart, is used. When another is added, the system is upgraded to *2n* channels. Several types of interleave filters have been proposed, but to the authors' knowledge, none have met all the crucial requirements. These requirements include a box-like spectral response and low chromatic dispersion (CD). Among the most promising filter candidates, a Fourier transform–based MZI [5] seems to be best because it predicts the potential of the arbitrary spectral response by using theoretical calculations. The authors propose and demonstrate a new Fourier transform–based interleave filter with a box-like spectral response and low CD.

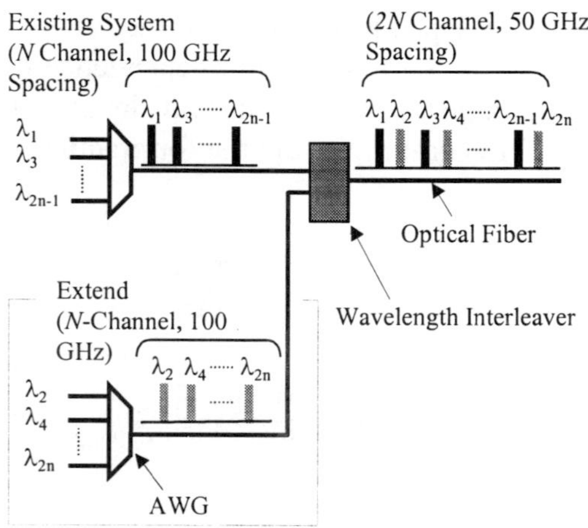

Figure 8: *Increasing the Number of Channels by Using a Wavelength Interleaver*

Figure 9 shows the structure and parameters of the basic Fourier transform–based MZI. It has four different couplers and three different delay lines. *Figure 10(a)* shows the calculated spectral response of the circuit in *Figure 9* for paths A–C, and A–D. *Figure 10(b)* shows the response for paths B–D and B–C.

As is shown, the spectral response of paths A–C and that of paths B–D are exactly the same. The same is true for paths A–D and B–C. *Figure 11* shows the calculated CD distributions for paths A–C and B–D. They show equal but opposite CD behavior. This behavior depends on input port selection (upper or lower) because the ports have a phase conjugate relationship. Consequently, if the circuits were cascaded and the optical signal was launched from a different input port in each circuit, CD compensation would be possible [6].

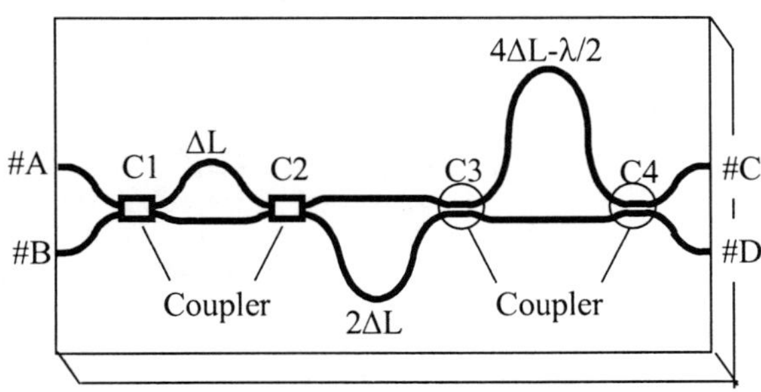

Figure 9: *Three-Stage Fourier Transform–Based MZI Circuit*
(C1: 50 percent, C2: 50 percent, C3: 2 percent, C4: 2 percent)

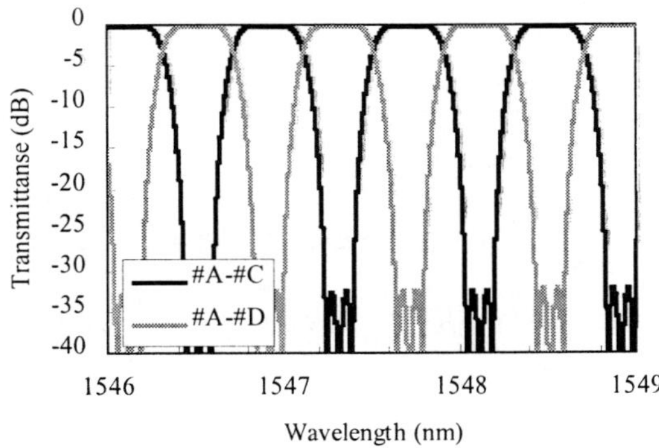

(a) Input from #A

Figure 10(a): *Calculated Spectral Response of Three-Stage Fourier Transform–Based MZI—*
Input from A

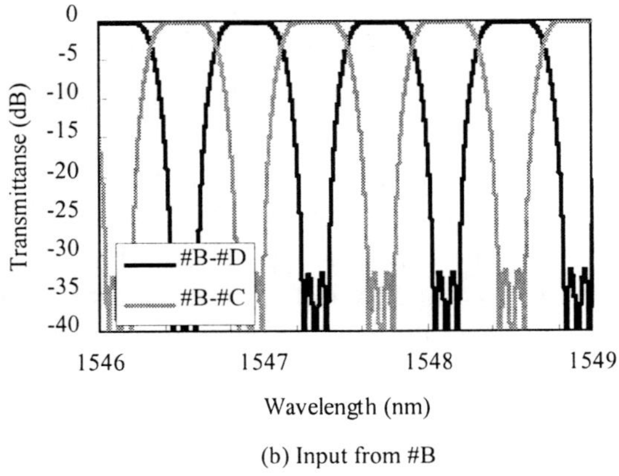

(b) Input from #B

Figure 10(b): *Calculated Spectral Response of Three-Stage Fourier Transform–Based MZI—*
Input from B

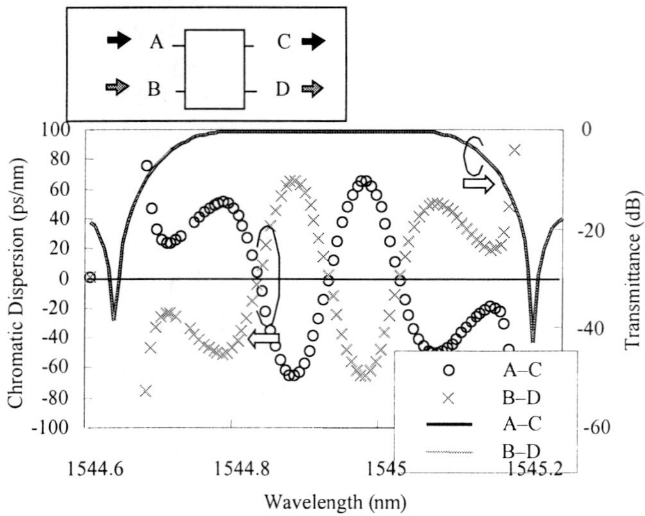

Figure 11: *Calculated CD Distributions*

Figure 12 shows the new PLC interleave filter circuit configuration. The filter consists of three identical circuits. Taking the phase conjugate relationship into consideration, the upper ports of circuit 1 are connected to the lower input port of circuit 2, and the lower port of circuit 1 is connected to the upper input port of circuit 3.

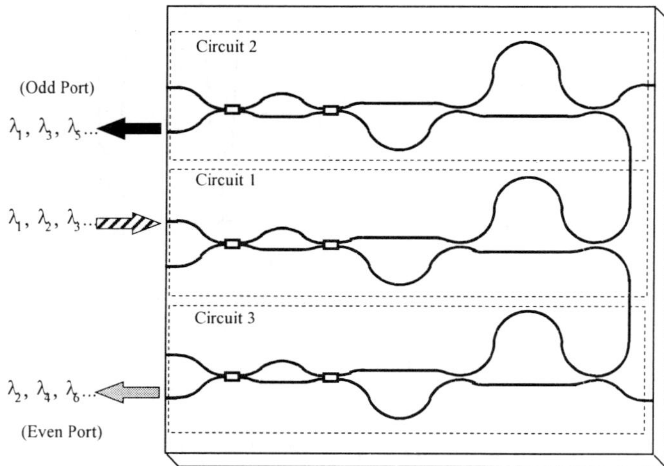

Figure 12: *Fabricated Optical Circuit with Configuration of Dispersion Compensation*

The optical circuit for the 50–100 GHz wavelength interleaver was fabricated (see *Figure 12*). *Figure 13* shows its spectral response. The insertion loss is around 2.2 dB, and the isolation is more than 25 dB. Box-like spectral responses were obtained with no exceptions. By comparing *Figures 11* and *14*, the degree of CD compensation offered by the new optical circuit configuration can be seen. This device performs sufficiently for use as a DWDM interleave filter. *Figure 14* shows the measured CD distribution in the pass band.

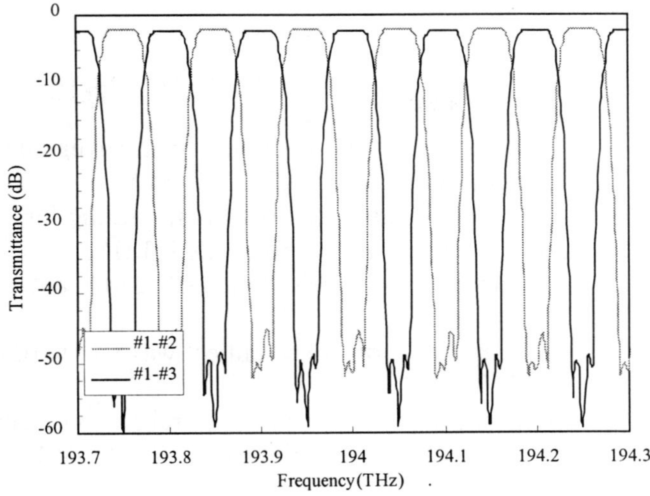

Figure 13: Measured Spectrum Response of Fabricated 50–100 GHz Wavelength Interleaver

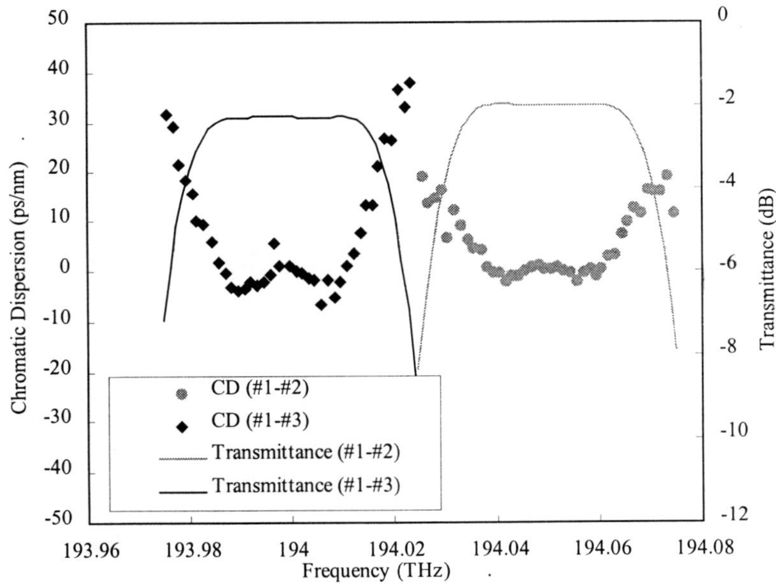

Figure 14: Measured CD Distribution of a 50–100 GHz Wavelength Interleaver

Conclusion

This paper has described the status of the components under development for the next-generation DWDM system. The authors have fabricated AWGs with 16–40 channels spaced 50–200 GHz apart. They have also developed an AWG that does not need a heater for temperature control. This device will be particularly useful in local-area networks (LANs). For 50 GHz and 100 GHz systems, an OADM consisting of two optical circulators and an FBG has been developed, as has a temperature-compensating structure for PLC grating. Because the wavelength shift at the temperatures from –5°C to 70°C is less than ±0.011 nm, this PLC grating is suitable for practical use. Moreover, the authors have developed a wavelength interleaver that consists of cascaded MZIs. The insertion loss is 2.2 dB, and the isolation is more than 25 dB.

References

1. Maru, K., et al. Athermal and center wavelength adjustable arrayed-waveguide grating. *OFC2000, Technical Digest*: 130–132.
2. Hill, K.O., et al. May 1978. Photosensitivity in optical fiber waveguides: Application to reflection filter fabrication. *Appl.Phys.Lett*, 32 no.10: 647–649.
3. Inoue, A. et al., August 1998. Optimization of fiber Bragg grating for dense WDM transmission system. *IEICE Trans.Electron.* E81-C no.8: 1209–1218.
4. Takasugi, T., et al. 1999. Temperature compensation package for optical add/drop filter using PLC. *ECOC'99* 1: 108–109.
5. Li, Yuan P., et al. 1996. Fourier transform-based optical waveguide filters and WDMs. *OFC '96 Technical Digest*: 97–98.
6. Arai, H., et al. 2000. Interleave filter with box-like spectral response and low chromatic dispersion. *NFOEC2000, Technical Proceedings* 2: 444–451.

Dynamic Signaling for Photonic Networks

Luc Ceuppens
Senior Director, Product Management
Calient Networks

Jonathan Lang
Senior Systems Engineer, Network Architecture
Calient Networks

Introduction

Communications networks have grown at tremendous rates in recent years, and nothing indicates that this growth will decline in the near future. In addition, rapid advancements in optical technology hold enormous promise for building high-capacity transmission networks. With wavelength division multiplexing (WDM) and Internet protocol (IP) as the dominant technologies, it is widely expected that the convergence of the IP and optical layers will be the defining theme in the next expansion phase of the Internet. At the optical layer, WDM offers enormous bandwidth and increasingly sophisticated reconfiguration capabilities. At the services layer, IP offers connectivity and service capabilities to end systems.

Unfortunately, these observations are in sharp contrast with the current state of the transmission networks. Transmission networks have traditionally been run very conservatively and have included the manual provisioning of bandwidth. While this worked fine in a world where the bandwidth consumption over time was not particularly high (voice and leased lines were the predominant communications services), it does not work very well in today's environment where the primary user of a carrier's bandwidth is the IP service layer. It becomes appealing to bypass intermediate-layer technologies, such as synchronous optical network (SONET)/synchronous digital hierarchy (SDH) and asynchronous transfer mode (ATM), and to move valuable intermediate-layer functions to either the optical or the IP layer. The emergence of the IP–multiprotocol label switching (MPLS) framework as a common control plane for data and optical layers is an early but definitive trend indicator. Furthermore, optical switching technology is rapidly evolving and will ultimately lead to the long-promised all-optical or photonic network, which naturally succeeds the current point-to-point dense WDM (DWDM) networks.

In today's world, the optimally designed carrier network offers dynamic bandwidth creation and deletion to higher-layer service infrastructures, primarily the IP network. This requirement, combined with the phenomenal pace of innovation and development in the optical transport layer, spurred a dramatic rethinking of how networks are constructed. The outcome will be two-layer networks that concentrate intelligence for resource management in the IP service layer, with optical resource management provided through a control plane shared by the IP layer and the photonic transport layer. The transport layer delivers managed multiple-gigabit bandwidth and provides reliable, wavelength-level, traffic-engineered network interfaces to the IP layer. This network architecture leverages the strengths of both IP and optical technology—utilizing the vast amount of research that has been done in network control and management in IP and the huge bandwidth available through WDM and photonic switching.

Rather than the well-known and familiar homogenous electronic infrastructure made up of routers, ATM switches, and SONET rings, the new networks will have transparent, service-agnostic, mesh-connected optical cores surrounded by routers. This has created a new divide that must be bridged by routing and signaling protocols that permit quick and easy provisioning of end-to-end services. This instigated work in the Internet Engineering Task Force (IETF) and has also caused the creation of the Optical Internetworking Forum (OIF).

While the work done in the IETF and OIF takes varying approaches, it all aims at allowing electrical devices such as routers, ATM switches, and SONET add/drop multiplexers (ADMs) to request dynamically the creation of point-to-point bandwidth between ports on an optical network. However, any dynamic protocol supporting faster provisioning of bandwidth in an IP–over-optical network must do the following:

- **Leverage existing work**
  Because the technical requirements for this protocol are similar to protocols already developed for data networks, this work should be leveraged with optical networking equipment. The MPLS work in the IETF has created a number of different signaling protocols as well as extensions to protocols such as open shortest path first (OSPF) and intermediate system–to–intermediate system (IS–IS). Any work on dynamic signaling between the electrical and optical domains should leverage the existing work.
- **Minimize application-specific extensions**
  Because dynamic signaling between the electrical and optical domains is different from the original application of the MPLS control plane, extensions will have to be made to existing protocols. However, these extensions should be done in the most generic way possible so that basic MPLS, as well as future applications, can leverage them.
- **Support both user network interface (UNI) (i.e., overlay network) and UNI+network-to-network interface (NNI) (i.e., peer network) capabilities**
  A key architectural decision that must be resolved before protocols can be specified is the nature of the interface between the optical core and the service-delivery platforms at the optical edge. Two models have been proposed: overlay networks and peer networks. In the (traditional) overlay model, data networks (e.g., IP, ATM, frame relay [FR]) have no topological visibility to the underlying transmission infrastructure. An example would be an IP router that is connected to another IP router without knowing how the transmission infrastructure is providing connectivity. In contrast, the peer model breaks down this distinction by merging the optical service-delivery layer with that of the optical core (at least from a routing and signaling perspective) by treating them as peers in the routing and signaling protocols. Because of visibility into the core structure, the service-delivery layer can make smarter routing decisions. In other words, the service-delivery devices—not the optical core—control the way the optical core is used.

Both architectural models must be supported. There will clearly be applications where the operator of the optical network will not want its customer to see the internals of the optical transport network. This illustrates the need for the overlay model (UNI signaling). However, when a carrier's IP services division requests bandwidth from its own optical infrastructure, it is desirable for the IP layer to be able to see the internals of the optical network. This visibility into the optical network improves the scalability (because it does not hide the physical topology from the routers as was done in the IP–over–ATM days) and allows the IP routers to use many of the recent traffic engineering (TE) capabilities to use bandwidth efficiently. This indicates a need for the peer model (UNI+NNI signaling).

Furthermore, by defining an NNI, a fundamental step is made to standardize the internals of the optical networks' control plane and to promote vendor interoperability. Finally, as it is difficult to predict exactly how the technology will play out and which model will be dominant, the protocol should be as open as possible and not preclude any mode of operation.

The generalized MPLS (GMPLS) protocol suite, currently being developed within the IETF, possesses all these characteristics and hence enjoys a wide group of supporters from both vendor and service-provider communities. Leading router and optical equipment vendors and service providers are collaborating to combine the control plane of MPLS with the point-and-click provisioning capabilities of the photonic transport network. While originally instigated under the name "multiprotocol lambda switching" (MPLambdaS), this work has recently been renamed GMPLS to illustrate its broadened scope and applicability. GMPLS is a natural evolution of MPLS, which groups the set of extensions to OSPF, IS–IS, resource reservation protocol (RSVP), and label distribution protocol (LDP) to support the routing of paths for TE purposes. Intrinsic to MPLS is the notion of separating the forwarding (data) plane from the control (signaling) plane. MPLambdaS described the concept of leveraging the MPLS control plane to support routing of wavelength paths. GMPLS is the realization of the MPLambdaS concept, created by extended MPLS to support not only devices that perform packet switching, but also the devices that perform switching in the time, wavelength, and space domains. The development of GMPLS requires modifications to current signaling and routing protocols. It has also triggered the development of new protocols such as the link management protocol (LMP). More recently, work was initiated to extend LMP to provide control and management interworking between photonic switches and the DWDM transmission systems (and between service-layer equipment and DWDM gear in the absence of a photonic switch). This interworking interface is currently known as the "optical link interface" (OLI).

This paper briefly presents the TE enhancements to the OSPF Internet routing protocol and IS–IS intradomain routing protocol, two popular routing protocols, to support GMPLS. It also discusses the LMP that can be used to make the underlying links more manageable. The paper then provides a brief introduction to the recently formulated OLI. Finally, it discusses a possible practical roadmap to dynamically signaled photonic networks.

Current Transport Networks

Delivering solutions that enable service providers to carry a large volume of traffic cost-efficiently has proven to be a challenge within the current data network architecture. Today's data networks typically have four layers: IP for carrying applications and services, ATM for TE, SONET/SDH for transport, and DWDM for capacity (see *Figure 1*). This architecture has been slow to scale for very large volumes of traffic, and is, at the same time, fairly cost-ineffective. Multilayer architectures typically suffer from the lowest common denominator effect where any one layer can limit the scalability of the entire network and can add to the cost of the entire network.

Effective transport should optimize the cost of data multiplexing as well as data switching over a wide range of traffic volumes. DWDM is a cost-efficient multiplexing technique that offers significant technical advantages. DWDM increases the bandwidth-carrying capacity of a single optical fiber by effectively creating multiple virtual fibers, each carrying multiple gigabits of traffic per second onto a single fiber. This provides a multiple-fold increase in bandwidth while leveraging the existing fiber infrastructure. Likewise, photonic switches are likely to emerge as the preferred option for switching multiple gigabit or even terabit data streams, as electronic per-packet processing is avoided.

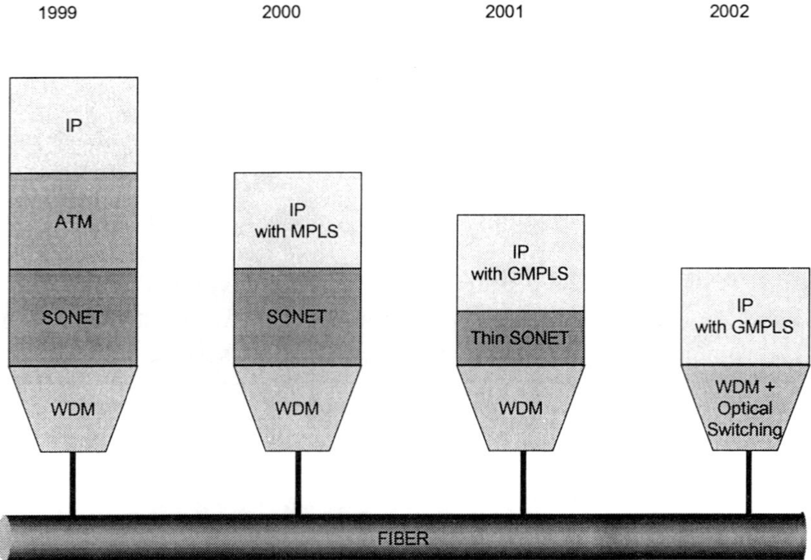

Figure 1: Evolution to Photonic Networking

It is widely expected that the predominant traffic carried over data networks will be IP–based, which suggests that the development of fast router technologies is essential for the aggregation of slower data streams into streams suitable for switching by photonic switches. Likewise, IP packet-based statistical multiplexing is likely to be the predominant multiplexing technology for data streams smaller than those suitable for DWDM. As the capabilities of both routers and photonic switches grow rapidly, the high data rates of optical transport suggest the distinct possibility of bypassing the SONET/SDH and ATM layers. To bypass these layers their necessary functions must move directly to the routers, photonic switches, and DWDM equipment. This will result in a simpler, more cost-efficient network that will transport a wide range of data streams and very large volumes of traffic.

Emerging Two-Layer Networks

In today's competitive business environment, any technical strategy for network evolution must ultimately provide competitive differentiation in service delivery. Scaling the network and delivering bandwidth and services when and where a customer needs them are absolute prerequisites for success.

However, the limitations of the existing network infrastructure are hindering movement to this service-delivery business model. A new network foundation that will easily adapt to support rapid growth, change, and a highly responsive service delivery is required. The answer lies in an intelligent, dynamic photonic transport layer deployed in support of the service layer.

More generally, the photonic network model divides the network into two domains: service and optical transport (see *Figure 2*). The architecture combines the benefits of photonic switching with advances in DWDM technology to create a network architecture that delivers multiple-gigabit bandwidth and provides wavelength-level, traffic-engineered network interfaces to the service platforms. The service platform includes routers, ATM switches, and SONET/SDH ADMs, which are redeployed from the transport to the service layer. The service layer relies entirely upon the photonic transport layer for the delivery of bandwidth where and when it is needed to connect to peers. In this model, bandwidth is provisioned at wavelength granularity rather than at timeslot granularity. To meet exponential growth rates, rapid provisioning is an integral part of the new architecture. While the first implementations of this model will support error detection, fault isolation, and restoration via lightweight SONET, these functions will gradually move to the optical layer.

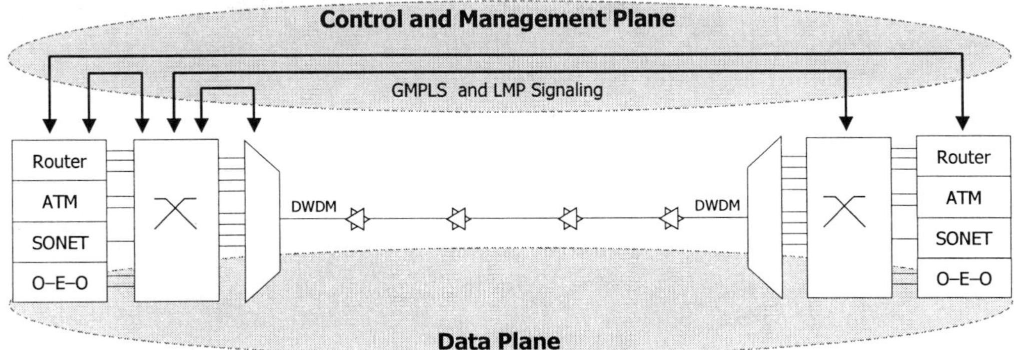

Figure 2: Photonic Network Reference Model

Operational Models

Operationally, this new optical architecture can be viewed from two vantage points, best described as an overlay model and a peer model. The overlay model hides details of the internal network resulting in separate control planes with minimal interaction between the two control planes (see *Figure 3a*). One control plane operates within the core optical network and the other between the core and the surrounding edge devices (called the UNI). The edge devices support light paths that are either dynamically signaled across the core optical network or statically provisioned without seeing inside the core's topology. This is very similar to current combined IP/ATM networks. The overlay model imposes administrative control boundaries between the core and edge by effectively hiding the contents of the core. The disadvantage of an overlay network is that for data forwarding, an $O(N^2)$ mesh of point-to-point connections must be established between the edge devices. Unfortunately, these point-to-point connections are also used by the routing protocols, producing an excessive amount of control message traffic, which in turn limits the number of edge devices that can participate in the network. A single link state advertisement (LSA) flooding event, for example, creates $O(N^3)$ messages on the point-to-point mesh.

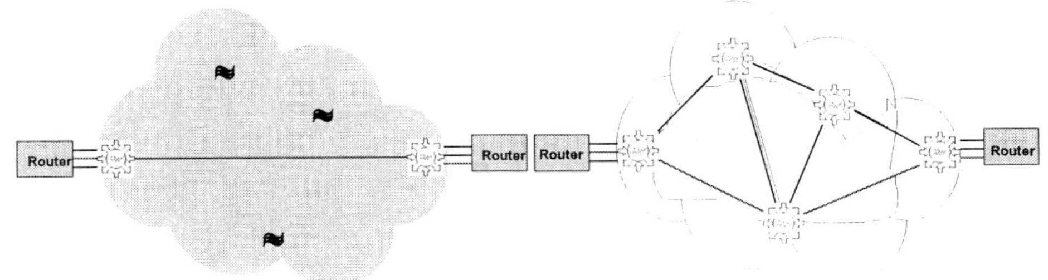

Figure 3a: Overlay Model (UNI)   *Figure 3b: Peer Model (UNI+NNI)*

In the peer model, a single instance of the control plane spans an administrative domain consisting of the core optical network and the surrounding edge devices (see *Figure 3b*). This allows the edge devices to see the topology of the core network. Although an $O(N^2)$ mesh of point-to-point connections is still required if full connectivity between the edge devices is needed, it is used exclusively for the purpose of data forwarding. As far as the routing protocols are concerned, each edge device is adjacent to the photonic switch to which it is attached, rather than to the other edge devices. With $O(N)$ routing adjacencies, a full mesh of $O(N^2)$ forwarding paths can be supported. This allows the routing protocols to scale to a much larger network.

Another approach is a hybrid model that combines both the peer and the overlay models. Some edge devices serve as peers to the core network and share the same instance of a common control plane with the core network. Other edge devices could have their own control plane (or a separate instance of the control plane used by the core network), and interface with the core network through the UNI. This represents the highly desirable solution of offering carriers and service providers substantial flexibility to deploy the most cost-effective model for their needs, be it a peer, an overlay, or some hybrid of these models.

Functionally, the peer model forms a superset of the overlay model. That is, the set of functions required to support the overlay model is a subset of the set of functions required to support the peer model. An overlay model can be derived from a peer model by administratively disabling topology sharing while preserving the connection signaling functions. Conversely, when the overlay model is used, it is not possible to open parts of the optical network without segmenting the network into multiple subnetworks. This observation suggests that rather than having one set of protocols to support the overlay model and a different set of protocols to support the peer model, one suite of control plane protocols with enough flexibility to support both models constitutes the most efficient approach.

Network Protocols in Support of Photonic Networking

Sustaining the advantages of flexible control models are the challenges of rapid provisioning, routing, monitoring, and efficient restoration. Reliability requirements make these challenges paramount to photonic networks. While several vendors are developing proprietary routing and signaling protocols to enable automatic provisioning, such implementations are unlikely to interoperate in multivendor deployments. Providers of "best-of-breed" optical solutions require the integration of photonic switches into a heterogeneous optical network, which will eventually combine next-generation equipment with legacy equipment. A common, standardized control plane must be used to communicate between the various elements. Fortunately, such a standard is well known in the industry.

Over the last few years, IP routing has evolved to include new functionality under the umbrella of MPLS, and recent work has been done on extending MPLS as a control plane that can be used not merely with routers, but also with legacy equipment (e.g., SONET, ADMs) and newer devices such as photonic switches. These efforts offer the necessary standardized common control plane, an essential component in the evolution of open and interoperable optical networks. First, a common control plane simplifies operations and management, which reduces the cost of operations. Second, a common control plane provides a wide range of deployment scenarios ranging from overlay to peer, where the overlay model is realized by using only a subset of the functionality provided by the peer model. A common control plane allows the choice of peer or overlay (or a combination of both) to be driven by business considerations, rather than being constrained by technology. At the same time, building the common control plane from proven signaling and routing avoids reinventing protocols, thereby minimizing risk while reducing time to market.

Some modifications and additions are required for the MPLS routing and signaling protocols to adapt to the peculiarities of photonic switches. These are being standardized by the IETF under the umbrella of GMPLS, which can be summarized as follows:

- A new LMP has been designed to address issues related to link management in optical networks. Recently, interworking with the DWDM systems in the optical transport layer has been added to this work.
- Enhancements to the OSPF/IS–IS routing protocols to advertise availability of optical resources in the network (e.g., generalized representation of various link types, bandwidth on wavelengths, link protection type, fiber identifiers, etc.)

- Enhancements to the RSVP/LDP signaling protocols for TE purposes that allow an LSP to be specified explicitly across the optical core
- Scalability enhancements, such as hierarchical LSP formation, link bundling, and unnumbered links

MPLS

MPLS allows a variable-length stack of labels in front of an IP packet. Label-switched routers (LSRs) forward incoming packets using only the value of the label on the top of the stack. This label, combined with the port on which the packet was received, is used to determine the output port and forwarding label for the packet. Connections established using MPLS are referred to as LSPs. MPLS makes use of extensions to existing routing protocols (OSPF and IS–IS) to exchange link-state topology, resource availability, and policy information required in the computation of paths for LSPs. It also uses extensions to signaling protocols (RSVP and LDP) to specify explicit paths for LSPs through the network and reserve resources.

GMPLS

There are several synergies between LSRs and photonic switches, and between an LSP and an optical channel trail. Analogous to switching labels in an LSR, a photonic switch switches wavelengths from an input to an output port. Establishing an LSP involves configuring each intermediate LSR to map a particular input label and port to an output label and port. Similarly, the process of establishing an optical path involves configuring each intermediate photonic switch to map a particular input lambda and port to an output lambda and port. As in LSRs, photonic switches need routing protocols such as OSPF and IS–IS to exchange link-state topology and other optical resource availability information for path computation. They also need signaling protocols such as RSVP and LDP to automate the path establishment process.

New Protocols and Protocol Extensions in Support of GMPLS

As mentioned above, several modifications and additions are required to the MPLS routing and signaling protocols to adapt to the peculiarities of photonic switches and networks. These modifications are discussed below.

Link Management Protocol

A consequence of generalizing MPLS to encompass non–Public Service Commission (PSC) links is that a label is no longer an abstract identifier, but must now be able to map to timeslots, wavelengths, and physical resources such as the ports of a switch. This requires that the association of these "physical" labels be created between adjacent nodes. For interior gateway protocol (IGP) scaling purposes, multiple links between nodes may be combined into a single bundled link. The LMP runs between adjacent nodes and is used for both link provisioning and fault isolation. A key service provided by LMP is the associations between neighboring nodes for the component interface IDs that may in turn be used as labels for physical resources. These associations do not have to be configured manually, a potentially error-prone process. A significant improvement in manageability accrues because the associations are created by the protocol itself.

Within a bundled link, the component links and the associated control channel do not need be transmitted over the same physical medium. LMP allows decoupling of the control channel from the component links. For example, the control channel could be transmitted along a separate wavelength or fiber, or over a separate Ethernet link between the two nodes. A consequence of allowing the control channel for a link to be physically diverse from the component links is that the health of a control channel of a link does not correlate to the health of the component links, and vice versa. Furthermore, because of the transparent nature of photonic switches, traditional methods can no longer be used to monitor and manage links.

Although LMP assumes that the messages are IP encoded, it does not dictate the actual transport mechanism used for the control channel. However, the control channel must terminate on the same two nodes that the bearer channels span. As such, this protocol can be implemented on any optical switch, regardless of the internal switching fabric. A requirement for LMP is that each link has an associated bidirectional control channel and free bearer channels that must be opaque for verification purposes (i.e., able to be terminated); however, once a bearer channel is allocated, it may become transparent. Note that this requirement is trivial for optical cross-connects (OXCs) with electronic switching planes but is an added restriction for photonic switches.

LMP is designed to provide four basic functions for a node pair: control channel management, link connectivity verification, link property correlation, and fault isolation. Control channel management is used to establish and maintain connectivity between adjacent nodes, and consists of a lightweight, keep-alive, "Hello" protocol that is transmitted over the control channel. The link verification procedure is used to verify the physical connectivity of the component links, which is paramount due to the all too often human error–prone cabling process. The LinkSummary message of LMP provides the correlation function of link properties (e.g., link IDs, protection mechanisms, and priorities) between adjacent nodes. This is done when a link is first brought up and may be repeated any time that a link is up and not in the verification procedure. Finally, LMP provides a mechanism to isolate link and channel failures in both opaque and transparent networks, independent of the data format.

LMP, in its current form, is defined between two GMPLS devices, such as service-layer devices and photonic switches. However, these devices may be separated by transmission systems that do not implement GMPLS. Currently, no standardized protocols exist between a photonic switch and the transmission systems to communicate link and system characteristics and performance-monitoring information. Such information can assist photonic switches and service-layer equipment in managing network resources more efficiently and reacting to faults quickly.

Recently, work was initiated to extend LMP to provide control and management interworking between photonic switches and the DWDM transmission systems (and between service-layer equipment and DWDM gear in the absence of a photonic switch). This interworking interface is currently known as OLI. OLI will enable the exchange of management and performance-monitoring information between the transmission systems, photonic switches, and service-layer equipment. This information can be used by the network elements (NEs) to manage the transport network and its resources more efficiently. Photonic switches and service-layer equipment will also be able to react quickly to faults or degradation in performance if such information is communicated to them by the transmission systems across the OLI. The extension of LMP to incorporate the OLI will unleash the full capabilities of photonic networks by allowing interworking between two critical devices in any transport system.

Three types of information are considered for communication over the OLI:

- Transmission system characteristics
- Performance-monitoring parameters
- Fault localization and troubleshooting information

LMP extensions are being defined to specify the exact format to exchange the above information.

OSPF Modifications
OSPF is a link-state routing protocol designed to run within a single area/autonomous system (AS). Each node in the area describes its own link states by generating LSAs. These LSAs are distributed to all nodes in the network using a process called "reliable flooding." This information is used to create a link-state database, which describes the entire topology of the area/AS. Once a network has converged to steady

state, all nodes will have identical link-state databases. As a result any node in the network can use its link-state database to calculate the best route to any other node in the network.

The TE and GMPLS extensions to OSPF add additional information about links and nodes into the link-state database. This information includes the type of LSPs that can be established across a given link (e.g., packet forwarding, SONET/SDH trails, wavelengths, or fibers) as well as the current unused bandwidth, the maximum size LSP that can be established, and the administrative groups supported. This allows the node computing the explicit route for an LSP to do so more intelligently. The concept of a "derived link" has also been added. For example, if a wavelength LSP is established, that wavelength LSP can then be advertised in the link-state database as a derived link which is capable of supporting SONET/SDH or packet-forwarding LSPs.

RSVP Modifications
In the mid 1990s, RSVP was developed as a response to the increasing demand for Internet applications (such as video on demand [VOD]) that require high levels of quality of service (QoS) from the network. Until recently, RSVP has been ignored for use in long-haul optical networks due to scalability problems associated with the message overhead, the lack of mechanisms to provide traffic management efficiently, and the fact that it is based on unreliable messaging. However, with the convergence of the Internet and telecommunications communities, extensions to RSVP are being developed to support MPLS and TE, as well as to address the scalability, latency, and reliability issues of the soft-state nature of RSVP. The RSVP–TE proposal provides a number of extensions to establish MPLS LSPs. The refresh-reduction extensions proposed address many of the drawbacks associated with the soft-state feature of RSVP for LSPs.

Support for provisioning and restoration of end-to-end optical trails within a photonic network consisting of heterogeneous NEs imposes new requirements of the signaling protocols. Specifically, optical trails will require small setup latency (especially for restoration purposes), support for bidirectional trails, rapid failure detection and notification, and fast intelligent trail restoration. The modifications proposed enhance the extensions of RSVP TE and the refresh-reductions draft to support the following functions:

- Reduction of trail establishment latency by allowing resources to be configured in the downstream direction
- Establishment of bidirectional trails as a single process instead of establishing two unidirectional trails in a two-step process
- Fast failure notification to a node responsible for trail restoration so that restoration techniques can be quickly initiated.

A Practical Roadmap to Dynamic Signaling between Routers and Photonic Networks

There is much interest from service providers and equipment vendors, both optical switch vendors and router vendors, to offer dynamic signaling capabilities for optical networks. There are, however, two challenges that must be overcome to make this a reality. First, two organizations (IETF and OIF) are working on protocol specifications, and although this work is stabilizing, it is not yet sufficiently stable for implementation. Secondly, implementing and testing these new features will take time to ensure that they are of a high enough quality to ship to customers.

Below is presented a practical roadmap to accelerate the development of these tools and provide them to customers as quickly as possible:

- Endorse and support the IETF work to ensure quick completion of the GMPLS protocol suite, supporting UNI+NNI signaling (overlay and peer model support).

- Endorse and support the OIF work to ensure quick completion of a subset of the GMPLS protocol suite specifying UNI signaling (overlay model). Note that there is a possibility that the IETF IP over optical (IPO) working group will co-standardize this UNI–only subset of GMPLS along with the OIF. This collaboration between the IETF and the OIF is a very positive development.

- In parallel with this specifications work in the standards bodies, actually implement the most current version of the specifications. While there may be changes to the specifications, those changes are expected to be marginal and not to require fundamental implementation changes. Early implementations will allow interoperability demonstrations and will be the most practical way to accelerate the completion of the specifications.

Practical implementation of two-layer (service and transport) networking is important to help service providers deal with the exploding traffic loads in their networks effectively. These implementations should focus on the following:

- The rapid and practical development of solutions that customers value
- Using intelligent optical NEs to create efficient mesh networks
- Bringing new best-of-breed products and technologies into the networks that can maintain existing services while establishing new service offerings

Given the number of choices to be made and the difficulty for some vendors to introduce a standards-based control plane quickly, implementations will most likely be phased. This phased approach will allow a gradual education of the service-provider community and its customers regarding the advancements and advantages created by the new "photonic paradigm." To accelerate the creation of best-of-breed, multivendor photonic networks, the following step-by-step approach could be considered:

Phase I

(a) Static configuration of an out-of-band control channel, such as Ethernet, between the router and the optical switch to demonstrate MPLS interoperability between router and a photonic switch. This phase supports the creation of LSPs using RSVP to set up LSPs via the control plane.

(b) Minimalist LMP between the router and the optical switch to demonstrate the operation of LMP. This phase includes the use of a control channel between the optical switches and bearer channel initialization and verification. It will also demonstrate the operation of GMPLS and includes the creation of LSPs using GMPLS.

(c) UNI–only signaling between the router and the optical switch based on the GMPLS–RSVP extensions. Features supported include dynamic routing, preferred paths, constraint-based path computation, and priority-level source reroutes.

Phase II

(a) This phase concentrates on optical path redundancy and introduces 1+1 network redundancy to provide fault detection and network restoration at SONET time scales. Other restoration mechanisms include 1:N path restoration and M:N bearer channel protection. This phase also introduces load-balanced path selection, an enhancement to the constraint-based path computation to improve load balancing among equal-weight paths.

(b) The final phase adds support for UNI+NNI signaling and introduces performance monitoring by extending LMP to include OLI. This interface allows optical switches and routers to communicate with DWDM transport systems and to retrieve health and status information from the optical cloud and make it accessible to edge devices for use in network management and path computation.

Conclusion

GMPLS defines a standard control plane and signaling framework across NEs, resulting in better operational control. It supports multiple connection types (packets, SONET, wavelengths, fibers, bundles, etc.) and allows flexible network implementations (overlay, peer, combinations). In essence, GMPLS builds on existing deployed standards and leverages 50 years of experience in design, signaling, routing, and management of communications networks.

GMPLS will be an integral part in deployment of next-generation data networks. It provides the necessary bridges between the IP and photonic layers to allow interoperable and scalable parallel growth in the IP and photonic dimensions. With GMPLS dynamically bridging the gap between the traditional transport infrastructure and the IP layers, the path is being paved for rapid service deployment and operational efficiencies, as well as increased revenue opportunities. The necessary provisions have been put in place to support a smooth transition from a traditional segregated transport and service overlay model to a more unified peer model. The functionality afforded by GMPLS, its associated generalized notion of an LSP hierarchy, and bundling create sufficient flexibility in support of either the segregation or unification of most any operational paradigm desired by an operator. By streamlining support for multiplexing and switching in a hierarchical fashion and combining the flexible intelligence of MPLS TE, the business value of optical GMPLS will prove to be an essential part of any solution that aims to enable large volumes of traffic cost-efficiently for service providers.

Combining existing control plane techniques with the point-and-click provisioning capabilities of photonic switches, GMPLS will be used to distribute optical transport network topology state information and to set up optical channel trails. The GMPLS control plane will support various TE functions and will enable a variety of protection and restoration capabilities, while simplifying the integration of photonic switches and label-switched routers. Specifically, GMPLS offers the following advantages:

- Faster service deployment with end-to-end provisioning utilizing a single set of semantics
- Elimination of unnecessary network layers by enabling two-layer networking
- Cost savings in network operations by using widely available IP management tools
- Cost savings accrued in training by using a common control plane for optical and service-management layers
- Service creation enabled by common network knowledge of service and transport-layer elements
- Open foundation protocol promotes innovation at the services layer.
- Promotes best-of-breed product selection for service providers
- Accepted protocol enables independent innovation curve within each product class and eliminates proprietary vendor "islands of deployment."

The resulting photonic network offers the opportunity to compress protocol layers, drop intermediary technologies, and eliminate functional overlap. With a dynamic, intelligent, scalable, and transparent photonic network that provides rapid provisioning and fast restoration of multiple-gigabit services, service providers can scale their networks to meet even the steepest increases in bandwidth demand, turning this new optical technology into a revenue spinner rather than merely a way to save money.

Given the number of choices that need to be made and the difficulty for some vendors to introduce a standards-based control plane quickly, implementations will most likely be phased. This paper has discussed a phased approach that will allow a gradual education of the service-provider community and its customers regarding the advancements and advantages created by the new "photonic paradigm."

Acknowledgements

The authors wish to thank John Drake, Ayan Banerjee, and Brad Turner from Calient Networks, and Kireeti Kompella and John Stewart from Juniper Networks, all of whom have contributed to this paper in the form of comments, discussions, and joint work on various projects.

References

Andersson, L., P. Doolan, N. Feldman, A. Fredette, and B. Thomas. August 2000. Label distribution protocol specification. *Draft-ietf-mpls-ldp*, Work in Progress.

Ashwood-Smith, P., A. Banerjee, L. Berger, G. Bernstein, J. Drake, Y. Fan, K. Kompella, J. P. Lang, E. Mannie, B. Rajagopalan, Y. Rekhter, D. Saha, V. Sharma, G. Swallow, and B. Tang. November 2000. Generalized MPLS—signaling functional description. Internet Draft, *draft-ietf-mpls-generalized-signaling-01.txt*, Work in Progress.

Awduche, D., L. Berger, D. Gan, T. Li, V. Srinivasan, and G. Swallow. August 2000. RSVP-TE: Extensions to RSVP for LSP tunnels. *Draft-ietf-mpls-rsvp-lsp-tunnel-07.txt*, Work in Progress.

Awduche,D., J. Malcolm, J. Agogbua, M. O'Dell, and J. McManus. Requirements for traffic engineering over MPLS. RFC 2702, IETF.

Awduche, D., Y. Rekhter, J. Drake, and R. Coltun. July 2000. Multiprotocol lambda switching: Combining MPLS traffic engineering with optical crossconnects. Internet Draft, *draft-awduche-mpls-te-optical-02.txt*, Work in Progress.

Berger, L., D.-H. Gan, G. Swallow, P. Pan, and F. Tommasi. June 2000. RSVP refresh overhead reduction extensions. Internet Draft, *draft-ietf-rsvp-refresh-reduct-05.txt*.

Braden, R., L. Zhang, S. Berson, et al. September 1997. Resource ReserVation Protocol—Version 1 Functional Specification. RFC 2205.

Callon, R. Use of OSI IS–IS for routing in TCP/IP and dual environments. RFC 11 95, IETF.

Kompella, K., and Y. Rekhter. September 2000. LSP hierarchy with MPLS TE. Internet Draft, *draft-ietf-mpls-lsp-hierarchy-01.txt*, Work in Progress.

Kompella, K., and Y. Rekhter. September 2000. Traffic engineering with unnumbered links. Internet Draft, *draft-kompella-mpls-unnum-02.txt*, Work in Progress.

Kompella, K., Y. Rekhter, A. Banerjee, J. Drake, G. Bernstein, D. Fedyk, E. Mannie, D. Saha, and V. Sharma. July 2000. OSPF extensions in support of MPL(ambda)S. Internet Draft, *draft-ompls-ospf-extensions-00.txt*, Work in Progress.

Kompella, K., Y. Rekhter, A. Banerjee, J. Drake, G. Bernstein, D. Fedyk, E. Mannie, D. Saha, and V. Sharma. November 2000. IS–IS extensions in support of generalized MPLS. Internet Draft, *draft-ietf-gmpls-extensions-01.txt*, Work in Progress.

Kompella, K., Y. Rekhter, and L. Berger. November 2000. Link Bundling in MPLS Traffic Engineering. Internet Draft, *draft-kompella-mpls-bundle-04.txt*, Work in Progress.

Lang, J. P., K. Mitra, and J. Drake. March 2000. Extensions to RSVP for Optical Networking. Internet Draft, *draft-lang-mpls-rsvp-oxc-00.txt*.

Lang, J.P., K. Mitra, J. Drake, K. Kompella, Y. Rekhter, D. Saha, L. Berger, D. Basak, H. Sandick, and A. Zinin. December 2000. Link management protocol. Internet Draft, *draft-ietf-mpls-lmp-01.txt*, Work in Progress.

Moy, J. OSPF Version 2. RFC 2328, IETF.

Oran, D. OSI IS–IS intra-domain routing protocol. RFC 11 42, IETF.

Rosen, E., A. Viswanathan, and R. Callon. July 2000. Multiprotocol label switching architecture. *Draft-ietf-mpls-arch-07.txt*, Work in Progress.

EDFA Gain Flattening Filters Using Tapered Erbium-Doped Fiber

Youngbok Choi
Access Network Laboratory
Korea Telecom

Introduction

In recent years, the gain flattening filter (GFF) of erbium-doped fiber amplifier (EDFA) has been a research issue for high-capacity wavelength division multiplexing (WDM) optical-communication systems. When a system requires only one wavelength, the gain variation is not much of a concern. However, with 8, 16, 32, and higher-count wavelength channels, the transmission problem arises because a conventional silica-based EDFA has intrinsic gain of non-uniformity. Usually, the problem can be solved using optical GFF. *Figure 1* shows the effect of GFFs.

In this paper, the author proposes a new type of GFF utilizing twist fused and elongated fiber with silica-based erbium-doped fiber (EDF).

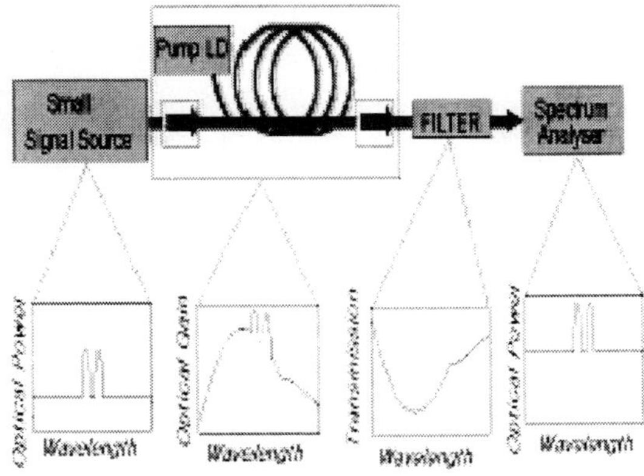

Figure 1: The Effect of GFFs

New Type of GFF Design

Figure 2 shows the out shape of the new GFF, which can be divided into two parts. One part is used by EDF in front, and the other part is used by the tapered single-mode fiber (SMF).

EDF Absorption Properties
Located in front, the EDF plays the role of an amplified spontaneous emission (ASE) rejection filter. The absorption property of EDF is compensated for at the highest gain peak point of EDFA in nearby 1530 nm

"

wavelengths. The absorption wavelengths of EDF, which are major absorption stages of 1530 nm, similarly coincide with the emission wavelengths of EDFA. The EDF length can control the amount of EDFA gain.

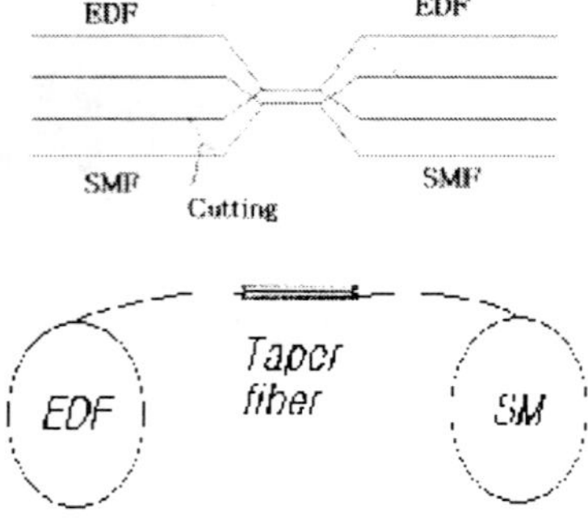

Figure 2: Scheme of the GFF

However, the EDFA emission spectrum does not coincide perfectly with the EDF absorption spectrum due to unwanted photon emission, or the energy gaps between the ground state and the higher states. Thus, only EDF does not use GFF. In *Figure 3*, line D represents ideal GFF, while line A represents EDF absorption loss.

Tapered Fiber Properties
A tapered SMF is obtained by local heating and stretching of the fiber. Tapered sections are produced on a specially adapted fused fiber coupler jig. The tapered sections are pulled by locally heating a length of fiber using a micro-torch with a flame size of approximately 900 microns and pulling the fiber apart using two high-precision motors. The taper angle is sufficiently small to ensure negligible loss of power so the tapering process does not affect the fiber's transmission. In the central region of the structure, the waveguide is ensured by the fused cladding, which plays the role of a core and in which air serves as the cladding. This process is repeated sequentially until the desired filter profile is attained.

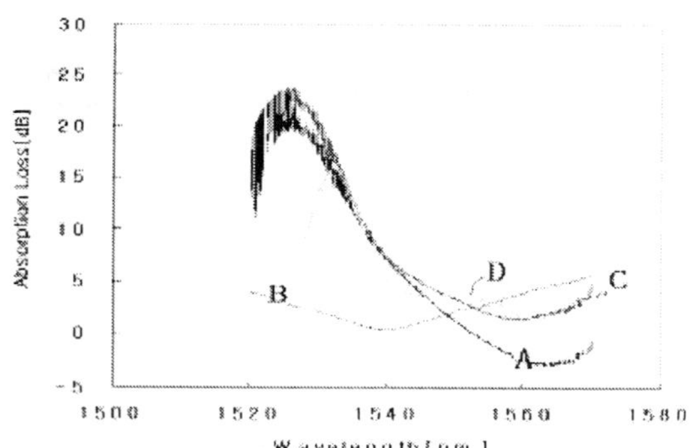

Figure 3: The Absorption Loss Curve
(A): EDF Absorption Curve (B): Tapered SMF Curve
(C): Tapered Silica-Based EDF Curve (D): Ideal GFF Curve

In *Figure 3*, line B shows single-mode tapered fiber's absorption property curves, which it is able to design specially.

Properties of Tapered EDF

Using tapered EDF, a GFF was fabricated. Line C in *Figure 3* shows the transmittance spectra of tapered EDF. Its spectra property can be used as a GFF. The transmitter spectra of the GFF—the C line—preferably coincides with the gain of the emission wavelengths of EDFA, as shown in line D.

Experiments and Test Setup

The testset was set up for measurement of the GFF's properties with a small-signal source. The testset consisted of an EDFA, a small-signal source, and the GFF (see *Figure 1*). The EDFA was aluminum-doped silica–based fiber pumped by 980 nm LD. The pump power launched into the EDFA was 100 mW. The EDF was 20 m long. Optical isolators were also used. The amplified signals were measured using an optical spectrum analyzer. There was a three-channel small-signal input power launched into the EDFA. LD power was –20 dBm, and wavelength differences among LDs were 16 nm.

Results

The resulting optical gain spectrum of the EDFA, including the GFF, is shown in *Figure 4*. In the case of the EDFA (line A) that did not use the GFF, gain excursion was 15 dB, which is the typical output spectrum of an EDFA. In *Figure 4*, lines B and C show the output spectrum of the EDFA that used only EDF and tapered EDF (the author's proposed GFF), respectively. Gain excursion was 8 dB and 1.5 dB, respectively. As shown in *Figure 4*, line C is the output spectrum of the gain-filtered EDFA with the three-channel input power of –20 dBm of LD. The linear gain slope was maintained within the 1530 nm – 1560 nm range.

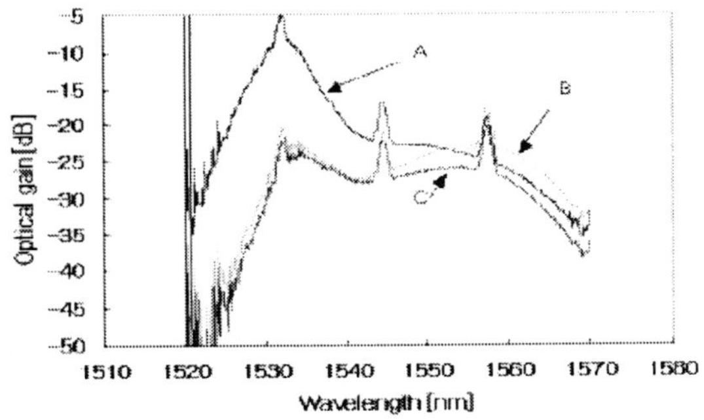

Figure 4: Transmittance Spectra of EDFA Excursion
(A): GFF Not Used, Gain Excursion 15 dB
(B): EDF, Gain Excursion, 8 dB
(C): Tapered EDF Gain Excursion, 1.5 dB

Conclusion

GFFs have been produced using a tapered fiber technique to enable the assembly of low–gain ripple EDFAs for WDM systems. The gain excursion is less than 1.5 dB for three-channel WDM signals over the ~30 nm (1530 nm – 1560 nm) wavelength range. The gain flatness and the total bandwidth can be further improved with a better fitting of the GFF.

References

Su, S. F., R. Olshansky, D. A. Smith, and J. E. Baran. 1993. Flattening of erbium-doped fiber amplifier gain spectrum using an acousto-optic tunable filter. *Electron. Lett.* 29 no.5: 477–478.

Vengsarkar, A. M., P. J. Lemaire, G. Jacobovitz Veselka, V. Bhatia, and J. B. Judkins. 1995. Long-period fiber grating as gain-flattening and laser stabilizing devices. *Pro. IOOC '95*, (Hong Kong) 5: PD1–2.

New Dynamics and Requirements for the Optical Network

Siraj Nour El-Ahmadi
Director, Product Management
Qtera Corporation

This paper will discuss the new dynamics and requirements for an all-optical network (AON). These new dynamics are occasioned by recent changes in the competitive landscape of the marketplace and in new optical technology development. The paper will address the future of the network and the network's efforts to satisfy key requirements that equipment or platforms through such a network must have. The discussion will begin with market dynamics and the evolution of transport networks. It will also cover the optical network requirements and the architectural implications of the changes. What are the technological requirements for architecture kept in the network?

Market Dynamics

In terms of market dynamics, there is an explosion of Internet demand; demand continues to exceed supply. New facility-based service providers emerge constantly. The landscape has changed drastically. In addition, the cost of bandwidth continues to fall.

The new focus for operators is to make the network scalable. Those who build networks once every six or 12 months could do so within one week. The only way to have a scalable network is to have a simple architecture. Nothing complex scales no matter what smart network operating system is used. If it is not simple, it does not scale.

Service deployment is also a hot topic today. Additional capacity must be deployed faster and faster. Operation involves lowering the life cycle of the network and the floor space or consumption. There have also been many changes in the network because of mesh and new services such as wavelengths. The challenge for the new optic infrastructure is to be future-proof. It would be advantageous to design an optic layer that is optimized for existing services, whether they are circuit, Internet protocol (IP), wavelength, or pipe, with flexibility inherent in the architecture.

Because dense wavelength division multiplexing (DWDM) technology and the new competitive landscape decrease costs, costs are a significant differentiator in the marketplace. New carriers cannot gain a bigger market-share simply by dropping the cost. They must also shorten their time to market by reducing their provisioning cycle. Banks and other companies will likely pay as much as carriers demand for an optical carrier (OC)–192c pipe between their key centers, provided they get the bandwidth in a timely manner because their business demands it.

Figure 1 illustrates network evolution. It shows a typical North American carrier as it appeared two to three years ago. There is a key characteristic. It takes many, many dollars to establish market presence.

The distance between most of these cities is 400 to 500 kilometers, and there was much back hauling and aggregation via electrical cross-connects. This was acceptable because carriers had a good 12-month static view of the network. Carriers knew exactly how much they would build the next year.

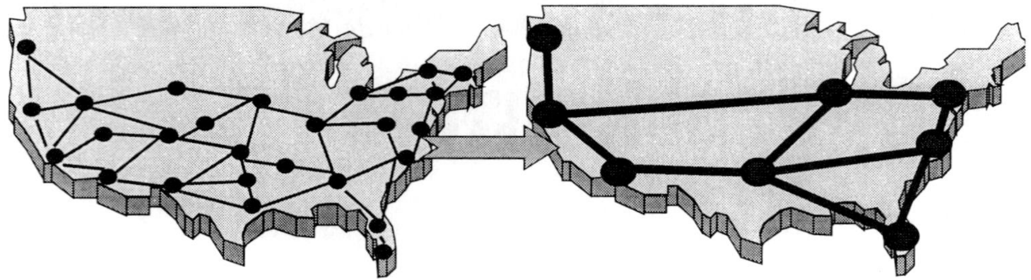

Figure 1: *Transport Network Evolution*

Bandwidth efficiency was paramount to the building of the network. In the absence of wide deployment of DWDM, bandwidth was a premium, so efforts were made to increase each wavelength utilization and efficiency. Carriers wanted their OC–48 or OC–192 wavelengths to be 80 or 90 percent full or utilized.

Today, the Internet drives evolution. New "super" points of presence (POPs) are emerging. These are at the large financial markets in New York, Chicago, Los Angeles, and San Francisco. There is a very long distance between them and abundant traffic. Cross-connects and aggregation are no longer necessary. There may not have been enough traffic to warrant a wavelength directly between these nodes two or three years ago, but that is no longer the case.

Optical-to-electrical (O–E) conversion is still needed at the edge to manage those wavelengths, but it should be totally absent along the wavelength path. Every time O–E has to be used, not only is significant capital cost incurred, but the carrier must also pay the price for the associated floor space, power consumption, and labor. More important, the deployment of O–E equipment along the wavelength path significantly increases the provisioning cycle and costs the carrier time to market.

Network growth is dynamic. There has been an interest in voice in the last 20 years (see *Figure 2*). Most are familiar with the 80–20 percent rule of voice traffic: Eighty percent of the voice traffic stays within

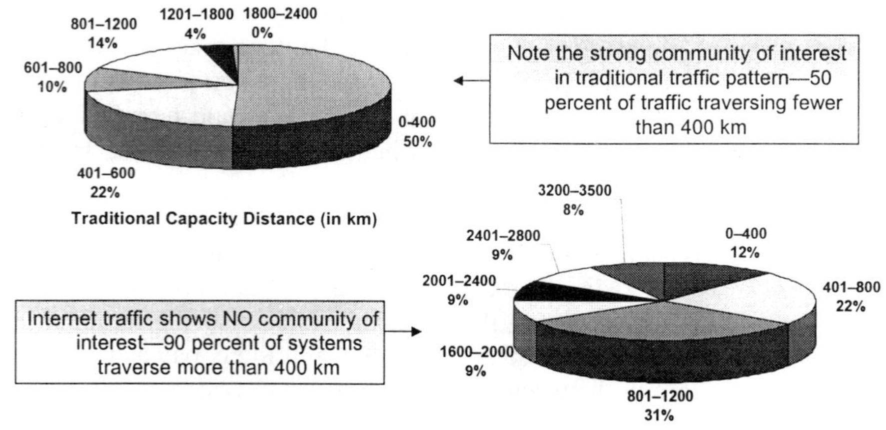

Figure 2: *Traditional versus Internet Traffic Pattern—Testimonial*

originating call community, while only 20 percent leaves and hits the long-haul network. The Internet has changed all that. With one click of a mouse, a user immediately can visit a Web site. The user does not care where that server is located. In a long-distance call, the distance is reflected on the telephone bill; distance is unimportant with today's data traffic.

Ultra-Long-Reach Transport Value Proposition

What is needed for this type of network (see *Figure 3*)? First, ultra-long reach is needed for long distance. The optical to electrical-to-optical (O–E–O) regenerators must be eliminated. No service provider in the mid 1990s had a transmission system capable of going thousands of kilometers because the requirement was not there. The requirement has emerged only within the last year, and it will take off (see *Figure 4*).

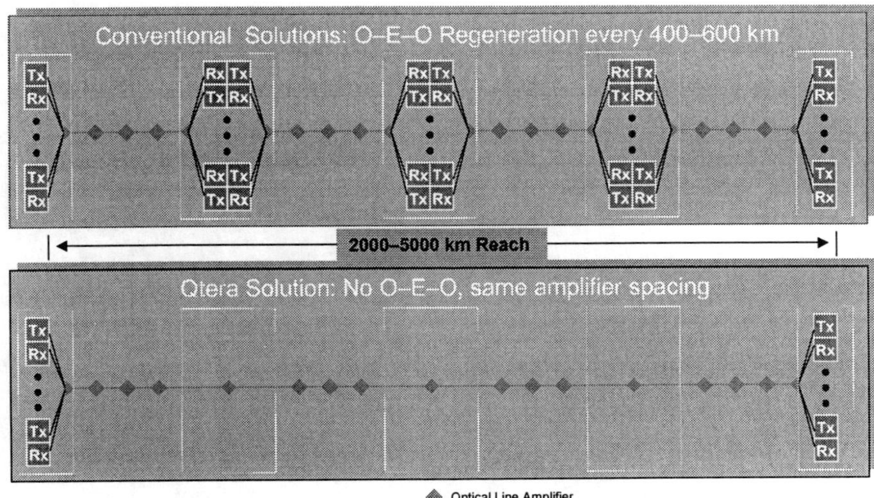

Figure 3: *Ultra-Long-Reach Transport—Value Proposition*

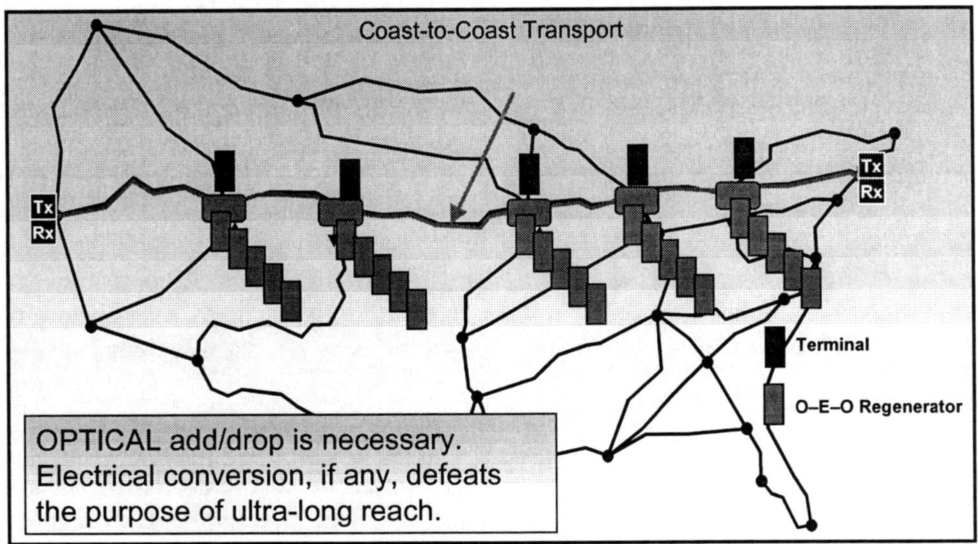

Figure 4: *Ultra-Long-Reach Transport—Requirement One*

The ultra-long reach is not the ultimate goal. One goal is to use the same amplifier infrastructure to service the express network as well as the local traffic and to drop wavelengths where needed. Optical connection management is also necessary. The chart in *Figure 5* explains this concept. To keep with the growth, management must be scalable and translated in the number of wavelengths and, more importantly, in the architecture. Certain architectures are more scalable than others. Given today's competitive landscape, real-time provisioning is also necessary.

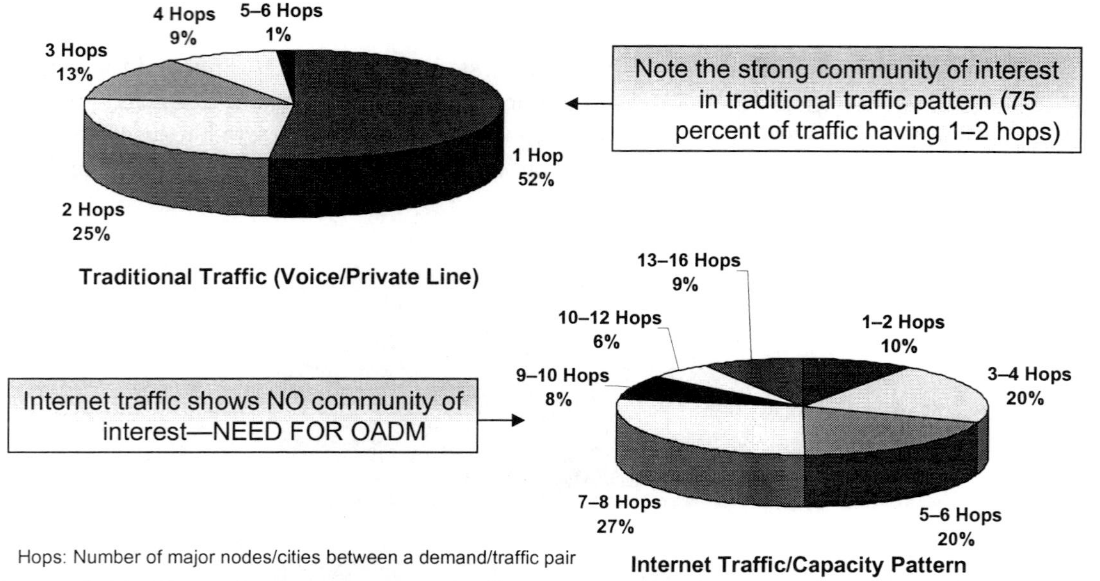

Figure 5: *Traditional versus Internet Traffic Pattern—Purely Photonic Add/Drop Requirement*

Construction of an OC–192 pipe between New York and Los Angeles today can take months. Could it be done in minutes or hours? Future success depends on that capability, not only on who drives the cost. There are ways to drive the cost, but operations require money. In the optical network, there are capital costs and operations costs; these costs are not simple. Architectural decisions cannot be made without consideration of operations costs.

Figure 2 shows that 50 percent of all the optical connections in the network travel a distance of 400 or fewer kilometers; another 22 percent travel only an extra 200 kilometers.

Most traffic stops at about 600 kilometers for two reasons: the community of interest associated with voice and too little bandwidth to warrant direct, express wavelengths between all node pairs. These wavelengths must be stopped and aggregated with other traffic to increase network efficiency. Again, bandwidth was expensive, so the network was built as efficiently as possible.

There is no community of interest with the Internet. Anything could change. A company could change its Web-farm location or add another data center to change everything because of the sheer amount of data traffic involved. The entire network changes. In addition, the distance the wavelengths must travel is in the long range. Thirty-one percent would like to increase to 1200 kilometers and the majority of traffic must now go 3000 kilometers.

This has established the first optical requirement (see *Figure 4*). Currently, there is a conventional synchronous optical network (SONET) system at 10 Gigabits. There are also O–E–O regenerators at 400 or 600 kilometers. Each time a channel or terminal is added, these sites must be revisited and the O–E–O regenerators added again. The sole purpose for the regenerators is signal beautification; the network does not need them. The regenerators keep the signal clean and manage dispersion, power, and timing.

Capital costs, life cycle costs, power requirements, floor space, and most importantly, the service provision of the network all factor into the overall network cost. If a channel is added, it must be added at the terminal, and the technician must drive to every site to install that equipment. This takes a lot of time.

Thus, the paramount requirement is ultra-long reach at multiple Gigabits per second, not only at 2.5 Gbps. Some have suggested 10 Gigabits as the minimum bit rate. If a company grows quickly, it needs numerous wavelengths and the higher bit rate. Forty Gigabits is probably the minimum going forward, and 80 is needed. There must be ultra-long reach without compromising the bit rate.

The solution is one in which transmitters and receivers are added only at the egress and ingress nodes; the middle is ignored. This lowers the cost tremendously and that, in itself, is very important. However, provisioning Gigabit pipes in minutes rather than in weeks or months is even more important to gain the market-share.

In the optical add/drop, if the wavelength knows it is traversing longer distances, it will see more nodes that might also need to be served. *Figure 5* shows that, traditionally, most of these wavelengths travel one hop; then, they stop.

Data traffic causes the number of hops a wavelength traverses along its path to increase tremendously. The majority sees three or four hops, and to keep the ultra-long reach without O–E–O regenerators, optical add/drop multiplexers (OADMs) must be at these nodes. Only then is it possible to glass-thru the intermediate nodes without going to electric conversion and still terminate the "local" traffic at these nodes. A second requirement is to cascade numerous OADMs without electrical regeneration, as shown in *Figure 6.*

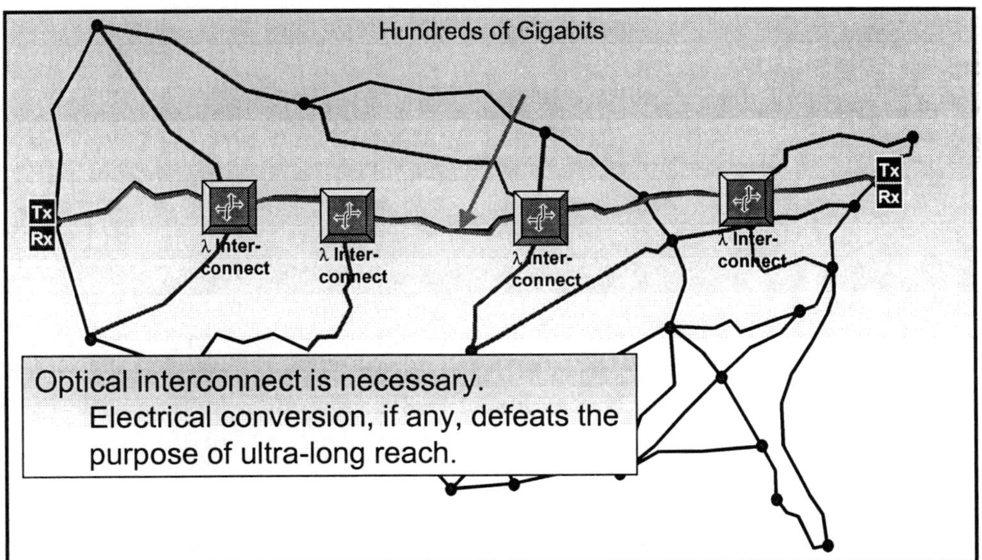

Figure 6: Ultra-Long-Reach Transport—Requirement Two

The third requirement is a corollary effect of the New York and San Francisco nodes talking to each other. Traditionally, an OADM has only two degrees of freedom. A transmission from the east can either travel west or be dropped. A multidimensional OADM is required to allow more than the two or three degrees of freedom that exist today. Again, because growth is unknown, community of interest is lost, and it is difficult to plan for these networks. It is uncertain from where traffic will come and where wavelengths will need to be terminated. Wavelength manipulation must be possible to allow interconnection at key OADM nodes. The application is for provisioning.

Another key requirement for OADM is that it must be possible to optical add/drop in any direction and to continue optically. If that is impossible, it defeats the purpose of the reach that the network needs today. These are some of the optical requirements that are driven by the data world and by the competitive landscape where service providers want to excel.

The result can be called a network paradigm shift (see *Figure 7*). The left side of *Figure 7* depicts traditional networks. The distance is an average of 440 kilometers—1.8 hops that a wavelength sees as it traverses the network—and there is a need for only a few channels. The provisioning cycle takes weeks, if not months. On the right is the future optical network: the data-centric transport network. Wavelengths must travel 1300 kilometers on the average and blast through numerous nodes without having to be terminated. Demand is astronomical, and the provisioning cycle must take only minutes.

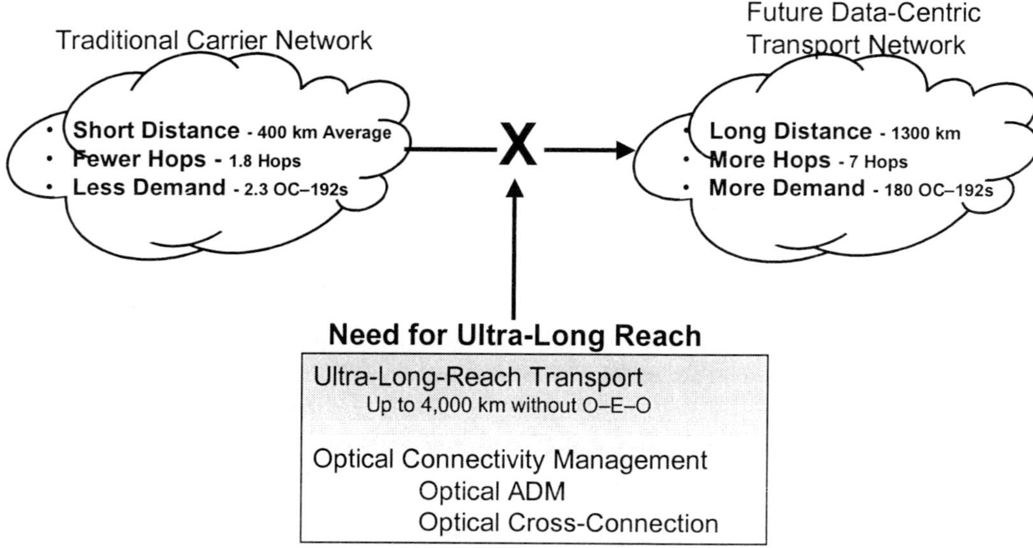

Figure 7: Paradigm Shift—Requirement/Solution

There are no longer only two, three, or four carriers. New carriers enter the marketplace constantly. The key competitive advantage is time to market. Obviously there is discontinuity, and the next-generation networking platform must bend to it. This is where that focus should be. The next-generation networking platform must be able to provision wavelengths for very long distances without regenerators. That allows for abundant service provisioning. However, that alone is not enough. To achieve that reach, optical connectivity management is needed, which involves two parts. One is the pure-optical handling of the channels. Second, a transparent, optical cross-connect (OXC) would be required to manipulate these wavelengths.

There are ways to extend the reach, but only at the expense of a lower bit rate. Today, it is 10 Gbps and moving toward 40 and 80 Gbps rates. Whatever new platform that will support this discontinuity should keep that bit rate and, in fact, should push it to 40 and 80 Gbps. Compromising one for the other is not in the service provider's interest.

Architectural Options

One way to build the AON is to continue what has been done all this time. However, it would be a tragedy simply to adopt the past and not try to leverage the new technology discontinuities.

Figure 8 shows a typical transport network carrier with a breakdown in terms of how much is spent on regenerators, optical terminals, amplifiers, etc. In this example, the regenerators and the amplifier would constitute the line equipment. Terminal or nodal optics is whatever is put at the node, so it could be a SONET OC–192 add/drop multiplexer (ADM) and/or an ultra-broadband cross-connect for bandwidth management.

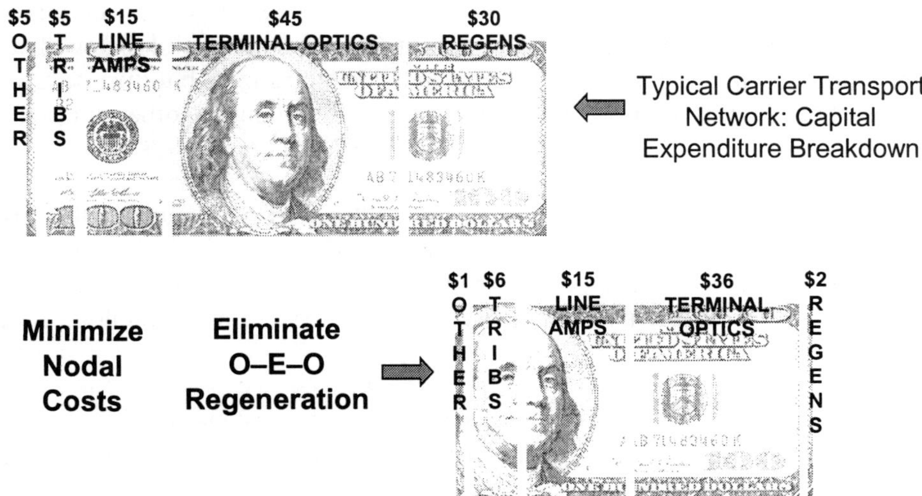

Figure 8: Distance Equals Cheap Bandwidth

Traditionally, line equipment cost constitutes about 70 percent of the total cost. Consequently, mesh networks make sense because they reduce the capacity or wavelength required, and hence decrease line cost at the expense of higher terminal or nodal cost. With the advent of optical amplifiers and DWDM, the cost breakdown became 50 percent–50 percent, and ring networks become more efficient because they reduce the nodal cost (see *Figure 9*).

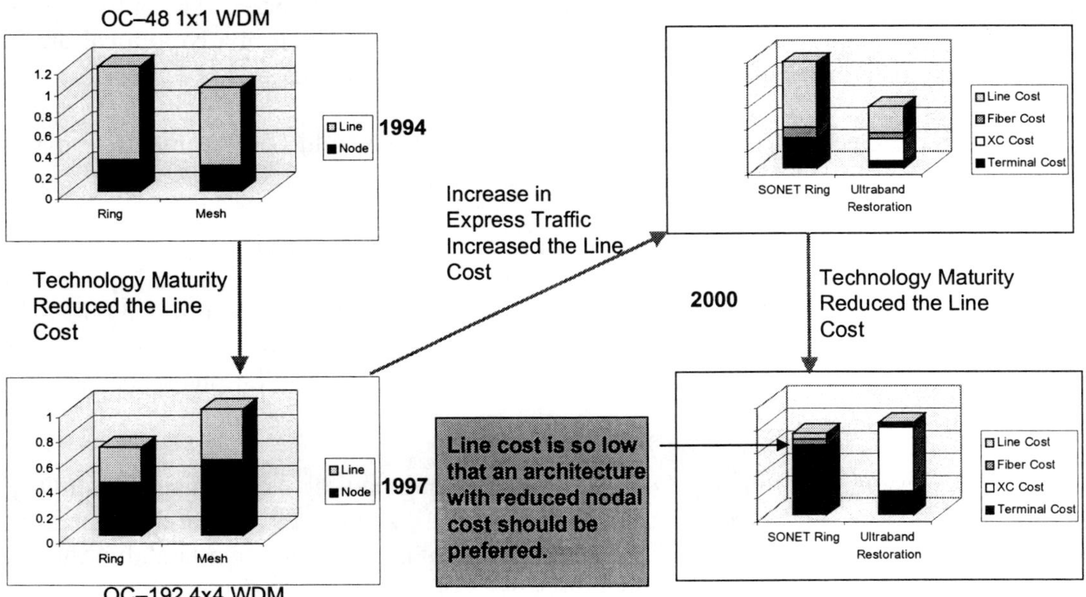

Figure 9: Key Dynamics—Node versus Line Cost

With the advent of ultra-long reach, line optics cost is significantly reduced, and the network cost is dominated by the nodal cost. Any new architecture must consider that. Architecture is chosen and designed either to save money or to allow flexibility. Does it still make sense to go to mesh to try to minimize something that does not have the line share of the pipe? The answer is unknown because the new discontinuities force new architectures and possibly a nonintuitive architecture.

Operations cost must also be remembered. A complex network will cost more to run; this is simply more operational cost. No one can quantify operational cost but the providers certainly know that it consumes a huge piece of their budget. Capital and equipment cost must be minimized, but the operations cost should not be ignored when choosing this architecture. If one is pursued without the other, the right solution may be missed.

In the final analysis, traffic patterns and the connectivity of the network show that there is a better way than a mesh network. However, numerous assumptions are involved. It is important not to ignore what is happening in technology. These issues should be considered when selecting this new architecture; the industry cannot continue to build the network it used to build. Emerging technologies will change the network architecture philosophy.

Reach involves cost savings and service velocity, and being able to plug-and-play at the end—plug in transmitters and receivers, and the process is complete. Every node does not need to be visited to install a regenerator. Again, reach cannot come at the cost of bit rate, which cannot be compromised simply to keep pace with the young dot-coms. Ten Gigabits is actually a minimum. The push should be for 40 Gigabits and beyond.

There are, even for the existing carrier, strong economics for a network designed hierarchically with both a super express network that takes wavelengths between the big data and financial centers in the country and a local network that serves all the tier-two and tier-three cities in between.

The optical platform should service both. Logically, it is a hierarchical network, where expressways of Internet traffic can cruise through while local traffic is appropriately terminated. The only way to do that is through ultra-long reach and wavelength connection management, which involve optical drop as well as being able to continue to manipulate wavelengths at those junction nodes.

Optical connectivity management is key. Transparent OXCs that would be able to scale to where the network needs them are required. Hopefully, the cost will be reasonable.

In summary, there is a new scalable, simple architecture empowered by ultra-long reach that may be as effective as mesh or ring networks. Consider the case for metro and long-haul networks, which involve different architectures. People build unidirectional path-switched rings (UPSRs) in the metro, while they build the more expensive bidirectional line-switched rings (BLSRs) in the long-haul because of the difference in the associated line cost. UPSR was simply not an option in the long-haul because of the cost of the O–E–O regenerators it requires.

Given that ultra-long reach is eliminating O–E–Os from the long-haul network, this discontinuity makes the long-haul ""look" increasingly like the metro for all practical purposes. The architectures that were successfully and optimally applied in the metro may or may not apply in the long-haul, but they certainly must, nevertheless, be considered given the convergence of both.

Amplified System Technology Trends and Their Impact on the Design of High-Capacity Networks

Madeleine Gagné
Optical Network Planning
Nortel Networks

Robert Gojmerac
Manager, Optical Internet Marketing

Marc Verreault
Brand Management, Next-Generation Amplifiers

Dimitrije Radakovic
Optical Network Planning

Introduction

Recently, there have been significant advances in the commercial deployment of terrestrial dense wavelength division multiplexing (DWDM) amplifier and transmitter technologies. These advances have delivered increased flexibility to network planners to design optimized optical-layer transport solutions and can be grouped into five thrusts: extended optical reach, increased spectral efficiency through high λ counts per fiber, increased spectral efficiency through higher line rates, improved optical add/drop functionality, and dynamic optical-layer λ management.

Extended optical reach, upwards of 4,000 km, is provided with new amplifier technology that uses techniques such as dynamic gain flattening, Raman amplifiers, dispersion management, and improved forward error correction (FEC) encoding. To achieve a very high wavelength channel count per fiber—in excess of 160 wavelengths—conventional (C)- and long (L)-band erbium-doped fiber amplifiers (EDFAs) are enabling fiber capacity cross sections beyond 1.6 Tbps for 10 Gbps line rate systems. Current amplifier technology has now positioned the industry at 10 Gbps for backbone DWDM optical systems, and new amplifier and DWDM technologies are emerging that can support 40 Gbps and 80 Gbps line rates, allowing fiber capacity cross sections in excess of 6 Tbps per fiber pair! Finally, new amplifier designs and dispersion compensation strategies are providing a high degree of flexibility for adding and dropping multiple wavelengths at intermediate sites along an optical link.

This paper considers how these advances in amplifier technology are changing the way that networks are designed today and the impact that this new technology has on network planning and on the ability to meet new service demands.

First, different amplifier technologies, including dynamic gain flattening filters (DGFFs), FEC, and Raman pump technology is discussed. This is followed by a discussion on the evolution of service demands, from which the authors have created a model that fits into two distinct data-transport layers that form part of a nationwide network, called backbone and regional.[1] The backbone model will satisfy service demands serving larger communities of interest with typical optical routes varying between 1,000 and 3,000 km. The regional model will cover shorter distances and service demands, typically up to 1,000 km.

Next, the paper will examine the key customer drivers for efficient and cost-effective networks and the impact that new amplifier technology has on the following:

- Maximizing the customer's return on investment
- Life cycle and operational considerations
- Enhanced revenue-generating capabilities

Finally, the interrelationship between these customer drivers—the service-demand models, additional network-planning considerations, and existing network architectures—is discussed. It is observed that the choice of amplified system technology must consider all of these elements to design efficient and cost-effective networks.

Advances in Amplifier Technology

Since the introduction of DWDM systems to terrestrial-based networks in the mid 1990s, an aggressive growth demand has fueled further investment into amplifier technology solutions to provide a more cost-effective and flexible optical layer–backbone capability. As it is well known, the challenges are numerous to reach the vision of a true optical, intelligent, and robust network architecture with rapid bandwidth-provisioning and reconfiguration functionality. This capability should be available within the next year or two. Discussed here are the recent developments in DWDM and amplifier technology that are likely to lead to this optimal vision of the next-generation optical layer–backbone solution.

FEC

Common in radio communication and undersea DWDM systems, FEC has also become a standard feature of terrestrial DWDM systems. Much of the FEC research done for the super-long-haul–undersea DWDM links is now being adapted to terrestrial systems. Using super-encoding algorithms, FEC can decrease raw bit-error rate (BER) by a multitude of factors of 10 to reach the industry-leading standard of 10^{-15} corrected BER. By leveraging this technology, network vendors can provide extended reach and increased channel density at the cost of very little overhead bandwidth. Employing enhanced FEC techniques in optical networks can potentially realize several dBs of gain per span, depending on fiber plant conditions. For this reason, work should continue on these techniques, especially in the area of out-of-band FEC and digital wrapping to enable open architecture compatibility.

---

[1] This model has been limited to the North American demographic space. Studies are also being undertaken for the European market.

Increased Spectral Efficiency

The EDFA spectral window is the spectral region of choice for vendors to design highly cost-effective cascaded optical amplifier systems. The trend will continue to maximize the use of this spectral region through L-band architectures.

Until recently, many DWDM solutions followed the evolutionary path of increased spectral efficiency through tighter channel density. This is achieved by adding more λ channels to a fixed spectral window; moving from 100 GHz channel spacing to 50 GHz and, ultimately, to 25 GHz. However, experience has shown that greater spectral efficiency gains could be achieved by relaxing channel density requirements and increasing line bit rates from 10 Gbps to 40 Gbps. This reasoning stems from the increasing challenge of managing nonlinear distortion on extended-reach DWDM systems with a high wavelength density. By easing channel density and extending the spectral window of operation it becomes much easier to manage distortion and, hence, to increase the bandwidth-distance product figure of merit.

DGFF Technology

This technique is well known in submarine DWDM systems and is now being implemented in terrestrial DWDM systems. The benefits are clear; network vendors can now potentially increase optical links from typical 4–7 span architectures to 10–15 spans, thus greatly increasing optical reach. By dynamically adjusting the per-channel gain characteristics at every amplifier stage, per-channel noise deviations are much better managed and permit each DWDM channel to travel through many more amplifiers than was possible before. Another advantage of DGFF technology is the ability to adjust and optimize DWDM system performance adaptively to enable users to add or remove bandwidth services without going through lengthy re-optimization procedures.

Raman Technology

This technology is gaining momentum in the industry. The principle is simple: Distribute the gain medium though the fiber plant in the opposing direction to the receive wavelengths to keep per-channel optical powers below nonlinear distortion thresholds. Distributed Raman technology also reduces the effective noise accumulation that is common with discrete-gain EDFA amplifiers. The net benefit is extended optical reach because the two primary degradation factors, noise and distortion, are reduced or controlled. The main challenge of Raman distributed amplifier deployment is related to the ability of the outside fiber plant to handle the high optical powers that may be required at amplifier sites. This can be addressed through improved methods and procedures, mainly related to the cleaning procedures and the quality of optical connectors. The authors believe that a hybrid combination of Raman and EDFA technology is the best approach to provide cost-effective and reliable optical-layer solutions for the next few years.

Dispersion Management Technology

Dispersion, both linear and nonlinear, is continually present in today's fiber plant and must be corrected regardless of the fiber plant used. In the last few years, it has become evident that gross correction to wavelength bands is not very effective in high-capacity DWDM systems. Contributing factors that necessitate per-channel management include increased channel counts, band separation, increased time division multiplex (TDM) rates, and fiber types, as well as fiber age. To meet these new challenges, network vendors are promoting per-channel dispersion-management solutions. Though initially unattractive as a solution, given the more complex configuration management, the introduction of smart passive optical components and optical path auto-discovery should simplify this aspect of network deployment.

Techniques under investigation include dispersion slope compensation and dynamic dispersion compensation. Also transmitter data formats are moving toward methods—such as return to zero (RZ) format, which is gaining acceptance over non–RZ (NRZ) formats on today's systems—that minimize channel susceptibility to dispersion. The net result of these techniques is a more controllable and accurate dispersion mapping on a per-channel basis. This is a requirement for high–TDM–rate systems that are being developed to operate at 40 Gbps and 80 Gbps. This technology can also be leveraged toward better managing optical add/drop within point-to-point optical links. Under variable per-channel optical-path distances, as with optical add/drop multiplexers (OADMs), per-channel dispersion management should allow the residual dispersion of each optical path to be corrected more easily and, later, to be managed dynamically.

The key question for network planners is, How much flexibility will be required in the customer's network today and in, say, two to three years? The technology developed today will definitely allow more flexibility in optical-layer growth and management. The problem is that the definition of flexibility is quite broad and involves many degrees of freedom. For example, full OADM functionality at all sites on an optical link may be possible today, but it may place restrictions on which wavelengths can be deployed for express or optical add/drop and thus limit long-term flexibility in network growth unless traffic growth can be well predicted.

Service Demands for Typical Nationwide Networks

Changing traffic requirements combined with different network growth rates have resulted in a layered optical-transport network. These layers have been traditionally designed as a combination of "collector" and "express" SONET rings. Collector rings must normally satisfy the demand for full traffic add-and-drop capability at all major sites on a dedicated ring and benefit from a highly flexible OADM solution with medium optical reach. Express rings are deployed to satisfy longer, coast-to-coast routes, adding and dropping traffic only at major traffic hubs, and benefit from an extended-reach optical solution with limited OADM capability.

In view of this architectural characteristic, the authors have devised a typical national network composed of a backbone and regional layer by considering the service demands of 143 traffic nodes representative of commercial North American data-transport networks. These nodes have been selected primarily based on expected traffic loads, bandwidth management requirements, and critical junctions. Analysis of this network topology has revealed a two-level traffic structure with the following characteristics:

- Within the backbone transport layer, nodes are typically comprised of large metropolitan traffic centers—such as New York or Los Angeles—or crucial fiber junctions—such as the Midwestern cities of Indianapolis and Kansas City. At present, the authors' analysis has identified 34 of these centers.
- Within the regional transport layer, nodes are typically comprised of medium-sized metropolitan centers with a relatively high traffic demand, but in all cases they are inferior to the demands imposed upon the backbone super-nodes. Normally, the regional-node traffic is hubbed toward the backbone super-nodes, particularly for long-distance (interstate) traffic needs or for traffic that is more efficiently transported between a combination of backbone and regional nodes, rather than solely through regional nodes. At present, the authors' analysis has identified 109 of these centers.

Traffic projections for the next two years can provide insight into the expected network loads, the optical reach requirements, and the topology requirements for both layers of the network. The annual growth rate has been based on the population size of nodes that communicate with one another and varies according to node size.

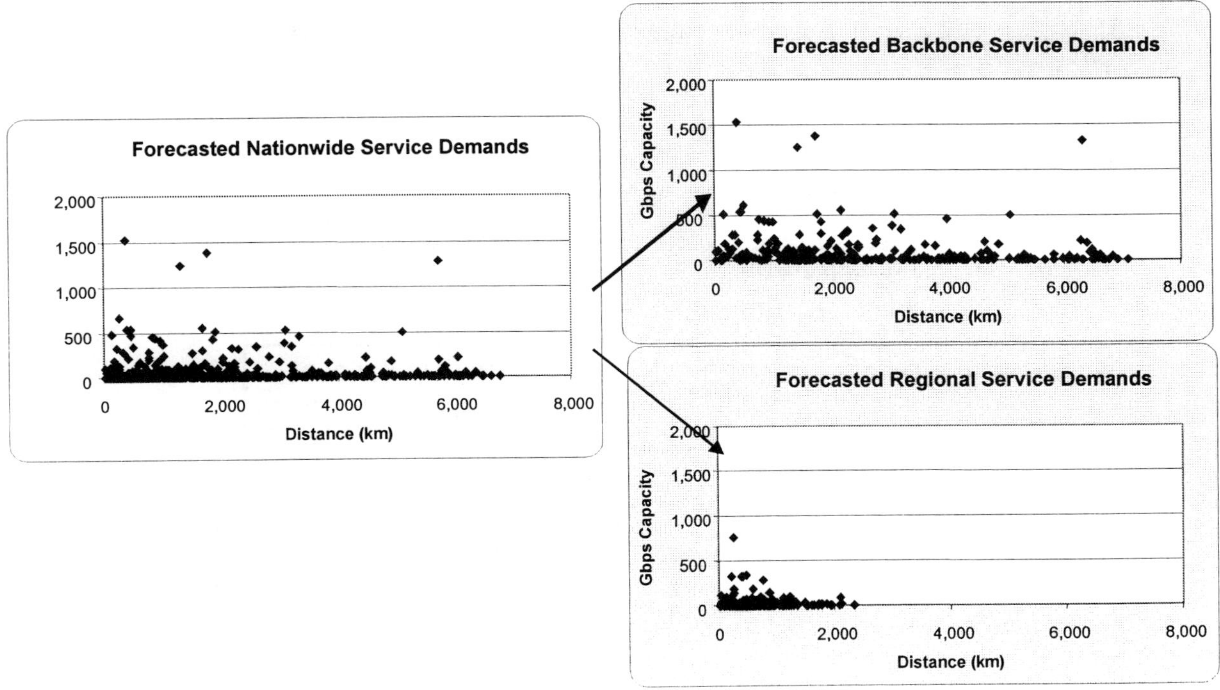

Figure 1: *Forecasted Service Demands of a Representative 143-Node Network*
These capacity forecasts can be differentiated into two layers of service demands based on nodal size and route length. Each layer possesses distinct optical feature requirements thereby influencing amplified system deployment considerations. (Note: The y-axis represents capacity in Gbps.)

As shown in *Figure 1*, the forecasted logical A–Z service demands of this 143-node network vary widely in both capacity and distance. However, with large backbone nodes the capacity and reach requirements are both high and comprise the majority of the total network load (for example, roughly 50 percent of the service-demand routes exceed 1,500 km). The regional network considers smaller, geographically closer nodes and contains shorter service-demand routes with the majority—almost 95 percent—fewer than 1,500 km. In both scenarios, routing logical service demands through fiber plants adds, on average, 1.6 times the air distance, placing further constraints on optical reach requirements.

Customer Drivers

Of the many variables to consider, there are three value propositions that customers use, each to varying degrees, when making a decision as to which amplifier or DWDM technology to deploy on their optical-layer network. These are return on investment (ROI), life cycle and operational effectiveness, and revenue-enhancing optical-layer capabilities. As amplifier technology improves and provides new ways of leveraging customers' investment in the network, these value propositions become more and more linked with technology (see *Figure 2*). Thus, the decision as to which technology to deploy requires more careful analysis. Consequently, the following sections present in more detail each of these customer drivers and how they are linked to amplifier and DWDM technologies.

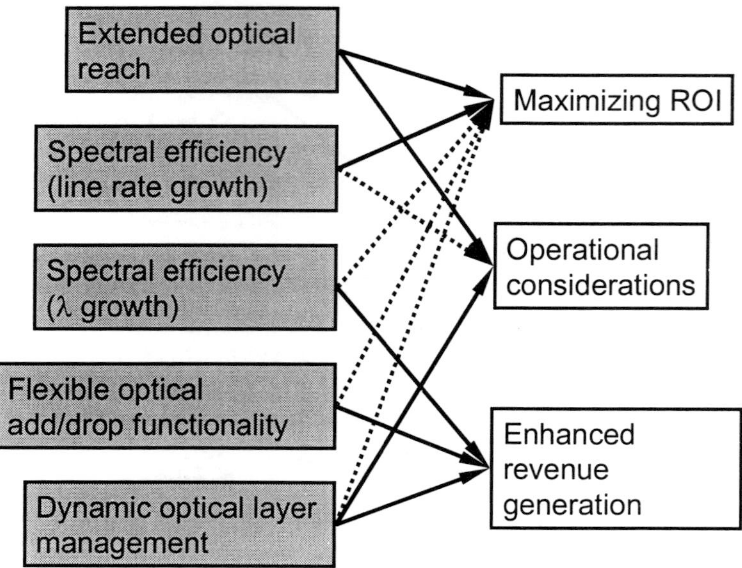

Figure 2: *Relationship between Thrusts and Customer Value Metrics*
The solid line represents a strong dependence; the dotted line a medium dependence. As an example,
flexible optical add/drop functionality and dynamic optical-layer management offer a strong return on
investment and, more importantly, offer excellent potential to enhance revenue generation through
bandwidth leasing and higher service granularity.

Maximizing ROI in the Backbone Layer

Typically, for the backbone layer, the ROI is driven by the lowest cost per bit. This can be correlated to a figure of merit called the "bandwidth-distance product" and is calculated by multiplying the maximum optical reach of a DWDM system with the total projected capacity cross section per fiber. The units are normally presented in Gbps $\times$ km, with typical values varying from 10^5 to 10^6 on commercial systems deployed today.

The bandwidth-distance product is dependent upon three design parameters: the total number of wavelength channels available per fiber, the maximum allowed optical reach enabling regenerator elimination, and the supported line rate of each DWDM channel. As a rule these parameters are interrelated—for example, maximizing wavelength count on a fiber may come at the cost of restricting the maximum allowed optical reach. For this reason the bandwidth-distance product provides an excellent overall measure of optical layer–system performance and allows customers to compare different commercial solutions easily, based on their total technical value. When deployment costs are factored in, the customer must determine which amplifier and DWDM technology actually delivers the best ROI for a required application space. It is at this point that the role of network planning is critical to balance the three design parameters with deployment costs so as to map the right technology to the right application.

Maximizing ROI in the Regional Layer

For the regional layer, though still important, the bandwidth-distance product should not be the sole decision metric when evaluating ROI. When determining the cost-effectiveness of a proposed solution for the regional layer, equally, if not more, important network attributes come into play. These include ease of system deployment and capacity upgrade, higher service granularity and flexibility through feature-rich on-ramp services (mixed synchronous optical network [SONET]/synchronous digital hierarchy [SDH], Ethernet, and asynchronous transfer mode [ATM]), and flexible optical add/drop functionality. Thus, given these new parameters, maximizing ROI in the context of the regional layer is not as obvious to quantify. This also means that these types of attributes can sometimes be neglected when evaluating the selection of an optically amplified DWDM network design. These concepts are addressed further in the following sections on operational effectiveness and revenue-enhancing optical-layer capabilities.

Other Factors to Consider When Maximizing ROI

Limited available power and real estate, the number of available fibers, and fiber type play equally important roles in determining the best fit between technology and ROI. As amplifier and DWDM technology evolves to support faster line rates and longer optical reach, customers are faced with the challenge of building new technologies and solutions over an existing mature technology base. As customers look to leverage their investment, the challenge for amplifier designers is to provide a logical evolution path.

Operational Effectiveness

While new amplifier technology enables customers to increase their speed of provisioning, the increase of system complexity must also be taken into consideration. Today, as the possibility of carrying different vendors' wavelength on the same DWDM infrastructure emerges, a new level of complexity arises in the management of each of these systems. This is predominant at major junction sites, where several systems, all carried onto an existing backbone network, must co-reside. Customers need efficient ways to monitor and manage each vendor's equipment. The emerging trend of bandwidth management performed at some major node sites will migrate toward the optical layer, while other node sites will retain synchronous transport signal (STS)–1 electrical granularity, resulting in streamlined, flexible management schemes. This differentiation between optical and electrical managed bandwidth nodes will depend upon the nature of the overall optical reach, the cost-effectiveness of moving from electrical to optical, and the required nodal flexibility. This translates into lower operational costs and moves the customer toward a more optimized lifec ycle solution.

Also, new long-reach amplifier technology provides the opportunity of regenerator elimination between add/drop sites, thus reducing the need to deploy support equipment and skilled personnel at remote sites for maintenance and repairs. With fewer regeneration points, service provisioning and activation will be greatly accelerated and will move toward greater automation. As the number of regenerator sites decreases, so does the high cost of managing regenerator component inventories for the overall network.

Revenue-Enhancing Optical-Layer Capabilities

Customers are demanding faster, easier service provisioning to increase time to generate revenues. Some of the amplifier technology enabling this includes lambda-leasing services, open optical interfaces for service transparency, and efficient interconnection of routers and switches to the optical layer. Greater service interface diversity, such as Gigabit Ethernet or Internet protocol (IP) optical carrier (OC)–48c's residing on the same terminal bay and managed by the same system, creates broader service portfolios operated on a common optical infrastructure. As the cost of ownership is driven down at the line, smaller businesses begin to appear that can now afford to build their own backbone network mediated by lowered cost/Gbps × km solutions. Alternatively, carriers may now offer OC–12c or OC–48c services to new, previously unserved market segments at competitive price points creating greater revenue opportunities.

How Does Selecting an Amplifier Technology Relate to Network Design?

The process of selecting the right amplifier technology for a customer's network is highly dependent upon the customer value metrics, the characteristics of the service-demand model, and the existing infrastructure of the customer network (see *Figure 3*). Further, a set of network planning considerations must be addressed to confirm the validity of this technology selection. These considerations are addressed in the following section.

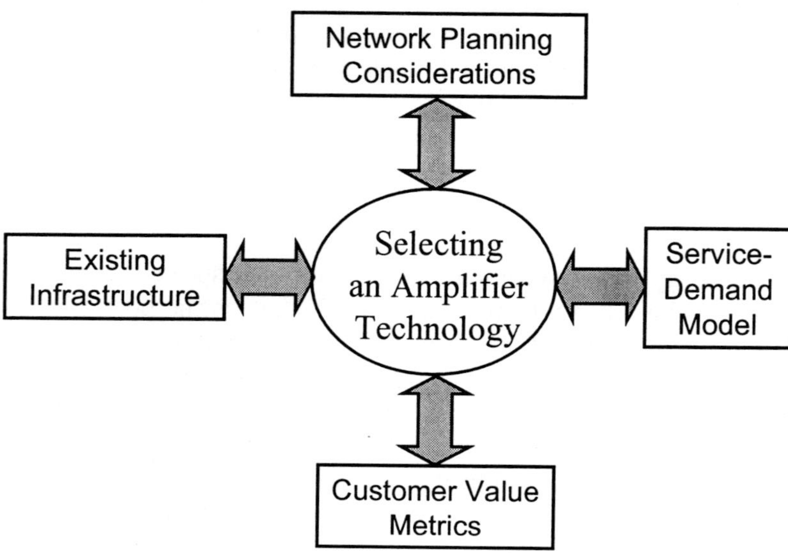

Figure 3: Interrelationship between Factors That Influence Amplified System Technology Selection

Network Planning Considerations

Because there are many factors involved in a network design, the amplifier of choice will depend on, but is not limited to, the following:

- Type of service (ToS) demand, community of interest (COI), and geographical coverage
- OADM applications
- Fiber types/dispersion compensation factors

ToS Demand, COI, and Geographical Coverage

Referring to backbone and regional networks, each model must accommodate the traffic characteristics using different solutions. In the backbone model, reach is a factor that must be considered with differing requirements for medium-reach demand routes and very-long-reach demand routes. In the medium-reach demand routes, amplifiers with larger capacity and shorter-reach capabilities will provide a better fit than would longer-reach optical systems at lower capacity, which are best suited for the very-long-reach demand routes. A spectrum of technologies beginning with typical NRZ–based high-channel C- and L-band solutions for medium-reach demand routes can be contrasted with RZ–

based DGFF solutions, which may or may not employ Raman pumps for very-long-reach demand routes. The differences reside in the degree of technology to be housed in an amplifier bay to serve a particular service-demand set. By increasing the functionality in an amplifier site, costs are also escalated but must be balanced with forecasted demand profiles to ensure that the most optimal customer metric criteria have been satisfied. The backbone model reveals very large bandwidth parameters, making the bandwidth management role much more dependent on amplifier selection. Large bandwidth management nodes will favor higher line rates, greater channel counts, sophisticated monitoring features, and open interfaces, to name only a few. On the other hand, regional solutions will reflect shorter-reach systems with high OADM flexibility and very scaleable capacity designs in the C- and L-bands as bandwidth is added at the local level. In short, network-design models must address the stringent service-demand attributes dictated by both backbone and regional networks resulting in tailored planning for traffic patterns, capacity, and reach.

OADM Applications

Advances in amplifier technology enable multiple-site OADM support for deployment flexibility on DWDM networks. This OADM functionality can also be designed, day one, to accommodate unpredicted traffic growth, where full terminal add/drop would be costly. This technology provides the ability to add or drop wavelengths simply by introducing OADM couplers at the mid-stage of the amplifier site while still allowing express channels to continue through. Initially, OADM technology was limiting the reach of the links, and high-loss penalties incurred restricted the number of wavelengths that could be added or dropped at each amplifier site. With new amplifier designs, using low-loss in-line passive optical components (e.g., band splitters) and multiple mid-stage access architectures, this restriction has been lifted, enabling 50 to 100 percent add/drop capability at OADM sites. Introducing banded OADM alongside per-channel OADM modules offers a wider solution set that can be engineered for differing add/drop loads throughout the optical layer.

Fiber Types and Dispersion Compensation Factors

Fiber types have always been an important factor in the design of the backbone network because of their different characteristics. Now more than ever, the fiber infrastructure will dictate the optical-layer solutions to be transported across the fiber plant with newer low-loss fibers better suited for higher line rates, C- and L-band wavelength support, longer reach, and more tolerant dispersion management. Today, the trend of fiber swapping between service providers is increasing, so the need for careful link analysis is even more important because the types of fibers exchanged vary dramatically. In particular, network designers must consider that ultra-long links may well see city-to-city links requiring the support of more than one fiber type. In the future, special considerations for dispersion compensation will balance the need for traditional passive, coarse compensators in favor of active, per-channel compensation devices. Finally, the trade-off between NRZ and RZ transmission formats will also enter the dispersion equation because each offers differing levels of pulse broadening. In brief, the need for more customized network design with service providers will ensure that fiber and dispersion issues have been properly addressed by amplifier and transmitter selection.

Conclusion

This paper has demonstrated that advances in amplifier technology—such as new transmission formats, DGFF, FEC, and Raman amplifiers—are introducing new challenges into the network-planning role. Evolution into higher line rates, moving from 10 Gbps to 40 Gbps and 80 Gbps systems, are impacting the planning of major hub sites. The enabling of higher capacity per fiber—well over a hundred wavelengths per fiber—demands careful evaluation of how these wavelengths will be managed in the network.

To design a cost-effective and efficient network, there is not one solution, but rather a mix of solutions arrived at through a closer, more collaborative relationship between planning teams and service providers. Networks are defined by a mix of characteristics with efficient network design best achieved by making the right choice of amplifier technology, thus maximizing customer value metrics, satisfying forecasted service demands, and leveraging the existing infrastructure.

Deep Fiber Architectures for Multiple-Service Revenue

John Gibbs
Director, Access Systems Product Marketing
Marconi Communications

Every consumer wants high-speed data, dependable voice service, and high-quality video services. Whether these are delivered by digital subscriber line (DSL), cable modems (CMs), or wireless architectures does not matter to the consumer as long as service is fast and dependable. This paper will discuss market drivers for high-speed data, end-user issues, services, and available infrastructures to meet these needs. It will include a high-level discussion of fiber-to-the-curb (FTTC) and fiber-to-the-home (FTTH) architectures and a glimpse of the pros and cons of deploying deep fiber solutions.

Market Drivers

The Internet explosion drives the access market for FTTC and FTTH. Everyone has seen the hockey stick curves illustrating the dramatic growth in Internet use. Data traffic is surpassing voice traffic in volume. There are currently numerous ways to get data out to the customers, including CMs and DSL.

Another market driver is increased competition. The number of competitive local-exchange carriers (CLECs) has grown rapidly in recent years, and this trend will continue as regulations and barriers to entry are increasingly relaxed. Another key factor is the increase in traditional cable television (CATV) providers entering the voice market and traditional telephony providers entering the video market.

There are, however, numerous complexities involved in rolling out these services. It is not easy to get a DSL line out to every customer. Load coils on the line, bridge taps, and truck rolls all make for difficulties. Hybrid fiber/coax (HFC) also offers no easy path to multiple-service revenue due to the complexities of powering and noise ingress when increasingly more data and telephony users are added to the system.

A final, quite significant issue concerns technology life cycles. What type of architecture is being put into the ground today? DSL may be great today, but will the network be able to handle video in the future? How will these problems be solved? Operators must put a network into the ground that is future-proof. That architecture is fiber optics—FTTC today, and FTTH in the near future.

One Network, All Services

"All services" means video, lifeline telephony, and high-speed data. To the end users, it does not matter if these services are delivered by CMs, DSL, or satellite feed. All that matters to consumers is that their services function rapidly and reliably. This applies to all broadband applications, particularly high-speed data and video. In the near future, the reality of video telephony will begin to take hold among consumers. This will be a major driver in getting nontechnical people to demand bandwidth into their homes. If a picture is worth a thousand words, a video is worth millions to the grandmother who lives across the country.

It is likely that we will see homes where two or three TVs show separate videos on demand while another person has a videoconference from the home office and someone else plays an interactive Internet game. This could all be done while the local police station monitors the outside of the home with a video monitoring system.

Two Basic Architectures

There are primarily two basic wireline architectures installed today. The traditional telephone architecture is mostly twisted pair (TP) copper cable, which was engineered to deliver voice. DSL may be used to enable high-speed data and very limited video capability.

The main upgrade option with TP copper is to add high–bit rate DSL (HDSL), asymmetric DSL (ADSL), and very-high–data rate DSL (VDSL) to get as much as possible out of that architecture. While some increases in bandwidth will be realized, the ability to deliver traditional broadcast video over these networks has not been proven. Through the use of VDSL, some providers have used switched digital video (SDV) to provide broadcast-type services, but the truth is that video has been delivered over TP copper in only very limited deployments, very small applications, and very select areas. Only one video channel is 3–4 Mbps. Simple math suggests the difficulties involved in delivering that kind of bandwidth over TP copper to the end user who has multiple TVs and personal computers (PCs) operating simultaneously. Video on demand (VOD) and videophones are the future. Operators should ensure that their networks are capable of handing large numbers of subscribers to these services on their networks.

The second architecture is HFC, which was originally engineered to deliver broadcast video services. Through the use of a return path, and data over cable service interface specifications (DOCSIS) modems, these architectures are now used by the traditional CATV companies to deploy video services, CM data, and to a limited extent, telephony.

HFC does not present the same bandwidth problems in dealing with video. There are no issues in broadcasting a hundred channels into the home today. There are, however, some issues associated with delivering lifeline telephony, high-speed data, and video upstream. There is also a quality of service (QoS) issue from the consumer perspective that must be overcome regarding the performance of the CATV companies in maintaining the networks. Also, the large node sizes in HFC architectures lead to bandwidth contention when increasingly more subscribers for data and voice are added to the network, which limits the level of penetration for voice and data services over the HFC network. To solve these issues with their HFC networks, service providers continue to decrease their node sizes by taking fiber-optic cable deeper into their networks.

The Ultimate Solution

Operators have the opportunity to select a topology from the available architectures that will meet all of their business requirements today and allow them to grow their revenue opportunities in the future.

The ultimate solution is to take fiber optics deeper into the network. Today, numerous service providers have installed fiber into their local loops. Deeper and deeper fiber architectures are being installed. First, there was fiber-in-the-loop (FITL), then FTTC, and now FTTH. Many companies are currently deploying

FTTH in limited quantities to test the technology. FTTC and FTTH are the network of the future. It is no longer a question of why—it is a matter of when operators will actually deploy these fiber-rich networks. With these deep fiber architectures, service providers can deliver high-speed data, reliable telephony service, and both broadcast and VOD services via Ethernet, CM, or DSL. With this type of service-delivery platform, operators can meet the needs of the home of today with enough scalability and capacity to meet the needs of the future as well.

Fiber-to-the-Curb Architecture

FITL is defined as fiber from the central office (CO) to a remote terminal (RT). Services are then delivered to the end user using coax or TP copper to the home or office. FTTC takes fiber from the RT into an optical network unit (ONU), and service is then delivered the remaining 500 to 1,000 feet to the home using TP copper or coax. In the FTTC architecture, service providers are able to deliver high-speed data, broadcast video services, and traditional telephony using CMs, Ethernet, or DSL. Typically, one RT will support between 24 and 56 ONUs. Each ONU will then support between 4 and 30 homes

A key concern in FTTC architecture is the power consumption of the electronics. For the active electronics at the ONU to function, they must be powered by direct current (DC) power. This can either be achieved by a local alternating current (AC) source converted to DC power at the ONU, or through network powering. Network powering is most often used and requires copper cable to be run from the RT to the ONU. Batteries at the RT provide the necessary backup for lifeline telephony service. The power requirements at the ONU directly affect the distance the ONU can be placed from the RT. Solutions currently on the market will allow the ONU to be placed up to 12,000 feet from the RT and still function on network power. This drastically reduces the cost of deploying the FTTC network.

FTTC architectures deployed today deliver voice, video, and high-speed data from a fiber-feed ONU located within 500 to 1,000 feet of the end user. CATV, Ethernet, DSL, CM data, and plain old telephone service (POTS) are offered. These systems supply services to millions of users.

There are numerous benefits to an FTTC network. First, the nature of the architecture will allow a service provider to deploy services selectively. One home can subscribe to voice services only while an adjacent home subscribes to voice, video, and data. This allows the service provider to scale the network to meet the demands of the customer without incurring additional costs. If a customer wishes to add services in the future, cards can be added at the ONU, deferring costs until revenue streams are realized.

A second advantage of FTTC architecture is the unlimited ability to upgrade in the future. Once the fiber is in the ground out to the curb, there will never be a need to dig up the streets to increase bandwidth delivery to the home. The capacity of a single fiber to deliver data using wavelength division multiplexing (WDM) and dense WDM (DWDM) has yet to be determined. Optical transport systems exist that are capable of delivering more than 400 gigabits on a single fiber, which, obviously, is more than adequate for the home user. The power of these fibers can be realized in the future by simply upgrading the electronics at the ONU and the RT.

The negative issues that are traditionally associated with FTTC should also be noted. If the fiber is not already in the ground, it must be put in, which means digging up the streets. Also, in sparsely populated rural areas where there are not many homes to pass, it becomes difficult to justify deployment financially.

An argument could be made that FTTC is prohibitively expensive to deploy. However, it is important to be clear that it is misleading simply to compare the costs of FTTC versus HFC or TP copper. Operators must ask themselves if they can really afford not to put in FTTC. With FTTC, companies can offer phone, CATV, and Internet service. Companies rolling out a TP copper network will not be able to put that whole package together and will encounter difficulties in attempting to provide video at all. So is it meaningful to say that a TP copper network is cheaper? Several operators have found the costs of FTTC to be comparable to the costs of traditional copper networks. Can an operator really afford to put in a TP copper network when that network will limit the operator's ability to offer desired services to customers? The first competitor that offers those customers phone, Internet, and video services in a single package for one monthly fee will win those customers. Whoever gets to those customers first with these types of fiber architectures and service offerings will become the incumbent. After that, competition will be based simply on price between only those providers capable of offering a bundled package.

Real-Life Experiences

FTTC has shown through real-life experience that it is simple to install, and it provides a dependable network even in the worst conditions. A network in Florida survived Hurricane Andrew not long ago. Voice service remained available during the hurricane. In a multiple dwelling unit (MDU) application, a system elicited not a single service call for CATV problems in 15 months. Typically, it seems that CATV is fairly failure-prone, so to not have a single service call in that period of time seems remarkable. This company has also seen significant reductions in service staff. Obviously if the CATV service calls dry up, service people are free to install service for other customers instead of spending time to maintain the network.

Turning these systems up has proven to be less complicated than first thought. Six hundred units for an MDU application were installed and turned up in two weeks in one network application. Training technicians is obviously one of the challenges with FTTC. People who have been working with copper networks must be trained to work with fiber and fiber-optic electronics.

In many cases it is necessary to overbuild existing networks. Unless an operator is fortunate enough to have a greenfield application or is going into a new apartment complex, it will be necessary to dig to put the fiber into the ground, which can cause numerous headaches for the service provider and end user.

Fiber-to-the-Home

Full-service access networks (FSANs) are a standard for FTTH and enterprise. To simplify considerably, FSAN specifies asynchronous transfer mode (ATM) over a passive optical network (PON) to the home. A PON system means that there are no active components between the CO and the home. The advantage of this is that the network does not have to be powered with DC power, and the maintenance requirements are drastically reduced. Through the use of passive optical splitters, services can be delivered to an end user located more than 30,000 feet from the CO or RT without using active components in the middle of the network.

In an FTTH system, equipment at the head end or CO is interfaced into the public switched telephone network (PSTN) using digital signal (DS)–1s. It is also connected into the data world through ATM or Ethernet interfaces. Finally, video services are brought into the system from the CATV head end or from a satellite feed. All of these signals are then combined onto a single fiber using WDM techniques and are transmitted to the end user. An ideal FTTH system would have the ability to provide all of the existing services that users pay for today, which are circuit-switched telephony, high-speed data, and broadcast video services (CATV and direct broadcast satellite [DBS]).

Approximately 30,000 feet from the CO, the optical signal is split by a passive optical splitter and sent to multiple homes. The split ratio can range from 2 users to 32 users and, again, is done with no active components in the network. The signal is then delivered another 3,000 feet to the home or enterprise over fiber.

At the home, the optical signal is converted into an electrical signal using an optical–electrical converter (O–EC). The O–EC then splits the signal into the services required by the end user. Ideally the O–EC will then have standard user interfaces and not require special set-top boxes to provide service. These interfaces would include registered jack (RJ)–11s for telephony, RJ–45s for high-speed data, and 75-Ohm coax ports for CATV and DBS service.

Because FTTH is a PON, the advantages to the service provider are immense. Not only can a single network replace two or even three architectures today, the FTTH architecture is virtually maintenance free in the outside plant. It is easy to see how reducing the number of networks to maintain, and thereby reducing the overall maintenance costs while simultaneously increasing the revenue potential, is the quickest way to maximize return on investment for the service provider.

Providing lifeline telephony is a basic requirement for the telephony network. Network powering in the FTTC architecture has already been discussed. One of the key advantages of an FTTH architecture is the elimination of active components in the network, although lifeline telephony must still be provided. There exist two ways to do this. One way is to run copper cable with the fiber cable to the end user to supply power from the network. However, this will increase the costs of the network, and it will also decrease the distance that the service can reach.

Another, more viable, option is to provide powering at the customer premises. This is accomplished by simply plugging the O–EC into a local AC power source. Lifeline service is maintained by a battery source, as simple as a C-cell battery from the hardware store, which would support one of the telephone lines in the event of a power failure. The O–EC would be equipped with a warning light to indicate when the batteries need to be replaced, or a rechargeable battery pack could be used. This solution allows the service provider to reduce the infrastructure costs and reap the full benefit of the PON architecture.

Another key issue in FTTH networks is the split ratio at the point of service (POS). FSAN specifies up to a 1:32 split using 155 Mbps optics. This implies that each customer-premises equipment (CPE) device must have a transceiver capable of supporting 155 Mbps optics. This will increase the cost of each CPE device considerably. Another alternative could be to use 25 Mbps optics and split the signal four times, which would allow the use of lower-cost optics but would increase the cost of fiber in the outside plant. Fewer splits would also simplify the installation for craftspeople in the field.

A final criterion with FTTH concerns the delivery of video to the customer premises. In an ideal world, the end user will still be able to subscribe to traditional CATV or DBS using the FTTH delivery system. In an environment using SDV, the number of channels that can be delivered to each home is limited by the bandwidth available on the data channel. In the scenario above using 155 Mbps optics and a 1:32 split ratio, the maximum bandwidth that can be guaranteed to the end user is fewer than 5 Mbps, or one compressed video channel. This is not enough bandwidth to deliver multiple video channels to each home. A broadcast video solution is much more cost-effective to deliver hundreds of channels to each user simultaneously.

Summary

End users want reliable, cost-effective, fast services. The demand for bandwidth continues to grow with no end in sight. It is easy to envision a scenario in which today's chat rooms, where several people type simultaneously, become video chat rooms where participants can see their friends via videoconferencing. Only the architectures that deliver this type of bandwidth—FTTC and FTTH—will survive. HFC and DSL architectures cannot offer this seemingly unlimited bandwidth delivery.

PONs are becoming a reality, and increasingly more service providers are installing these systems. The ability to deliver high-speed data, broadcast video, SDV, DBS, and high-quality telephony are key to the future of the networks and to the survival of incumbents as well as competitive carriers.

Deep fiber architectures offer the only solution for delivering bundled services to the end user over a single network and for maximizing return on investment for the service provider.

DWDM, Gigabit Routing, and Optical Switching in the Carrier Networks

Kevin Groom
Data Network Engineer
AT&T

Frank Groom
Professor, Graduate Center for Information and Communication Science
Ball State University

Introduction

National and metropolitan carriers are rapidly investing in optical equipment and gigabit switches/routers to increase the capacity of embedded fiber plant, to allow customers to participate in transmitting gigabit traffic over the network, and to provide a potential for carriers to reduce their costs in carrying both high-bandwidth and traditional traffic.

Dense Wavelength Division Multiplexing

The technology that is contributing to the major drop in transmission costs and bandwidth expansion is dense wavelength division multiplexing (DWDM). With this technology, the light beam from a light transmitter is divided into its separate frequency waves, and separate independent transmissions are sent over each of the waves. DWDM is made of the following components:

- A bank of lasers, each of which receives a separate information stream and is tuned to a separate frequency
- A coupler, which receives a separate transmission from each laser—tuned to a specific frequency (and different for each laser)—and directs the combined set of frequencies on a single light beam over a fiber-focused and tuned light transmission
- A fiber, which carries the light and frequencies
- Optical amplifiers every 60 to 100 miles so that the frequencies and their content can be deciphered at a destination DWDM box with a bank of filters to isolate the individual wavelengths and the transmissions they each carry

Figure 1 illustrates these DWDM endpoint components.

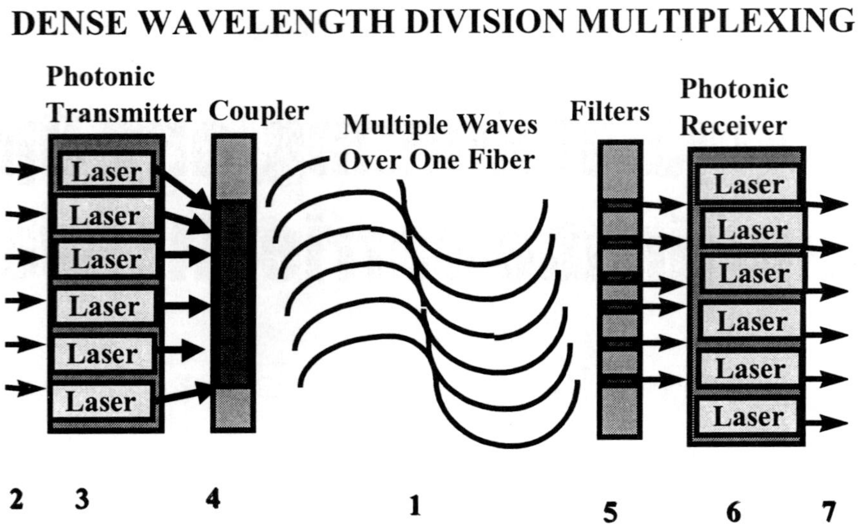

DENSE WAVELENGTH DIVISION MULTIPLEXING

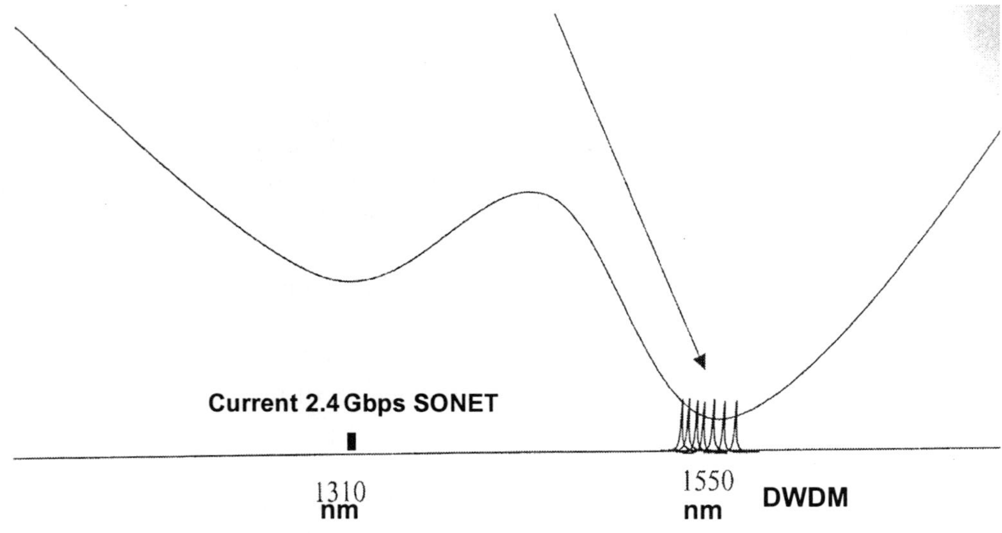

Figure 1: *Transmission of Separate XXs over Individual Frequency Waves of a Single Light Beam*

Light is composed of many frequencies, some of which can be displayed as colors with the usable spectrum between 1 and 1710 nanometers (nm). As portrayed in *Figure 2*, there are two points on a curve representing the amount of noise in the frequency spectrum. Those two points are considered the lowest-noise "sweet spots." The single frequency 1310 nm has long been used as the frequency for single-mode synchronous optical network (SONET) transmission in both metro and national backbone networks because it is a low-noise spot in the spectrum. In the last half of the 1990s, DWDM equipment used a higher portion of the spectrum, between 1500 nm and 1600 nm, because it is another low-noise area of the spectrum. The overall frequency spectrum and these two "sweet spots" are portrayed in *Figure 2*.

Figure 2: *The Single-Wave and Multiple-Wave Portions of the Spectrum*
Wavelengths are tightly packed together; therefore, spectral width must be tightly maintained, i.e., one clock frequency and one modulation schema.

The frequency band between 1500 nm and 1600 nm can be further divided into the lower portion or C-band (up to 1560 nm) and the L-band (from 1560 nm to 1600 nm) (see *Figure 3*). In fact, the C-band itself can be divided further into a lower blue-band and an upper red-band. Current deployments of DWDM equipment have concentrated on the C-band frequencies. There are 40 nm available in the C-band. Early deployed equipment has divided this band into 40 waves with 1 nm between each wave. Others have used only 32 of the frequencies, also with 1 nm between the waves.

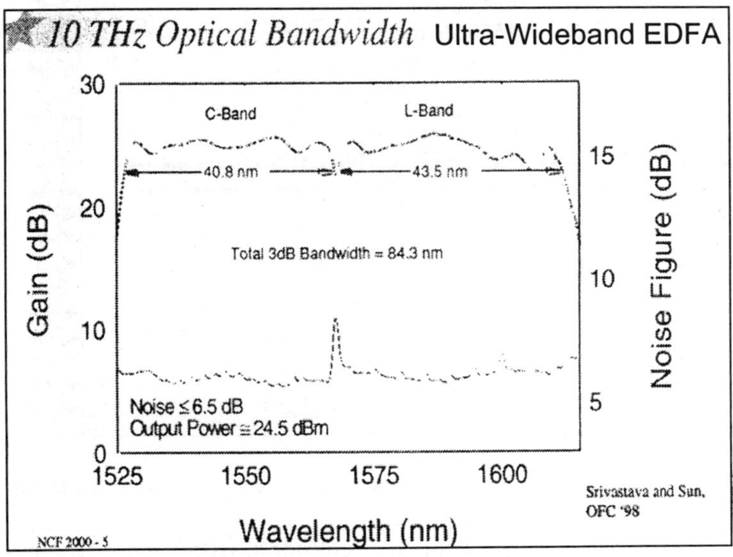

Figure 3: The C-Band and the L-Band in the 1500 to 1600 Frequency Range

More advanced equipment reduces the separation between the waves to only .5 nm. With this, 80 waves can carry transmissions if all the frequencies from 1520 nm to 1560 nm are used (or 64 waves if only the frequencies between 1528 nm and 1560 nm are used). *Figure 4* shows 80 waves (the points) and the spacing between the waves. Further experimentation is underway to use the 43.5 nm of the L-band for additional frequency waves to carry transmission.

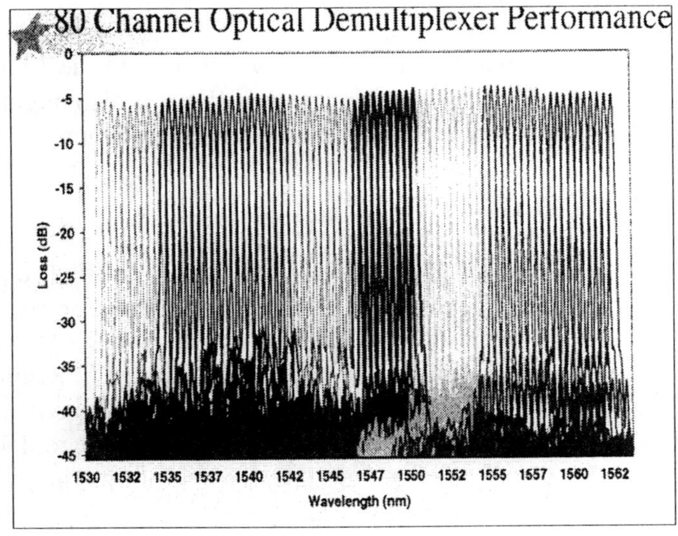

Figure 4: Individual Frequency Waves and Spaces between the Waves

Figure 5 portrays some of the individual waves of a light beam over a single fiber. The individual waves are approximately .8 nanometers apart, and each carries a separate transmission. These displayed waves are from 1542.14 nm to 1560.61 nm, although they likely extend down to 1528 nm or 1520 nm.

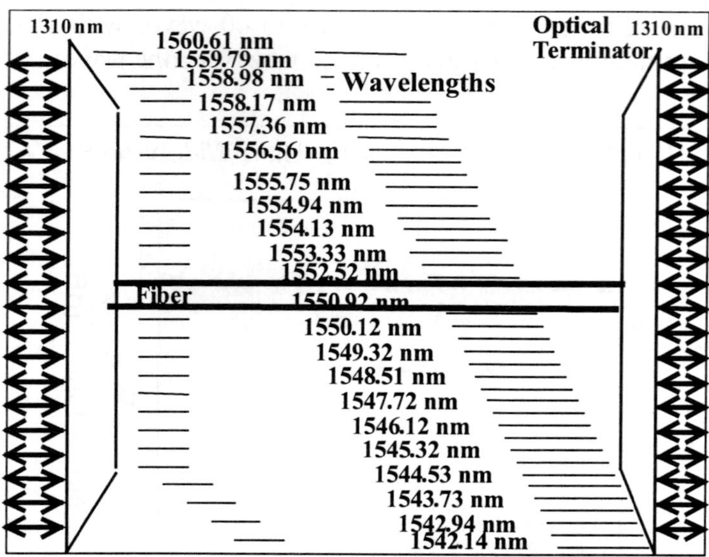

Figure 5: *A Number of These Frequencies at Their Appropriate Nanometer Spacing*

Optical switches offer the ability to provision and switch traffic at sub-second speeds from remote operations support systems (OSSs).

Currently, the most advanced and commercially available DWDM equipment employs a set of 80 waves in the C-band but will soon offer many hundreds of waves that each carries traffic from a separate laser. Each laser is tuned to a particular frequency, such as 1556.56 nm, and carries an individual transmission stream. Each laser carrying that transmission stream is focused on a coupler, which then places that stream on the associated frequency wave (such as 1556.56 nm) of the single fiber strand. Amplifiers are then placed along the fiber path every 30 to 100 miles to ensure that that frequency and its information are detectable at the reception point. The receiver has a filter that isolates each frequency and places it on a path to a new laser for further transmission.

Further work is underway to carve a number of separate waves around the 1310 nm sweet spot of low noise where traditional single-wave fiber transmission occurs. Work is also ongoing to extend the number of usable frequencies in the L-band up to and beyond the 1600 nm frequency.

Gigabit Routing over the Carrier Network

Gigabit routing enables carriers to transmit high-bandwidth customer traffic (up to 1 Gbps and eventually in the 10 to 30 Gbps range) over the metropolitan, regional, and national networks while off-loading the management of the submitted, high-bandwidth traffic to the customer. A router at the edge of the submitting customer's enterprise network will provide such management functionality.

Customers can thus deliver a gigabit backbone network throughout one location and extend that transmission speed to a distant location by submitting it to a carrier through the gigabit routers placed at the edge of their corporate network. Traffic can then be forwarded to a telephone central office (CO) and transmitted either over a SONET ring or over a frequency wave of a SONET–free system.

Thus a customer's high-bandwidth, intersite traffic can be delivered to one of the lasers in a DWDM box in the carrier network. It can be transmitted by that laser over a single-wave frequency across the network and then forwarded to an associated gigabit router at a destination. These routers can either be at the customer premises or at the carrier office—the point of presence (POP)—at the edge of the network.

Figure 6 portrays customers submitting high-bandwidth traffic to a gigabit router in a telephone CO. That router then forwards the traffic to an add/drop multiplexer (ADM) for placement of a SONET ring around a city or across the country.

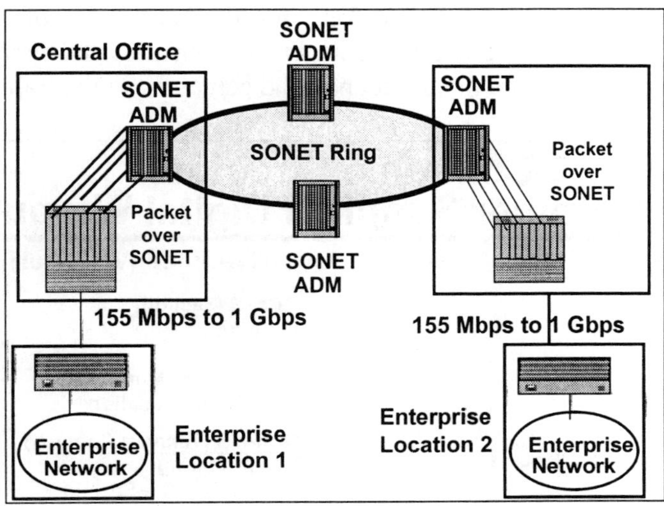

Figure 6: *Customers Submitting High-Bandwidth Traffic over Carrier SONET Rings*

Thus, intersite gigabit private-line traffic can be transferred over a separate SONET ring or over a SONET transmission over a DWDM–supplied frequency wave. However, the deployment of optical equipment allows transmission without the deployment and use of expensive SONET equipment. Each transmission can be placed directly on one of the frequency waves provided by a DWDM box and extracted by a filter at the destination CO where it can be routed to the customer's location. *Figure 7* illustrates this process.

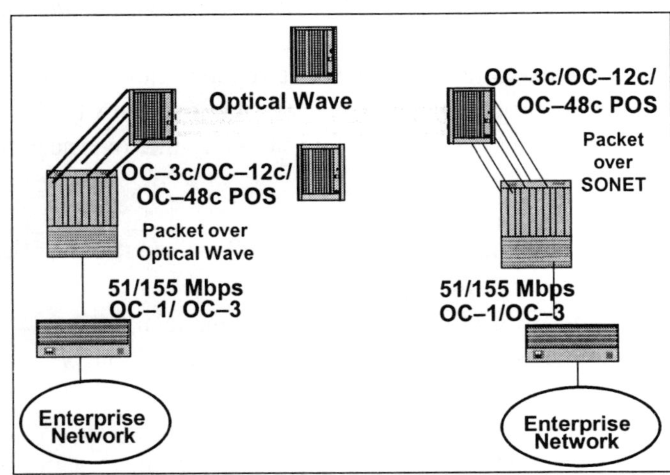

Figure 7: *Customers Submitting High-Bandwidth Traffic over Individual Waves*

One of SONET's principle functions, beyond the timing of the node-to-node transmission, is to provide a framing structure for the managed transmission of a bundle of packets. This is performed by creating a large packet frame that will carry the offered packets to be transported and some header information to manage node-to-node transmission and end-to-end transmission. However, because no SONET equipment exists in the path that the customer's traffic is taking in *Figure 7*, the customer's gigabit edge routers must assume the task of managing the node-to-node and path transmission in a fashion similar to the SONET approach.

One of the techniques developed to achieve the SONET–like management function is that of using a digital wrapper to "wrap" a collection of submitted packets in a SONET–like container frame. This frame not only carries a number of the transmitted packets but also adds a set of management information placed in the frame header. *Figure 8* presents an image of such a digital wrapper, with the overhead wrapper portion in the frame header and the packet payload contained in the frame.

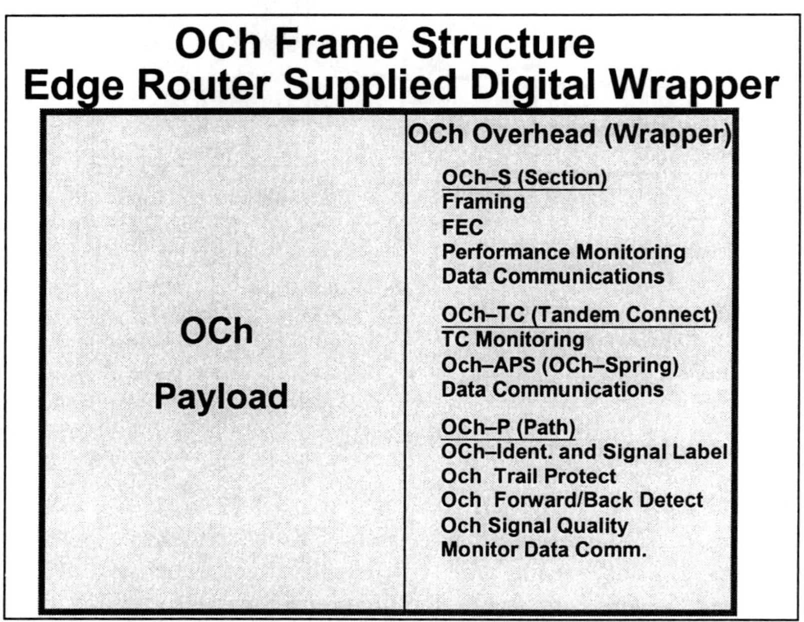

Figure 8: Customer-Provided SONET–Like Framing to a Bundle of Packets

Digital wrapper technology uses existing opto-electronic regeneration points in DWDM to add or extract information from the header. The process uses a time division multiplex (TDM) wrapper around the optical channel (OCh) client (the gigabit router) to add SONET–like overhead to the router client's signal. This wrapper (header) provides the following:

- Operations, administration, and maintenance (OAM) information for the optical network
- Optical-layer performance-monitoring information
- Forward error correction (FEC) for system margin enhancement
- Ring protection and network restoration on a per-wavelength basis
- Independence from the input signal format

Wrapped in their own digital wrapper frames with appropriate management and monitoring information, separate transmission streams can be brought to the individual lasers of the DWDM box for focus on the waves of a single fiber and transmission across a network without SONET equipment in the connection.

DWDM Deployment with SONET

A DWDM box can receive separate traffic streams into each of a bank of lasers. These lasers can then be tuned to a specific and different one of the 40 or 80 available wavelengths in the C-band of the optical spectrum and transmitted over a single fiber to a downstream DWDM box (see *Figure 9*).

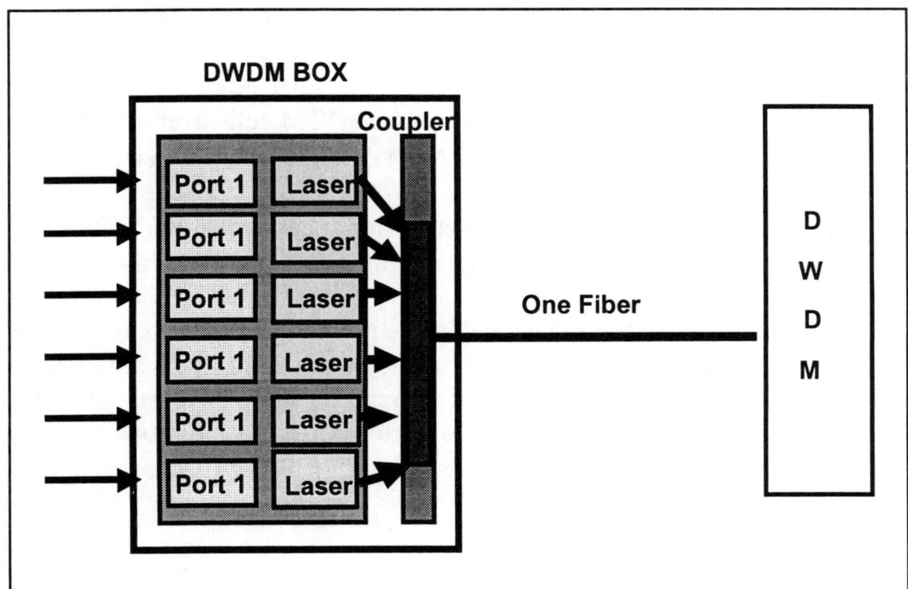

Figure 9: DWDM Receiving Multiple Transmissions and Placing Them in Parallel on a Single Fiber

Each of the separate streams that enter a DWDM box can be a separate SONET path (a portion of a SONET ring). As illustrated in *Figure 10*, many parallel SONET rings can be constructed by placing

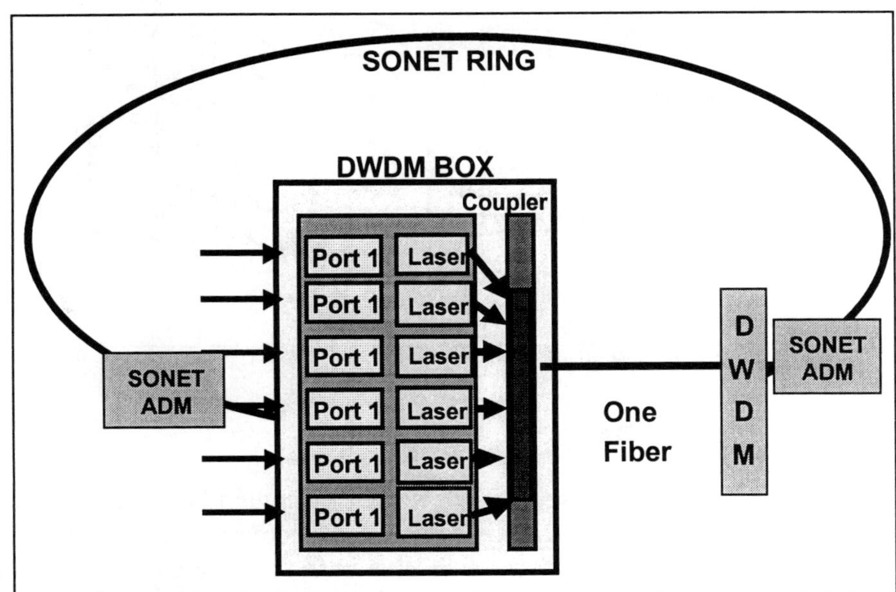

Figure 10: Multiple SONET Rings, Each Carried by a Different Wave of One Fiber

DWDM equipment in an existing SONET ring. SONET is usually deployed as a set of four fibers, only one of which actively carries information; the other three are for restoration and backup. By adding DWDM equipment inward of the SONET ADMs, what once was one SONET ring over one fiber can soon be 40 or 80 parallel rings, each a separate wavelength transmitted over the single fiber.

By deploying DWDM equipment in metropolitan and regional networks, many diverse protocols can be carried simultaneously over a single fiber. The following can be seen in *Figure 11*:

- Asynchronous transfer mode (ATM) switches sending ATM cells over a SONET transmission
- Internet protocol (IP) packets segmented into ATM cells and placed on a frequency wave without SONET
- IP packets placed into SONET frames for transmission (packet over SONET [POS])
- IP packets placed directly on one of the many gigabit waves carried by the single fiber

DWDM equipment allows a carrier to support each of these protocol formats simultaneously over a single fiber. Each transmission is carried over a different frequency wave. Thus the single fiber can carry many traffic formats and protocols. This eliminates the need to create separate physical networks for each of the protocols and protocol combinations that customers wish to transmit at high-bandwidth speeds between enterprise locations.

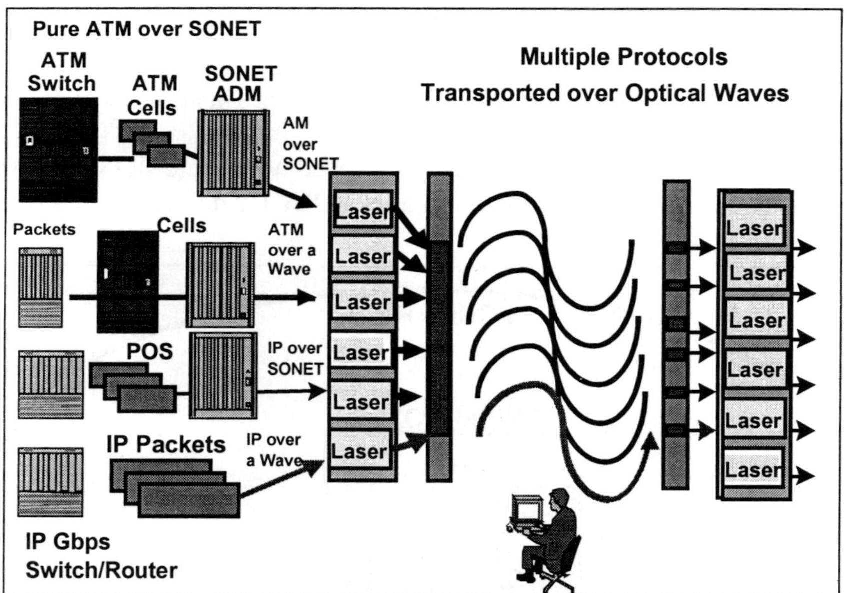

Figure 11: Multiple Protocols Carried by Separate Waves over a Fiber

Optical Switching in the Carrier Network

Carriers can now carry diverse traffic over separate waves over a single fiber. Each wave can carry unique traffic protocols, and these protocols have different reliability, recovery, and congestion characteristics. Carriers managing these optical networks with such diverse traffic characteristics, each carried in parallel with the others, are faced with the difficult challenge of how to reroute or divert one wave of traffic individually without affecting the others. Each wave must be treated as if it were a physically separate transmission facility, although, in reality, these share the same fiber media and DWDM equipment. *Figure 12* presents an optical switch that can be placed along a fiber to divert selected waves when the situation arises.

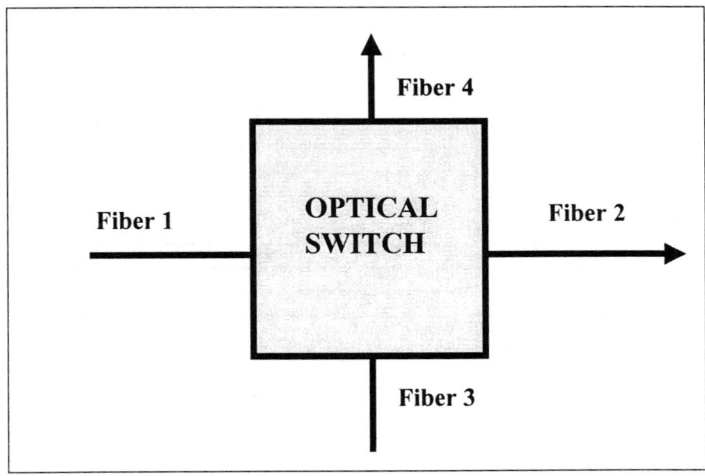

Figure 12: *Four Fibers Entering and Leaving an Optical Switch*

An optical switch can be placed along the fiber at strategic locations. Currently, laboratory optical switches can handle up to 1000 waves. However, most deployed switches carry 4, 32, 40, 64, or 80 waves for carrying parallel transmissions. The switch passes the optical waves through the switch without any interference. Microscopic mirrors or bubbles are placed slightly below each optical frequency (wave). The mirrors are in a flat position allowing the wave to pass undiverted, entering on a fiber on one side of the switch and exiting on a fiber on the other side of the switch. *Figure 13* portrays such a switch with a bank of microscopic mirrors placed underneath the light beam of the entering fiber on the left and exiting on the right.

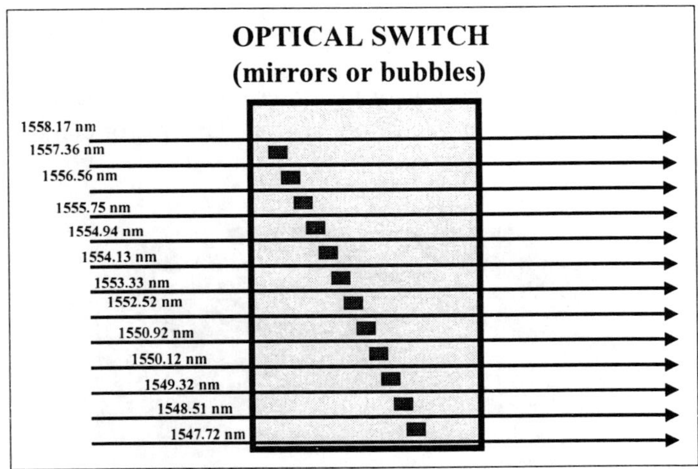

Figure 13: *A Single Fiber Carrying Multiple Waves across a Bank of Mirrors*

The mirrors can be raised individually, based upon a signal from an OSS to divert an individual wave to another fiber that also passes through the switch. The wave stays at the original frequency on the new fiber: If it enters at 1556.56 nm, for instance, it is placed on the new fiber at the same 1556.56 nm frequency. *Figure 14* portrays traffic on 1558.17 nm, 1554.94 nm, and 1549.32 nm frequencies being diverted to the fiber that exits from the top part of the switch.

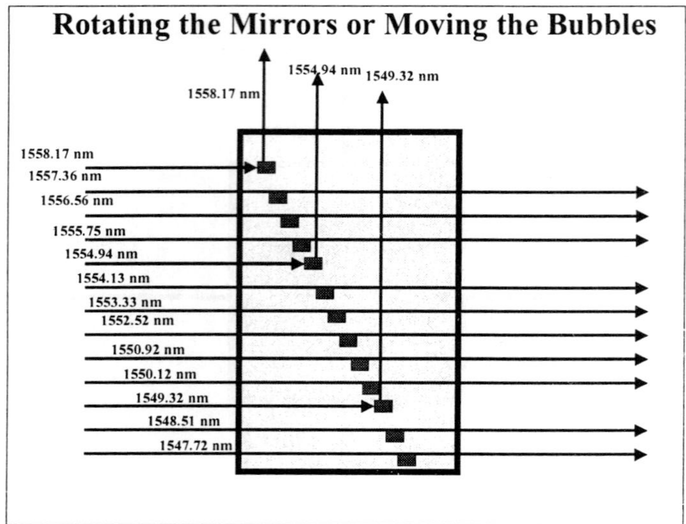

Figure 14: Switch Selects Frequency Waves by Tilting Associated Mirrors

Each wave can be separately switched by elevating its associated mirror. Because individual traffic streams are brought to each laser separately over one of the many wavelengths carried over the fiber, the nature of such transmissions can be quite dissimilar. The fiber itself and the DWDM box, which places the individual traffic on one of the fiber's wavelengths, are unaware of the particular traffic protocols each wavelength carries. Thus, many diverse protocols can be carried simultaneously, each on its own particular wavelength of the fiber. Such protocols and transmissions might be as follows:

- Telephone voice over SONET
- ATM over SONET (carrying voice IP, or IP in ATM cells)
- IP over SONET
- IP from a gigabit router using point-to-point protocol (PPP) placed directly on a wave

Figure 15 offers an example of an individual mirror in a micro-electromechanical systems (MEMS) optical switch. These mirrors are usually sculpted from silicon. The mirror can be seen tilted upward, with the bottom portion elevated.

Figure 15: Individual Lucent Mirror in an Optical Switch

Many of these mirrors—currently 40 or 80, but soon hundreds—are placed in a bank of mirrors (see *Figure 16*).

Figure 16: A Bank of Mirrors for a Lucent Optical Switch

The light beam entering an optical switch passes through a filter that separates the individual wavelengths. The wavelengths are then passed over the mirror bank with the possibility that individual wavelengths might be rerouted by tilting the associated mirror up under that wave. *Figure 17* presents a wavelength entering an optical switch on a fiber from the left and exiting on another fiber to the right. A second pair of fibers enters from the bottom and exits from the top. This second pair can be employed if wavelengths need to be rerouted over the path of the second fiber.

Figure 17: Four Light Beams Crossing a Bank of Optical Mirrors

Strategies for Optical Switching in the Carrier Network

The first implementation of optical switches will be in the national backbone network. These switches will be placed within the network, in-board from the DWDM boxes that separate out the frequency waves for the individual transported streams. They will initially be used to reroute non–SONET ATM or gigabit router traffic through alternate routes to the wavelength's destination.

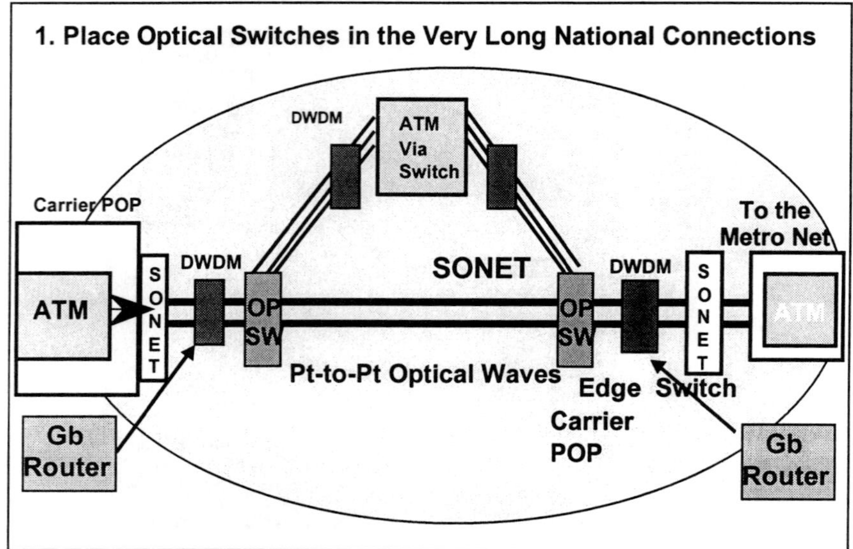

Figure 17: Optical Switches in the Long-Haul National Backbone Networks

The next implementation of optical switches will be in the national backbone and regional networks to switch SONET traffic (see *Figure 18*). These switches will be placed within the core network after SONET boxes (and gigabit routers) at the edge of the backbone have been deployed and after the DWDM boxes that separate the frequency waves for the individual transported streams. Switching SONET traffic is a much more difficult proposition because SONET provides a narrowly synchronized, node-to-node transmission. If a SONET stream over a DWDM wave is to be switched to an alternate route, the stream will have to be resynchronized before it will be able to operate as SONET. Alternatively, where optical switches are deployed in SONET systems, the synchronization can be relaxed significantly to tolerate the alternate routing of the transmission over a longer path to the next SONET box.

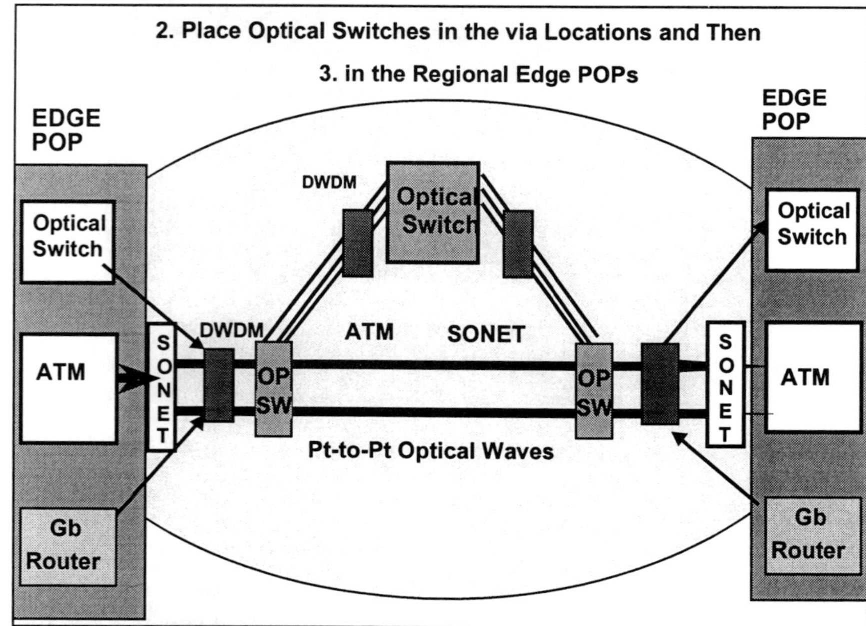

Figure 18: Optical Switches Replacing ATM Switching in the National Backbones

Optical switches will subsequently replace ATM switches in the network. These optical switches can switch both waves transporting SONET transmissions or point-to-point IP transmission over other wavelengths of a fiber. The transmission on a wavelength can be sent to an alternate wavelength by the optical switch and then on to the destination optical switch via an alternate fiber route, DWDM equipment, and possibly a SONET ADM; it is then on to a gigabit router at the destination. Where SONET transmissions are optically switched to alternate routes to a destination, new approaches will have to be deployed to resynchronize the SONET node-to-node timing when rerouting occurs. Alternatively, a looser timing must be deployed in the node-to-node SONET synchronization.

The final implementation of optical switches will be in metropolitan areas (see *Figure 19*).

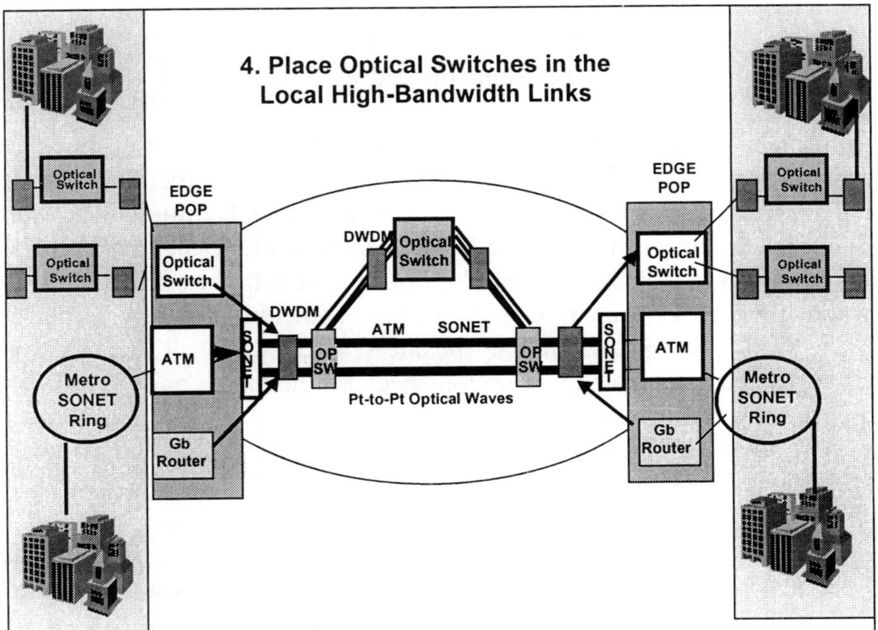

Figure 19: Optically Switched Access Networks in the Metropolitan Area

Parallel, optically switched access networks will transport gigabit traffic streams from corporations to an optically switched network to optical components in the carrier's POP. At the POP, they will be placed on optical waves by DWDM equipment for transport over a regional or national optical backbone network. This corporate, high-megabit and gigabit traffic will then be transported and switched as purely optical waves to be delivered to a destination DWDM box and optical switch with no slowdown due to electronic switches. From the destination carrier POP, they will be locally delivered over a gigabit, optically switched, local-metro network connection.

Optical Switches

Three basic approaches have been employed as the earliest designs for creating optical switches:

- Combination optical-electronic switches using standard electronics for performing the switch with input and output ports, which are connected to the fibers carrying the various frequency waves of transport
- Optical switches using a bank of mirrors over which the light carrying frequency waves passes, with pure optics end to end through the switch
- Switches using liquid bubbles that perform similarly to the mirrors for switching frequencies of light individually

The following are some examples of optical-switch vendors and the switching technology they employ:

Manufacturer	Optical Technology
Xros (Nortel Networks)	Optical micro-mirrors (MEMS)
Lucent Technologies (LambdaRouter)	Optical micro-mirrors (MEMS)
Optical Micro Machines	Optical micro-mirrors (MEMS)
Calient Networks	Optical micro-mirrors (MEMS)
Tellium	Electro-Optical
Sycamore	Electro-Optical
CIENA	Electro-Optical
Agilent Technologies	Liquid bubbles-optical
Corvis	Not yet announced
Brightlink	Not yet announced
Qtera	Not yet announced

Some of the emerging new-entrant carriers are among the first to place these optical switches in their networks. These new carriers can attract new business customers by constructing networks specifically to carry IP packet traffic. They can offer IP transport over dedicated facilities under contract to special customers; the remainder of their unsold bandwidth can be offered wholesale to other carriers. Thus, these new-entrant carriers can be in two businesses simultaneously and with the same facilities—in the primary carrier business where they provide their own integrated bandwidth and service packages to selected customers, and in the wholesale bandwidth–only business where they sell the pipes only, without the supporting services. These carriers have these options because they have no legacy facilities or services that they must support; they can offer whatever has a high customer demand and looks to be profitable. Accordingly, they can grow their business profitably as they acquire capital and can place their facilities to support only their growing customer base.

Thus, these new carriers do not face the problem of incumbent carriers—that of integrating optical switching in their existent ATM–over–SONET backbone networks. They can offer IP over pure optical networks with strategically placed optical switching as a primary service.

Conclusion

As customers migrate to high-bandwidth access connections and media-rich applications, the current SONET–oriented metropolitan and national backbone networks will quickly congest. Carriers are continuing to lay new fiber to increase their capacity. But the real solution to the coming bandwidth crunch is to employ DWDM equipment that expands the number of wavelengths for each fiber to at least 80 waves, with each wave carrying traffic streams at 10 Gbps (four times the total capacity of a complete fiber in the recent past).

The speed of this kind of transmission will require fast switching of heavily multiplexed wavelengths, and this must be accomplished cost-effectively. Optical switches with a pure optical design—those that do not need to slow down the signal by conversion to electronic form along the path—and a set of event-triggered and policy-guided "expert" OSSs will be required to provide the alternate routing and quick switching necessary for the expected traffic patterns and volume of the next decade.

Opto-Electrical Cross-Connects (OEXCs) with Virtual Transparency of Optical Networks

Jun-Ho Koh
Senior Engineer, Optical Communication R&D
Samsung Electronics

Byung-Jik Kim
Engineer

Sang-Ho Kim
Engineer

Han-Lim Lee
Engineer

Seong-Taek Hwang
Senior Engineer

YunJe Oh
Senior Manager

Introduction

The rapid growth of the optical network raises questions about how to route high–bit rate data traffic efficiently and how to design a cost-effective optical network by reducing the function of intermediate equipment such as synchronous optical network (SONET)/synchronous digital hierarchy (SDH) multiplexers and asynchronous transfer mode (ATM) switches. Optical cross-connects (OXCs), which are fully transparent and independent of bit rate and format, are a good candidate. To utilize fully the function of an OXC, an optical wavelength conversion technique must be used. However, the optical wavelength conversion technique to date is immature and expensive. Thus, instead of using an immature optical wavelength converter, much research has been conducted on the so-called "opaque optical network," in which both optical-to-electrical-to-optical (O–E–O) converters and optical switches are used. However, the opaque optical network has some shortcomings. The O–E–O converters with 3R are not bit-rate transparent, and optical switching devices are not mature.

This paper demonstrates a low-cost and compact OEXC, which has virtual bit-rate transparency, by using the proposed bit rate independent (BI) receiver/transmitter (Rx/Tx).

Configuration of an OEXC

Figure 1 shows the block diagram of an OEXC, which consists of Rx/Tx blocks, a switch block, and a control block. The BI–Rx block consists of a PIN–PD module; a limiting amp; the programmable clock and data recovery (CDR) block, which regenerates the input signal corresponding to the selected signal; and a bit rate identification (BID) block (patent pending) [1]. The BI–Tx block consists of an LD driver, an automatic power control (APC) block, an automatic temperature control (ATC) block, a BID block, a

"

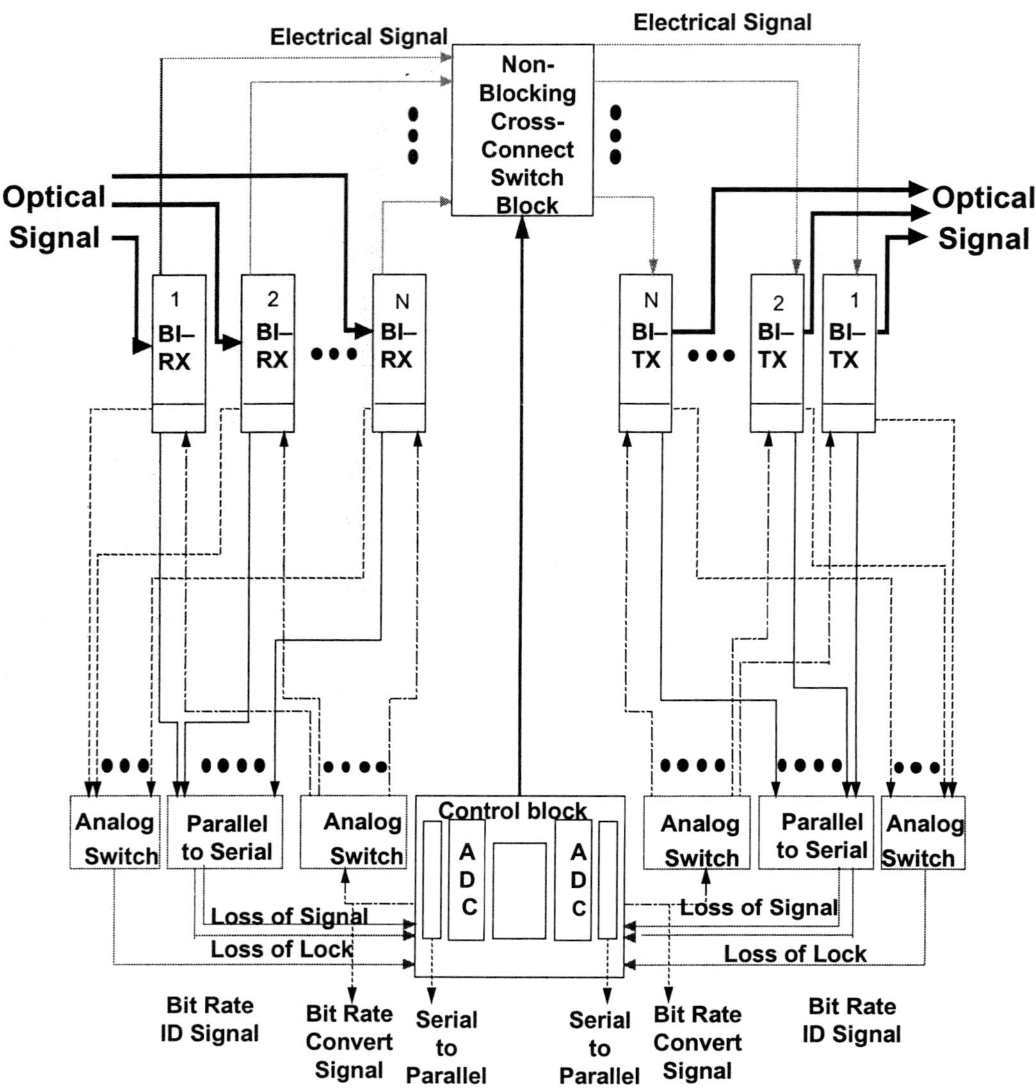

Figure 1: Block Diagram of an OEXC Independent of Bit Rates

distributed feedback laser (DFB)–LD, and the programmable CDR block, which retimes the signal to improve signal skew, jitter, and noise. The switch block indicates a nonblocking 16×16 digital crosspoint switch. The nonblocking architecture uses 16×1 multiplexers, which allow each output port to be programmed independently to any input port. All switch ports are programmed within 80 ns. The control block automatically detects a loss of lock (LOL) signal from the BI–Rx/Tx block. That indicates either a change in the bit rate input signal or signal loss. If there is signal loss, the loss of signal (LOS) detector sends the information to the control block. Also, it identifies the bit rate of the input signal, which receives BID signals, from BI–Rx/Tx block and returns a selected signal to the programmable CDR block. The CDR extracts and regenerates the input signal corresponding to the selected signal. The regenerated signal is routed and sent to the Tx block to activate the laser diode. The laser output is stabilized by use of APC and ATC blocks to maintain the pre-assigned International Telecommunications Union–Telecommunication Standardization Sector (ITU–T) recommended frequency.

Receiver Block

The BI–Rx block consists of the PIN photo detector, the limiting amp, the programmable CDR block, and the BID block. *Figure 2* shows a block diagram of the receiver block. The PIN photo-detector module contains a transimpedance amplifier. The pigtail of the receiver is multimode fiber to receive the signal from data equipment such as fiber distributed data interface (FDDI) or Ethernet. It often uses a multimode fiber transmission. This output is designed to drive 50 Ω impedance. These differential outputs are ac-coupled through capacitors to the package data output pins. The sensitivity is –24 dBm at 2.5 Gbps, 2^{23} – 1 PRBS, and the overload is 1 dBm at 2.5 Gbps, 2^{23} – 1 PRBS. The limiting amplifier is manufactured using gallium arsenide. The device is a wideband amplifier with differential inputs and outputs. It has a 3 GHz bandwidth and provides a typical gain of 34 dB using differential data with a 50 Ω impedance line.

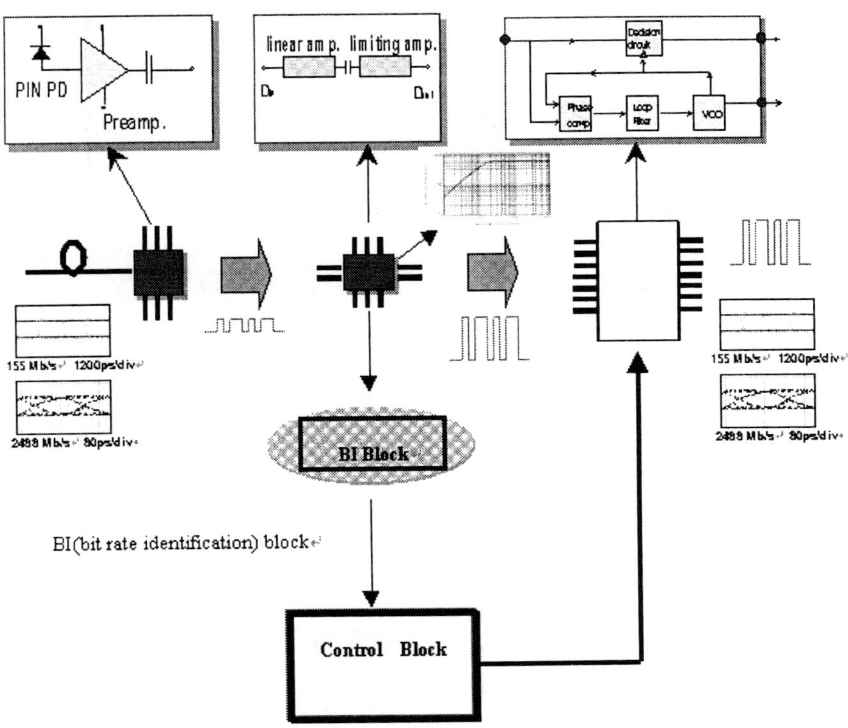

Figure 2: Block Diagram of the Receiver Block

The inputs are internally terminated in 50 Ω to simplify connection to the input source, and the outputs drive 50 Ω load impedance efficiently. The programmable CDR block consists of a phase-frequency detector (PFD) with a charge pump. The voltage-controlled oscillator (VCO) is controlled by the reference clock frequency and the selected division ratio. Clock generation is performed through a frequency multiplier phase locked loop (PLL) with source as reference. The CDR is an integrated circuit that retimes 45 Mbps data to 2.5 Gbps non–return-to-zero (NRZ) data. *Figure 3* shows a charge pump PLL. The closed loop transfer function is shown by *Equation 1*, where K_{osc} is VCO conversion gain, ω_n is natural frequency, and ξ is damping factor [2: 65–79].

$$H(s) = \frac{I/2\pi \times (RC_pS+1)K_{osc}}{S^2 + 2\xi\omega_nS + \omega_n^2} \qquad (1)$$

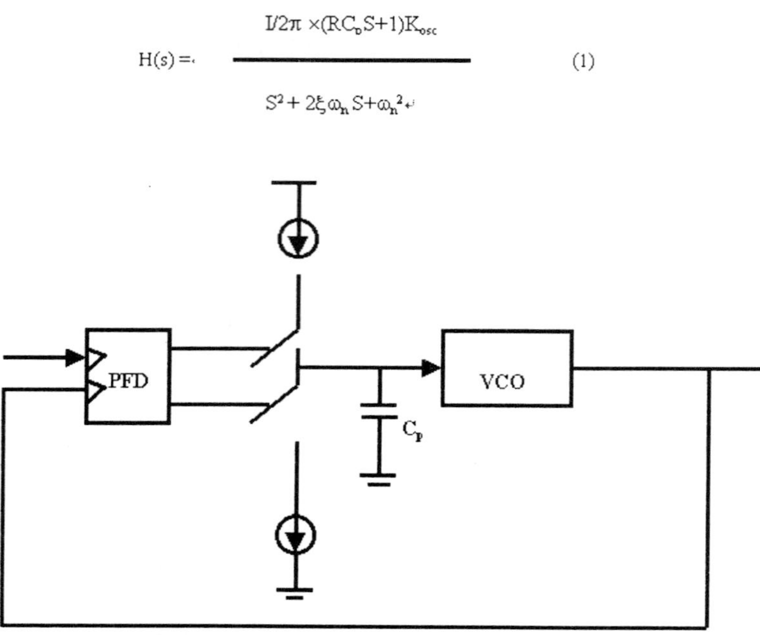

Figure 3: Block Diagram of a Charge Pump PLL

It is desirable to maximize the loop bandwidth, which is usually proportional to ω_n. If I and K_{osc} are increased, both ξ and ω_n can be increased. The BID block is a bit rate–identifying unit that detects the input signal data format. The information that identifies bit rate is sent to the central control block, which selects a reference clock generator and programmable CDR. The clock generator generates a reference clock signal according to the detected bit rate. The CDR recovers the clock signal and data from the input signal according to the reference clock signal. *Figure 4* shows the OEXC–Rx block.

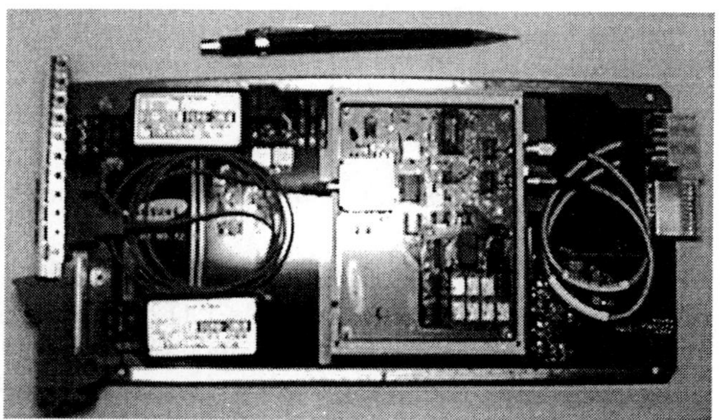

Figure 4: The OEXC–Rx Block

Transmitter Block

Figure 5 shows a block diagram of the transmitter block. The APC feedback loop is incorporated to maintain a constant average optical power to compensate for the change of laser threshold current over temperature.

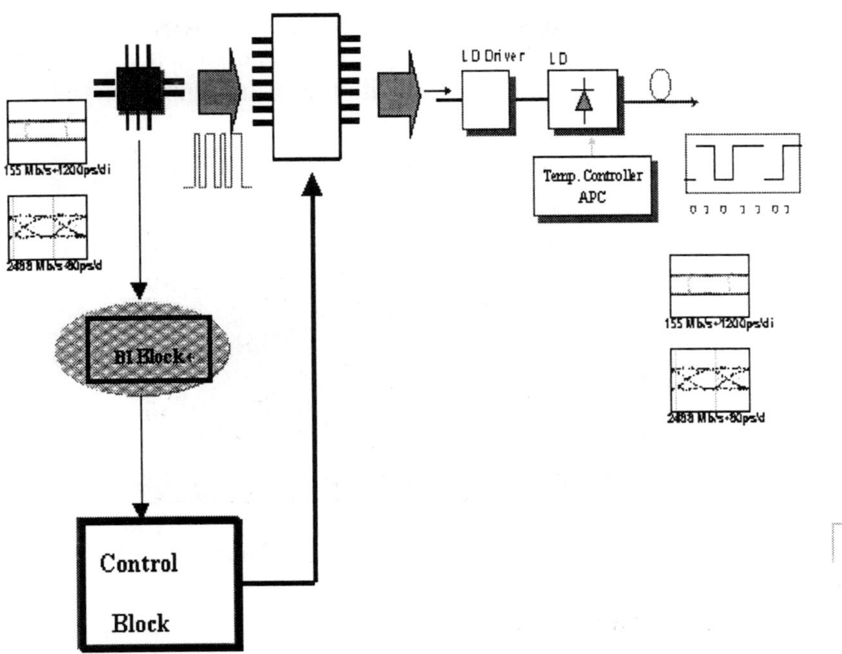

Figure 5: Block Diagram of the Transmitter Block

A back-facet photodiode mounted in the LD module is used to convert the optical power into a monitor current. The APC loop adjusts the laser bias current so that the monitor current is matched to a reference current set by the reference bias. The time constant of the APC loop is determined by PI circuit factors. The LD driver is capable of driving laser-diode bias currents up to 80 mA. The ATC controls wavelength with the LD temperature, which is implemented using a DFB–LD, which satisfies the ITU–T recommended frequency requirement. The Tx block is installed in an aluminum case that is 100 mm × 80 mm × 16 mm. *Figure 6* shows the OEXC–Tx block. In this system, eight bit rates corresponding to digital signal (DS)–3 (45 Mbps), FDDI (125 Mbps), enterprise systems connection (ESCON) (200 Mbps), fiber channel (1.062 Gbps), Gigabit Ethernet (1.25 Gbps), and SONET/SDH (155 Mbps, 622 Mbps, and 2.5 Gbps) are implemented. More bit rates also can be implemented.

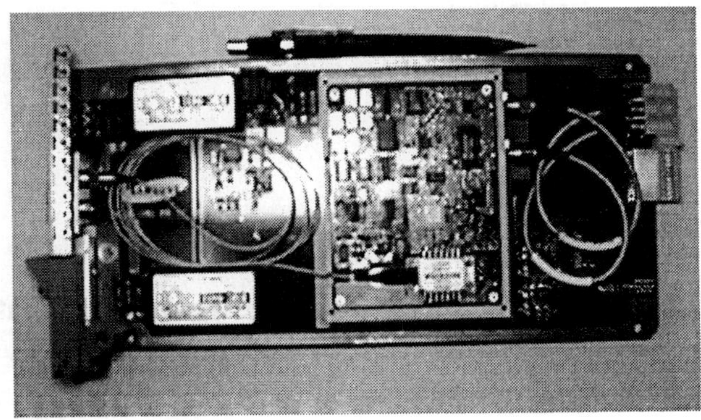

Figure 6: OEXC–Tx Block

Electrical Switch Block

The switchboards that accommodate all the elements—the Rx/Tx modules, the control module, and the digital cross-connect (DXC) switch—are connected simply using push-pullback connectors. The high-speed data line is designed to match 50 Ω line impedance and to reduce crosstalk (XT). *Figure 7* shows an OEXC 16 × 16 switchboard. The switch uses 16 fully independent 16:1 multiplexers that allow each output port to be programmed independently of any input port. The switch is configured by sequentially loading the four-bit program latch of each multiplexer with the desired input port address [3].

Figure 7: OEXC 16 × 16 Switchboard

Figure 8 shows OEXC switchblock architecture

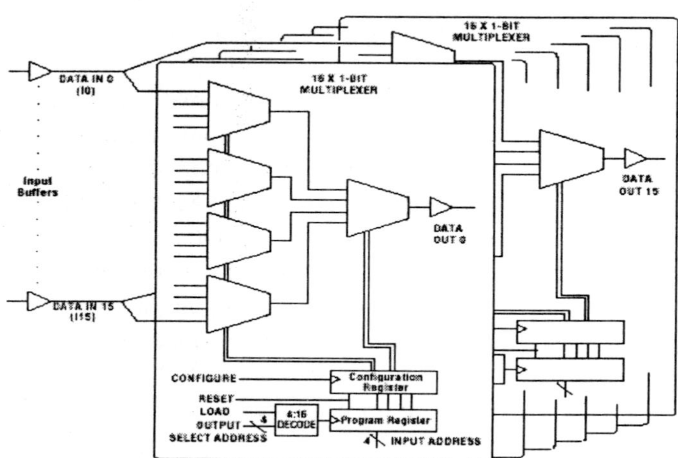

Figure 8: OEXC Switchblock Architecture

Central Control Block

The central control block performs several functions:

- Judgment of data from BI–Rx that detects the input signal data format
- Selection of reference oscillator to match input data format
- Programming of CDR to suit input data format
- LOL check, which means either change of input signal bit rate or signal loss
- Programming of each output port and input port to route paths

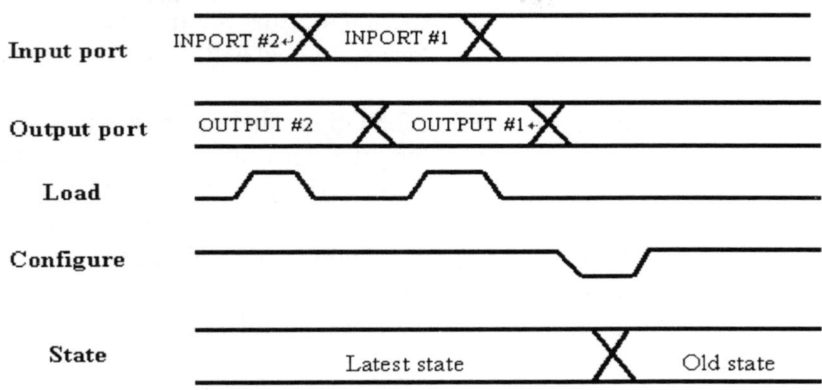

Figure 9: Timing Diagram to Route

Figure 9 shows a timing chart to program switch paths. *Figure 10* shows a photograph of the OEXC central control block.

Figure 10: OEXC Central Control Block

Measurement Results

The OEXC with virtual transparency systems is designed to operate over eight bit rates corresponding to DS–3 (45 Mbps), FDDI (125 Mbps), ESCON (200 Mbps), fiber channel (1.062 Gbps), Gigabit Ethernet (1.25 Gbps), and SONET/SDH (155 Mbps, 622 Mbps, 2.5 Gbps), which are the rates implemented at

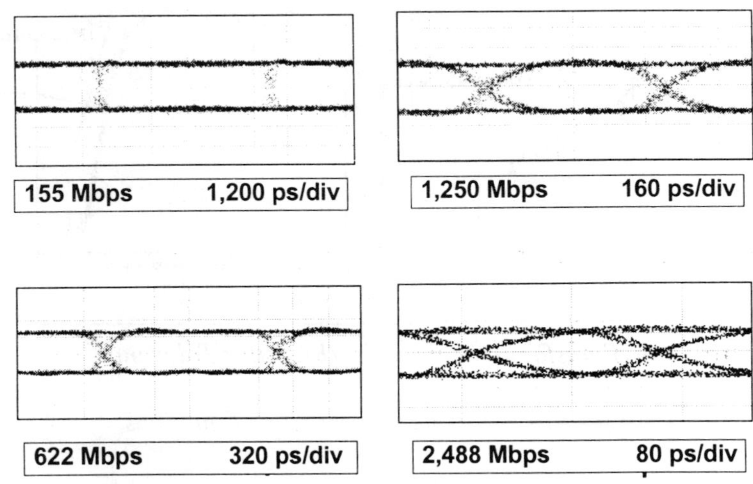

Figure 11: Eye Diagrams of Optical Output after Switch

present. *Figure 11* shows representative eye diagrams of optical output after switch. The optical output quality of the OEXC is tested with a transmission experiment using the lightwave CDR from HP (model 83446A) and the bit-error rate (BER) detector from Anritsu (model MP1764) at 2.5 Gbps.

Figure 12 shows the BER performance measured at 2.5 Gbps. The transmission penalty is 1.0 dB. The OEXC satisfies the ITU–T G.958 recommendation for jitter performance at each bit rate. It is shown at 2.5 Gbps in *Figure 13*. It was tested using the SONET/SDH analyzer equipment from Anritsu (model MP1552B). The jitter transfer characteristic of the OEXC is defined as the ratio of the output jitter to the applied input jitter as a function of frequency. The OEXC performed well, as shown in *Figure 13(a)*. The jitter tolerance characteristic of the OEXC is defined in terms of the sinusoidal jitter amplitude, which causes a designated degradation of error performance. The jitter tolerance is a function of the amplitude and frequency of the applied jitter.

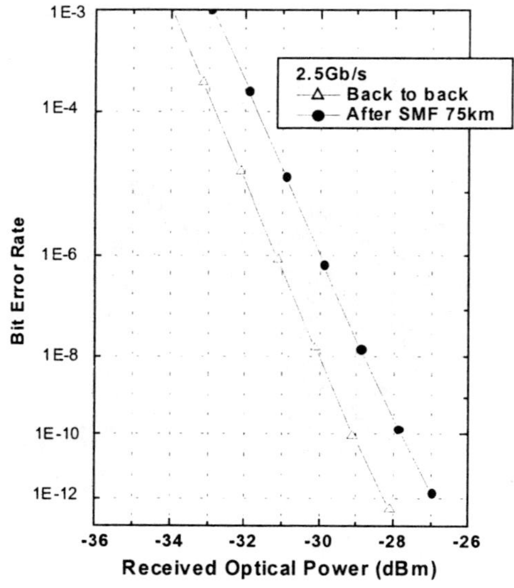

Figure 12: *BER for Back to Back and the Output from the OEXC*

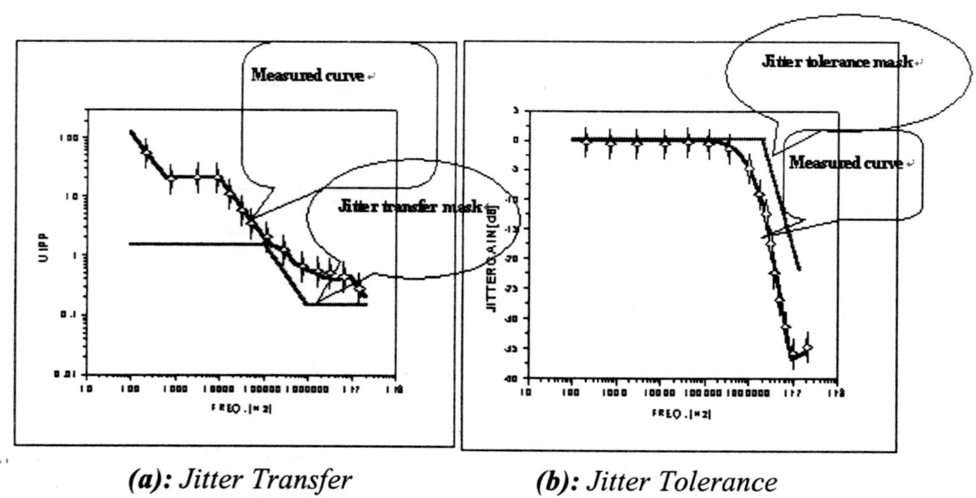

(a): *Jitter Transfer*   **(b):** *Jitter Tolerance*

Figure 13: *Measurement of Jitter Functions*

The jitter tolerance requirements are specified in terms of jitter templates, which cover a specified sinusoidal amplitude/frequency region. Jitter templates represent the minimum amount of jitter that equipment must accept without causing the designated degradation of error performance. The BER penalty criterion for jitter tolerance measurements is defined as the amplitude of jitter at a given jitter frequency. That duplicates the BER degradation caused by specified signal-to-noise ratio (SNR) reduction [4]. Again, the OEXC performed well, as shown in *Figure 13(b)*. The operating monitor program displays detected signals and alarms and routes input signals to desired output ports. *Figure 14* shows an operating monitor program. *Figure 15* shows a prototype of a 16 × 16 OEXC.

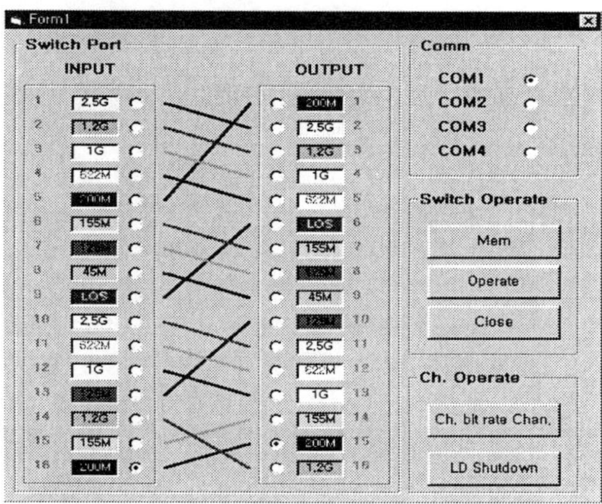

Figure 14: Operating Monitor Program

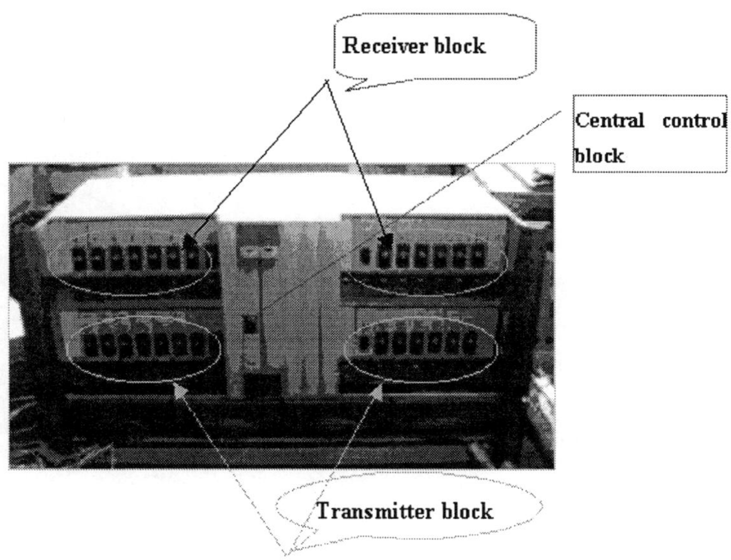

Figure 15: Prototype of a 16 × 16 OEXC

Conclusion

In this paper, the authors demonstrated for the first time a low-cost and compact OEXC, which has virtual bit rate transparency. In this system, eight bit rates corresponding to DS–3 (45 Mbps), FDDI (125 Mbps), ESCON (200 Mbps), fiber channel (1.062 Gbps), Gigabit Ethernet (1.25 Gbps) and SONET/SDH (155 Mbps, 622 Mbps, 2.5 Gbps) were implemented. More bit rates also can be implemented.

Acknowledgments

We would like to acknowledge Dr. Bongsin Kwark and Dr. Moon-Soo Park for the contribution of their ideas.

References

Yang, K.J., Jun-Ho Koh, Bongsin Kwark, and Moon-Soo Park. 1999. Optical transponder with bit-rate independent clock and data recovery (BICDR) for DWDM systems. *NFOEC '99:* 151–157.
Wolaver, Dan H. 1991. *Phase Locked Loop Circuit Design*. Upper Saddle River: Prentice-Hall.
TriQuint Semiconductor. 1998. *Triquint Data Sheet*. pp. 1–4.
ITU–T Recommandation 0.171. July 1997. pp. 4–8.

Optical Waves

Sunita Krishna
Senior Product Manager for Optical Waves
Williams Communications

Optical Wave Service

In introducing optical waves, it is important to highlight some details about what optical waves actually are. In layman's terms, an optical wave is a color of light. If light is passed through a prism, different colors or frequencies of light are seen on the other side. Each color can be called a wave. Transponders, which provide transparent, concatenated bandwidth to customers requiring high-bandwidth solutions, enable companies to sell colors of light.

It is called transponder-based service because it is the transponder that allows one to provide wave services. It is a concatenated service, which means that it provides high efficiency, high throughput, and less overhead. Transparency allows customers to complete synchronous optical network (SONET) rings. Optical waves make it possible to take half a ring to complete a SONET ring.

Today, customers in the telecommunications industry want to focus on the core competencies. Customers do not believe in the entire vertical integration. They want to own the areas in which they are the best, not the entire value chains. By buying the wave service, the network provider manages the core, and the customers manage the edge. Thus, it is a true model of e-construction, and it really drives outsourcing.

The entire industry is in the middle of a growth spurt, and the demand for bandwidth is exploding. The wave service currently provides optical carrier (OC)–48 services. It will provide OC–192 and, most likely, OC–768 in the future. Thus, it represents the perfect solution for high-bandwidth requirements.

Customers and emerging carriers are becoming more savvy. There are also new companies and new application service providers (ASPs) that want quick time to market, that want to get their initial public offerings (IPOs) fast, and that want to see their business models running quickly. Customers desire an easy way to build a scalable network, but they do not want to build it themselves. They want it as fast as possible so that they can earn revenue as fast as possible.

Mature carriers also want to complete SONET rings. These customers have experienced the demand for bandwidth, and they want to expand their networks to provide more bandwidth. The optical wave service can cater to both types of customers.

True Wave Applications

Waves can be used to complete SONET rings and diverse routes, to construct entire network infrastructures, and to build asynchronous transfer mode (ATM), frame relay, Internet protocol (IP), or SONET services over the infrastructures. Transparency is an important feature for the optical waves because it allows completion of SONET rings.

There are cases in which a carrier desires to complete a SONET half-ring with some level of bandwidth. Compatibility issues prevent completion of SONET rings by traditional private line. However, the optical wave service is transparent and can be used to complete SONET rings. That is one of the major applications for the wave service.

Transparent waves can also be used to complete diverse routes (see *Figure 1*).

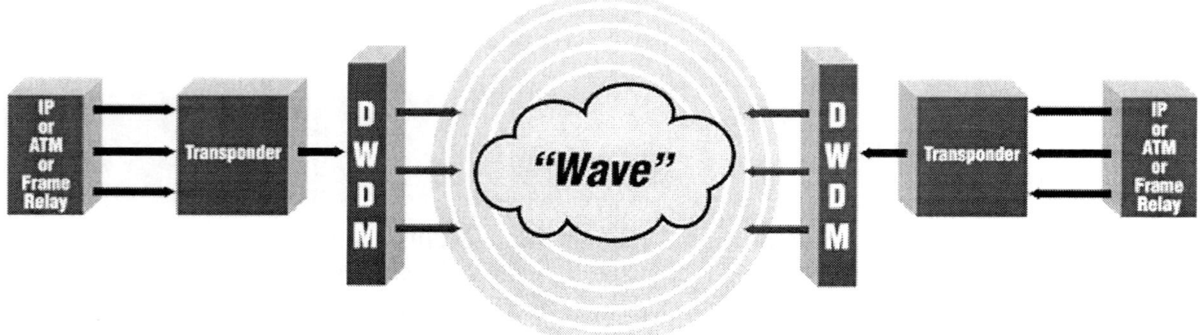

Figure 1: Transparent Waves

Wave customers typically do not buy one or two waves from suppliers. Instead, customers buy 20 to 30 waves because they have to build an entire wave network infrastructure and then add services such as ATM and IP over it. Carriers can buy waves as single-strand waves, as equipment-protected, dual-strand waves, or as fiber-protected, dual-strand waves.

Why would an Internet service provider (ISP) need the optical wave service? ISPs can use the service to expand capacity and to achieve a low-cost network that is scalable and transparent. Waves, then, are also a perfect solution for ISPs.

To reiterate, wave service is an ideal solution for customers who do not want to deploy fiber or optronics in the ground and that want simply to manage the edge solutions and not deal with the core. A network provider can handle the core for them, and the carrier can manage only the edge solutions.

The Future

The future is exciting. While providing OC–48 wave services, companies have already experienced demand for OC–192 and OC–768 services. One can imagine how quickly customers will require fat pipes.

The industry is also moving toward quality of service (QoS) within the optical domain, which represents unprotected wave services because they form a single strand of service. QoS also represents mesh protected–wave services and ring protected–wave services. This means that services are customized to the customers' requirements.

The service features of optical switching equipment enable companies to provision the different levels of restoration. This is an evolving and innovative technology where almost everything can occur within the optical layer and that essentially divides communications into three layers: a dense wavelength division multiplexing (DWDM) layer, an optical layer, and a service layer.

Wave service allows an ideal means of transport. Until today, if a company needed bandwidth for data traffic, they had to buy the most expensive bandwidth with high restoration. Mesh technology, however, will allow data carriers to buy different levels of restoration to fit a customer's needs.

Conclusion

Ultimately, then, wave service is the perfect transport to provide customized solutions for customers. Wave service allows customers to manage what they are best at. Customers can direct the edge solutions and can leave the headaches of the core to the network provider. Logical optical circuits will provide operations and simplicity and, thus, quicker time to market and quicker provisioning of services.

Routing of Wavelengths in Switched Optical Networks Based on Optical-Domain Noise

D. Levandovsky
Lead Engineer
Tellabs

S. Makam
Staff Engineer
Tellabs

V. Sharma
Research Engineer
Tellabs

Introduction

One important aspect of a dense wavelength division multiplexing (DWDM)–based optical network is its scalability. This usually means the ability to grow network capacity (e.g., adding more wavelength division multiplexing [WDM] channels to some fibers and providing additional nodes and connectivity) in small increments. Scalability of a network to very large capacities is made possible by the presence of optical cross-connects (OXCs) and wavelength converters, which permit a very large number of possible network topologies. The selection of an optical channel route through such a network must depend on the following factors:

1. The source and destination addresses or ports
2. The type of protection and auto-recovery that the network must provide in case of equipment and/or link failure
3. Quality of service (QoS) expected (in terms of bandwidth size, bit-error rate [BER], etc.)
4. State of the network (i.e., network availability and connection-blocking probabilities)

This paper will concentrate mainly on item three above, which deals with physical constraints limiting the possible light paths that may be provided by the network. One way to guarantee QoS is to design a network topology in such a way that every physically realizable light path delivers an adequate level of performance with no impact on any other light path. However, the resultant network design is often too inflexible and expensive to be considered for a practical implementation on a large scale. An alternative method would be to establish a set of guidelines (i.e., engineering rules) based on the desired QoS and on the physical limitations of the network elements (NEs) and fiber link parameters, which will be used in selecting a possible light path through a connection-rich network. The amount of recent activity in attempting to define a standard way for rapid, on-demand establishment of light paths through an optical multiwavelength network indicates the need for these rules to be built into the very protocol that is used to discover and provision such light paths.

This paper proposes a set of engineering rules that will guide the possible light circuit establishment based on the specified QoS. It also discusses some of the proposed signaling protocols for establishing dynamic paths through optical networks from the point of view of their capability to implement these engineering rules.

The ASON

Recently, a lot of activity in the optics industry has been directed toward enabling optical networks to perform automatic, end-to-end wavelength routing. The goal of such an automatically switched optical network (ASON) is to allow client network devices to request bandwidth dynamically from the optical network on an "as needed" basis [1]. The optical network consists of transmission and switching equipment that can provide point-to-point connection service to the attached client terminals. These connections can be of various sizes, such as a whole wavelength (supporting arbitrary signal encoding), a synchronous optical network (SONET) or Gigabit Ethernet–encoded leased line, or a time slot (i.e., a fraction) of a SONET line. Typical client network devices requesting services from such a network include SONET add/drop multiplexers (ADMs), asynchronous transfer mode (ATM) switches, and Internet protocol (IP) routers. Such dynamic control of the optical network is becoming feasible owing to the imminent availability of the transparent optical switching devices, which may be used as the intelligent routing elements in such a network. This paper is concerned with transparent wavelength connections across the ASON, as shown in *Figure 1*. It does not attempt to address the problem of optimum resource allocation in the ASON, such as minimizing the channel request blocking rates, etc. Those issues have recently been investigated in numerous papers [2, 3, 4]. This paper is concerned only with establishing whether or not a given light path, however it might be chosen, is capable of delivering the performance expected of it by the client devices.

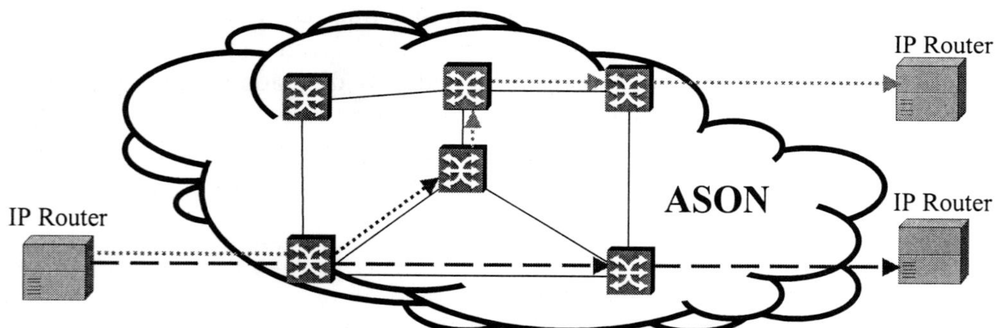

Figure 1: ASON with Several IP Routers Connected by Switched Wavelength Connections

Islands of Transparency

One way to satisfy physical constraints in the ASON is to divide it into a set of smaller subnetworks, as shown in *Figure 2*.

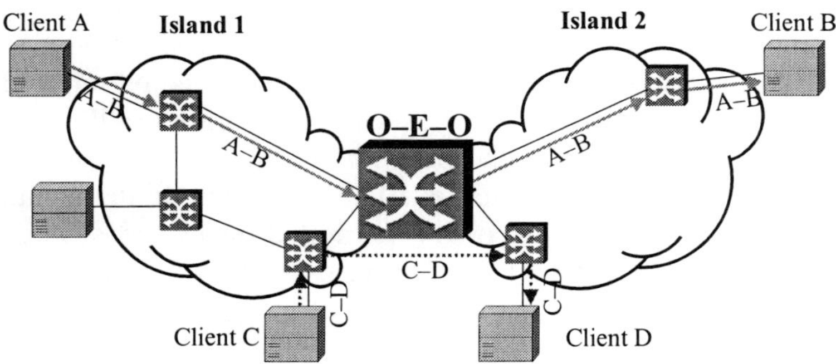

Figure 2: *ASON with Two "Islands of Transparency" Connected by Means of Optical-to-Electrical-to-Optical (O–E–O) Regeneration*
The figure illustrates how clients C and D, which are geographically close to one another, would benefit if a transparent connection C-D were made available.

Every subnetwork constitutes a small "island of transparency," where every path that does not exceed a certain maximum distance and a certain number of maximum hops can deliver some universally acceptable QoS. The subnetworks are interconnected via O–E–O cross-connects (XCs), which are capable of regenerating the signals in the electronic domain, thereby eliminating the undesirable optical transmission impairments [5]. An ASON based on the "islands-of-transparency" metaphor is very simple to plan, but the resultant network design will be costly due to the large number of O–E–O devices that are needed. This is because all network clients—including the ones that may be close to each other geographically—that reside in different "islands" will have to establish a path through the O–E–O, as shown in *Figure 2* (path A–B).

Because an all-optical switching technology is expected to be far more economical in routing wavelengths inside an ASON, it is desirable to allocate O–E–O functionality in a more flexible manner. Namely, the route selection software must be capable of figuring out whether or not a given path between two network clients can be established transparently and with the desired QoS. In case no such transparent path is possible, the chosen route shall have to traverse an O–E–O device. However, if such a path exists and is available (path C–D in *Figure 2*), then the O–E–O device may be bypassed. This method is much more flexible than the "islands-of-transparency" one, as it permits one to use fewer O–E–O devices, which may be placed at strategically selected locations throughout the ASON. One may still choose to break down the ASON into multiple subnetworks, but the choice will now be based on administrative or business considerations rather than on physical constraints. Hence, the paper shall concentrate on the methods that may be employed in estimating the QoS for a transparent connection through such an ASON.

Physical Constraints as the Basis for Route Selection

Preliminary Assumptions
In evaluating a potential wavelength channel, one must consider the effects of self-phase modulation (SPM), group velocity dispersion (GVD) and polarization mode dispersion (PMD) as these tend to broaden (or sometimes compress) the propagating optical pulses, which eventually leads to neighboring

bits interfering with each other and thus causing additional nonlinear mixing effects. Such intersymbol interference is usually avoided by careful dispersion and optical power management along the transmission path. In this treatment, it is assumed that the intended light transmission path is properly dispersion-managed. This means that the fiber GVD and/or PMD is periodically compensated to ensure an acceptable level of pulse shape distortions at the receiver. It is typically accomplished via dispersion compensating devices at (some of) the amplifier sites. The accuracy and periodicity with which dispersion is required to be balanced at every such site is dependent on the desired bit rate and the type of fiber used on the links involved. It is assumed that the route selection algorithm will track the dispersion and its management on every transmission link in the path for the purposes of pulse-shape management. In particular, for a given bit rate (i.e., pulse power level) and wavelength, the maximum distance between the dispersion compensating devices, the maximum permitted dispersion deviation (and the associated dispersion power penalty), and the maximum distance between electronic regenerators must be established and made available to the route selection software.

In addition, it is assumed that the optical network operates in a steady state. This means that the effect of optical power transients (usually caused by various wavelength channels turning their powers "on" or "off") on gain parameters of the optical amplifiers in the system is not considered. In practice this means an additional physical constraint on the transient suppression performance (i.e., time constant) of the optical amplifiers. The required transient recovery time is inversely proportional to the number of amplifiers in the chain and, therefore, must also be validated by the route selection software [6].

All of the above mentioned constraints are considered verified (by the route selection software) and satisfied before the algorithm described in the rest of this section is applied.

Optical QoS Attribute

The main idea is to use a simplified method for estimating the QoS for a given tentative connection before it is admitted into the switched optical network. In a circuit-switched network, such as the ASON, where a light path is expected to carry a digital payload only, the QoS attribute of interest is the BER. Consequently, the authors' task is reduced to proposing a scheme that provides a fair estimate of the BER across the network for a light path with a given optical bandwidth. This approach reflects the fact that fiber attenuation and dispersion, being compensated for by the optical amplifiers and dispersion compensating devices, is not the dominant factor in analyzing the performance of a given light-wave circuit.

In general, in a static optical network, such as exists today, the optical design parameters for a given route are chosen depending on the fiber type, the available amplifier sites, the number of wavelengths, and the desired bit rate per wavelength. The prospective design of the optical transmission line must be carefully simulated on a high-speed digital computer to determine if the desired BER can be achieved.

Such a custom route–optimized approach is infeasible in a switched-network environment because such simulations require extensive use of computational resources, making the approach slow and prohibitively expensive to be used on a call-by-call basis. Moreover, even if such careful simulations could be carried out for every circuit setup, a lot of the design parameters are not known because in a dynamic network they may change as soon as another light circuit is added on the same physical link or switch node in a network. This would force one to plan for the worst possible network scenario in the analysis software (e.g., highest possible optical power losses, largest allowable average powers in every link, etc.), which negates the whole purpose of such careful simulation in the first place.

In this proposal, such unpredictability is taken into account by assuming the worst-case scenario in some cases (e.g., power levels) and by using accurate estimates in others (power loss, etc.), as will be shown below.

BER Estimation Algorithm

Before a circuit is admitted into the network one must estimate the BER for the connection. The BER will depend on the amount of power launched into a fiber on a particular carrier wavelength λ_k, the amount of noise added by the transmission and switching facilities, and the quality of the receiver on the other side of the connection.

First consider the noise added by the transmission and switching facilities. It is assumed here that a transparent wavelength connection is desired between two points in the network. These points may be customer sites, O–E–O regenerators placed between separate administrative domains in the network, or a combination of the two. The sources of noise in the optical switching facilities (transparent optical switches are assumed here) may be broadly classified as the following:

- Crosstalk (XT), which is added to the signal by light scattered from other channels present in the OXC
- Optical loss, which is the reduction of signal power upon transmission through the XC, and which is also detrimental to the BER

The switching facilities referred to above include transparent OXCs and optical ADMs (OADMs) that are used to route wavelengths through the network.

The transmission facilities typically include optical fiber and optical amplifiers, such as the ones based on erbium-doped fiber amplifiers (EDFAs) or on Raman amplifiers. The principal sources of noise in optical fibers are due to nonlinear scattering and wave mixing [7]:

- Stimulated Brillouin scattering (SBS)
- Stimulated Raman scattering (SRS)
- Cross-phase modulation (XPM)
- Four-wave mixing (FWM)
- Noise-induced timing jitter [9]

In addition, one must consider the effects of dispersion in evaluating the net contributions of these nonlinear impairments because the amount of noise added via some of these effects depends on the interaction length between multiple simultaneous bits that may be traveling on different wavelengths in the same piece of fiber (e.g., as in a WDM system). This is in addition to the dispersion management considerations affecting the signal pulse shapes that were discussed earlier.

The paper will now discuss how one may approach the task of estimating the BER along such a route. The principal quantity of interest along the path is the optical signal-to-noise ratio (SNR^o), given in photon number units as the following equation, where $n(z)$ denotes the total number of photons at a point "z" along the path, with $<...>$ indicating a time average over one bit time interval, $n(z)_{EXE}$ representing the total number of photons in the excess noise associated with a given signal, and σ^2 denoting the variance (i.e., the noise power) measured during the interval [8]:

$$SNR^o(z) = \frac{\left[\langle n(z)\rangle - \langle n(z)\rangle_{EXE}\right]^2}{\sigma^2} = \frac{G^2(z)\langle n(0)\rangle^2}{\sigma^2} \tag{1}$$

The optical power gain or loss at a given point along the path relative to input $z=0$ is denoted by $G(z)$. This equation, therefore, corresponds to an SNR of a single bit, which at this point remains unspecified (a "1" or a "0"). Squared photon number units are in deference to electrical SNRs, which use electrical powers, thus being proportional to optical powers squared. Use of photon numbers instead of optical powers is more convenient because it avoids making the treatment dependent on the bit rate and the photo-detector integration constant. For signals with Poisson statistics (so-called shot noise–limited signals), $<n(z)> = \sigma^2$, and $<n(z)>_{EXE} = 0$, which makes $SNR^o = <n(z)>$. The principal idea, therefore, is to treat the entire process of signal propagation from $z = 0$ to $z = L$ as a quasi-linear distributed amplification process (for a given fixed set of optical powers along the transmission path) where the net contribution of all the nonlinear interactions (and the spontaneous emissions inside optical amplifiers) are reflected in the degradation of the SNR^o. In this representation, therefore, the entire signal path is a chain of lumped or distributed amplifiers with lossy elements in between, or a single fiber amplifier, whose emission and absorption coefficients follow a periodic (or even discontinuous) distribution along the length.

A loss (as in fiber during propagation) is possible in this model, as $G(z)$ will be allowed to dip below the value of 1. To guarantee the validity of this approach, a certain set of constraints must be met on every link in the transmission path, as is discussed below. In other words, this paper treats the entire noise propagation problem (normally done via detailed computer simulations) via linearization approximation around the relevant transmission system parameters that are likely to prevail in the network. The results of the simulations will be needed to establish the values of the aforementioned transmission system parameters. If the parameter estimation process is accurate, the approach is likely to result in a good estimate of the BER.

The excess noise, represented here by $n(z)_{EXE}$, refers to an additional noise-generation process inside a transmission or switching device. Its exact probability distribution is not necessarily known except for its main features—the mean and the variance. This excess noise itself may consist of various contributions, some stemming from the amplified spontaneous emission process inside the EDFA, others due to FWM in the fiber. The measure of excess noise in an optical signal is given by the Fano factor, which is equal to 1 for the shot noise–limited signal:

$$F(z) = \frac{\sigma^2(z)}{\langle n(z) \rangle} \tag{2}$$

Every device in the transmission path, be it passive fiber, active optical amplifier, or an OXC, is characterized by its associated noise figure, conventionally defined as this equation (here $z = 0$ is assumed to correspond to the device input):

$$NF^o(z) = \frac{SNR^o(0)}{SNR^o(z)} \tag{3}$$

Presumably, a set of noise figures would be computed for every device in the network for several operating scenarios. Let M be the total number of modes detectable by the photo-detector at the end of the line (spatial, polarization, or frequency), so that $<n(z)>_{EXE} = MN(z)$. Only one out of M modes carries the desired signal. The noise, however, can be introduced via all M modes [9]. It is convenient to introduce the equivalent input noise factor, given by this equation, which corresponds to the equivalent average number of noise photons on the input per mode:

$$n^{eq}(z) = \frac{N(z)}{G(z)}$$

Consider first an EDFA. For the case of linear amplification of a single wavelength channel, its NF^o may be shown to be of the form

$$NF^o = \frac{SNR^o(0)}{\langle n(0) \rangle}\left[F(0) - 1 + 2n^{eq} + \frac{1}{G(z)} + \frac{Mn^{eq}}{G(z)\langle n(0) \rangle}\left(1 + n^{eq}VG(z)\right)\right] \tag{4}$$

where $n^{eq} = n^{sp}(z) - 1/G(z)$. n^{sp} is commonly referred to as the spontaneous emission factor, which is equal to 1 for an ideal amplifier (with complete medium inversion), and $N(z)$ takes on the meaning of the per-mode amplified spontaneous emission photon number. The V parameter is equal to 1 for a linear amplifier, and its meaning shall be explained later. Note that for a shot noise–limited input $SNR^o = <n(0)>$ and $F(0) = 1$. Furthermore, typically, in a realistic situation, $G(z) >> 1$ and $G(z)\langle n(0)\rangle >> N(z)$. Consequently, $NF^o(z) = 2n^{sp}(z)$ (i.e., the commonly known 3 dB limit). The expression in Eq. (4) can be used as the basis for estimation. For "ideal" (i.e., linear) lossy fiber spans where $G(z) < 1$, $n^{eq} = 0$, so that for a shot noise–limited input, $NF^o = 1/G(z)$, as expected. In a realistic situation, however, fiber (including gain fiber) is not linear and, therefore, in this linearized treatment, adds excess noise. The noise referred to includes FWM, cross-gain modulation in an amplifier with multiple wavelength channels, SBS, SRS, etc. The equivalent noise power, represented here by $n^{eq} \neq 1$, depends on the optical power per wavelength, the number of wavelengths simultaneously present, and fiber dispersion parameter (which depends on pump power and signal power for gain fiber). Because the number of wavelengths, and hence the total power, is subject to change at any time in a switched optical network, one should use worst-case scenario (maximum number of wavelength channels present) to obtain n^{eq} for every transmission element. Eq. (4) implicitly assumes that this additional noise has an amplified spontaneous emission (ASE)–like statistical nature, (i.e., a variance as in the Bose-Einstein distribution). This is also a very conservative estimate. If it is felt to be too conservative the V parameter may also be adjusted for a given transmission element (i.e., for an optical amplifier with no photons on the input $\sigma^2(z) = N(z) + VN^2(z)$, thus setting V = 0 makes it closer to Poissonian, whereas setting it equal to 1 makes it like Bose-Einstein). One may also extend this method to distributed amplifiers, where the signal is amplified as it propagates along the transmission fiber. This is relevant to the case of Raman amplifiers, which are gaining acceptance in the ultra-long-haul transmission market segment. In that case the equivalent noise factor may be shown to be the following equation, with the noise figure for Poissonian input as before: $NF^{o\pm} \approx 1 + 2n^{eq\pm}$:

$$n^{eq}_+ = \frac{a_0}{\alpha'_k} \exp\left(-\frac{a_0}{\alpha'_k}\right)\left\{ Ei\left(\frac{a_0}{\alpha'_k}\right) - Ei\left(\frac{a_0}{\alpha'_k}\exp(-\alpha'_k z)\right)\right\} \tag{5}$$

$$n^{eq}_- = 1 - \frac{1}{G} + \frac{\alpha'_k}{a_0}\left(\exp(\alpha'_k z) - \frac{1}{G}\right) \tag{6}$$

The two values of n^{eq} in Eqs. (5) and (6) correspond to forward and backward pumping configurations with α_κ'—a fiber absorption coefficient, which is assumed approximately the same here for the signal and the pump wavelengths—and $Ei(x)$ is the exponential integral function.

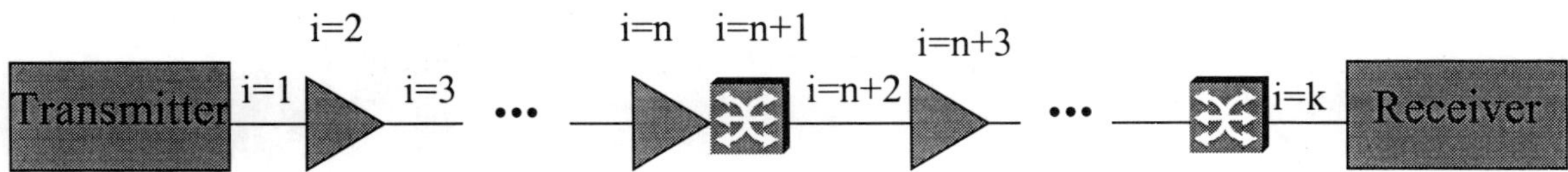

Figure 3: *Chain of Optical Transmission Elements*

Let us now consider a chain of such gain/loss elements as our complete transmission path (see *Figure 3*). It is evident from the definition of the noise figure that the total noise added by such a system can be characterized by $NF^{Total} = NF_1\, NF_2\, ... NF_{k-1}\, NF_k$ for a chain consisting of k elements. The definition of the noise figure, however, depends on a state of the signal at the input to the device (characterized by $SNR^o(0)$), which evolves in this situation from one element in the chain to the next. Defining the noise figure with respect to a Poissonian input state, therefore, would make it independent of its position in the chain and useful as a surrogate for the devices' aggregate noise performance. Setting $SNR^o(0) = <n(0)>$ and $F(0) = 1$ in Eq. (4), we write for the "i[th]" transmission element in the following chain, where n_i^{eq} is the equivalent input noise factor for that particular element, V_i is a parameter for the element as discussed earlier, and $<n_{i-1}>$ is the input number of photons:

$$NF_i^{Poisson} = 2n_i^{eq} + \frac{1}{G_i(z)} + \frac{Mn_i^{eq}}{G_i(z)\langle n_{i-1}\rangle}\left(1 + n_i^{eq}V_iG_i(z)\right) \tag{7}$$

With this definition the noise figure for a chain of transmission elements may be shown to be

$$NF^o(k) = \frac{\langle n_0\rangle}{\sigma_0^2}\left\{F_0 - 1 + NF_1^{Poisson} + g_1\frac{NF_2^{Poisson} - 1}{G_1} + g_1g_2\frac{NF_3^{Poisson} - 1}{G_1G_2} + \dots + g_1g_2\cdot\cdot g_{k-1}\frac{NF_k^{Poisson} - 1}{G_1G_2\dots G_{k-1}}\right\} \tag{8}$$

with

$$g_i = 1 + M\frac{n_i^{eq}}{\langle n_{i-1}\rangle} \tag{9}$$

and

$$\langle n_i\rangle = G_1G_2\dots G_i\langle n_0\rangle + M\left[G_1G_2\dots G_in_1^{eq} + G_2\dots G_in_2^{eq} + \dots + G_in_i^{eq}\right] \tag{10}$$

It is obvious from Eqs. (8) and (9) that it is very important to have low values of NF_i and g_i for smaller values of "i" (i.e., it is important to have a low-noise transmitter with sufficient power and low-noise fibers/amplifiers early in the chain). The latter should be the major consideration in the routing software's algorithm, which is used to select the physical medium for the signal transmission. In the case where the signal to each element dominates the noise (i.e., $\langle n_{i-1}\rangle \gg n_{eqi}$ and $g_i \approx 1$) and the input signal has Poisson statistics, the cumulative noise figure may be written as the following equation, which is the well-known Friis' formula widely used for electronic amplifier chains:

$$NF^o(k) = NF_1^{Poisson} + \frac{NF_2^{Poisson} - 1}{G_1} + \frac{NF_3^{Poisson} - 1}{G_1G_2} + \dots + \frac{NF_k^{Poisson} - 1}{G_1G_2\dots G_{k-1}} \tag{11}$$

Once the overall noise figure is computed, one may estimate the BER at the photo-detector. The process involves accounting for all the noise sources at the photo-detector and depends on the design details. The BER can be computed on the basis of the Q parameter (widely known as Personick's Q-factor), defined as the following, where $<i_1>$ and $<i_0>$ refer to average photo-currents corresponding to bits 1 and 0, respectively, whereas σ_1^2 and σ_0^2 refer to the corresponding variances:

$$Q = \frac{\langle i_1\rangle - \langle i_0\rangle}{\sqrt{\sigma_1^2} + \sqrt{\sigma_0^2}} \tag{12}$$

The latter depend in general on the optical noise power, the detection responsivity, and the ratio of the optical to electrical bandwidth, as well as the thermal noise added by the photo-detector/amplifier of the receiver circuit. The BER is easily obtained from the Q factor via this equation:

$$BER = \frac{1}{2}erfc\left(\frac{Q}{\sqrt{2}}\right) \tag{13}$$

The value of $Q = 6$ is required for a BER of 10^{-9}. For a simple case of identical noise variances $\sigma_0^2 = \sigma_1^2$ for zeros and ones (which may happen in an unlikely case where thermal noise dominates all other noise sources), the SNR is given by $SNR = 4Q^2$. In any case, every optical receiver will have some required SNR to achieve the value of $Q = 6$.

Network Control Considerations

Based on the results of the previous section one may estimate the BER for a given connection. Given the desired bit rate, the acceptable BER, and receiver parameters (e.g., ratio of the electronic to optical bandwidth, receiver power penalty, etc.) one determines the optical SNR required before the detection as well as the M parameter (which depends on the bit rate that will be on this particular channel). Subsequently, the expression in Eq. (8) is used to verify that the desired performance is attained. This could be arranged in one of two ways, either via extending the topology distribution protocol or through signaling protocol extensions.

The first method involves distributing all the physical constraints when the network is assembled and powered up. Because an IP style of control is currently being adopted for optical networks, the existing link advertisement protocols, such as open shortest path first (OSPF) or intermediate system–to–intermediate system (IS–IS), need to be extended to associate the physical parameters with every piece of transmission equipment in the optical network. These parameters will include a range (or ranges) of optical powers together with the associated $[n_i^{eq}\ V_i,\ G_i]$ for every piece of transmission equipment in the network, as well as an M parameter associated with every receiver for any given bit rate. As mentioned before, the equipment must be assumed to be carrying the worst-case load on all wavelengths except for the one being considered for this connection.

Consequently, a source node will be able to use Eq. (8) to determine if a given connection meets the performance requirements and should be admitted to the network. $[n_i^{eq},\ G_i,\ V_i]$ must be chosen carefully for the expected average input power levels for every link so as to result in realistic (erring on the conservative side) estimates for the noise figure. In the cases where average input optical powers are not known ahead of time for a given link, one can use "k^{th}" order polynomial expansion of the following form, where X_i is one of $[n_i^{eq},\ G_i,\ V_i]$:

$$X_i = X_i^0 + X_i^1 \langle n_{i-1} \rangle + X_i^2 \langle n_{i-1} \rangle^2 + ... + X_i^k \langle n_{i-1} \rangle^k \tag{14}$$

In these cases $k = 5$ or 6 is expected to be more than adequate in most situations. Simulations must be used to assign values to the aforementioned parameters for all different types of devices in the ASON.

The second way involves a signaling protocol, such as an appropriately extended version of resource reservation protocol (RSVP) or constraint-based routing–label distribution protocol (CR–LDP), to validate the quality of the proposed connection. The extensions are necessary to enable hop-by-hop distributed route selection. In this scenario, the aforementioned physical constraints are not distributed to every routing element in the network but rather are stored in the transmission equipment involved.

For instance, an OXC or an OADM may keep the parameters for itself and every fiber link attached to its ports in its own internal configuration memory. In this case, the algorithm will take advantage of the recursive nature of Eq. (8), with every device responsible for calculating its own optical noise figure and updating the "running total," using Eq. (8), which is passed along to the next routing element in the chain. The parameters to pass in this case are the noise figure so far—NF_i, the average number of photons $<n_i>$, and the cumulative products of gains and excess noise coefficients: $g_1...g_i$ and $G_1....G_i$. Using these four numbers and the M parameter, the NF_{i+1}, $<n_{i+1}>$, $g_1...g_{i+1}$ and $G_1....G_{i+1}$ can be determined. This procedure must terminate when a receiver is reached or when the resultant optical SNR gets outside the desired range. If the latter condition is reached, the signaling protocol will have to backtrack some number of hops and start exploring another possible route for the connection. This is similar to the signaling capabilities of the private network-to-network interface (PNNI) protocol specified for use in the ATM networks. The advantage of this approach is that the noise-figure calculation for a given element is done locally and could employ a more sophisticated and precise method than the one employed in Eq. (7). This method may be entirely proprietary and implemented internally by an equipment vendor. Also, certain OXC designs will have different physical attributes (e.g., optical losses) depending on which pair of ports is picked for the connection.

With this method, actual numbers could be used instead of worst-case estimates for such OXCs. Moreover, the deterioration in noise performance (e.g., increased optical losses due to aging, etc.) could also be kept in check locally with greater simplicity (in the first approach they would have to be advertised periodically to every routing node in the network). The down side of this method, of course, is the fact that it requires a more sophisticated signaling protocol that permits such hop-by-hop route reservation with the added attributes.

The ultimate decision on whether a given path satisfies all the performance requirements must be made on the receiving end, where the final BER is computed. The receiver, therefore, is expected to send the final reservation message, which is used to turn on the connection at every intermediate routing node.

References

1. See for instance <http://www.odsi-coalition.com/odsi_docs/odsiFunctSpecApril21.doc>
2. Ramamurthy, B., D. Datta, H. Feng, J. P. Heritage, and B. Mukherjee. Oct 1999. Impact of transmission impairments on the teletraffic performance of wavelength-routed optical networks. *Journal of Lightwave Technology* 17 no. 10: 713–723.
3. Alanyali, M., and E. Ayanoglu. April 1998. Provisioning algorithms for WDM optical networks. *Proceedings of the IEEE INFOCOM 1998*: 910–917.
4. Sabella, R., et al. Nov. 1998. Impact of transmission performance on path routing in all-optical transport networks. *Journal of Lightwave Technology* 16 no. 11: 1965–1971.
5. Saleh, A. A. M. July 1998. Islands of transparency—an emerging reality in multiwave optical networking. *1998 IEEE/LEOS Summer Topical Meeting on Broadband Optical Networks and Technologies*: 36.
6. Sun, Y., A. K. Srivastava, J. L. Zyskind, J. W. Sulhoff, C. Wolf, and R. W. Tkach. Feb. 13, 1997. Fast power transients in WDM optical networks with cascaded EDFAs. *Electronics Letters* 33 no. 4: 313–314.
7. Kaminow, I. P., and T. L. Koch. 1997. *Optical Fiber Telecommunications*. Academic Press, ISBN: 0123951704 (v. A); 0123951712 (v. B).
8. Agrawal, G. P., 1997. *Fiber-Optic Communication Systems*. (2nd ed.). New York: Wiley, ISBN: 0471175404.
9. It may also be desirable to keep track of the path averaged excess noise (an additional physical constraint parameter) because in an ultra-long-haul communication system these random intensity fluctuations will couple (via SPM) to carrier frequency noise, which results in additional pulse timing jitter at the output. This parameter is defined as follows:

$$\overline{\langle n(z) \rangle}_{EXE} = \frac{1}{L} \int_0^L \langle n(z) \rangle_{EXE} \, dz \tag{F1}$$

The timing jitter induced by amplifier ASE on optical pulses called solitons is known as the Gordon-Haus timing jitter and can be an important source of bit errors.

Ringbusters: With Optical Mesh, Carriers Can Stop Going in Circles for "New-World Bandwidth"

Nitish Mandal
Manager, Strategic Markets
Tellium

The spread of optical networks from long-haul to metropolitan rings and now to customer premises is creating an ocean of bandwidth for voice and data communications users. At the same time, there has been a phenomenal growth in demand for bandwidth due to the increased use of the Internet. To meet this growing bandwidth demand, established carriers have built their optical networks with synchronous optical network/synchronous digital hierarchy (SONET/SDH) ring architectures in North America and in other locations around the world. However, as we enter the next-generation network era, a question must be put to carriers and manufacturers: Are SONET rings enablers of new-world bandwidth and a successful future?

In many ways SONET has earned its stripes. The SONET ring architecture is a proven technology; incumbent local-exchange carriers (ILECs) and others have invested tens of billions of dollars in equipment to support SONET rings; and besides offering effective performance, rings have demonstrated advantages in operating through faults in the network. In addition, as Lawrence Gassman, president of Communications Industry Researchers (CIR), points out, restoration of disabled rings is fairly rapid.

But SONET ring–based networks have limitations, and the time may come when we thank SONET for its contribution to the public network infrastructure and embrace the solutions that will take us to the next-generation networking environment.

The Bandwidth Tidal Wave

Recent advances in optical technology have resulted in an exponential increase in the traffic-carrying capacity of optical equipment. From only 16 wavelengths in 1996, dense wavelength division multiplexing (DWDM) technology has progressed so rapidly that equipment capable of supporting several hundred wavelengths is on the horizon. This, coupled with advances in fiber cable technology, creates a bandwidth tidal wave that crashes on the shores of carrier core networks. The rapid growth of the Internet, on the other hand, together with newer applications and services—such as leased lambdas, data mirroring and backup, video-based applications, e-commerce, and other high–bit rate services— make it imperative that this bandwidth be quickly and effectively harnessed.

New-world bandwidth must be rapidly provisioned and must have flexible quality of service (QoS) mechanisms. Moreover, exponential growth in Internet traffic and increasingly unpredictable demand patterns make the need for a scalable network architecture paramount. SONET ring architectures that were well suited for the old-world networks—where most of the bandwidth was dedicated to voice—now lay siege on a carrier's ability to unleash the power of dynamic, flexible, just-in-time bandwidth.

According to telecommunications consultant Alan Stewart of Network Interface Corporation in Crystal Lake, Illinois, "Higher-rate optical channels, such as optical carrier (OC)–48s, need multiple frequencies, not time slots; hundreds, not tens, of wavelengths; switch-integrated add/drop multiplexer (ADM) functionality, not individual ADMs; mesh networks tailored to Internet protocol (IP) data flows, not point-to-point connections designed for predictable voice traffic. Implementing such techniques by overlaying stacked SONET rings with point-to-point wavelength division multiplexing (WDM) creates a nonscalable architecture."

Consequently, carriers are forced to look for other topologies. In growing numbers, carriers are finding a solution in optical mesh design. According to a recent forecast from RHK, the market for optical cross-connects (OXCs), on which mesh networks are built, is slated to jump to $1.8 billion by 2003.

A Closer Look at Mesh Topologies—Enabling the New-World Bandwidth

While rings are well understood by the current digital literati, much less is broadly known about mesh. This section provides a brief review.

In mesh topologies, theoretically, each point in a network has a direct pathway to every other point in the network. Mesh networks, built on new-world OXCs, exploit a combination of IP– and SONET–based mechanisms to provide dynamic provisioning and fast deterministic shared protection. The benefits of mesh topologies for optical networks fall into five categories:

- Migration and scaling
- Deployment speed
- Capacity utilization
- Network restoration
- Operating costs

Migration and Scaling
Rings can be an inflexible architecture. The typical SONET OC–192 bidirectional line-switched ring (BLSR), for instance, is unable to transport OC–192 Internet backbone trunks. While an increasing number of router vendors offer OC–192 interfaces, carriers are unable to place the OC–192 circuits on their existing ADM–based rings. Thus, the highest available bandwidth is never made available to the client-layer elements, and the carriers must wait for the next-generation technology to be available to support BLSRs. This can present problems as networks migrate to Internet architectures to meet the exploding demand in IP traffic.

Mesh topologies, on the other hand, provide services at the highest line rate and can work seamlessly with IP routers. They enable routers dynamically to request and release bandwidth for relief on congested routes.

Deployment Speed
The rapid growth of Internet traffic is also increasing the need for long-haul capacity at a fast pace. With the Internet, users are as likely to be checking a site that is hosted halfway across the country as one that

is hosted locally. In a ring-based network, a typical long-haul circuit must be interconnected through many rings using expensive and inefficient back-to-back tributaries. This requires the availability of bandwidth simultaneously on all rings to provision the long-haul circuit.

In reality, however, the bandwidth may not be available in all rings, and the carrier must wait until the next buildout for the circuit to be provisioned. In the meantime, the available bandwidth is reserved on the other rings. The end result is a very slow provisioning process—typically measured in months. Carriers are forced to waste working capacity on their rings and are unable to respond quickly to rapid shifts in demand. Clearly, this is not the "bandwidth-on-demand" capability that newer Internet applications require.

Utilizing dynamic mechanisms that are inherent in the IP, mesh architecture enables point-and-click bandwidth solutions. This capability becomes increasingly important for two reasons:

- As bandwidth prices decline rapidly, carriers need other means to differentiate themselves.
- With unpredictable demand patterns that are getting tougher to forecast, carriers must be able to configure bandwidth on demand to prevent severe congestion in parts of their networks.

Capacity Utilization

For fully protected traffic, rings are unable to achieve 50 percent capacity utilization. Capacity may be available only in certain portions of a ring and not over the entire ring, thus forcing the carrier to provision an entirely new wavelength, while it is unable to exploit the unused portions of the ring. Thus, not all working capacity can be utilized to carry traffic. However, carriers are still forced to dedicate protection bandwidth for non–traffic bearing capacity as well. Mesh architectures, on the other hand, can offer utilization rates of 70 percent or more. Mesh also provides intelligent and flexible protection sharing.

BLSR protocol was specified with an additional capability: extra traffic that was meant to improve the bandwidth utilization of rings. Extra traffic was specified to use the dedicated protection bandwidth available in the ring and was preempted in the event of a failure anywhere on the ring. However, with no network database capability to support provisioning for end-to-end extra traffic circuits, preemptive traffic service is a no-show in ring networks. With end-to-end dynamic provisioning support, mesh networks, on the other hand, provide a unique capability to exploit preemptive traffic. For instance, a router requesting temporary bandwidth for congestion relief may provision a preemptive circuit on demand. The short-duration preemptive traffic can utilize the idle protection capacity without any significant impact on reliability. The likelihood of preemption is extremely low. In addition, the user network interface (UNI) specification currently being defined in the Optical Internetworking Forum (OIF) allows fast provisioning of short-duration preemptive traffic.

Network Restoration

The mesh architecture also supports QoS features that make more sense for the mixed voice, data, and Internet traffic that characterizes today's transmission needs. Circuits can be provisioned while taking into account restoration priorities and restoration times. This flexibility avoids the one-size-fits-all approach taken by ring-based networks. Mesh also allows the carrier to provide restoration times comparable to those of stacked rings for voice traffic.

Operating Costs

With rapidly falling prices for bandwidth, ring-based architectures are fast reaching a point where the cost of building a network exceeds the revenue they generate. Operational savings through optical mesh topology can be as high as 60 percent compared to ring structures, according to CIR's Gassman. A ring network, with its dedicated backup bandwidth, uses between two and four times the number of line cards as a mesh network of similar capacity (depending on the degree of connectivity of the nodes and the demand patterns).

Circle the Wagons...and Stop Circling

With the rapid spread of e-commerce and an exponential growth in Internet traffic, networks are clearly on the threshold of a new era. Customer demand for bandwidth and for new, more sophisticated services is soaring. Enabling new-world bandwidth requires a network characterized by rapid point-and-click provisioning, QoS–based restorations, and a least-cost-per-bit mindset. Newer service providers are offering consumers and businesses choices they never had before. With optical mesh, a carrier is not held hostage to purchasing a stack of equipment for enabling another wavelength on the stacked ring.

Optical mesh is a key enabler of the new-world bandwidth paradigm. SONET rings deployed now could well be the noose around carriers' necks. Carriers must evaluate their network architectures and decide whether or not the path to a successful future follows the traditional ring architecture or the optical mesh topology. Choices made now will have a major impact on how carriers compete in the future.

Why Is Turning Up and Maintaining a Wavelength So Difficult in a DWDM System?

Raymond Moyer
Senior Marketing Manager, Long-Haul DWDM Marketing Department
Fujitsu Network Communications, Inc.

Turning up a new wavelength while maintaining the integrity of existing wavelengths is becoming more and more difficult as today's state-of-the-art dense wavelength division multiplexing (DWDM) systems increase in size and complexity. Provisioning a new wavelength can require multiple technicians and up to several hours or days—even if the system is operational with live traffic carried on other wavelengths. What's worse, much of the DWDM technology deployed today requires that an entire system be taken out of service while the new wavelength is turned up.

Additional problems occur if a wavelength is lost for any reason. When a system is turned up, optical power balancing and gain/tilt parameters must be set to optimize performance. If a wavelength goes away, all other wavelengths change their power level. Wavelengths react differently—increasing or decreasing in power. Wavelengths that decrease in signal level will not be propagated successfully to the other end of the fiber. This is a serious problem: When one channel drops out of service, it affects or eliminates the ability of several other channels to reach the other end of the fiber.

Today's DWDM systems support 160 to 176 channels/wavelengths. Each wavelength can carry traffic capacity up to optical carrier (OC)–192, the equivalent of 129,024 phone calls. Every hour of every day that a wavelength does not carry traffic translates into many dollars of lost revenue. DWDM equipment requires a power-monitoring capability to maximize amplification. Today's products offer a variety of power-monitoring options, but not all of these methods will work in next-generation DWDM systems.

These same technologies can also be used to reduce ongoing maintenance and related downtime. As the fiber and other optical components age, their characteristics change. This change generally results in a shift of optical gain/tilt. Many of today's DWDM systems require out-of-service optical power balancing and gain/tilt correction as part of routine maintenance. Because this is a time-consuming procedure, it is costly from both a technician-expense and a lost-revenue perspective. Anything that reduces or eliminates the time required for out-of-service maintenance will improve a carrier's bottom line.

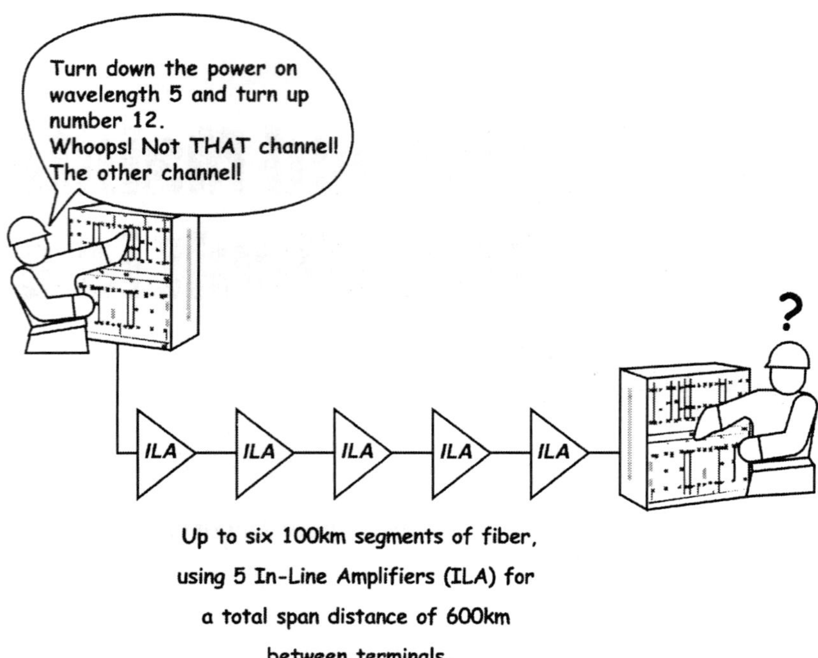

Up to six 100km segments of fiber,

using 5 In-Line Amplifiers (ILA) for

a total span distance of 600km

between terminals.

The Problem

Today, a typical DWDM deployment requires two technicians to set the optical power and gain/tilt on a wavelength. Each technician bears an average loaded cost of about $200 per hour. They must travel to terminals at opposite ends of the fiber, set up a spectrum analyzer to look at the received signal, and load the information into a computer to calculate the correct attenuator values to level the signal. This process introduces human error. Some DWDM systems employ external fixed attenuators; others use internal manual attenuators, or microprocessor-controlled variable attenuators. Initial settings must be transmitted to the technician at the opposite end of the fiber. In most cases, the technician manually sets these values, waits for the system to stabilize, then checks to ensure that the settings provided the correct results. This procedure does not always work the first time; unexpected results may occur because as one wavelength's signal strength is increased, others could decrease. Consequently, the process tends to be very iterative, especially if there are intermediate in-line amplifiers (ILAs) that also require a visit and manual adjustment.

One solution for simplifying the process requirements uses microprocessor-based variable attenuators. When spectrum analyzer data is obtained at the receive end of the fiber, a transaction language 1 (TL1) command is sent back to the transmit terminal to adjust the variable attenuator. This automatic adjustment reduces the time required for optical power balancing from several days to one hour; it also compresses necessary manpower needed to a single technician—saving time and money. Because time is money in the long-distance optical-transport business, the result is increased revenue and reduced cost.

Figures 1 and *2* present the output of a spectrum analyzer connected to a terminal at the receive end of the fiber. *Figure 1* shows that every wavelength initially carries a different signal or power level. There is a total power level threshold, above which nonlinear effects generate enough noise to make the signals undistinguishable.

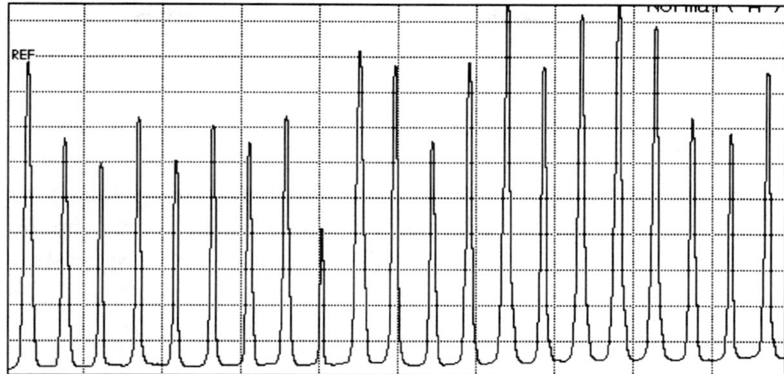

Figure 1: Initial Power Levels across the Spectrum

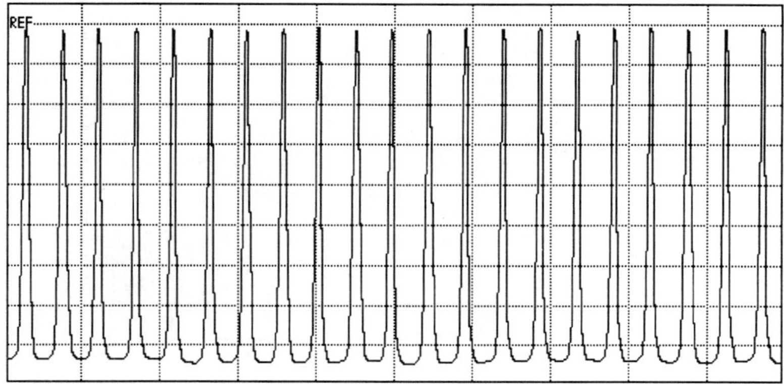

Figure 2: Final Power Levels after Power Balancing

If the signals in *Figure 1* are amplified to the maximum level allowed by the nonlinear threshold, each signal is increased in amplitude by that same percentage. This means that the large signals get larger while the small signals increase in amplitude but remain much smaller than the larger signals. These small signals may not have sufficient power to reach the other end of the fiber. When enough amplification is applied to the aggregate signal to allow the smaller signals to reach the other end of the fiber, the increased power level of the larger signals could damage the receiver or other optical components.

Figure 2 demonstrates the effect of an optical power-balancing procedure. In this example, attenuators on each of the optical inputs are properly adjusted to within 2 dB. The signal can be amplified to the nonlinear effects threshold, and all of the signals have an equal piece of the aggregate power. This equalization allows all the signals to be received successfully at the other end of the fiber.

Unfortunately, there is no easy way to set the attenuator on a new signal to ensure that it will work successfully with existing wavelengths. Each wavelength requires a different amount of power to reach a given level, and each wavelength will exhibit a unique reaction when others are changed or added. In most DWDM systems this means that the entire power-leveling-and-balancing procedure must be performed whenever a new wavelength is added.

Today's Solution

Most of the DWDM systems in operation today provide some type of optical power-monitoring capability that prevents the aggregate signal from being amplified into the nonlinear region. Because these systems typically address the aggregate signal level, they cannot make the adjustments required to balance the signal without a technician's help.

Systems with built-in microprocessor-based variable attenuators automatically make adjustments for added wavelengths without requiring outside intervention. Only a couple of DWDM systems offer this capability. In addition to the microprocessor-based variable attenuators, the system must measure the power level on each individual signal while the aggregate signal is in the transmitting terminal's amplifier. This function can be performed only with a built-in spectrum analyzer or with a spectral gradient and power meter (see *Figure 3*).

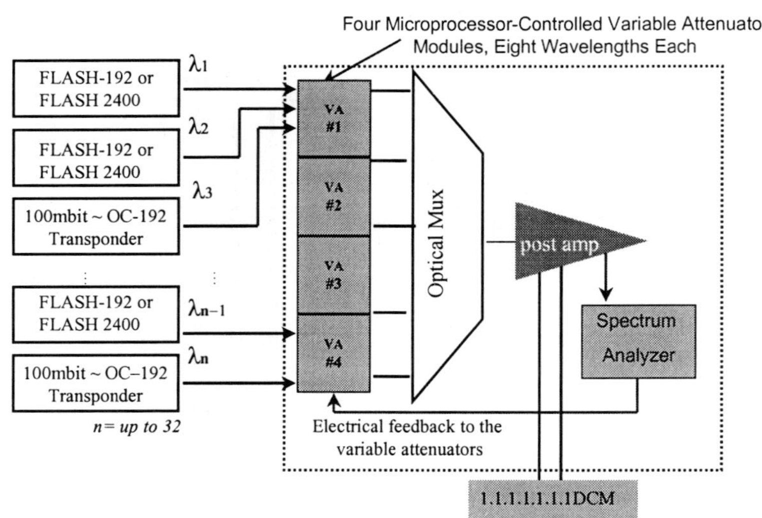

Figure 3: DWDM Terminal with Built-In Spectrum Analyzer

The spectrum analyzer digitizes the optical waveforms and determines what, if any, adjustments to make. A spectral gradient and power meter accomplish a similar function at a lower cost, but also with lower performance and reliability than a spectrum analyzer. The spectral gradient splits an aggregate signal into its individual light components on the power meter input. The power meter then measures the power of each wavelength simultaneously. This measurement is less accurate because the power meter must be uniform across the input surface or errors will be introduced. If a signal generates high power levels at initial turn-on or during operation, the power meter can be damaged at that position, or completely eliminating readings for that wavelength. In this situation, the entire power meter would have to be replaced to correct the problem once the error is isolated.

In the second situation, where a wavelength is suddenly lost or removed, a typical system would require technicians to travel to the system to re-balance optical power and restore wavelength conformity. For efficient operation, the specific problem should be pinpointed and repaired before the power-balancing procedure is performed. Otherwise, excess time is spent performing initial power balancing, then repeating the process when the lost channel is turned back on. Time added to the process by repetitive procedures results in longer system downtime, mounting repair bills, and lost revenue.

Systems with built-in spectrum analyzers or spectral gradients and power meters can automatically correct changes in power levels caused by a lost signals. When a signal disappears, the spectrum analyzer or spectral gradient and power meter detect the resulting power changes. The equipment then corrects the incoming power levels by sending a signal to the variable attenuators. Using a closed feedback control loop between the analyzer and the variable attenuators, a power balance can be restored within seconds. Data corruption is minimized, and system downtime is reduced from days to milliseconds. This automatic correction effectively monitors and controls any loss of revenue-generating traffic and eliminates the need for repair technicians to balance the system manually. Technicians can focus their efforts on restoring the signal.

Once the signal is restored, savings begin to multiply. Because the analyzer hardware will automatically re-balance the power when the wavelength is regained, the system will not require an out-of-service condition (that could last for several days) to effect a manual re-balance. Automatic re-balancing restores the wavelength faster and virtually eliminates the need for downtime on the entire system. The revenue-generating capability of the system is maintained throughout a limited catastrophic failure situation.

As a DWDM system ages and weather conditions change, the equipment will not pass light with the same efficiency or characteristics as it did at initial turn-up. This continual degradation requires that wavelength optical power be readjusted at regular intervals to optimize system performance. Many systems require a full out-of-service optical power balancing to address the problem. As discussed above, out-of-service conditions are expensive, time consuming, and hard on revenues.

A built-in spectrum analyzer or a spectral gradient and power meter continuously performs the optical power adjustments necessary to maintain optimum performance and does not require outside intervention. This capability greatly reduces maintenance requirements and eliminates the effects of slowly degrading system performance.

The Next-Generation Solution

As DWDM systems become more complex, the number of channels increases and the optical signal-to-noise ratio (OSNR) becomes a larger issue. Because the aggregate power-level threshold for nonlinear effects stays constant as the channel count increases, lowering the amplitude of each signal reduces the amount of power available to each signal. When the amplitude of the signal is reduced, the OSNR is also reduced.

Larger numbers of ILAs are required with longer total distances. Each ILA introduces noise into the composite signal and amplifies the noise already present. Noise increases in direct proportion to the distance a signal travels. The longer the distance a signal must travel, the more noise is injected into the composite signal. This increase in noise level acts to reduce the OSNR.

Therefore, as the number of channels and the distance increase, the OSNR becomes smaller and smaller. Why is that important? With a smaller signal and more noise, it is difficult to distinguish the signal from the noise (see *Figure 4*).

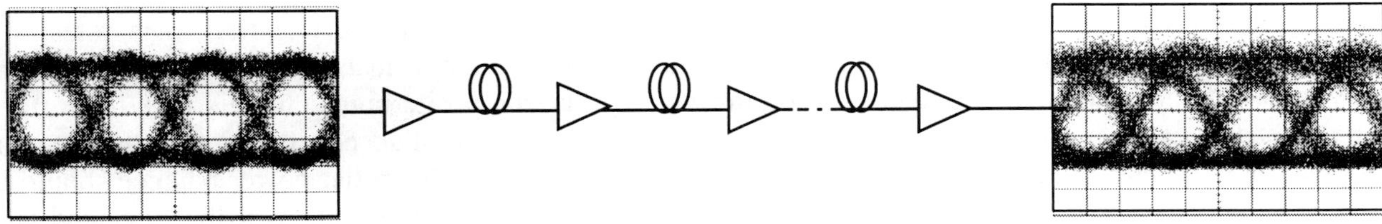

Figure 4: Eye Patterns Showing the Composite Noise Increasing from Transmitter to Receiver

Because the signal amplitude is reduced as the number of channels increases, the importance of maintaining a correct optical power balance grows. In a DWDM system with a good optical power balance, it is possible to amplify all the signals to their maximum amplitude at each amplifier stage. Because each amplifier introduces noise, fewer amplifiers operating at greater amplification push the signals over long distances and generate less noise through the system.

In high–channel count systems, the spectral-gradient and power meter measurement hardware does not work as well. This equipment must monitor and make adjustments to as many as 160 channels instead of the 32-channel systems it was designed to monitor. The increased number of channels means that spectral sensitivity must be uniform over a surface that is now five times larger, and the meter must perform five times the number of readings at the same time.

For next-generation DWDM systems, a spectrum analyzer is the only way to maintain maximum performance and move large quantities of data over the longest distance. At longer distances and higher channel counts, the signal must be analyzed at both the receive end and the transmit end of the fiber to optimize the performance.

The optical power-balancing process can be fully automated by connecting a built-in spectrum analyzer to both the transmit and the receive fibers (see *Figure 5*). Therefore, instead of analyzing the power at only the transmitting terminal, the spectrum analyzer also analyzes the optical power from the receiveend of the fiber. This eliminates the need for an external spectrum analyzer at the receive end of the fiber and for manpower to set up test equipment at either terminal. Simply push the button and system software sets up its own optical power balancing at turn-up. In addition, 50 percent of the in-service channels can be added or removed at once without repeating the optical power-balancing procedure. And, as the system and fibers age or change characteristics, the spectrum analyzer automatically makes the needed corrections.

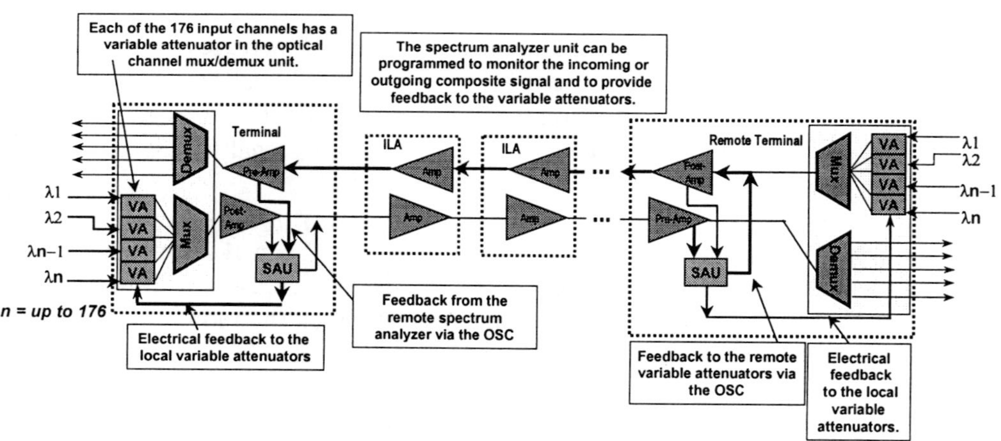

Figure 5: Spectrum Analyzers at Both Ends of the Fiber to Maximize Performance

Setting the correct signal tilt becomes even more critical as the number of channels, data rates, and distances increases. If all signals are equally balanced to maximize amplification at the transmit terminal, the signal at the receive end will not be optimized. As the composite signal travels down the fiber, nonlinearities cause individual wavelengths to move at slightly different speeds, creating a sloped signal instead of a flat one. To maximize the amplification across the fiber, the initial power balance should be set with the opposite slope so that the signal will become flat about halfway down the fiber and then increase after that point. When the spectrum analyzer is added to the ILA sites, balancing and tilting functions can access additional information to optimize system performance. This information also allows maximum amplification based on the tilt of the signal at intermediate sites. Because the tilt must be set by adjusting the individual channels, the ILA sites cannot influence the tilt, but only the aggregate signal (see *Figure 6*).

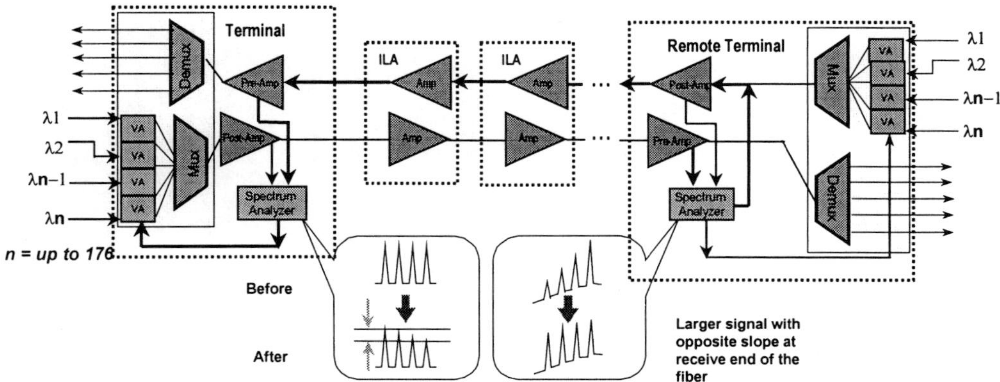

Figure 6: Effects of Optical Gain/Tilt on the Composite Signal

The latest-generation systems with built-in spectrum analyzers automatically optimize the critical signal-to-noise ratio (SNR) for maximum performance. The spectrum analyzer enables more signals with correspondingly smaller amplitude signals to be transmitted farther with less human intervention.

The larger number of signals transmitted simultaneously results in greater bandwidth per fiber. Longer distances offer engineering flexibility and decrease the number of expensive regeneration sites, resulting in a reduction of the total system cost. Spectrum analyzers reduce the ongoing expense of a system by reducing the need for highly skilled field technicians and engineers, and therefore fit the long-haul carriers' requirement for transmitting information over the longest distance with the most bandwidth at the lowest cost per bit.

With the energy business emerging as a new entrant in the long-haul market, bandwidth has become a trading commodity. This new market segment produces a need for faster provisioning, more bandwidth, and most of all, lower costs. The explosion of data and video traffic has generated a growing demand for more bandwidth to carry more data at higher speeds using more wavelengths. Built-in spectrum analyzers will continue to play an important part in system optimization and cost reduction as this immense growth continues.

Simplification of the Optical Layer in the Core Backbone Network by Use of Packet-Based Technology and Wavelength Multiplex–Level Protection

Takao Okamawari

Information and Communication Laboratories
Japan Telecom

Introduction

The exponential growth of data traffic accelerates the deployment of wavelength division multiplexing (WDM) transmission systems that provide cost-effective bandwidth. By using today's WDM technology, it is possible to carry more than one Tbps signal in single optical fiber. Moreover, Internet protocol (IP) routers have evolved rapidly in terms of routing speed and data interface rate. The routing speed of such IP routers increases to several hundred Gbps or higher, and the data rate of the interface card will reach to tens of Gbps in the near future.

These innovations impose several challenges. One challenge is related to the impact of a network failure. The failure of the optical transmitter/receiver or the interface card of the IP router will affect wavelength channel level, which results in the loss of the several tens of Gbps. Wavelength multiplex–level failure is caused by fiber cut, cable cut, and optical amplifier failure. Because single fiber can support about 100 wavelengths in today's WDM transmission system, the impact of the multiplex-level failure may reach to several Tbps. To construct a robust core backbone network, an effective and scalable restoration system is needed for both wavelength channel–level and multiplex-level failure.

Another challenge is related to total cost reduction of the core backbone network. The high-speed IP routers can connect directly to the WDM transmission system. This network architecture is referred to as IP/WDM, where the intermediate-layer technologies or systems, such as frame relay (FR), asynchronous transfer mode (ATM), and synchronous optical network (SONET), are bypassed. This is very attractive from the viewpoint of cost reduction and simplification of the backbone network. One question in this scenario involves implementing the restoration scheme for network failures because the intermediate-layer technologies such as SONET and ATM provide traditional restoration functions.

Several approaches have been reported to equip the restoration system in IP/WDM model. One approach is referred to as "optical-layer restoration" [1]. Both wavelength channel–level and wavelength multiplex–level optical restoration are possible. Basically, this approach is circuit switch–based restoration where the working traffic is switched to the predefined protection wavelength or to fiber, in the case of failure. Another approach evolved from the traditional IP restoration scheme. Current IP inherently has a degree of survivability because it updates the routing information table in distributed IP routers. This packet-based restoration method has a problem with restoration speed. In the worst case, it takes several minutes to restore the failed condition. However, new technologies such as multiprotocol label switching (MPLS) or space reusing protocol (SRP) improve the restoration speed within several hundreds of milliseconds or faster [2, 3, 4]. This means that the packet-based restoration will be one of the candidates for the restoration system in IP/WDM architecture.

In this paper, a novel network architecture using this packet-based restoration technique is proposed. In the proposed architecture, packet-based restoration is used for wavelength channel–level failures. At the same time, wavelength multiplex–level optical restoration is applied to the network to protect the wavelength multiplex–level failure. It is noteworthy that wavelength channel–level optical restoration system is not required (or optional) in the proposed architecture. The lack of the wavelength channel restoration system means the simplification of the optical layer and leads to the total cost reduction of the backbone network because complex wavelength routers or optical cross-connect (OXC) systems are not required.

Simple Optical Layer for a Data-Centric World

This section details the configuration of the proposed network architecture. The proposed architecture is compared with conventional layered architecture, where the network functions of the optical layer are independent of those of the service layer (e.g., IP, ATM, and SONET) [5, 6, 11]. The optical layer of the proposed architecture is very simple compared to the layered architecture because the optical layer functions are reduced with the help of the service layer's technology.

Layered Architecture
Figure 1(a) shows the main concept of the layered architecture. In this architecture, service-layer signals are accommodated in an optical path (OP)—a more general concept of the wavelength channel—by frame mapping or wrapping. The main purpose of this frame mapping/wrapping is the performance monitor and alarm signaling of each OP for the OP–level restoration. OP–level restoration is accomplished by switching from a working OP to a reserved one within several hundred milliseconds. Therefore, full redundant equipment is required not only in the optical layer but also in the service layer (e.g., working and protection interface cards are required in each IP router). Special hardware, such as high-speed optical switches and OP–level fault detection systems, is also required to realize fast OP switching. Additionally, OPs must be cross-connected for the purpose of add/drop, cut-through, and wavelength reuse. To realize the OP–level restoration and the cross-connect (XC) function, a complex OP XC system (OPXC) is required in the layered architecture, as shown in *Figure 1(a)*. These OPs are simply transported by point-to-point–WDM transmission systems (PP–WDMs). A remarkable point is that there is no restoration function in PP–WDMs.

Proposed Architecture
Instead of today's voice services, it is predicted that the data services carried by IP packet will be dominant in the near future. Additionally, the packet-based restoration technique is greatly improved, as already mentioned. To match these situations, the authors propose the novel architecture shown in *Figure 1(b)*. The network has three main features. First, the service-layer signals are accommodated in the OP by direct mapping. The term "direct mapping" suggests that the service-layer signals are transported without any special frame or wrapper for the optical layer, similar to the way in which today's PP–WDM simply

carries SONET signals without an optical-layer frame or wrapper. The second feature of the proposed architecture is that the OP–level restoration is realized at the service layers' function such as SONET restoration or IP/MPLS restoration. For services that do not have their own restoration scheme, such as Gigabit Ethernet (GbE) or wavelength rental service, some protocol for restoration may be optionally applied per wavelength basis. These types of services are shown as "others" in *Figure 1(b)*. The third feature is optical multiplex section (OMS)–level optical restoration system. OMS is the more general concept of wavelength multiplex level. The OMS–level restoration function is not used for OP–level failure but only for OMS–level failure. Consequently, neither frame mapping/wrapping nor OP–level restoration is required in the optical layer in the proposed network. Therefore the function of the OPXC in this architecture will be simpler than that in layered architecture. Actually, the OPXC function should be implemented to enhance optical-layer functions such as add/drop, cut-through, and wavelength reuse, and some kind of OPXC will be required in the proposed network.

Even in this situation, the function of the OPXC in the proposed model can be simplified. Basically, the design of the OPXC depends on the speed and frequency of the OP switching [7]. In the most complex case, the OPXC needs optical switches with millisecond-order switching time, OP–level fault detection function, and fast OP calculation function to realize OP–level restoration. The layered model described previously requires this type of complex OPXC. In the proposed model, fault detection function and fast OP calculation function are not required in the optical layer due to the lack of OP–level restoration. On the other hand, in the case of the proposed architecture, optical switching devices with several seconds of switching time are enough to XC the OP for fast OP provisioning. In the simplest case, OPXC may be replaced by manual connection at the fiber distribution frame for a semipermanent OP connection. Again, the complexity of the OPXC can be greatly reduced in the proposed model.

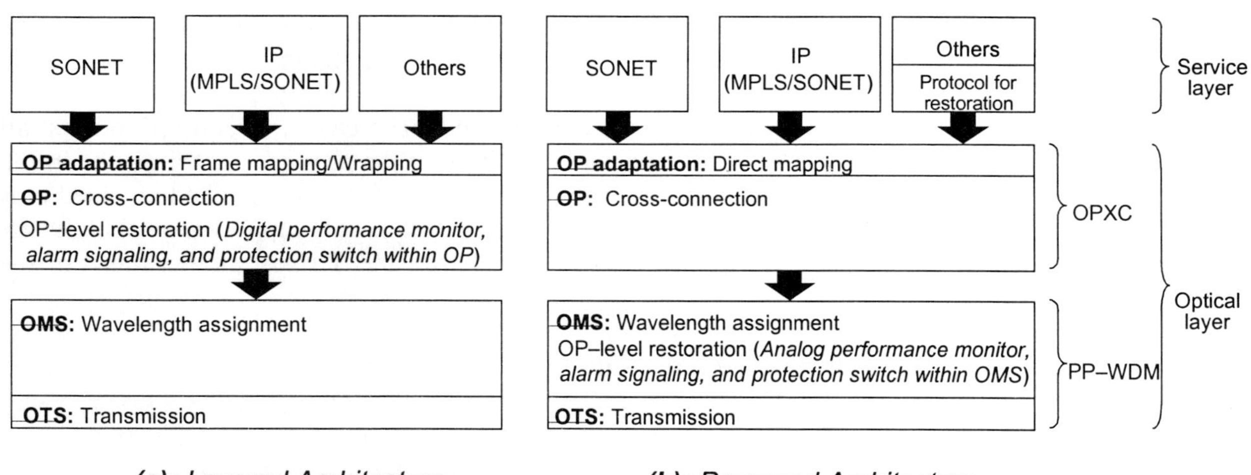

Figure 1: Comparison of Network Architecture

Comparison and Discussion

One of the most important differences between the layered architecture and the proposed architecture is the OP–level restoration scheme. In the proposed architecture, OP–level restoration is not implemented in the optical layer but in the service layer. This means that OP–level restoration is done by IP/MPLS in the IP–centric case. The elimination of optical restoration will be useful both for avoiding the overlap of the restoration function between layers and for reducing the total cost of the network. Another difference is that OMS–level restoration is effective for fiber cut, cable cut, and optical amplifier failure. In the situation where the IP services are dominant, the combination of packet-based restoration and OMS–level restoration will be a very effective solution.

To explain, an example of a physical ring network with three nodes is given in *Figure 2*. In this example, three nodes are connected by a logical, full-mesh topology. Each working OP has physical diverse wavelength as a protection OP for the reliability. In the case of OP–level restoration in the layered architecture, working OP (or wavelength) is used in the normal condition, and dedicated protection OP with a physical diverse route is prepared for the network failure. In the failure condition, the working wavelength is switched to protection (see *Figure 2(a)*). The packet-based restoration in *Figure 2(b)* can use not only working but also protection OP at the normal condition [2]. In other words, there is no distinction between working and protection OP; they are used at the same time in normal conditions. The restoration is accomplished by updating the routing information table in some IP routers, and the packets carried in failure OP are rerouted to the other OPs. Here, if the total bandwidth (or the number of active OPs) of the two architectures shown in *Figures 2(a)* and *(b)* are compared, it can be seen that the packet-based restoration in *Figure 2(b)* can support two times more bandwidth than the OP–level restoration in *Figure 2(a)* in the normal condition.

Another remarkable point is that OP–level restoration keeps its bandwidth after the failure, as shown in *Figure 2(a)*. However, *Figure 2(b)* shows that the bandwidth is reduced after the failure in the packet-based restoration. If there are numerous diverse routes in the network, the reduced bandwidth caused by single OP failure is so small that packet rerouting easily restores the failure. However, in the case of OMS–level failure, half of the bandwidth is immediately lost in the ring topology network (see *Figure 2(b)*). This example shows that the OMS–level failure in packet-based restoration significantly degrades the throughput of the packet-based network. To avoid this situation, the reserved protection fiber or bandwidth should be prepared for OMS–level restoration, as shown in *Figure 2(c)*. *Figure 2(c)* represents one of the examples of the proposed architecture shown in *Figure 1*, where both packet-based restoration and OMS–level restoration are used. Switching the bundle of wavelengths to the protection fiber at once effectively restores the OMS–level failure (see *Figure 2(c)*). This is why the author chose the combination of packet-based restoration and OMS–level restoration.

It is noteworthy that the redundancy of the interface card of the IP router is not required and that all of the interface cards are basically active at the normal condition in the proposed architecture In the layered architecture, half of the interface cards are inactive in the normal condition and are used only in case of network failure. This means that the proposed architecture can support more bandwidth than the conventional layered architecture can with the same hardware. This is another merit of the proposed architecture.

The above example compares the number of active OPs under the same number of total OPs. However, each model shown in *Figure 2* requires additional hardware. For example, complex OPXCs are necessary in *Figure 2(a)*, and additional fibers and optical switches for OMS protection are required in *Figure 2(c)*. Therefore the cost comparison is very difficult because the cost of the hardware greatly depends on topology, traffic pattern and quantity, required quality of service (QoS), and so on. Moreover, flexibility is needed to extend the network for future demand and requirements.

The most important feature is the possibility of the elimination of complex OPXC from the core backbone network. Complex OPXC is difficult to manage and control, and it is extremely expensive. On the other hand, the price for IP/MPLS routers and PP–WDM has been greatly reduced due to years of mass production. The author believes that the proposed architecture will be one of the most cost-effective and reliable network architectures for the future network.

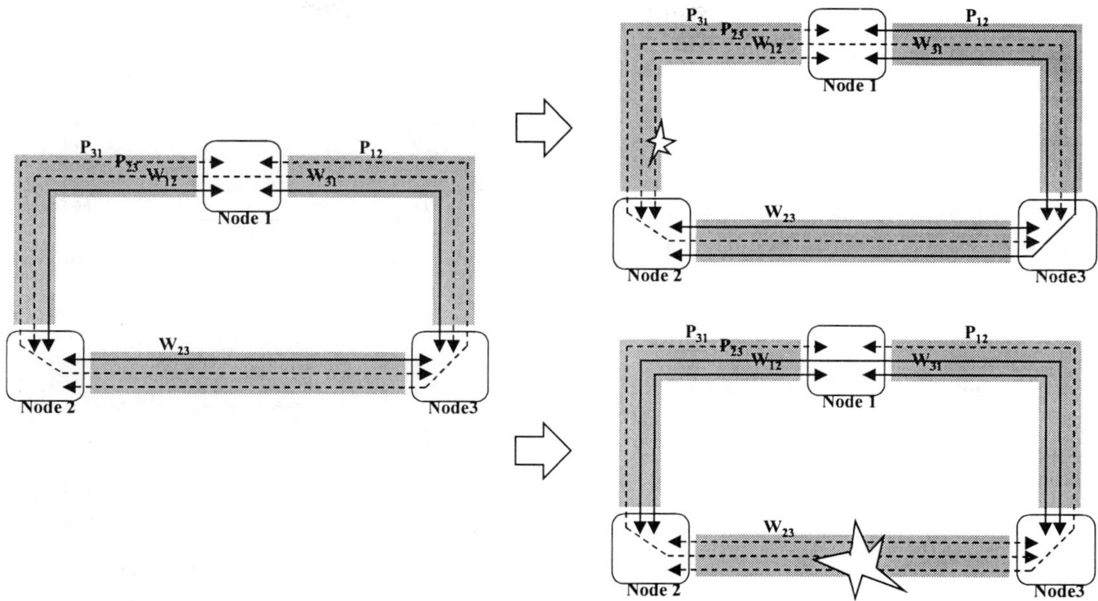

Figure 2: *Comparison of Restoration Scheme (Left: Normal Condition, Upper right: OP–level failure, Lower right: OMS–level failure)*
(a): OP–Level Restoration

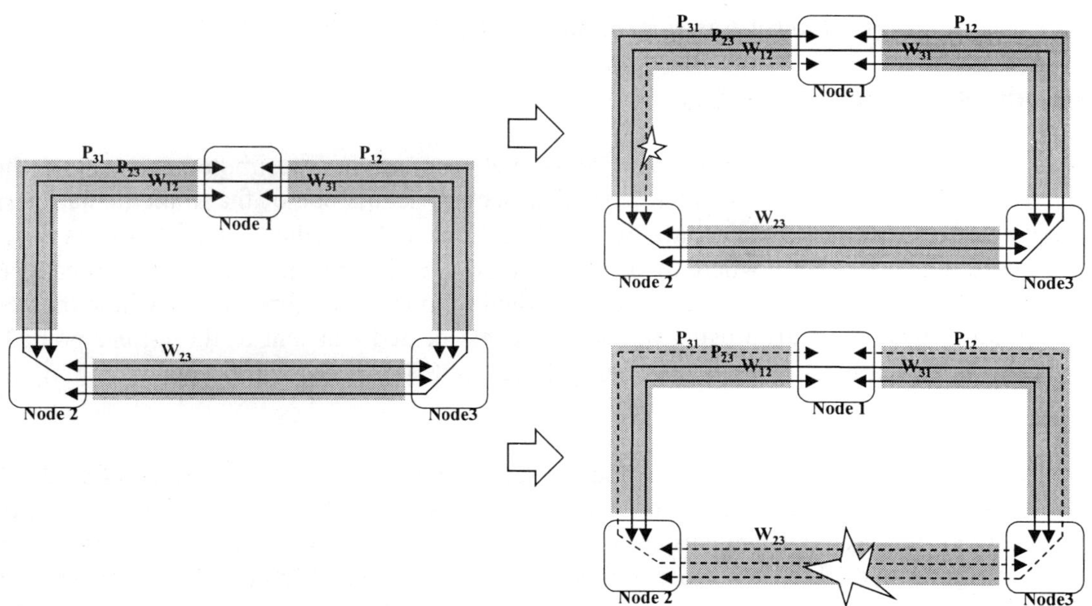

(b): Packet-Based Restoration

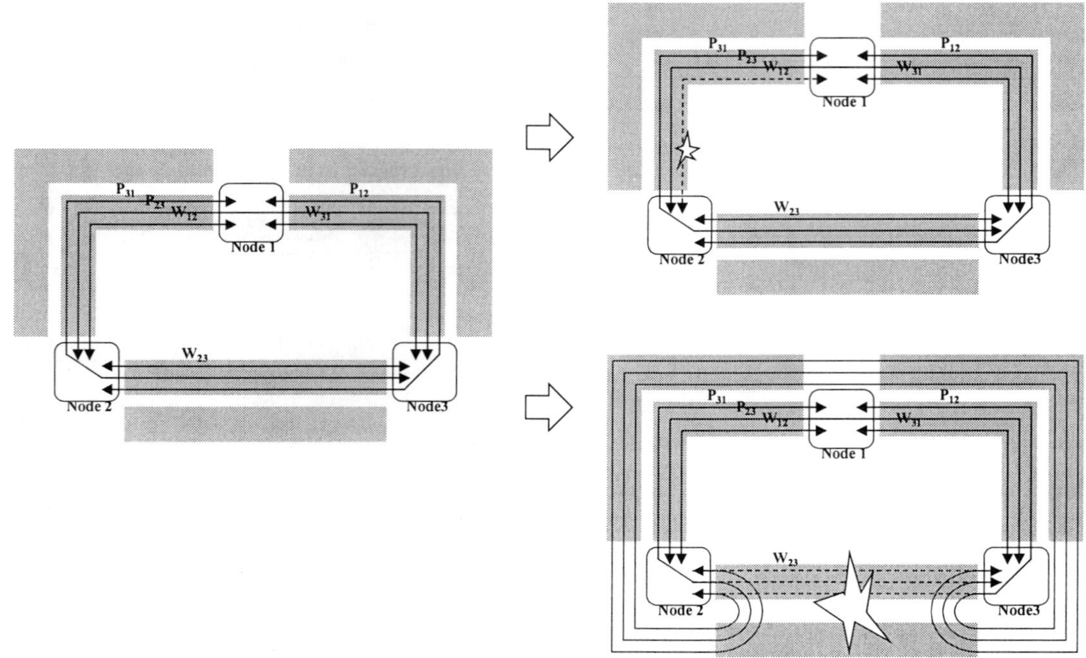

Solid line: Active OP W_{xy}: Working OP between node x and y

Dashed line: Inactive OP P_{xy}: Protection OP between node x and y

Hatched area: PP–WDM transmission system

(c): Packet-Based Restoration + OMS–Level Restoration

Implementation of OMS–Level Restoration

The proposed network architecture requires OMS–level restoration. The implementation of this OMS restoration system has not yet been standardized. This section details one of the implementation methods of OMS–level optical restoration. There are two main features in the method. The first one is the point-to-point (PP)–based network (PBN), where the network consists of PP systems that are regarded as network elements (NEs). Another feature is the restoration method proposed by Ellinas et al., where the network is restored by folding back the traffic using optical switches located adjacent to the failure point [9]. This section explains these features in detail, and then it discusses OMS–level restoration.

PBN

Optical network products are divided into node systems and PP systems. Consequently, the possible network is classified into two categories: the node-based network (NBN) and the PBN (see *Figures 3(a)* and *(b)*, respectively). A SONET ring system is an example of the NBN, and the node systems, regarded as NEs, are simply connected to each other with long-reach optics (see *Figure 3(a)*). An example of PBN is the PP–WDM transmission systems that are connected back-to-back. In this network, PP–WDM systems, which are also treated as NEs, are connected at the optical channel (OCh) level with short-reach optics. In the proposed OMS–level restoration method, the concept of PBN is applied, and each PP system in the network is regarded as an NE.

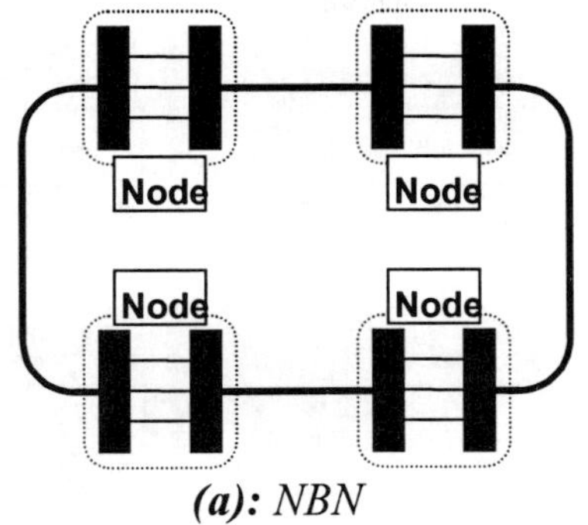

(a): NBN

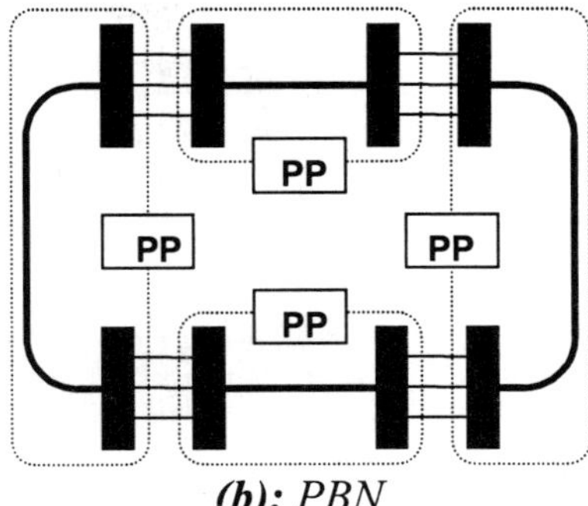

(b): PBN

Figure 3: *Classification of the Network*

Detail of the Restoration Method

To simplify the discussion, four-fiber bidirectional dedicated protection ring (4F/BDPR) is used as an example of the network [13]. The example is basically an opaque network where the noisy optical signal is cleaned up by optical-to-electrical-to-optical (O–E–O) conversion [14] Examples of 4F/BDPR with four nodes in terms of NBN and PBN are shown in *Figure 4(a)* [15] and *Figure 5(a)*, respectively. The OMS switch (OMS SW), which is adjacent to the failed point (see *Figure 4(b)* [15] and *Figure 5(b)*), does the restoration for both models. Though it is similar to today's SONET–bidirectional line-switched ring (BLSR) protection system, this type of restoration method is easily applicable to the network with mesh topology. Ellinas et al. showed that the network with plane graph topology could restore single-network failure by folding back the OMS SWs adjacent to the failed points. The detail of the plane graph and the mathematical background is given by reference [9].

The main difference between an NBN and a PBN is the location of the OMS SW. In NBN, the OMS SW is integrated into a node system. In a PBN, the OMS SW is divided into two parts within a node, and each OMS SW belongs to a distinct PP–WDM system. This difference affects the physical optical transmission property in the network. Generally, the optical properties of fiber spans are different from each other (e.g., the length of the span, the wavelength band, the characteristics of optical fiber, and so on). In such a situation, the WDM transmitter/receiver in the NBN must maintain two fiber spans (i.e., the fiber span that is used by a WDM transmitter/receiver before failure is different from the one after the failure) (see *Figures 4(a)* and *(b)*). Therefore, the transmission capacity in an NBN may be limited because it is very difficult to optimize the WDM transmitter/receiver to optical properties for several fiber spans. In a PBN, high-capacity transmission will be possible because the WDM transmitter and receiver can be optimized for the optical property of a single fiber span even after the failure (see *Figures 5(a)* and *(b)*). This means that a PBN is preferable for constructing a high-capacity network.

Another remarkable difference is the alarm signaling method between the NEs. In an NBN, to activate the restoration function, all the node systems in the network have to communicate for alarm signaling. Therefore some standardized protocol for the alarm signaling will be required in a multivendor environment. In a PBN, the alarm signaling is done independently within each PP system where the failure has occurred. *Figure 6* shows an example of the communication process for alarm signaling in a PBN. Some photo-detectors are placed in a PP system to detect an OMS–level failure. If the power level is below the threshold, the switching function for OMS restoration will be activated. These processes are done within each PP system independently. Therefore, a PBN allows use of a multivendor PP system in the network without a standardized communication protocol for alarm signaling.

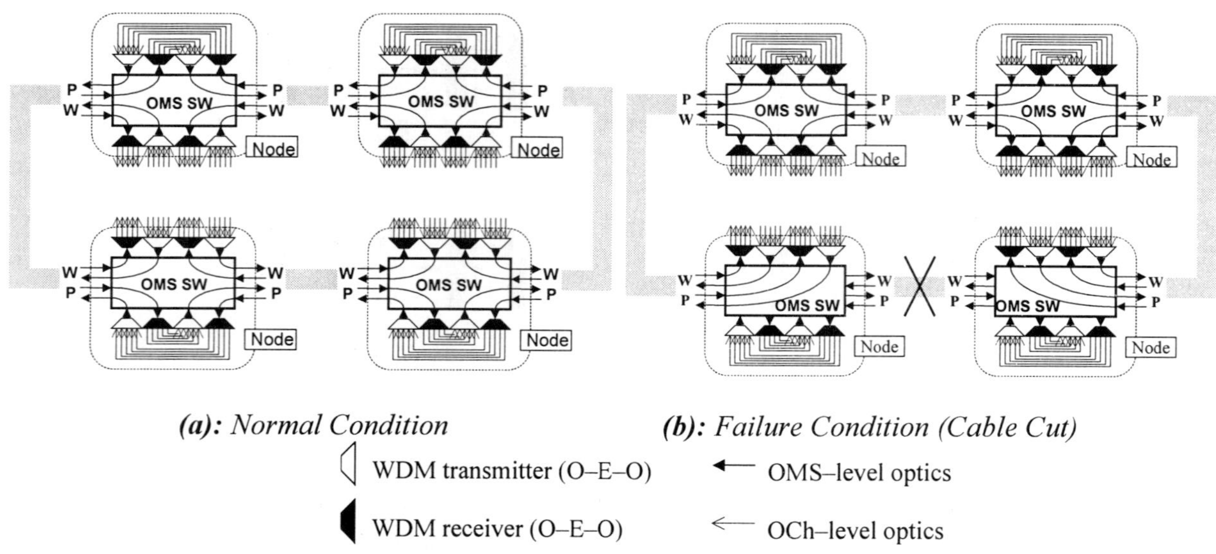

Figure 4: 4F/BDPR Configuration with NBN
("W" and "P" represent working and protection fiber, respectively)

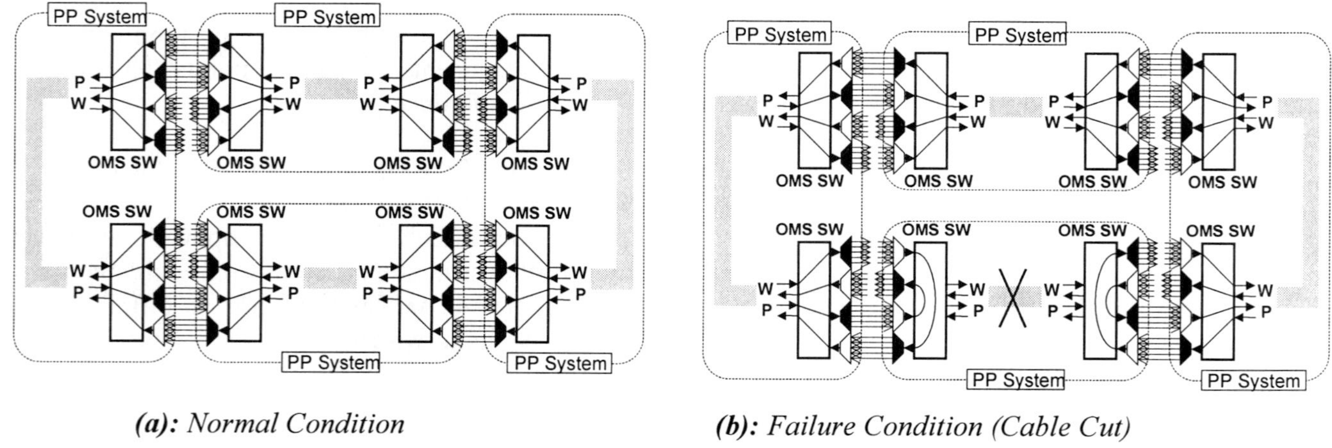

Figure 5: 4F/BDPR Configuration with PBN

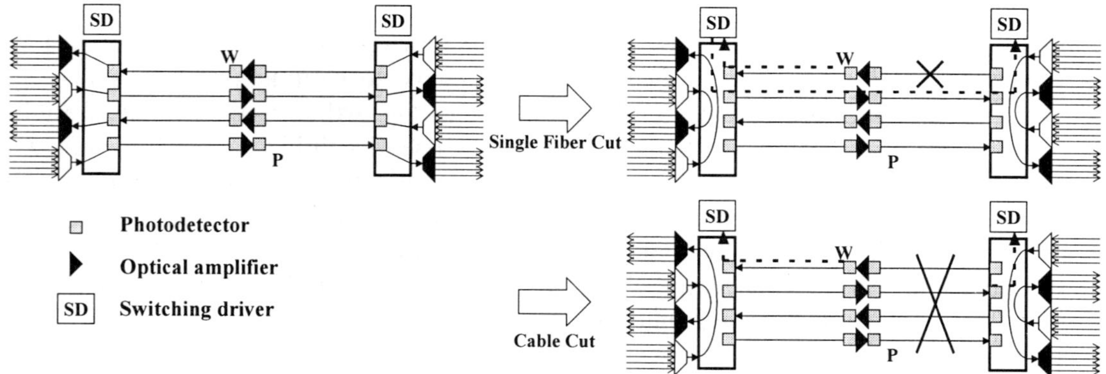

Figure 6: Signal Flow in the Case of Failure

A protection fiber in BDPR does not carry any traffic in a normal condition, but a protection fiber carries low-priority traffic in a bidirectional shared protection ring (BSPR) [13]. To add or drop the low-priority traffic, an add/drop switch (ADS) must be used (see *Figure 7(a)*). In case of failure, in addition to the OMS switching, all ADSs must switch into a "through" state (see *Figure 7(b)*). Here, the ADSs must communicate with the PP systems for alarm signaling in 4F/BSPR, and the switching speed for protection will become slower than that of the 4F/BDPR. Standard communication protocol between PP systems and ADSs will be required for the deployment of multivendor systems.

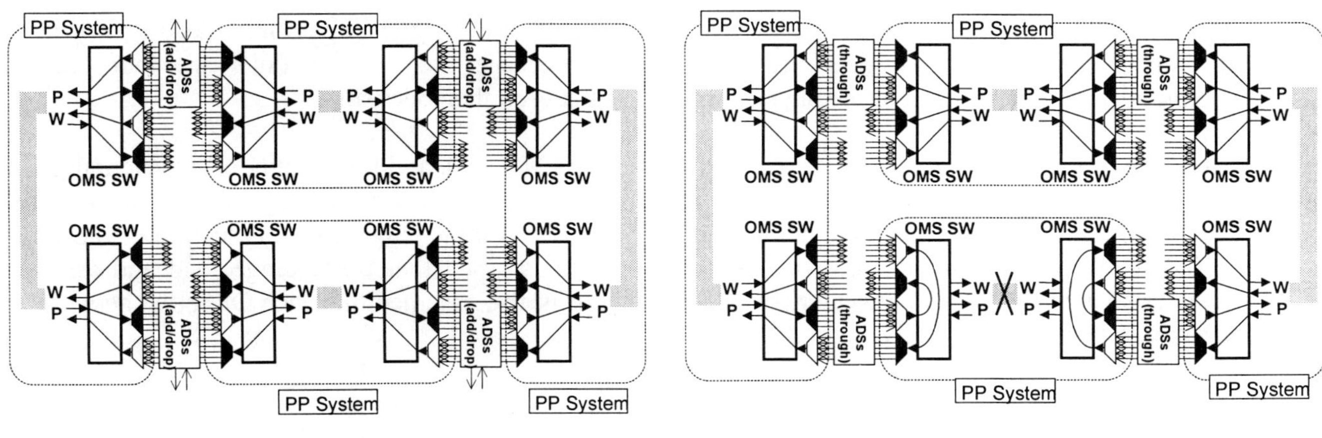

(a): Normal Condition   *(b): Failure Condition (Cable Cut)*

Figure 7: *4F/BSPR Configuration with PBN*

Discussion

This section presents a novel concept for the implementation of OMS–level restoration. The PBN concept is very attractive for multivendor environments and for high-capacity signal transmission. Though the example used in this section is ring topology, this concept is easily extended to the mesh topology if the topology of the network is a plane graph [9]. This OMS–level restoration method can be one of the candidates for the network architecture proposed in a previous section. But the proposed network architecture may adopt the other candidates of OMS–level restoration. More discussion is needed before OMS–level restoration is implemented in the network.

Packet-Based Restoration in the Large-Scale Network

The above section proposed a novel network architecture with packet-based restoration and OMS–level restoration. The key technology of the proposed architecture is the packet-based restoration scheme for OP–level failures. The packet-based restoration is done by fault detection, alarm signaling, optimum route searching, and updating of the routing table. To accomplish high-speed restoration, it is desirable to precalculate or predefine the optimum route for the restoration before the failure. To develop the packet-based restoration method for broadband optical network, this section introduces two simple rerouting methods for the packet-based restoration and compares them using numerical simulation. First, the ideal network model used in this analysis is given. This model has abundant bandwidth provided by optical technology. Then, two alternative restoration methods for single OP–level failure are introduced. Finally, the results of the comparison by means of the numerical analysis are shown with discussions and comments.

Network Model

The network model analyzed in this section is a physical ring network with N nodes. The physical ring consists of two fibers dedicated to clockwise or counterclockwise signal transmission, and the nodes are placed with equal spacing along the ring. Each node in the network has an IP router, and each pair of IP routers in the ring network is directly connected by OPs along both sides of the ring using the cut-through function in the optical layer. This means that each router pair has two physically independent bidirectional links. As a result, thanks to the optical technology, logical, full-mesh, duplex connections between routers can easily be obtained on the physical ring network. *Figure 8* shows an example of physical and logical topology for $N = 6$. The logical connection in *Figure 8(b)* corresponds to the topology of the overlay IP layer. Here, care must be taken that the logically parallel links do not have the same distance or the same delay time in general because the parallel links are independently accommodated on both sides of the physical fiber ring between the node pairs.

In this network model, the number of bidirectional links is $N(N-1)$, and the number of wavelength in the single unidirectional fiber is $N(N-1)/2$. Of course, two additional fibers are required to realize OMS–level restoration. However, they are not considered here, because the purpose of this section is to evaluate the OP–level failures. In the following discussion, it is assumed that the average load and statistical characteristics of the traffic between nodes is the same. It is also assumed that each OP in the network has the same transmission capacity.

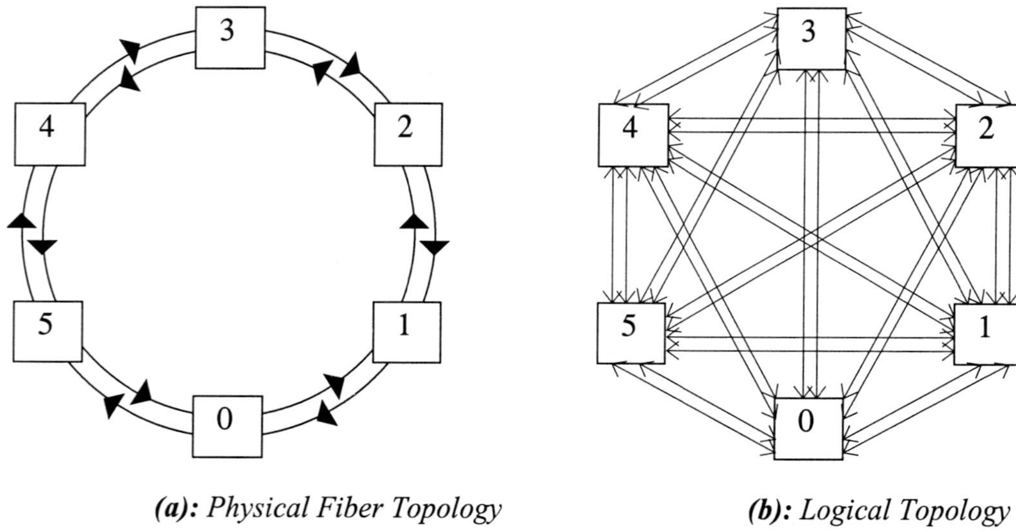

(a): Physical Fiber Topology　　　　*(b): Logical Topology*

Figure 8: *Network Model for the Analysis (Six Nodes)*

Rerouting Method

In the model presented in the previous subsection, packets can traverse the network with single hop using directly connected OPs between ingress and egress nodes (or IP router). Both of the logically parallel OPs are simultaneously used to carry the traffic. Single OP–level failure means that one of the parallel OPs is down. To restore such a situation, the authors introduce the following two alternative packet-rerouting methods:

(a) Use the remainder route.

> In this rerouting method, the remainder of the default single-hop routes carries all the traffic affected by the failure (the OP, which runs logically parallel to the failed OP) (see *Figure 9(a)*). In the figure, the OP–level failure occurs at one of the links between nodes 0 and 1, and the remainder route shown in the solid line carries all the affected traffic (i.e., the traffic

between nodes 0 and 1). As a result, it is predicted that the throughput of the affected traffic will be reduced significantly because the link capacity for the affected traffic is reduced by half. On the other hand, unaffected traffic will keep the throughput even after the failure.

(b) Use the remainder and the two-hop routes.

To avoid significant throughput reduction of the affected traffic, the affected traffic is rerouted to both the one-hop remainder route and the two-hop diverse routes (see *Figure 9(b)*). The traffic is equally distributed to these possible routes. If this technique is used, the throughput of the affected traffic is improved. If it is not, the throughput of the unaffected traffic is slightly degraded because the link capacity of two-hop routes must be shared by affected and unaffected traffic. In this scenario, it is easily predicted that the throughput will be enhanced for larger N due to the increase of two-hop routes. The number of diverse two-hop routes is given by $2(N-2)$.

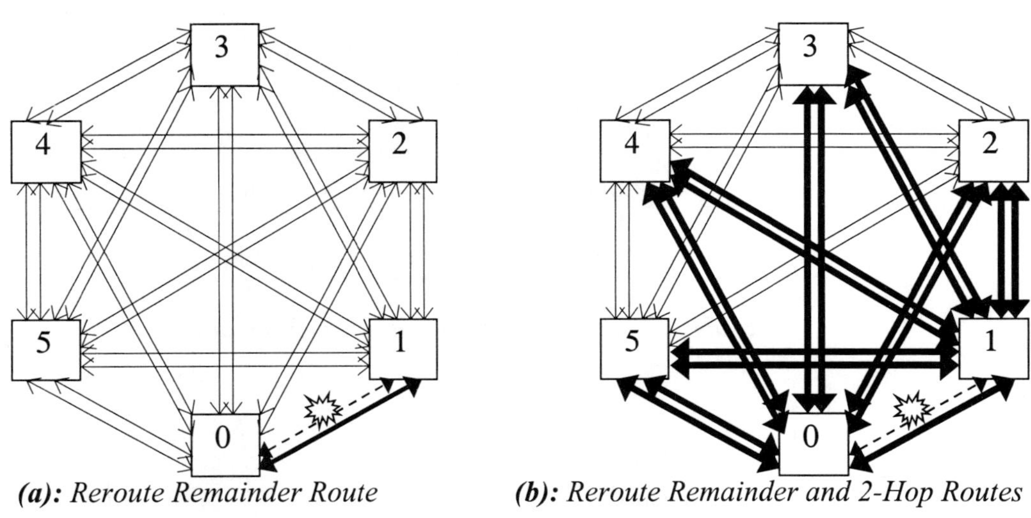

(a): Reroute Remainder Route   *(b): Reroute Remainder and 2-Hop Routes*

Figure 9: *Comparison of Rerouting Methods (Six Nodes)*
(Dashed Line: Failed Path; Solid Line: Path for Affected Traffic)

Numerical Simulation

To evaluate the throughput in the case of single OP–level failure, numerical simulation has been done with the network model shown in the "Network Model" section. *Table 1* summarizes the parameters used in this simulation. The throughput of the network was evaluated by the average packet delay of the affected traffic in the function of the normalized traffic load. In the calculation, total packet delay is obtained by the sum of transmission delay and queuing delay. Here, the term "average packet delay" means the average queuing delay from ingress to egress, which is obtained by subtracting transmission delay from averaged total packet delay. Normalized traffic load is defined as the average bit rate of the affected traffic divided by its link capacity.

First, rerouting method (a) was analyzed. The solid lines in *Figure 10* show the result, where the black circle and square represent the average packet delay before and after the failure, respectively. As predicted in the previous subsection, it is observed that the throughput degrades by about one-half before the failure. Second, rerouting method (b) was evaluated, shown by the dashed lines in *Figure 10*. The white circle and square represent the average packet delay for $N = 3$ and $N = 6$, respectively. The throughput is improved as the increase of the node number N. The throughput for $N = 3$ and $N = 6$ is about 65 percent and 81 percent of the normal condition, respectively. If the average traffic is set as a smaller value than this predicted value, traffic can be effectively restored in the case of single OP failure.

One of the drawbacks of rerouting method (b) is the ineffective use of the remainder route at the failure condition. The load of the remainder route is comparatively low because the route carries only equally divided affected traffic. On the other hand, the load of the two-hop route is heavy because its capacity is shared by affected and unaffected traffic. The flow of the affected traffic should be controlled to balance the load. Another drawback is the transmission delay at the failure condition. There are numerous two-hop OPs in the network for the failed OP. Here, the transmission delay depends on the physical route of the OP. This means that the packet propagates about twofold of the circumference of the ring in the worst case (which is why the throughput is evaluated by queuing delay). The worst delay corresponds to $(2N-3)$ msec for the simulation model. For a large-scale network, this value greatly affects the total delay of the packets. To improve these drawbacks, it is suitable to implement the traffic engineering (TE) technique provided by MPLS [10].

Network	Node number: N
	Distance between nodes along the ring: 200 km
	Circumference of the ring: N x 200 km
Traffic	Statistics: Exponential on-off
	Packet size: 250 byte
	Average on/off time: 1 msec
	Sending rate at on state: variable
OP	Capacity: 100 Mbps
	Transmission delay: five μsec/km

Table 1: *Parameter for the Numerical Simulation*

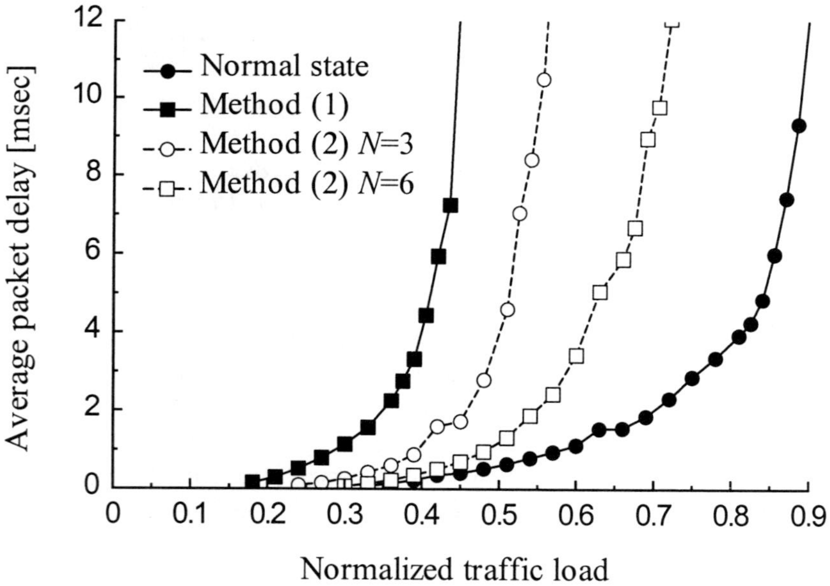

Figure 10: *Throughput in Failure Condition*

Conclusion

In this paper, the author proposed the novel network architecture for a data-centric core backbone network. The proposed network has both packet-based restoration for OP–level failure and OMS–level optical restoration for OMS–level failure. The basic idea of this architecture is a simplification of the optical layer or the OPXC by eliminating OP–level optical restoration function. As a result, the cost for OPXC will be significantly reduced, and the control plane for the optical layer will be very simple. Moreover, this network architecture is highly reliable for both OP and OMS–level failure.

One of the important technologies in the proposed network is OMS–level optical restoration. It is difficult to implement and has not yet been standardized. One of the examples for OMS–level optical restoration was discussed in terms of the PBN, which is very useful for the multivendor environment and for optical transmission system design.

Another important technology for this architecture is the packet-based restoration for OP–level failure. This paper evaluated the throughput of the network using numerical simulation. The author made the simulation model, where using optical technologies provided abundant bandwidth. The result of the simulation showed that the throughput of the network is improved by rerouting the packet to the multiple diverse routes. This means that the easy way for the effective restoration of a packet-based network is to make a number of diverse routes by using optical technology and to use these routes simultaneously at the failure condition. Moreover, the author believes that the throughput or delay of the network can be improved much more by using the traffic-engineering technique provided by MPLS technology.

The next-generation network should be optimized to the data-centric world. To realize such a network, the integration of optical and packet technology is essential [11, 12]. Optical technology provides abundant bandwidth, and packet technology provides powerful intelligence. The best combination of these advanced technologies must be found. The author hopes that the network architecture shown in this paper will be one of the candidates of the core backbone network architecture in the near future.

References

1. Afferton, T. S. 1999. Optical layer restoration—an operations perspective. *Proc. OFC'99*, (IEEE/OSA, San Diego) TuK1: 148–150.
2. St. Arnaud, B. 1999. Optical networks for the rest of us. *Proc. ECOC'99*, (SEE, Nice) 1: 2–4.
3. Dynamic Packet Transport Technology and Applications Overview. 1999. <http://www.cisco.com/warp/public/cc/cisco/mkt/servprod/opt/dpt/dpta_wp.pdf>
4. Hjalmtysson, G., P. Sebos, G. Smith, and J. Yates. 2000. Simple IP restoration for IP/GbE/10GbE optical networks. *Proc. OFC2000*, (IEEE/OSA, Baltimore) PD36.
5. Berthelon, L. 1999. Management of WDM networks. *Proc. ECOC'99*, (SEE, Nice) 2: 94–97.
6. Okamoto, S. 1998. Photonic transport network architecture and OAM technologies to create large-scale robust networks. *IEEE J. Select. Areas Commun.* 16: 995.
7. Kobrinski, H., and D. Zuckerman. 1999. Applications of automated cross-connection in the optical layer. *Proc. ECOC '99*, (SEE, Nice) 1: 64.
8. Okamawari, T., Y. Shimabayashi, and Y. Mukai. 2000. Optical multiplex section level protected network based on point-to-point WDM system. *Proc. OFC2000*, (IEEE/OSA, Baltimore) FE5: 90.
9. Ellinas, G., and T. E. Stern. 1996. automatic protection switching for link failures in optical networks with bidirectional links. *Proc. GLOBECOM'96*, (IEEE, London) 152.
10. Kuo, G., ed. 1999. Multiprotocol label switching. *IEEE Commun.* 37 no. 12: 36.
11. Luciani, J., B. Rajagopalan, D. Awduche, B. Cain, and B. Jamoussi. March 2000. IP over optical networks—a framework. Internet draft, *draft-ip-optical-framework-00.txt*.
12. Greenberg, A., G. Hjalmtysson, and J Yates. 2000. Smart routers—simple optics a network architecture for IP over WDM. *Proc. OFC2000*, (IEEE/OSA, Baltimore) ThU3: 292.
13. Johansson, S., A. Manzalini, M. Giannoccaro, R. Cadeddu, M. Giorgi, R. Clemente, R. Brandstrom, A. Gladish, J. Chawki, L. Gillner, P. Ohlen, and E. Berglind. 1998. A cost-

effective approach to introduce an optical WDM network in the metropolitan environment. *IEEE J. Select. Areas Commun.* 16: 1109.

14. Tkach, R.W., E. L. Goldstein, and J. A. Nagel. 1998. Fundamental limits of optical transparency. *Proc. OFC'98,* (IEEE/OSA, San Jose) WJ1: 161.

15. Fang, X., R. Iraschko, and R. Sharma. 1999. All-optical four-fiber bidirectional line-switched ring. *IEEE J. Lightwave Techonol.* 17: 1302.

Improving Network Efficiency with ATM–Based Optical Transport

Vish Ramamurti
Senior Member, Technical Staff
SBC Communications

This paper endeavors to answer the question of how to evolve from today's time division multiplexing (TDM)–based synchronous optical network (SONET) transport to a more efficient and cost-effective transport mechanism that can handle TDM as well as "data" traffic. The paper will briefly discuss current developments, the present method of operation, and the drivers for packet-based transport. It will compare SONET transport to transport solutions based on asynchronous transfer mode (ATM) technology. It will discuss industry trends toward the development of data-aware transport products. The paper will specifically focus on ATM transport products because they are ahead in the development cycle compared to other data-aware transport products, and because mature implementations are currently available. This paper concludes with a brief look at applications for ATM–based transport products.

Drivers for Packet-Based Transport

It is well known that in the last few years there has been an exponential growth in data traffic. In 1999, it is said that data traffic exceeded the voice traffic in carrier networks. SONET transport was originally conceived to carry voice and TDM circuits at the rates of digital signal (DS)–0s, DS–1s and DS–3s. Today, carriers face an ever-increasing demand for transport of "data" services such as Internet access, frame relay (FR), transparent local-area network (LAN), and cell relay. While the existing SONET transport network may be able to carry these services, it is not optimized to do so. However, packet technology has matured over the years and standards now exist, especially in the case of ATM, for efficient transport of TDM as well as data services. New services are being planned and deployed at companies such as SBC. Two such initiatives are asymmetric digital subscriber line (ADSL) rollout and voice trunking over ATM (VToA). These initiatives will generate abundant ATM traffic, so an ATM–based transport solution would be an efficient and cost-effective way to transport these services, at least at the edge of the network.

Figure 1 shows a high-level view of today's network architecture. SONET transport provides point-to-point bandwidth to connect different network elements, but it does not care about how efficiently the bandwidth "pipe" that it provides is utilized. Traffic within a "pipe" may be bursty in nature, so it is very possible that the "pipe" may not always be at its full capacity. SONET is oblivious of this fact. However, a data-aware transport solution, ATM in this paper, addresses this issue. *Figure 2* shows a schematic of a present-day local-exchange carrier (LEC) interoffice network. The boxes with broken lines represent central offices (COs). These offices contain cross-connect systems, voice switches, and possibly ATM and FR switches. SONET rings interconnect these COs.

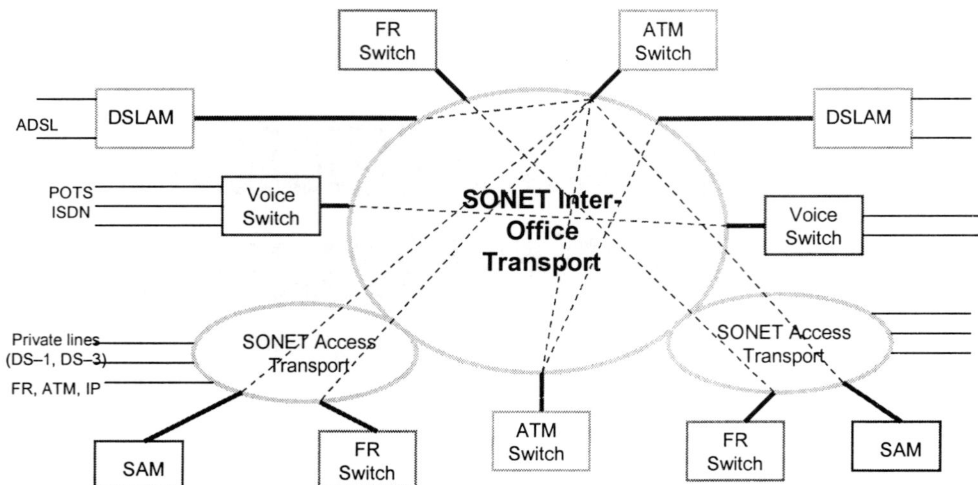

Figure 1: Current Network Architecture

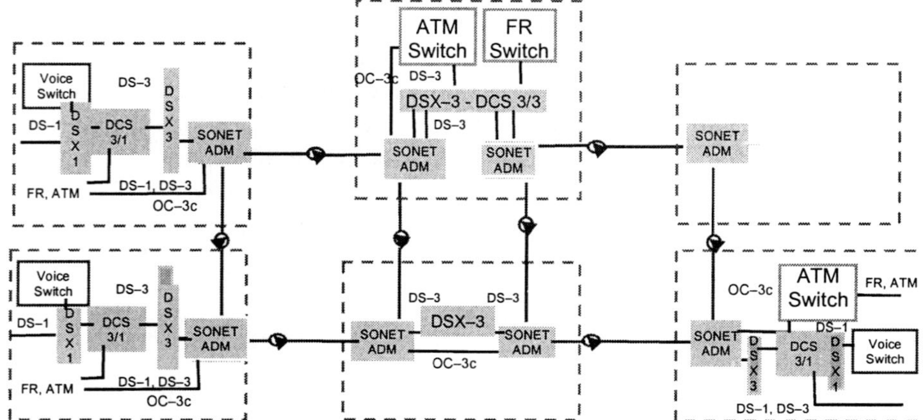

Figure 2: SONET–Based Transport Network

SONET versus ATM Transport Networks

SONET transport is a TDM–based approach. A number of multiplexing stages must be passed through to go from DS–0 to optical carrier (OC)–48 or OC–192. In each of the multiplexing stages, inefficiency is introduced if the higher-rate signal at any stage is not fully packed with the lower-rate signals. This is what happens typically in practice. DS–1s are not usually filled with the 24 maximum possible DS–0 signals. Similarly, a DS–3 is typically only half-filled with DS–1s. Higher-rate SONET rings have a granularity of synchronous transport signal (STS)–1, which makes partially filled STS–1s go wasted around the ring. In the case of an ATM transport model, multiplexing could happen in only one stage. An entire SONET OC–48c (concatenated OC–48) or OC–192c payload can be filled with ATM cells in a single multiplexing step, thereby occupying the SONET payload more efficiently.

ATM transport supports statistical multiplexing of data traffic. Different data interfacesadapted to ATM cells can be simultaneously mapped on to the same SONET higher-layer payload, and the averaging effect of the burstiness of the sources can be taken advantage of.

The International Telecommunications Union's (ITU's) I.630 standard specifies ATM–virtual path (VP) protection mechanism and error restoration within 50 ms. VPs at any arbitrary bandwidth rate can be protected on the ring unlike with SONET's fixed bandwidth protection at the STS–1/nc level.

The ATM–VP protection mechanism provides protection all the way up to Layer 2, whereas the SONET protection mechanism protects only SONET– and physical-layer defects.

Finally, while SONET protection mechanism requires dedicating 100 percent spare bandwidth for failure scenarios, the ATM–VP 1:1 protection mechanism allows traffic also to use the protection bandwidth. In case of a failure on the working side, higher-priority traffic takes precedence while using the protection bandwidth. Such a feature is not available on SONET TDM rings, although SONET bidirectional line-switched rings (BLSRs) allow "extra" traffic to ride on the protection channels that would be pre-empted in the case of a working-side failure.

Most of the benefits described above with the ATM VP ring (VPR) can be obtained by the use of other technologies such as Ethernet at Layer 2. However, products and standards in the use of these technologies for transport purposes are still not mature.

ATM Transport Developments

Current vendor developments have emphasized increasing the overall ATM switch fabric capacity and the adoption of higher network-side capacity (OC–12→OC–48) in the ATM add/drop multiplexers (ADMs). The aforementioned I.630 standard allows ATM layer protection mechanisms to achieve established SONET survivability performance levels. Development of ATM layer advanced protection switching protocols is underway to permit protection channel sharing by classes of service. This capability could potentially be used to carry unspecified bit rate (UBR) traffic on the protection side, to use the full-protection bandwidth under normal working conditions, and to use only available bandwidth on the protection side during a working-side failure scenario. Also, the forthcoming adoption of advanced ATM routing protocols (e.g., private network–network interface [PNNI]), will supplement ATM–VP layer protection for certain service classes and architectures.

Transport network elements are becoming more data-aware. Traditional ADM vendors are incorporating data switch fabrics into their products to create hybrid network elements. The data switch vendors, on the other hand, are also making their switches transport-aware by building transport functionality into their switches.

A pure ATM VPR adds and drops only at the ATM–VP level without TDM bandwidth management. A hybrid ring performs both VP and TDM (STS–1 and/or virtual tributary [VT]–1.5) bandwidth management

Figure 3 shows the different layers in an ATM VPR. Consider an OC–3 ring and logical STS–1s within the OC–3 bandwidth. The figure shows the arrangement of VPs and virtual circuits (VCs) within the

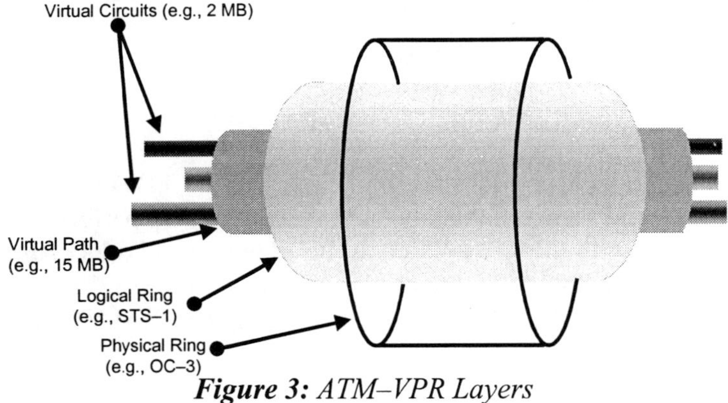

Figure 3: *ATM–VPR Layers*

STS–1s. *Figure 4* shows how one can obtain port and service consolidation through the use of an ATM VPR by setting up multiple VCs within a VP.

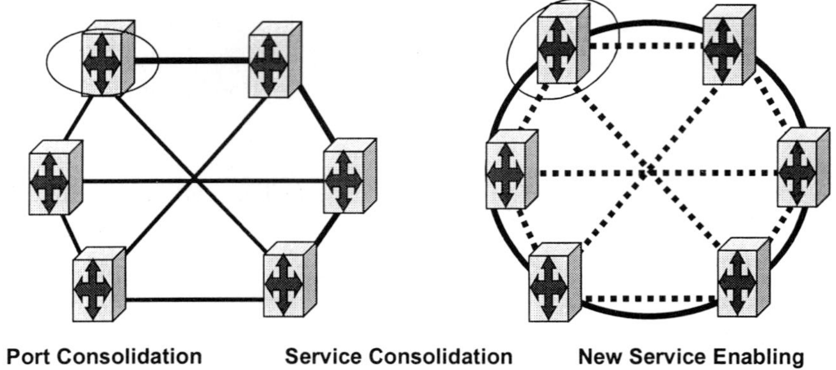

Figure 4: VPR Rationale—Flexible ATM Transport Bandwidth

Applications

This section examines how ATM VPRs would be useful in the VToA architecture (see *Figure 5*). A core of ATM switches takes the place of a voice tandem switch in the VToA application. In *Figure 5*, the trunk interworking function (TIWF) converts the voice circuits from the Class-5 switch to ATM "circuits" via circuit emulation and also communicates with the control signal–IWF (CS–IWF). In the figure, "EO" stands for "end office" and more specifically the Class-5 switch within the EO. The CS–IWF performs the interworking between the signaling system 7 (SS7) network and the ATM network.

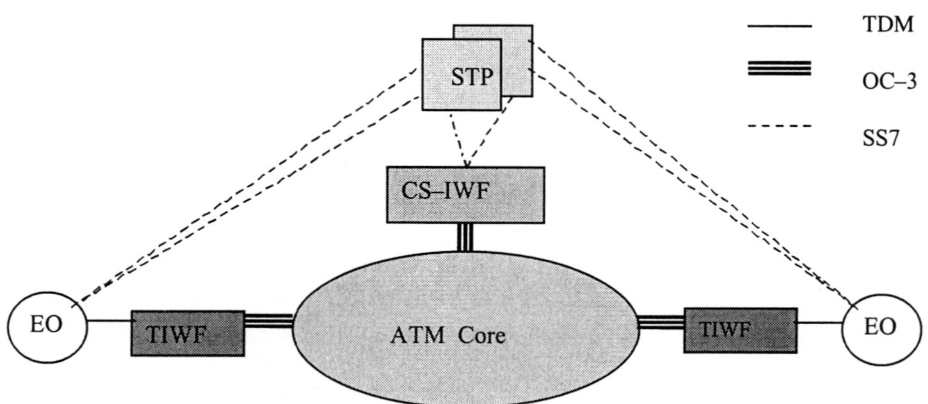

Figure 5: VToA Architecture

Figure 6 shows how an ATM VPR provides transport between the TIWFs and the ATM switches in the near-term architecture. It should be noted that an ATM ADM can also perform "circuit emulation." If the functionality to communicate with the CS–IWF could be added, the TIWFs could potentially be replaced in the future. Private-line traffic could also be carried in this architecture, again via circuit emulation at the ATM ADM. With so many ATM traffic-generating sources, the ATM VPR is an appropriate choice to provide transport functionality.

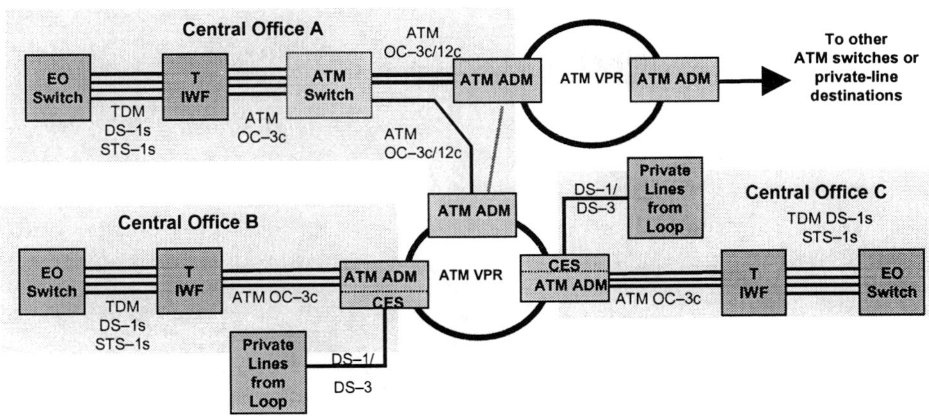

Figure 6: Near-Term VToA/PLoA Transport

If the switch fabrics in the ATM ADMs were large enough, the ADMs could replace the core ATM switches. The long-term architecture then would look very elegant (see *Figure 7*). Integrated access, switching, and transport (IAST) is the term coined for an ATM ADM that has all of the above capabilities.

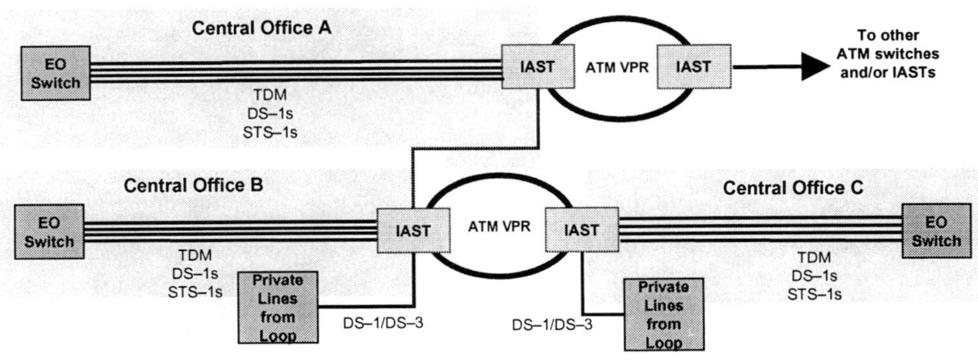

Figure 7: Long-Term VToA Architecture

Another application relates to the ADSL rollout. DSL access multiplexers (DSLAMs) at the CO and the digital loop carriers (DLCs) at the remote terminals terminate ADSL traffic (see *Figure 8*). Lightly loaded ATM trunks are back hauled from the DSLAMS and DLCs to an ATM switch at a distant CO

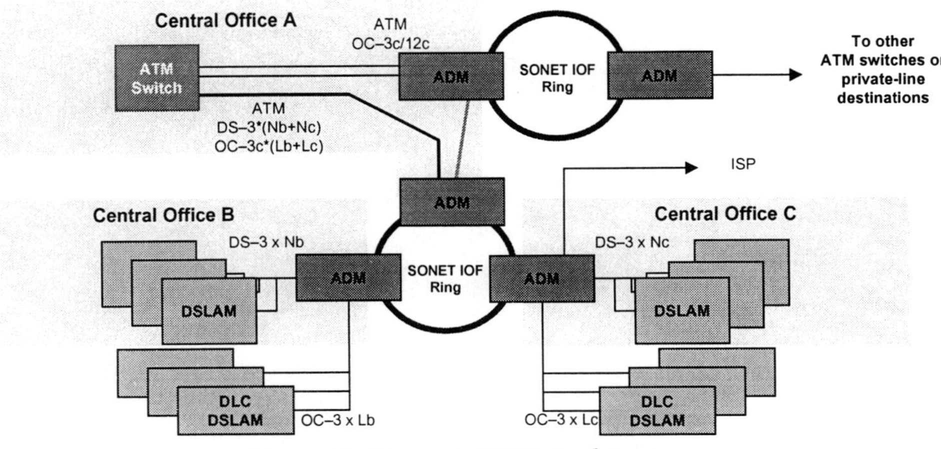

Figure 8: Current ADSL Architecture

before the groomed traffic is sent to the Internet service provider (ISP). If the ADMs in the architecture could be replaced with ATM ADMs, the lightly loaded trunks could be aggregated and the interoffice bandwidth could be more efficiently utilized. If the ATM switch fabric sizes in the ATM ADMs are sufficiently large, the backhaul to the core ATM switch could be avoided and traffic could be directly transported to the ISP.

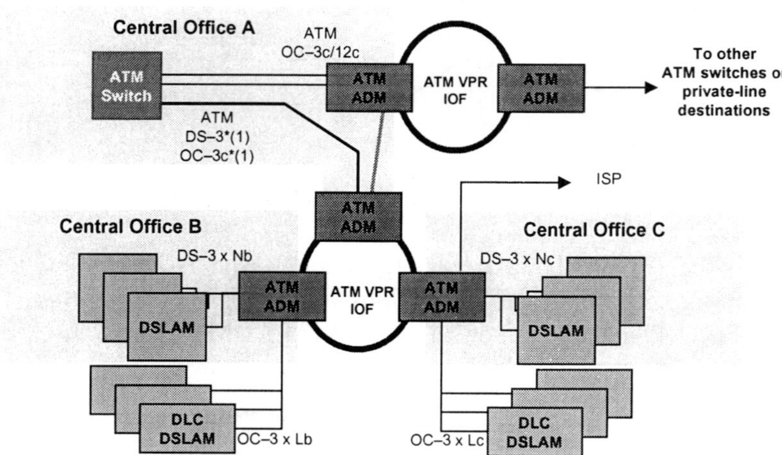

Figure 9: *ADSL Transport Architecture Evolution*

Optical Q-Factor Measurement— What's It All About?

Roland Bach
Project Manager, Fiber Optics Division
Acterna

Wolfgang Mönch
Senior Product Manager, Fiber Optics Division
Acterna

The optical Q factor is a parameter that directly reflects the quality of an optical communications signal. Instead of standing for "quality," Q is the standard mathematical symbol for the Gaussian error integral. The Q factor is determined by monitoring the amplitude and phase of the (analog) signal at the electrical level. The process involves intentional mis-sampling to simulate a poor signal-to-noise ratio (SNR) and increased bit-error rate (BER). Using statistical techniques, it is then possible to determine the Q factor corresponding to optimal sampling conditions. The Q factor is a property of the signal and not of the communications system. The procedure is thus fundamentally different from a BER test (BERT). Still, there are ways of relating the Q factor and the BER, assuming any signal impairments are stochastically distributed (white amplitude and phase noise) (see *Figure 1*). In case of discrete interference, there is no such correspondence. In any case, however, the Q factor is a measure of the true signal quality.

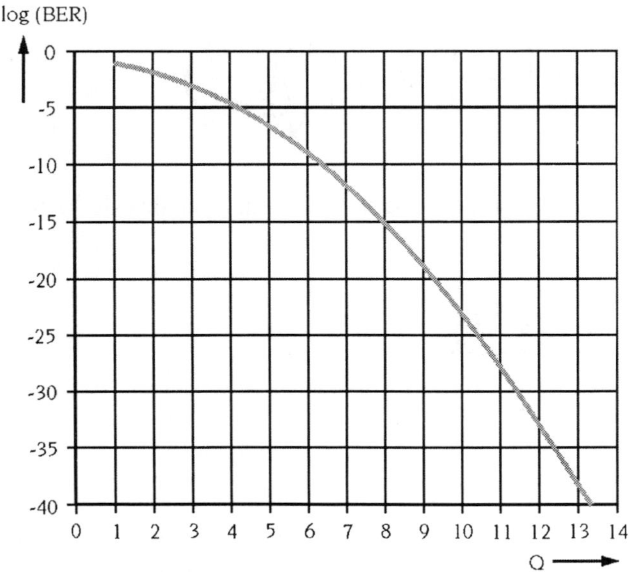

Figure 1: *Relationship between Q Factor and BER in Signals with White Amplitude and Phase Noise*

How Does One Determine the Q Factor?

Figure 2 shows a typical eye pattern, which is produced by overwriting multiple traces of the signal with the proper phase. The larger the eye opening, the more reliably the "1"s and "0"s will be detected, making for a lower BER.

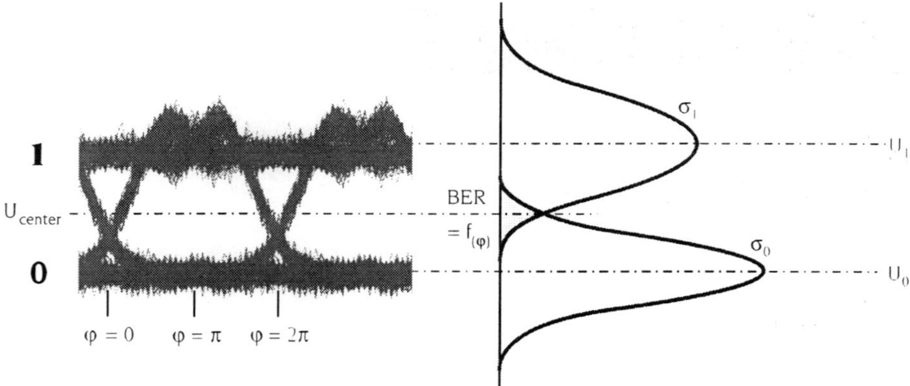

Figure 2: *Amplitude Distribution Density and BER for Signals with Superimposed White Noise*

The Q-factor meter takes several samples and computes the average value U_1 and the standard deviation σ_1 for the "1" state, and also U_0 and σ_0 for the "0" state (see *Figure 2*). For the sake of simplicity, this is done in the center of the eye ($\varphi = \pi$). Based on these samples, the meter builds two Gaussian distribution functions. The Q factor can then be determined using the following simple formula:

$$Q = \left.\frac{U_1 - U_0}{\sigma_1 + \sigma_0}\right|_{\varphi = \pi}$$

The probability of bit errors thus corresponds to the area where the two distributions overlap—again assuming stochastic interference (white noise), which is the basis for assuming a Gaussian distribution:

$$BER = \frac{1}{4}\, \text{erfc}\left(\frac{U - U_0}{\sqrt{2}\,\sigma_0}\right) + \frac{1}{4}\, \text{erfc}\left(\frac{U_1 - U}{\sqrt{2}\,\sigma_0}\right)$$

The minimum BER is obtained by setting the decision-making threshold for "1" or "0" so as to equally split the "error area":

$$U_{center} = \frac{\sigma_0\, U_1 + \sigma_1\, U_0}{\sigma_1 + \sigma_0}$$

Don't Forget the Phase!

Until now, it was assumed that sampling always takes place in the center of the eye. In networks with low transmission speeds and, therefore, "clean" signal edges, the Q factor determined in this manner is sufficient to assess the quality. The reason is that, here, it is mainly amplitude noise that disrupts the signal. In high-speed networks operating at 10 Gbps and faster, this result is only half of the picture because phase noise, dispersion effects, and interference resulting from, say, conversion of phase into amplitude modulation, play a larger role and cause additional bit errors. *Figure 3* shows the relationship between the Q factor and sampling phase.

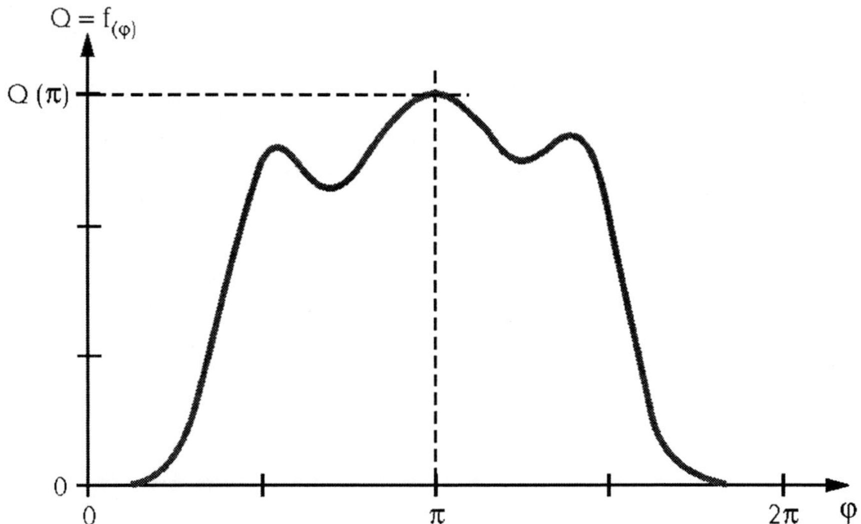

Figure 3: The Q factor, although usually defined for the center of the pulse, is also a function of the sampling phase. The resulting curve is not necessarily symmetric: Differing edges in the eye pattern produce asymmetric results.

The overall relationship between the BER, amplitude, and phase is thus three-dimensional. *Figure 4* shows the BER—log(BER)—as a function of amplitude threshold and sampling phase. It is plotted for a little more than 1.5 pulses of the signal. The cross-section at 3π helps to visualize the two-dimensional dependency between the BER and amplitude threshold.

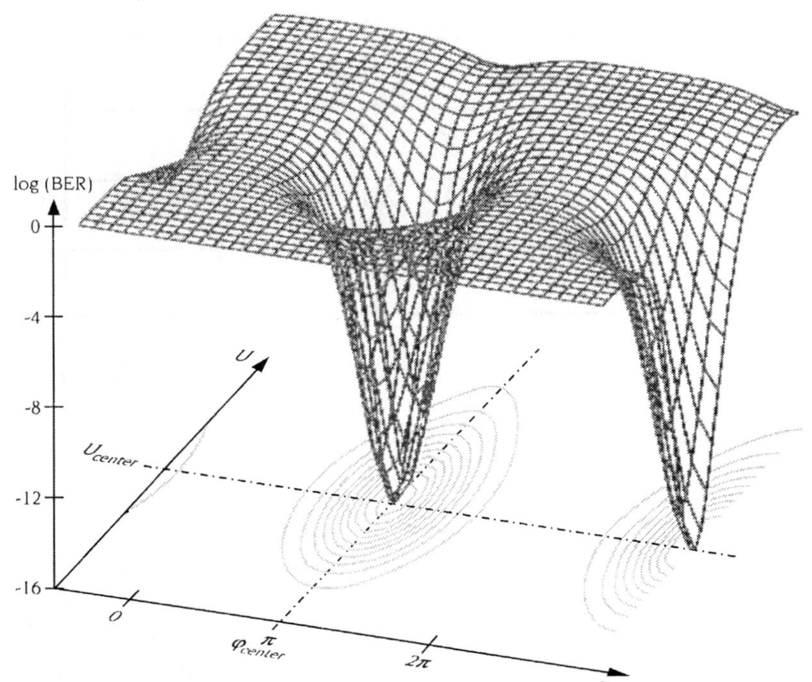

Figure 4: BER as a Function of Amplitude and Phase

Projection of the bit-error probability on the amplitude-phase plane gives the lines of constant BER. The surface of the log(BER) construct asymptotically approaches the plane BER = 0.5.

Figure 4 shows that it is impossible to extrapolate only the BER results versus amplitude without taking the phase into account. To find the lowest point of a funnel—i.e., the lowest possible BER and thus the actual Q factor of the signal—the sampling instant must be properly chosen.

In the two-dimensional representation given in *Figure 5*, the BERs are plotted versus amplitude for different sampling phases φ. The curves thus correspond to different funnel sections in the amplitude direction for constant phase. Again, asymmetric behavior must be assumed at the pulse start and end— i.e., for positive and negative values of φ with respect to the nominal sampling phase φ_{center}. If the minimum bit-error values found in *Figure 5* are plotted versus phase, the continuous curve in *Figure 6* is obtained. Through interpolation the "correct" sampling phase φ_{center} can be found. Here, only one interpolation is possible. In computational terms, a higher-order polynomial (e.g., fourth order) provides good results.

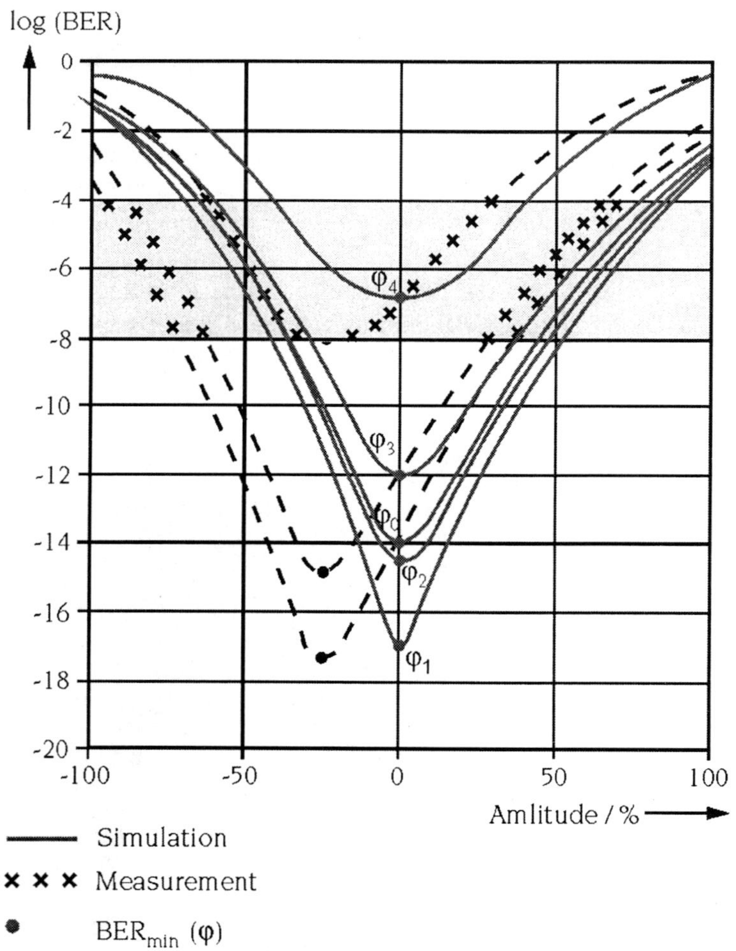

Figure 5: *Comparison of Measured and Simulated Values for BER = f(φ)*
The measurement range of the Q-factor meter (10^{-4} to 10^{-8}) has a gray background. The offset between measurement and simulation toward the amplitude axis is due to hardware effects. This graph shows why a Q-factor meter produces results so fast: Measurements are made at high BERs to estimate low bit-error values.

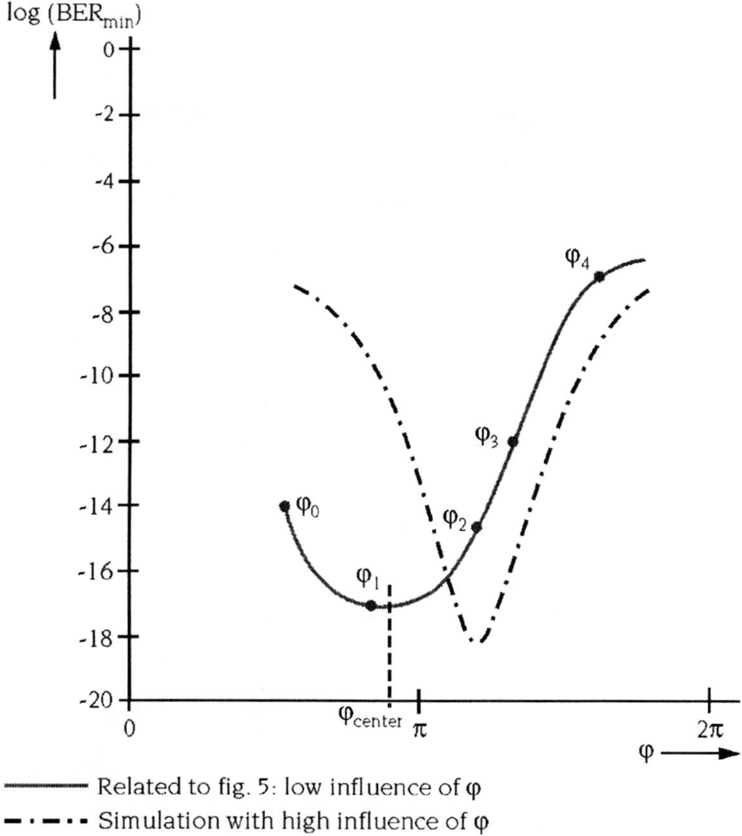

Figure 6: *Interpolation of the Measured Values for BER$_{min}(\varphi)$ Provides the "Correct" Sampling Phase φ_{center}*

Whereas extrapolation of the measured bit error–amplitude curve provides reliable results (assuming a Gaussian distribution), the same does not hold in general for the bit error–phase curve. Here, the results vary considerably as a function of the impairment type, meaning jitter, dispersion, PM–AM conversion, and so on. *Figure 6* illustrates this with a curve that arose through simulation of strong phase jitter. Outside the minimum, it is much steeper than the curve obtained from the measured values in *Figure 5*. In this case, it is necessary to determine carefully the optimum sampling phase using densely spaced measurements.

Circuit Design

The basic principle of the Q-factor meter involves sampling of the eye pattern. This is more convenient and effective than with an oscilloscope, which in the best case detects BERs in the range to 10^{-2}. The Q-factor meter can measure BERs from 10^{-4} to 10^{-8}, making the extrapolated values for the actual Q factor much more precise, although the measurement time remains the same.

The core of the device works as follows: In a dual-path detector, one comparator acts as a "center sampler" with a fixed amplitude threshold and fixed phase, while the other, acting as a "measurement sampler," can vary these two parameters. The results from the two comparators are compared in an exclusive OR circuit, and any differences are counted as bit errors. No special bit pattern is needed in the signal. This enables measurements at any time during system operation on live traffic. The limit is reached only when the signal quality degrades so severely that the circuit can no longer synchronize—i.e., at a BER of about 10^{-2}.

Clock recovery is broadband for all bit rates up to 10 Gbps. In other words, the measurement works at any bit rate. This is due to the use of a phase-locked loop (PLL). The phase detector and the synthesizer in *Figure 7* are regular components of any PLL. The only difference is the edge counter that controls the frequency divider in the synthesizer depending on the signal frequency. It takes advantage of the fact that the data signal—a non–return to zero (NRZ) signal—contains equally as many positive edges as negative edges in the center and that the frequency of the edges is equal to a quarter of the bit rate:

$$SP/t = SN/t = \tfrac{1}{4}\,BR$$

SP/t: Number of rising edges per unit time
SN/t: Number of falling edges per unit time
BR: Bit rate

The bit rate is thus automatically determined so the Q-factor meter can tune to it. Wandel & Goltermann has patented this process.

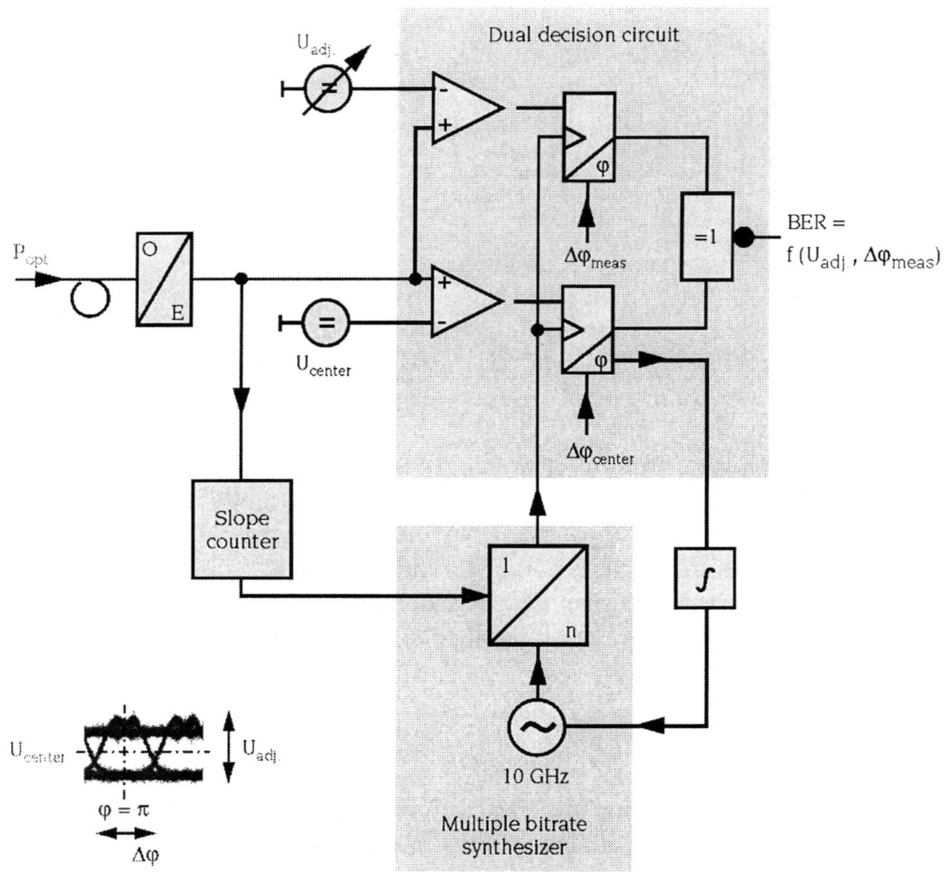

Figure 7: *Block Diagram of the Q-Factor Meter for Bit Rates from 25 Mbps to 10 Gbps*
The only prerequisite for operation is an NRZ signal at the optical input.
The bit rate and bit pattern have no effect.

The actual measurement involves shifting the comparator thresholds in the detector circuit so that the sampling moves to the edge of the eye. The Q-factor meter generates an artificially increased BER in this manner. Measurements in the 10^{-4} to 10^{-8} BER range have proven adequate.

The increased BER is the basis for the short measurement time (approximately one minute). On the other hand, discrete impairments at BER = 10^{-8} can still be detected so that extrapolation for the minimum BER still produces usable results (see *Figure 8*).

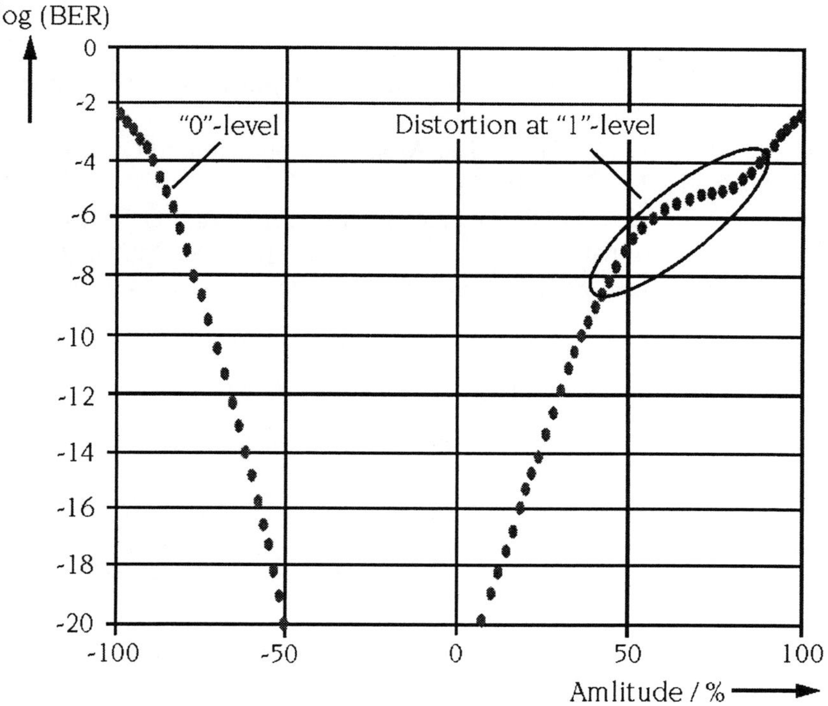

Figure 8: *Discrete interference such as crosstalk distorts the funnel curve on the side of the "1" state in the 10^{-4} to 10^{-8} range. However, reliable results are still obtained through extrapolation.*

So far the paper has discussed measured and extrapolated BERs. But what about the actual result: the Q factor? This is a purely computational task for the measuring equipment, involving average values and standard deviations.

Q, BER, SNR, OSNR—Different Measures of the Same Thing?

The paper has discussed in detail the relationship between the Q factor and the BER. Of course, there is also a relationship to other quality parameters. First, there is the electrical SNR. According to [3], the Q factor can be converted directly to an electrical SNR value:

$$\text{SNR (dB)} = 20 \log Q = 10 \log Q^2$$

Many different sources add to the electrical noise power level in optical communications systems, including thermal noise and shot noise of receivers, and signal-spontaneous beat noise and spontaneous-spontaneous beat noise of optical amplifiers, which results from mixing products in photodiodes.

The relationship to the optical SNR (OSNR) is, unfortunately, not so simple. It is defined only for systems with optical amplifiers as a ratio of the optical signal power to the amplified spontaneous emission (ASE) of the amplifiers. If the signal-spontaneous beat noise happens to be much greater than the thermal noise, however, one may use this approximation:

$$\text{OSNR} = 4 \, (B_e/B_o) \, Q^2$$

B_e and B_o are the bandwidths of the electrical and optical signals, respectively. To make a precise OSNR measurement, however, an optical spectrum analyzer is still needed.

References

[1] Bergano, N. S. March 1993. Margin Measurements in Optical Amplifier Systems. *IEEE Photonic Techn. Letters* 5 no. 3: 304–306.

[2] New ITU–T Rec. G.976 (ex G.OSS3) on "Test Methods Applicable to Optical Fibre Submarine Cable Systems," COM15R68 (TSB, 7. Nov. 1996).

[3] Fussgaenger, K., and S. Cascelli. Per Channel Bit-Error-Ratio, Q-Factor, El. SNR and OSNR Relations for Single- and Multi-Channel Long-Haul Systems with Optical Amplifiers in Draft Recs. G.691, G.692 and G.onp. *ITU–T Study Group 15—Contribution 31* (COM 15–31-E, Nov. 1997).

Optimization of Optical Networks Testing Resources

R. James Bateman
Director, Systems Engineering
Net-HOPPER Systems, a Spirent Communications company

Ashok Kumar
Director, Strategic Planning

Bert Lindgren
Chief Technical Officer

Introduction

In recent years, the telecommunications industry has seen sharp increases in market demand for both network access and advanced functionality. More specifically, today's telecommunications service providers are primarily driven by the following six market factors:

1. The need to scale rapidly to meet customers' increasing bandwidth demands wherever needed
2. The customers' demands for multiple broadband services
3. The provider's desire to maximize network equipment and operations support systems (OSSs) by providing multiple services over one network infrastructure
4. The need to offer services with partner carriers while maintaining service-level agreements (SLAs)
5. The provider's desire to leverage technical advancements, especially in all-optical networking, that increase network efficiency by reducing reliance on distinct network protocol layers
6. The need to optimize network operations support infrastructures and resources, as providers' budgets are growing at a much slower pace than their network's size, traffic, and users

This paper is an analysis of the testing and diagnostic component of the last factor—what it means to optimize network testing and diagnostic efficiency, and considerations that must be made when striving for maximum efficiency.

While other efficiency factors are considered in this paper, the central goal of network diagnostics is, of course, to be able to respond to trouble reports as quickly as possible. Without exception, network customers insist on high quality of service (QoS). When networks fail, quick restoration of lost services is most critical for ensuring continued customer satisfaction. The customer is not concerned about the service provider's problems or issues that delay network troubleshooting—it is expected that the service provider will be able to isolate and resolve network anomalies quickly and effectively. In the end, quick and effective problem resolution significantly reduces a key measure of customer service success: mean time to repair (MTTR). In addition to improving customer satisfaction, reduced MTTR means that network administrators may spend less time on network troubleshooting and more time on longer-term maintenance and development issues, thus avoiding many types of short- and long-term network problems.

The pressures that face network operators, especially when the operators are trying to optimize their diagnostic efforts, appear in *Figure 1*.

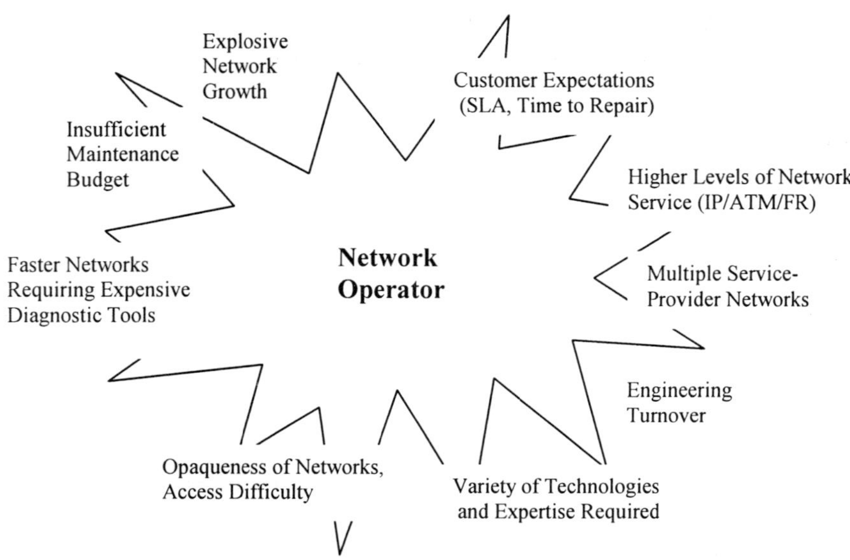

Figure 1: Network Operator's Pressure Points

With the current network bandwidth explosion, the need for network engineers and technicians trained in problem isolation, testing, and troubleshooting is growing faster than the networks themselves; qualified engineers are in short supply. In addition, the integration of network technologies often requires different sets of diagnostic tools capable of isolating and analyzing anomalies on network segments using different technologies. The expense of placing such a variety of suitable pieces of test equipment on every network segment is prohibitively expensive and requires highly trained technicians for their placement and maintenance.

Faced with continuous pressures to meet MTTR commitments and manage maintenance budgets that typically fail to keep pace with network growth, service providers must optimize their testing resources. One step toward test resource optimization lies in the capabilities offered by optical network test access systems (ONTASs). A test access network allows administrators to move remotely located test equipment to different network segments, increasing the test tool's network visibility and maximizing the usability of existing test equipment.

The complexity of today's optical technologies creates additional challenges in designing efficient and effective testing strategies. In examining optimal network testing, this paper will describe and motivate efforts to achieve network testing efficiency. The paper continues with an examination of factors that inhibit diagnostic efficiency, including trends in next-generation optical networks. Approaches to regain any efficiency lost due to these trends are also discussed.

Description of Network Testing Efficiency

The primary factors that must be considered in any discussion on improving network testing efficiency include the following four goals:

1. Minimizing the time needed for fault isolation
2. Minimizing investment in diagnostic infrastructure without sacrificing goal number one
3. Minimizing the number of personnel required to diagnose a network fault
4. Minimizing the training those personnel require to contribute to diagnostic efforts

The importance of rapid and effective network testing is evident in the financial realities of today's growing network systems. The exponentially expanded bandwidth and faster systems required to meet the growing market demand translate directly into higher risks. In fact, loss of service can cost as much as $10,000 per channel per minute (or roughly $19.2 million per hour) for a 32-wavelength system. Additional SLA penalties incurred during interruptions amplify the financial impact. Therefore, as networks serve more customers, provide more services, and become more complex, the need to maintain the system and respond instantly to trouble reports becomes substantially more critical.

Unfortunately, the rate of network growth significantly exceeds that of maintenance budgets. For example, it is estimated that networks are growing fivefold each year; maintenance budgets, however, are growing only twofold, at best. The cost of managing a large network often dominates the cost of network deployment, including investment in equipment. An inability to diagnose a large network efficiently can significantly increase a company's expenses.

Efficient network testing means maximizing the tools and personnel responsible for network monitoring and troubleshooting. In traditional networks, technicians must locate an appropriate diagnostic tool for testing on a specific network segments. After traveling to the trouble site, the technician depends on experience and expertise to attach the equipment correctly and begin analysis. Multiple issues arise: Is the correct piece of test equipment available? Is a knowledgeable technician available? Can the company afford repeated outages at distant sites? The technical complexities of optical networks exacerbate these issues.

A diagnostic system that allows a single group of test equipment to be used on a number of network segments without physical relocation would significantly minimize diagnostic tool investment. Furthermore, the ability to perform diagnostic procedures from a central or arbitrary location reduces time spent off-site, allowing technicians to concentrate on network diagnostic and not on transportation and physical access issues. Finally, consistent, value-added interfaces for managing the network system reduce investment in personnel training, allowing technicians to begin contributing to the bottom line more quickly. How service providers address these issues greatly impacts their profitability.

The next section discusses industry trends that work against efficient network diagnostic. In addition, factors that enable providers to maintain efficiency while managing legacy and next-generation networks are examined.

Overcoming Barriers to Optimal Network Testing

Thus far, this paper has described the key factors that should be optimized for achieving efficient testing mechanisms and methodologies. This section will discuss the following business and technical factors that inhibit optimal network testing and diagnostics:

- Explosive network growth
- Industry trend toward higher-speed transport technologies
- Ability to test next-generation networks
- Trend toward combining multiple services and technologies
- High engineer turnover and shortage of skilled technicians
- Dense wavelength division multiplexing (DWDM)
- Loss of operational network telemetry

After discussing these efficiency-reducing factors, approaches will be presented that help prevent their time- and money-wasting effects.

Explosive Network Growth

First among the issues facing any network engineering or operations staff is the impressive growth of the network that they design and maintain. Examples of the rate of growth abound, including the estimates that Internet traffic doubles every 100 days and that provider networks have recently grown, and will continue to grow, fivefold each year. Of course, these increases are a major motivating factor for efforts to improve network diagnostics efficiency. More network elements (NEs), more traffic, or more users often mean more problems, which benefit from efficient trouble resolution. Beyond providing a simple increased need for testing efficiency, network growth actually tends to increase MTTR, to require more staff to isolate anomalies, and to cost more to diagnose.

For example, when network growth results in increased distance between the troubleshooter and problem location, the time taken to isolate a remote network fault could quickly increase. Additionally, well-known difficulties exist when attempting access to co-location facilities. Therefore, management's commitment to remote diagnostics often follows major outages; network testers and protocol analyzers are then deployed to remote network segments. In the face of network growth, however, increasingly more of the operation's budget would be allocated to buying additional testers to diagnose this year's new network segments; five times more network could easily require twice the amount of test equipment to maintain coverage. Because operational budgets are always behind the growth of the network itself, visibility gaps frequently develop within the network fabric, which result in significant troubleshooting delays in the newly expanded parts of the network.

A more scalable test equipment deployment can be achieved by avoiding direct attachment of testers and analyzers to network segments. Instead, by attaching test equipment to reconfigurable electrical or optical matrix switches, the diagnostic equipment can be utilized remotely across the entire network; therefore, network growth can be managed without additional purchase of expensive test equipment. Such a test access system offers the benefit of overcoming several efficiency-reducing aspects of network growth.

Migration to Higher-Speed Networks

Given a substantial increase in customer demand, service providers must upgrade their optical infrastructure's capacity. This typically requires one or more of the following: First, the single-channel transmission rate can be increased (e.g., optical carrier [OC]–12, OC–48, OC–192, OC–768, etc). Second, more DWDM channels can be packed within a given wavelength range with tighter spacing (e.g., 25 GHz, 12.5 GHz, 10 GHz, etc.). Third, the wavelength ranges available for DWDM channels can be extended into the L-band and other regions. While these three mechanisms present different problems for test infrastructures, only the first of these will be discussed here. The remaining mechanisms will be covered in a following section on DWDM concerns.

Similar to most any other electronic device, the latest and greatest testing tools are more expensive than tools designed for older-generation networks. This certainly exacerbates the test infrastructure upgrade costs and, luckily, is also mitigated by the same test access infrastructure. Factors also exist beyond the pure increase in cost of the faster test equipment, most notably the typical path network providers follow for their upgrade. Providers rarely replace an existing infrastructure; instead, legacy and modern technologies coexist as customers are gradually migrated off the legacy system. In fact, in some instances older customers remain on the legacy infrastructure, and only new customers are provisioned on the new system. In any case, the point is that test equipment is concurrently needed for the previous and current technologies. Therefore, providers would see greatest benefit from testing systems (testers and test access) that are simultaneously compatible with multiple technologies.

Modern Networks Resist Diagnostics

Several technologies designed to improve networks' capacity, resiliency, flexibility, and adaptability have side effects that inhibit network testing efforts and, therefore, delay fault identification and isolation. For example, while all-optical networks (AONs) have traditionally been deployed in ring topologies that have very predictable data flows, there is an increasing trend toward mesh network deployments, primarily because the leaps in customer bandwidth and service demands are occurring in an unpredictable and nonuniform manner. Unfortunately, traffic through a mesh typically has several routing options (a ring topology has only two); therefore, a target network flow can prove difficult to find, analyze, and diagnose at the heart of a modern network. In such cases, integration between the testing and the network's provisioning OSSs are required. In the end, network operators describe the flow or circuit they wish to test, queries are made to the provisioning system to locate the traffic within, and the testers are then placed on the appropriate network segments by the test access hardware.

Additionally, despite improving a network's performance and resiliency, the use of traffic engineering or dynamic routing protocols—open shortest path first (OSPF), multiprotocol label switching (MPLS), user network interface (UNI) switched virtual circuits (SVCs)—exacerbates a mesh network's resistance to analysis because the provisioning system is no longer omniscient. Instead, where traffic is engineered or routed dynamically, the testing OSS must now expand on the provisioning information by performing on-line network queries to identify the current path assigned to circuits and flows.

Until now, it has been assumed that only a single path of the network's mesh is utilized by a traffic flow, and the difficulties enumerated above describe what could be involved in locating that path. Load balancing introduces yet another dimension; visibility may require deploying multiple tools simultaneously and aggregating their inputs because only together can the entire flow be seen and diagnosed. Therefore, penetrating analysis of a load-balanced network requires multiple testing tools that offer aggregation of their inputs as well as a test access network capable of simultaneously deploying these tools where needed.

A different approach to overcoming the barriers to efficient analysis within the core of modern networks is to focus the analysis at the edge before the flows get arbitrarily routed, engineered, or load balanced. All the same concepts of test access–attached test equipment apply here, except that the test infrastructure deployment must include the "edge" segments before they are aggregated.

Variety in Network Services and Technologies

Whether for historical reasons or because of customer growth in new areas, providers' networks are rarely homogeneous architectures designed for a single service. If a provider's network is rather homogenous, it is typically because the provider was able to make a fresh start recently; networks are already gradually losing that homogeneity. More often, however, multiservice networks utilize many different protocols and technologies, including a combination of synchronous optical network (SONET), asynchronous transfer mode (ATM), Internet protocol (IP), Ethernet, frame relay (FR), and others. In addition to this hybrid of telecom and datacom technologies, providers are moving up the service ladder from providing strictly transport services to providing value-added services (e.g., IP, virtual private network [VPN], QoS) or even content (e.g., application service providers [ASPs], voice over IP [VoIP]). This variety of services delivered over a variety of technologies has the potential to reduce testing efficiency drastically.

Primary among approaches that help avoid the potential time- and money-wasting pitfalls is centralized troubleshooting—utilizing a universal testing infrastructure that enables the rapid deployment of appropriate tools. Besides the desire to avoid wasting time traveling to remote network sites, the need for centralized troubleshooting is extended because of the range of expertise needed to track and isolate a high-level network difficulty. A centralized group of engineers with knowledge of the application service, network security, policy routing, and network transport will be able to address any issues that may arise.

Organizationally, various responsibilities of providing high-level services are often divided among different departments. Where appropriate, testing infrastructure shared among all departments is substantially more efficient than having multiple diagnostic access points (for example, one for the transport network group and one for the application-providing group). A final requirement exists for this shared test infrastructure: It must contain good diagnostic tools for the different technologies and different services. For instance, the tools needed to isolate a cabling fault are certainly different than those needed to diagnose unexpected performance behaviors within the queuing policies. With an appropriate network testing system, troubleshooters throughout the company can quickly and safely deploy the right tool onto the right network segment for the problem at hand.

Extending beyond the human-centric diagnostic procedures described above, the complex networks of multiservice providers benefit most from procedures to manage, troubleshoot, isolate, and correct problems. When combined with procedures to test before and during service deployment, issues with network health can be isolated very efficiently. Even more, the standard processes (i.e., those done during service establishment) should be as mechanized as possible, thereby reducing the human error factor and increasing efficiency. Toward this end, providers should consider both test access and test equipment whose operations can be automated.

Engineering Turnover

At SuperComm 2000 it was said that "Good people are tough to find and expensive to keep." With the current network bandwidth explosion, the need for network engineers and technicians trained in testing, troubleshooting, and diagnosing problems is growing faster than supply. Probably all providers have either lost valuable operations personnel or are looking to hire or train more. The shortage as experienced in recent years is detrimental to network testing efficiency primarily in the delay required for training new engineers. Because of the type and amount of training needed, new engineers may only become useful weeks or months after they are found and hired.

Maximizing the skills of incoming engineers as well as minimizing unique characteristics of the testing environment may achieve a reduction in this delay and training investment. Even troubleshooters with average skill sets will initially possess some experience with, and knowledge of, network protocols, generic network architectures, and some types of test equipment. If, somehow, these new engineers could be assigned a problem requiring a test tool that has already been deployed, they would be able to contribute immediately.

The barriers involved in troubleshooting every network are often where in the network the engineer should look for a given problem and how the engineer deploys a desired tool to this location. With brief introductory training focused on the network's architecture and well-integrated provisioning, monitoring, and testing OSSs, both the "where" and "how" problems can be drastically simplified. In fact, new operations hires have been seen troubleshooting before they have been issued a badge or found their mail slot.

Dense Wavelength Division Multiplexing

By demonstrating dramatic improvements in network capacity—nearly doubling capacity every nine months—DWDM has become a welcome advancement. AONs are rapidly moving into production and will be carrying up to 80 high-speed channels per fiber. At some level, two DWDM NEs will appear to have 80 fibers between them. The difference between channels and fibers becomes clear when one is faced with diagnosing a network flow within one of these channels: The entire investment in diagnostic tools cannot be used. Certainly, a DWDM–capable test device could also be deployed, and it would be able to observe the desired channel. However, the need for a variety of deployable tools has already been established (i.e., the importance of applying the right tool for the job). Replacing all existing tools with DWDM–compatible versions would be prohibitively expensive. Furthermore, most higher–protocol layer tools probably do not have DWDM versions.

Instead, a more systemic approach should be taken: Use a DWDM–enabled test access infrastructure that is capable of tool deployment onto DWDM channels as if they were single-wavelength optical links. In this manner, diagnostic procedures on DWDM channels offer complete symmetry with the tools and procedures used before DWDM technology.

Loss of Operational Network Telemetry

Two factors have caused providers to move away from SONET infrastructures, including a need for nonuniform, ad hoc bandwidth scalability and a mismatch between IP's packet-frame orientation and SONET's time division multiplex (TDM) orientation. Unfortunately, in their rush away from SONET toward routed, switched, or transparent optical networks, providers lose 20 years of experience in working toward interoperable supervisory and management device services. Eventually, DWDM networks will probably introduce similar services. However, in the short term, basic alarms (e.g., signal loss) are anticipated with limited analysis capabilities. Problem detection on individual DWDM channels will likely be limited to the edges where the channel becomes electrical. Intermittent failures, such as excessive bit-error rates, are also likely to be detectable only at the circuit's endpoints.

Thus, an increasing number of diagnostic procedures will require deployment of dedicated test devices instead of today's management queries to the NEs themselves. All the approaches and considerations discussed within this paper are, therefore, even more pertinent to the upcoming AONs.

Summary of Efficiency-Enhancing Aspects of a Diagnostic Infrastructure

Throughout this paper, various approaches to optimization of testing resources were presented. This section will summarize the key recommendations.

First, the overall arrangement of components in an efficient testing infrastructure is presented in *Figure 2*. As shown, the ONTAS serves as the intermediary between the optical circuits and the test equipment. As an intermediary, the ONTAS leverages the test equipment by giving each test tool visibility across the entire region of the optical network. Controlled through a centralized test access OSS, the ONTAS also safely, securely, and rapidly moves test equipment between optical circuits, putting the necessary tools where they are needed.

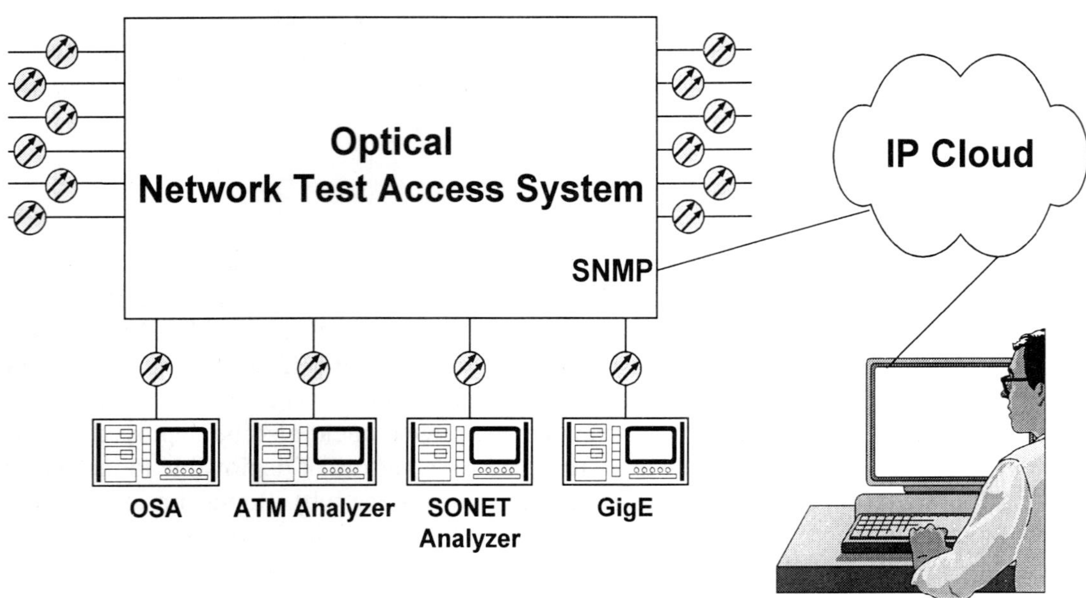

Figure 2: *Network Diagnostic Infrastructure Block Diagram*

Several considerations that will aid in choosing efficient test equipment have been presented. *Table 1* summarizes key efficiency-improving aspects of test tools.

Aspect of deployed test equipment	How network testing efficiency is improved
Remotely accessible	Allows immediate use of the tool for diagnostic of the attached network segment without transportation costs or delays for the engineer
Expandable to new technologies	Supports both the new network technologies as well as the migration time when both the legacy and new technologies are simultaneously used within the network
Supports data aggregation from multiple test equipment ports	Offers a unified view of multiple pieces of test equipment, providing consistent observation of load-balanced traffic as it is transported by the network
Appropriates variety of test equipment for technologies and protocols used within network	Ensures that the right tool is always available for different types of problems

Table 1: *Efficiency-Improving Test Equipment Access*

Similarly, *Table 2* summarizes key efficiency-maximizing requirements for the ONTAS hardware and OSS software.

Aspect of optical test access system	How network testing efficiency is improved
Fully photonic optical switching hardware	Provides a consistent access infrastructure that is network-technology, bit-rate, and wavelength independent; easily supports the migration from one networking technology to another
Multiple test devices attachment support	Ensures that the right tool is always available for different types of problems
DWDM wavelength extraction	Provides the same test equipment access to optical channels as to optical links
Centralized test access OSS server	Coordinates testing efforts throughout the network; centralizes management efforts while keeping test features widely available
Lightweight test access OSS Web-based clients	Provides a highly available tool that can be quickly and easily used to deploy testing resources; Web browser supports test device deployment from most convenient workstation
Access control and user authentication	Guarantees that only authorized operators are utilizing monitoring and diagnostic equipment

Table 2: Efficiency-Maximizing Requirements for Optical Network Test Access

Conclusion

The inherent complexities in today's optical networks require service providers to optimize their investment in testing resources. ONTAS can prove to be an important and valuable solution that saves money while minimizing troubleshooting efforts and time. The result is a drastic reduction in MTTR because technicians no longer need to be dispatched to distant sites with a test set and because it leverages several different technologies of test equipment over multiservice networks.

There is a need to optimize the investments made in testing resources used for optical networks. Network test access systems and planning tools designed up front in an optical network achieve this goal by accelerating the overall process of testing, analysis, and diagnosis for fiber networks.

References

Butler, Robert K., Ph.D. 2000. *Sprint perspective on next generation multi-service transport and switching networks*, Session TuE 3-1. Sprint, OFC-2000.

Kaufman, Darin, Saad Asif, Abdul Kasim, and Mike Mahoney. 1999. *The optical Internet: A new network architecture*, OTN planning session (D4). Sprint Transmission Engineering, NFOEC 1999.

Lindgren, Bert, and Ashok Kumar. *The Risks of Network Growth: Are We Ready?* LAN-Hopper Systems, Inc., Future Comms 2000, Sterling.

Palissery, Ram. 1999. *True interoperability in optical networks*, Network planning tools session (D3). Telcordia Technologies, NFOEC 1999.

Ramaswami, Rajiv, and Kumar N. Sivarajan. *Optical Networks: A Practical Perspective.* Morgan Kaufmann, 1998.

Optical Management from a Service Provider's Perspective

Robert Branton
Advisory Engineer, Network Technology and Planning Group
MCI WorldCom

In the effort to manage the evolution to an all-optical network (AON), network technology and planning groups are looking at the new technologies that will be embraced. They are focusing on what is necessary to manage those technologies as they are brought into the network. They are also examining how deployment life cycles can be shortened to facilitate moving new devices into the network quicker. Standards groups provide one avenue to evolve management techniques. However, the carriers are looking for other ways to integrate these technologies into the network.

It helps, sometimes, to regard a situation from a different perspective—in this case, network management oriented from an engineering point of view. Carriers are looking at methods of using engineering more effectively to automate network management. One method to accomplish this is through developing prototype management platforms. These platforms are used to test the viability of new management concepts and ideas early in the element development cycle. Using the management prototype concept, it is possible to test new management techniques that may or may not prove feasible for production management systems. This is accomplished quickly with a minimal development cost. Also, prototyping aids in discovering deficiencies in vendor element managers or craft interfaces before adoption into the carrier's network. In MCI WorldCom's case, for example, this technique has proved very effective with synchronous optical network (SONET) technology. A SONET prototype management system was developed six years ago, and this is pertinent because of the lessons learned from SONET—if things are not effectively managed in the beginning, then they are not likely to be managed effectively in the future. The vendors that supply SONET element managers have not been able to satisfy some of the carriers' network management needs, even over an extended period of time. This is a situation that must not be repeated in the AON.

The Need to Manage Technologies Together

As carriers move into an Internet protocol (IP) environment, it is not enough to manage the network aiming for the traditional "almost error-free" goals of telecom network management. It is necessary to manage points of network congestion, anticipate the problems, and reroute automatically and effectively. Bandwidth utilization and identification of expansion needs must become part of a proactive management system for a carrier's network. Carriers also must be able to provision new services automatically, to sectionalize problems, and to manage differently based on different service classes. Management systems must manage across multiple technologies. Finally, management systems must start to limit alarm reporting to only those portions of the network carrying traffic. This is currently a problem with many vendor supplied–element managers. They report everything, regardless of whether traffic rides on devices or not.

Managing Optical

Turning the focus to optical technology, the current problem is that there is little to manage except some laser readings. Standards for optical performance and alarm reporting must be fully defined and adopted quickly. The many independent standards organizations addressing the optical universe must communicate to develop consistent standards. A simple example of the detail that must go into standards is the need for the development of consistent time stamp rules. This is critical for network management and correlation across the different technologies that will exist in an optical network. Minor steps standards are being made, but the industry has far to go. At an International Telecommunications Union (ITU) meeting, a digital wrapper standard was adopted that will allow channels that are both wrapped and unwrapped to go across the network at the same time. This provides, at least for the short term, the capability to have multiple classes of service for optical. Thus, there is standards movement, but it is not sufficiently fast.

The carrier community is on the verge of seeing massive optical network growth for which management plans must be made. The current number of managed optical networks is minimal compared to what is expected in the next three or four years. This leads to the concern over forethought, as far as the vendors are concerned. Many vendors have effective element managers for small networks, but these element managers have their scaling limits. They must be enhanced to be able to scale the management with exponentially growing numbers of elements. In addition, the merger situation among carriers means that even when a good plan is adopted, element managers may fall short. Carriers are looking toward networks with elements from multiple vendors for each technology type. Past that, for each technology type there probably will be multiple communication protocols to manage across, such as simple network management protocol (SNMP), Q3, transaction language 1 (TL1), or common object request broker architecture (CORBA). All these protocols and other new ones that may emerge must be managed consistently. This is the key goal for managing optical networks.

Additionally, elements need to be network-aware. They must know where they are and to what they are connected, and they must be able to report it. Self-discovery must be in place so that new elements automatically report their existence to the network management system.

Timing Is Nearly Everything

Why is this push to plan and adopt standards quickly so important? Consider some experiences with SONET systems. Recently, carriers started to put path/section trace into their networks. A standard was written and adopted 10 years ago for this function. However, it has taken this long for vendors to get it into their products. It would be an immense help if vendors embraced the optical standards that are adopted quickly and implemented them where carriers and systems designers can use them.

Carriers Need Full View of Network Elements

Vendors must also provide full access to network elements (NEs), either through their element manager or by direct connection to the NEs. They usually provide only a partial view of what is actually occurring on the element. Network management, however, requires an encompassing view. A limited point of view will not suffice. If a carrier is going to develop a superior network management system to become a better carrier, then it must find ways to manage its network elements uniquely, directly, and consistently.

Things can often be accomplished that are craft-interfaced with SONET that would be impossible with an element manager. By talking directly, these things can be configured consistently. For example, when the networks of two companies (both of which have SONET equipment from the same vendor, but with different configuration settings) are merged, one might have threshold-crossing alert states that go up and stay up forever, while the other might have alerts that go up and away in 15 minutes. Suddenly, these elements are merged, and the new, commingled network must be able to unify the configurations so that the performance threshold–crossing alert means the same thing across 20,000 NEs.

The Need for Standards

Data should be managed consistently, particularly in alarm performance. A particular alarm must mean the same thing, regardless of the vendor, and performance must be controlled. What has been learned from SONET is that data can be overwhelming. For example, one network may have 20,000+ SONET elements broadcasting performance data every 15 minutes, including information at OC–192 and 192 synchronous transport signal (STS)–1s in both directions. To help manage the network, carriers are looking for ways to control that data. The SONET world has shown that performance data is nearly all zeros. If, in a 15-minute period, an OC–192 pumps out all zeros, why throw up all the performance data for 192 STS–1s in both directions? Instead, a carrier should throw up only the performance data that had events in it. This is particularly true in the case of an alarm storm because the moment there is an alarm storm, most of the element managers cannot keep up with it. When performance data is needed, mangers are too busy doing alarms. Thus, a way to filter such data must be discovered.

Alarms must be consistent when crossing technologies, and different vendors will put different names on alarms. In some cases, two alarms may be equivalent from one vendor to another, and there may be ways to manage them generically; consistent configurations are the key. If the alarm has a laser bias, optical power, reading alert, it must be within a certain threshold.

Optical Management Issues

In summary, the optical portion of the network requires management in whatever form it evolves. Problems in the optical network must be resolved without affecting the customers. Currently, there is little to manage. However, standards are anticipated—perhaps too many. Different standards organizations have different agendas, and every standards group wants to own the territory. Hopefully, they can get together quickly and develop a consistent set of standards. That will help vendors to know which management path to take. Carriers will have a standards framework from which to approach optical management that will lead to quick implementation and management. Carriers can assume that vendors' element managers will be unable to cover all the ways needed to manage elements. Any optical element managers that vendors develop must provide full control of the elements, either directly (able to get into the elements automatically) or by means of a second path directly to the element. With the right set of standards for a full set of optical alarms, performance and configuration information, and fully functional element managers, carriers have something upon which they can begin to build needed management platforms.

Full-Band, 40 Gbps, Long-Haul Transport Enabled by HOM Dispersion Management

Yochay Danziger
Chief Technical Officer
LaserComm

Doug Askegard
Product Planning
LaserComm

Introduction

As demand for bandwidth continues to increase at an astounding pace, chromatic dispersion is emerging as one of the primary limiters on optical network capacity and distance. Fulfilling the ever growing demand for bandwidth in long-haul optical networks requires a corresponding growth in the capacity of the networks. The most economical way to provide the expansion needed in long-haul networks has been to increase the number of channels per fiber and the bit rates per channel. Today's more advanced dense wavelength division multiplexing (DWDM) systems feature up to 160 channels with spacing as narrow as 50 GHz. These systems fill the entire C-band of erbium-doped fiber amplifiers (EDFAs) (1530–1565 nm) with channel bit rates of up to 10 Gbps. Unless chromatic dispersion is correctly compensated for in each channel and across the entire band, the span distances between regenerator sites will be limited. With transmission rates increasing to 40 Gbps and beyond, dispersion becomes an even greater challenge, especially with the expansion into the L-band portion of the spectrum (1570–1610 nm).

Sometimes, by implementing complicated system designs, older dispersion compensation techniques have been made to work for the new systems. However, these conventional devices are unable to manage dispersion slope across the entire bandwidth that is critical in the high–channel count DWDM systems. Traditional solutions cannot correct dispersion uniformly; they correct to a "center wavelength," while the wavelengths at the upper and lower ends of the C-band spectrum receive too much or too little compensation. New dispersion management devices (DMDs), based on high-order-mode (HOM) technology, feature a broad spectrum, continuous-slope compensation capability that corrects all C-band wavelengths. With this full-band dispersion management, next-generation ultra-long-haul systems can have large distances between regenerators and transmission bit rates to 40 Gbps and beyond.

Limitations from Chromatic Dispersion

The cause and effects of chromatic dispersion are well known and understood. Although optical signals, either non–return to zero (NRZ) or return to zero (RZ), appear to be laser light either on or off, they actually contain a narrow spectrum of wavelengths. To achieve very narrow spectral bandwidth, 10 Gbps laser transmitters are now continuous output lasers that are externally modulated. Even if an ultra-pure laser light source were available, any isolated pulse has a small spectrum by virtue of its modulation. As these pulses move through an optical fiber with positive dispersion (+D), the shorter wavelength components travel slightly faster than the longer wavelength components. This chromatic dispersion effect, more accurately known as "group velocity dispersion" (GVD), is due to the wavelength dependence of the group velocity of light in fiber and results in the broadening of the pulses as they travel through the fiber. Chromatic dispersion (D) is measured in units of ps/nm-km and is the amount of broadening in picoseconds that occurs to a pulse with a spectral bandwidth of 1 nm while propagating through 1 km of fiber [1].

Chromatic dispersion causes intersymbol interference that results in bit errors when the effect exceeds the receiver's ability to discriminate between light pulses. Over the last few years, fiber manufacturers have developed a number of new fibers that have lower dispersion than the older, conventional single-mode fiber (SMF). These nonzero–dispersion-shifted fibers (NZ–DSFs) have a range of dispersions in the C-band and a variety of slopes (see *Table 1*).

Type of Fiber	Typical Dispersion At 1550 nm	Typical Slope at 1550 nm	C-Band Dispersion Range 1530 to 1565 nm	L-Band Dispersion Range 1570 to 1620 nm
Conventional SMF	17 ps/nm-km	0.057 ps/nm^2-km	15.9 to 17.8 ps/nm-km	18.1 to 21.0 ps/nm-km
NZ–DSF (early)—type 1	2.6 ps/nm-km	0.067 ps/nm^2-km	1.3 to 3.6 ps/nm-km	3.9 to 7.3 ps/nm-km
—type 2	3.5 ps/nm-km	0.067 ps/nm^2-km	2.2 to 4.6 ps/nm-km	4.8 to 8.2 ps/nm-km
NZ–DSF (large effective area)	3.8 ps/nm-km	0.1 ps/nm^2-km	1.8 to 5.3 ps/nm-km	5.8 to 10.8 ps/nm-km
NZ–DSF (reduced slope)	4.4 ps/nm-km	0.045 ps/nm^2-km	3.5 to 5.1 ps/nm-km	5.3 to 7.5 ps/nm-km
NZ–DSF (new large effective area)	4.2 ps/nm-km	0.085 ps/nm^2-km	2.6 to 5.5 ps/nm-km	5.9 to 10.1 ps/nm-km
NZ–DSF (new "light" fiber)	8.0 ps/nm-km	0.057 ps/nm^2-km	6.8 to 8.9 ps/nm-km	9.1 to 12.0 ps/nm-km

Table 1: *Typical Dispersion and Slope Values in Various Types of Fibers*

The wide variety of optical fibers listed in *Table 1* is evidence of the need to reduce the amount of chromatic dispersion and thereby reduce its effect on the signals being transported. The NZ–DSFs have lower chromatic dispersion, but because most have increased dispersion slope (change in dispersion as a function of the wavelength), the need to provide the correct amount of dispersion compensation is even more critical, especially in high-speed, long-haul DWDM systems. For some of these fibers, dispersion at the longer-wavelength end of the C-band is nearly three times that experienced by the signals in the channels at the low end of the band. In many deployments, the accumulated dispersion is compensated every 80 km to 100 km along the SMF link. The issue of providing correct dispersion compensation becomes even more acute when the transmission rate increases to 40 Gbps, as is illustrated by the deterioration of the eye diagram (shown at the transmitter in *Figure 1a*) after 80 km in NZ–DSF (early) type 1, where the dispersion measured +210 ps/nm at 1550 nm (see *Figure 1b*).

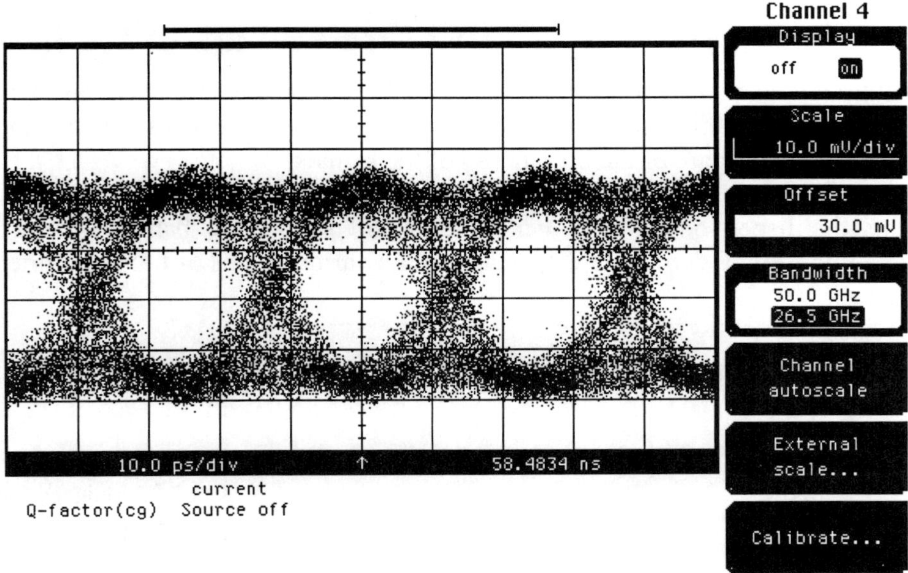

Figure 1a: 40 Gbps Signal at the Transmitter

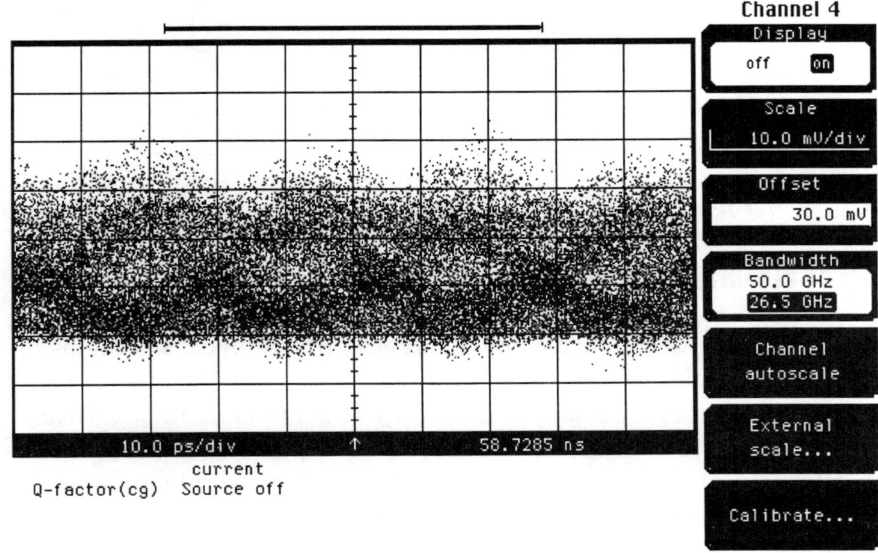

Figure 1b: 40 Gbps Signal after Traversing 80 km of NZ–DSF Fiber without Compensation (Total Dispersion = +210 ps/nm)

Dispersion management is the process of balancing positive and negative dispersion over the length of the transmission fiber so that as the optical signals travel through the network, they will always experience some chromatic dispersion, but when they reach the receiver (either at a terminal or from an optical add/drop node), the total dispersion is near zero, or within an acceptable limit. For 40 Gbps transmission, this acceptable limit is very small and hence, precise dispersion compensation must be provided.

The maximum allowable accumulated chromatic dispersion for a 1 dB penalty is a function of the bit rate (R) per channel and can be approximated by

$$DL \text{ (ps/nm)} < 10^5/R^2$$

This means that 2.5 Gbps signals can tolerate 12,000 to 16,000 ps/nm of dispersion; a 10 Gbps signal about 1000 ps/nm; and 40 Gbps can tolerate only about 60 ps/nm because dispersion tolerance is inversely proportional to the square of the bit rate (R) of the signals.

When plotted using the dispersion at the longer wavelength end of the C-band, *Figure 2* illustrates the distance limitations imposed by chromatic dispersion on the various types of transport fibers in use today as the bit rates increase.

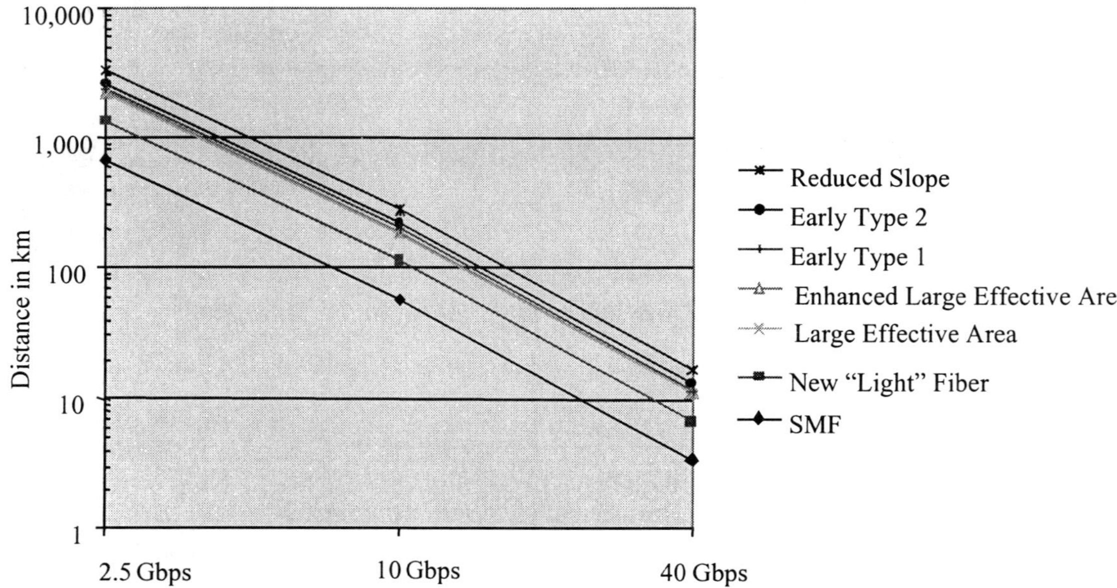

Figure 2: Chromatic Dispersion Distance Limitation

Traditional Chromatic Dispersion Management Technique

Dispersion management can be accomplished by numerous different means. Lengths of transport fiber of opposite dispersion characteristics (+D and –D) can manage the dispersion at a center wavelength—this is commonly done in submarine cable deployments. However, this presents a difficult design and build problem, and because the slopes are not complementary, the outer channels experience large amounts of dispersion. It also offers no solution for installed fiber. The most commonly used traditional method is the conventional dispersion compensation fiber (cDCF) compensator.

This special type of fiber has a typical dispersion of –70 to –90 ps/nm-km. Spools of the cDCF commonly hold 10 to 15 km of the fiber required to counteract the dispersion of the installed fiber (i.e., about –1360 ps/nm at 1550 nm for 80 km of conventional SMF). One characteristic of cDCF is that its dispersion is nearly constant (no slope) across the operational waveband. Thus, while the cDCF can neutralize the effect of chromatic dispersion at the specified wavelength, the steep dispersion slope characteristic of the transport fibers means that the DWDM transport channels away from the central wavelength may receive inadequate—or, sometimes, too much—negative dispersion compensation.

In a laboratory demonstration of this phenomenon, an 80 km length of early type 1 NZ–DSF was used as the transport medium for 40 Gbps NRZ signals. This fiber had a dispersion of approximately +210 ps/nm at 1550 nm. The cDCF compensator was trimmed to –210 ps/nm of compensation at 1550 nm so that the total chromatic dispersion experienced by the signal was near zero, as is demonstrated by the high quality of the eye diagram obtained (see *Figure 3*).

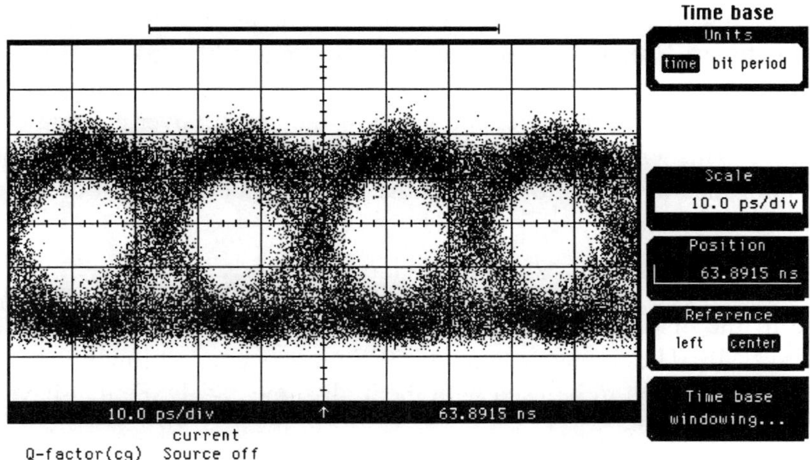

Figure 3: *cDCF Compensation for 40 Gbps Signals after 80 km NZ–DSF at 1550 nm*

However, when the wavelength of the signal was changed to 1530 nm, the quality of the eye diagram changed dramatically (see *Figure 4*). The cDCF compensator performed exactly as expected because at the 1530 nm wavelength, the dispersion of the 80 km length of NZ–DSF is only about +105 ps/nm. With a –210 ps/nm compensation from the cDCF, the signal experienced a net dispersion of about –105 ps/nm.

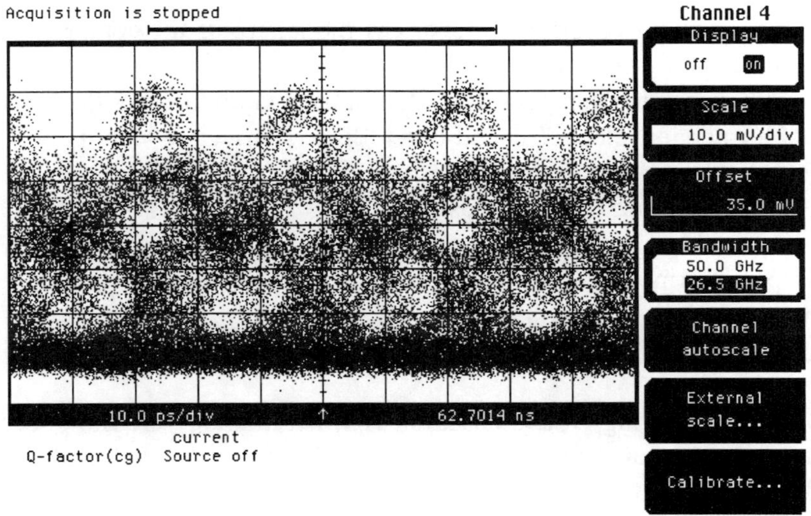

Figure 4: *cDCF Compensation for a 40 Gbps Signal after 80 km NZ–DSF at 1530 nm*

Other Chromatic Dispersion Management Techniques

Dispersion compensation can also be provided by gratings that are based on Bragg grating technology, where a chirped pattern index of refraction changes is impressed into the fiber core by ultraviolet light, effectively creating a set of wavelength-selective mirrors. Even though linear Bragg gratings are sometimes used as narrowband filters in optical networking systems, it has taken years of development to produce dispersion gratings that are only now becoming available for narrowband (1 to 2 nm) dispersion compensation applications. Broadband compensation gratings (40 nm), required for long-haul fiber compensation, have proven very difficult to bring to market. Group delay ripple, which manifests itself with rapid deviations from the mean dispersion slope of the grating over a range of wavelengths, is the most serious problem. This erratic deviation from the expected dispersion compensation at discrete wavelengths causes portions of pulses to shift in time with respect to the rest of the pulse train, resulting in serious distortion of the signal.

Recently, other new techniques have been demonstrated in lab experiments. Through further reduction in the A_{eff} of the core of the dispersion compensation fiber (DCF), some slope correction has been made possible for standard SMF. These new designs of DCF must operate at low optic power, and they have too much negative compensation to be used with the high-slope NZ–DSFs that have low dispersion (see *Table 1*).

Interferometric techniques have also been demonstrated in lab experiments to provide dispersion compensation by providing wavelength-dependent paths of different lengths for different spectral components of the transported signal. These have yet to be made into commercial products.

New Broadband Dispersion Management Based on High-Order-Mode Technology

In view of the preceding information, ideal chromatic dispersion management solutions should have the following characteristics:

- Strong negative dispersion coefficient (shortens the length of the compensation fiber if fiber is used as the compensating element)
- Broad bandwidth dispersion compensation (40 nm bandwidth to cover the C-band)
- Correction of dispersion slope (second-order dispersion) on a continuous (as opposed to a discreet) basis
- Low attenuation to minimize insertion loss
- Minimal or no introduction of nonlinear effects
- Low or no dispersion ripple
- Polarization independent
- Easy to manufacture to meet demand

Broadband high-slope-compensating DMDs based on HOM technology are now becoming available. It has been shown that chromatic dispersion can be managed effectively by HOM fiber with devices to transform the signal energy from the basic LP_{01} mode to a desired HOM and back to basic mode [2]. The conversion process is proprietary and is not discussed in this paper. HOM–based broadband DMDs (schematic shown in *Figure 5*) have low insertion loss, while they tolerate higher levels of optic input power, thus raising the high-power threshold prior to exciting nonlinear effects [3].

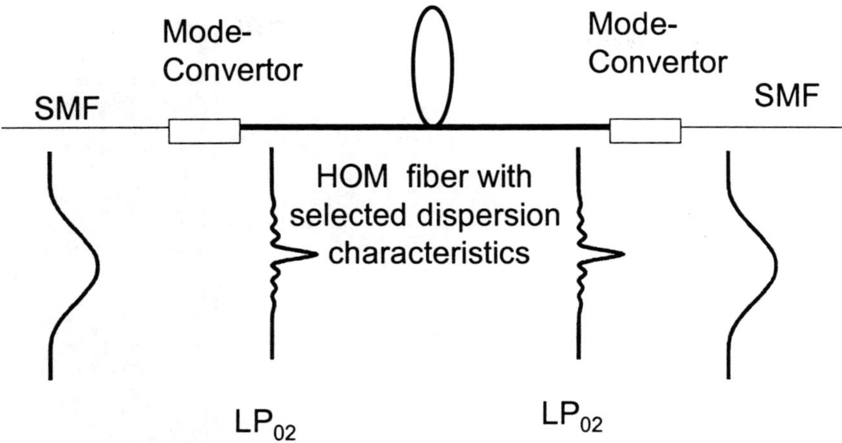

Figure 5: *HOM DMD*

The plot of the performance of a typical broadband high-slope-compensating DMD has a nominal dispersion of –270 ps/nm at 1550 nm and a slope of –5.6 ps/nm^2 over the C-band (see *Figure 6*).

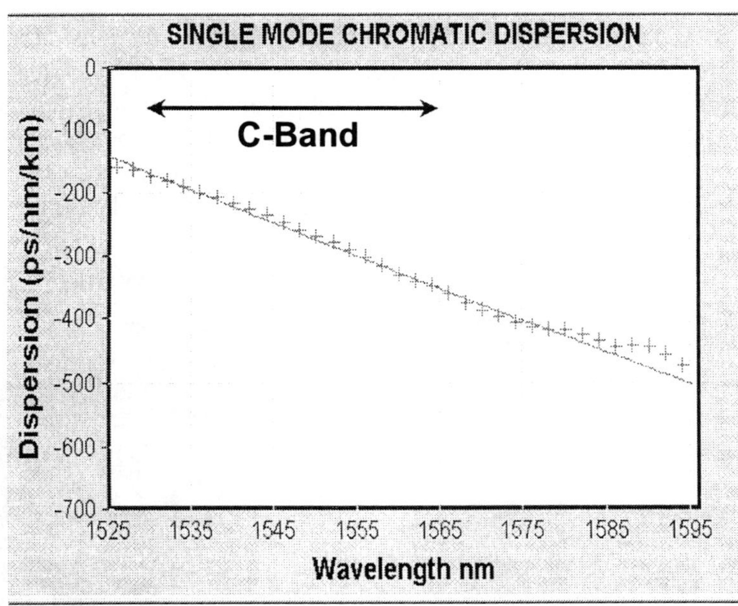

Figure 6: *Dispersion Compensation Performance of HOM–Based DMD [4]*

Broadband high-slope dispersion management means that the full bandwidth of transport fibers can be utilized with amplifiers and regenerators spaced at much greater distances than with other dispersion compensation techniques. The HOM–based DMDs provide the correct amount of dispersion compensation needed for each of the many channels used to fill the erbium band and beyond.

In contrast to the performance of the cDCF, broadband slope-compensating DMDs based on HOM technology provide the correct chromatic dispersion compensation to all the channels across the bandwidth of the transmission fiber (see *Figure 7*).

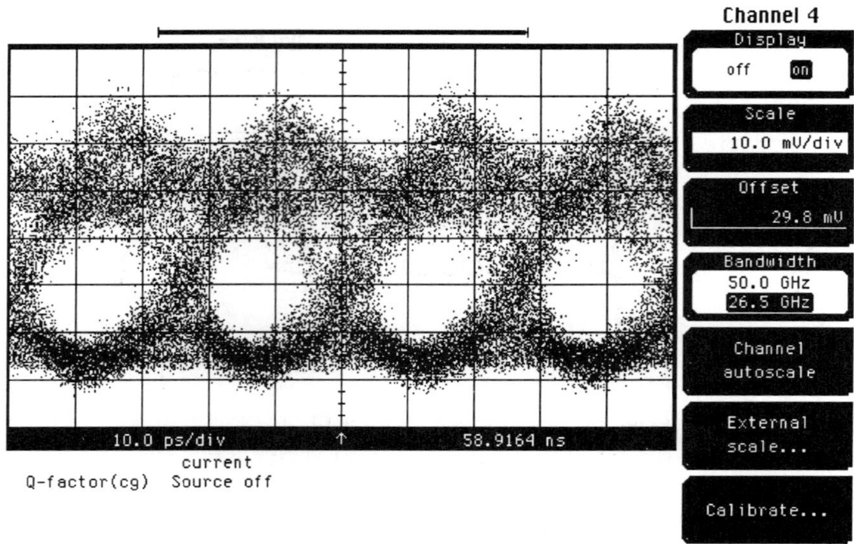

Figure 7: Broadband Slope Dispersion Management at 1530 nm with HOM Device

A More Dramatic Demonstration

A more dramatic illustration of the benefits of a broadband DMD is demonstrated by applying three of the devices to a 240 km span of TrueWave Classic fiber. The fiber used in this evaluation consisted of three 80-kilometer lengths with a total chromatic dispersion of 608 ps/nm at 1550 nm. EDFAs were used between each length to boost the signal. To provide a comparison with a cDCF compensator, the compensation was connected at the end of the 240 km span in the mid-stage access of a two-stage EDFA, so that the cDCF module and the HOM–based devices could be swapped in and out without disrupting the rest of the test bed.

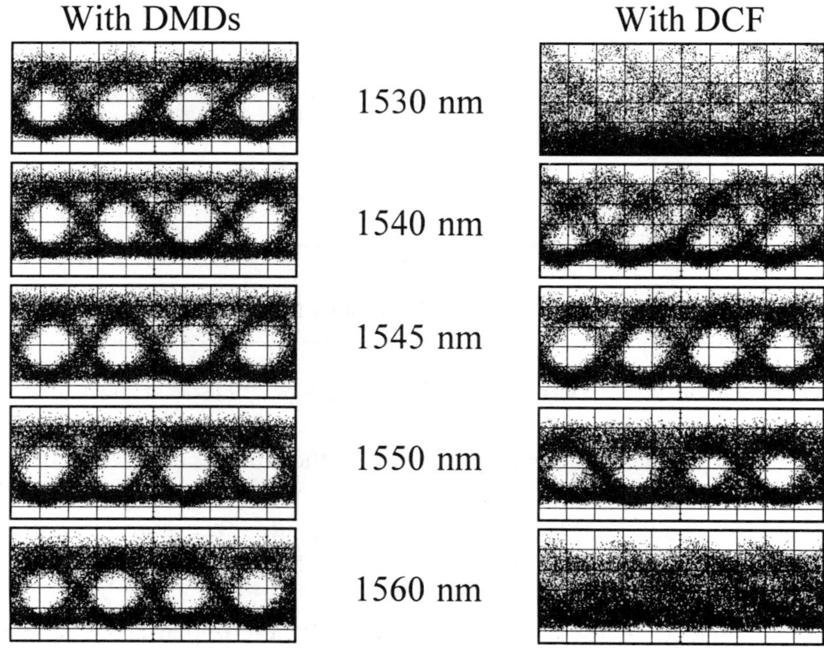

Figure 8: Comparison of High-Slope (DMD) and No-Slope (DCF) Dispersion Compensation [5]

As *Figure 8* shows, both techniques provide the correct amount of compensation at the center wavelength (1545 nm), but the cDCF permits only a small portion of the C-band to be used before chromatic dispersion limits are exceeded. On the other hand, the broadband high-slope-compensating DMD based on HOM technology enables the full C-band to be used for this test configuration of 240 km of NZ–DSF.

Summary

Chromatic dispersion has been one of the main constraints in the design of 10 Gbps and 40 Gbps DWDM transport systems. Although current techniques can provide dispersion compensation for a few wavelengths, they do not provide the correct compensation across the transport band, thereby limiting the span distance between regenerators. Expensive, complicated designs have been used to provide full-band compensation in some instances, but they are very cumbersome to implement and deploy.

On the other hand, broadband high-slope-compensating DMDs based on HOM technology have full bandwidth slope correction, low loss, and minimal or no nonlinear effects. The continuous slope compensation means that more channels can be accommodated with longer spans between regenerators. HOM–based DMDs open the bandwidth of deployed conventional SMF and new NZ–DSF for current 10 Gbps and next-generation 40 Gbps transport systems.

Credits

Special thank-you to Dr. Lara Garrett and Alan Gnauck of AT&T Research Labs in Red Bank, New Jersey, for their assistance making the 40 Gbps measurements.

References

1. Kaminow, T. Koch. 1997. "Optical Fiber Telecommunications IIIA."
2. Askegard, D. July 1999. New chromatic dispersion compensation technology opens the bandwidth for next-generation DWDM systems. *AMTC'99*.
3. Danziger, Y., and Askegard, D. June 2000. Full fiber capacity realized with high order mode technology. *IEC Annual Review of Communications 2000* 53: 62–68.
4. Danziger, Y., and Askegard, D. July 2000. Full band chromatic dispersion management. *Lightwave Magazine*.
5. Gnauck, A., et al. March 2000. Dispersion and dispersion-slope compensation of NZDSF for 40-Gb/s operation over the entire C-band. *OFC'2000 post-deadline paper*.

Issues Relating to Network-Design Sweet Spots for Reducing Cost

Robert J. Feuerstein
Senior Engineer
Level 3 Communications

Introduction

This paper will focus on the costs relating to network-design sweet spots. It will cover the network as it exists today and what is hoped for the future. Of course, everyone hopes for reductions in equipment costs. The constraints and design that will help lower costs, new forms of transport technology protection and restoration, and a vision of what a new broadband switch and router might look like will also be part of the paper, in addition to a small discussion of multiprotocol label switching (MPLS).

The network depicted in *Figure 1* is subject to debate. Some people say that such a network exists, while others argue that it will not be possible until 2001, the year many recognize as the true first year of the new millennium. In 2001, the new era of communications will be entered as well—when data traffic will certainly exceed voice traffic. However, as many have suggested, that does not mean that the revenue from data will exceed voice—only the traffic. The market is still such that voice revenue greatly exceeds data traffic revenue.

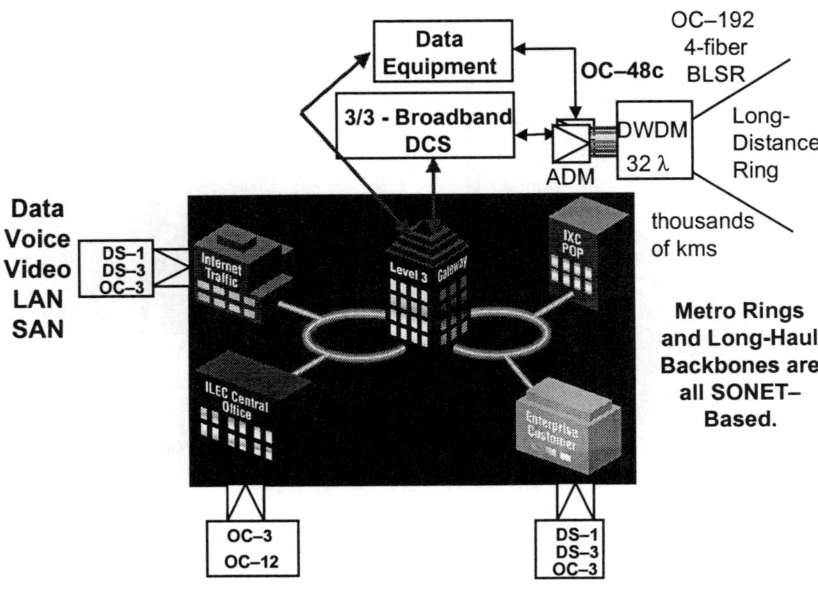

Figure 1: *Networks Today*

Today, local access comes in at various bit rates. Most of the traffic certainly begins via dial-in modems for data traffic. It is then aggregated up to T1s and digital signal (DS)–3s and moves into digital access carriers (DACs) for transport in the local area and switching around the city. Of course, if it is intended to go out to the long-distance network, broadband DACs are involved. In addition, most major carriers today have dense wavelength division multiplexing (DWDM) systems and multiple wavelengths; then the traffic is transmitted out into the world.

A major point is that most systems today are synchronous optical network (SONET)–based, which tends to make them rather expensive. In comparing the cost of a SONET optical carrier (OC)–48 transmitter card with a Gigabit Ethernet card, there exists a rough factor of 20 cost-difference versus a factor of 2.5 in the actual data rate. This, then, is an area in need of improvement. It is also hoped that systems will move toward transporting Ethernet signals and toward reducing dependence on SONET.

An interesting point about the youth of America, and probably youth throughout most of the world, is that they are used to seeing prices of electronics drop or dramatic improvements in price performance. A theory called *silicon economics* proposes that they will expect this in communications as well. It is expected that the price of bandwidth will drop dramatically over the next few years.

One of the ways to help achieve this is to bet on technology, and this is something Level 3 is doing with Williams Communications and Worldwide Fiber. These companies are burying multiple conduits, as opposed to burying a fiber-optic cable directly into the ground *(see Figure 2)*. There are currently many conduits, and it is likely that only one conduit has an actual fiber-optic cable in it with 100 to 600 fibers. As the technology changes, these companies will be able to upgrade to the new generation of fiber because there have been some changes; this will be discussed more thoroughly later in the paper. The conduit is the fixed asset; previously, though, the fiber and the conduit were the fixed assets. People, of course, want to take advantage of their fixed assets, but they will experience some limitations based on the particular fiber technology.

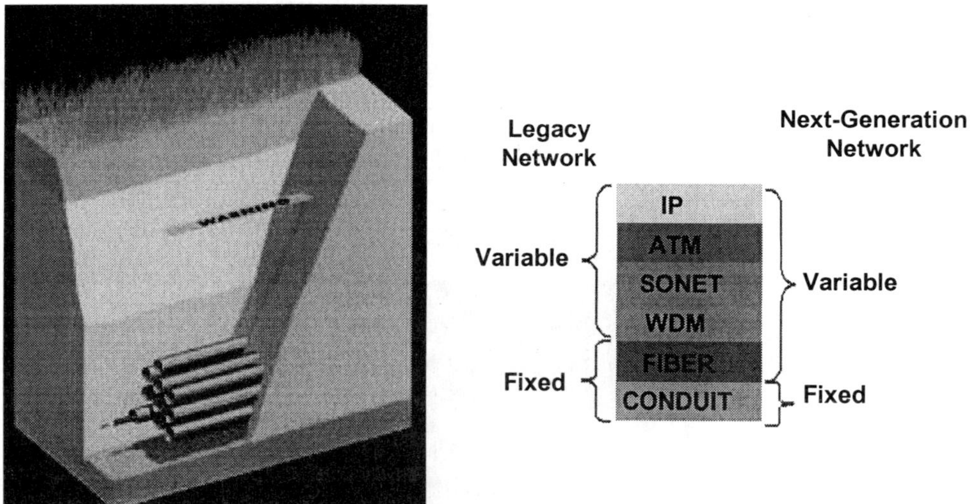

Figure 2: New Paradigm for Network Evolution

What will happen in a world where there are order-of-magnitude increases in availability and dramatic decreases in the price of bandwidth? What services will be offered? What if every building has an OC–48? Most companies hope that every building will have multiple OC–192s. Because of the drop in prices, spending on communication services, particularly data services, is expected to increase dramatically. Another critical issue is the network operating system. One issue with networks is maintaining control over provisioning and tracking the circuits that are provisioned—the available bandwidth and the capacity utilization. Is a circuit being used at 10 or at 80 percent of capacity? It would be advantageous to provision circuits without having a truck roll every time. Obviously, it can be done if there is equipment sitting out in the network ready to be provisioned, but a balance must be struck between truck rolls and stranded, unused capital.

All this must be considered when purchasing equipment. Again, operational and provisioning expenses are large components of a carrier's business. Carriers want to lower their costs. This is a very important part for a vendor offering. If they can provide an idiot-proof graphical user interface (GUI) network operating system that is idiot proof; companies will be very interested.

After all these "what-if" questions, people might ask, "Well, what if pigs could fly?" That is a fair question, but is it really wise to bet against technology? Only several years ago, everyone said voice over Internet protocol (VoIP) was years and years in the future. Consumers already have it in their homes. People also said that MPLS systems would not be installed for many years; Level 3 expected to have it in its network sometime during 2000. The company is betting on technological advances.

Figure 3 is a calculation to project some sense of the expected cost reductions. It is important to be aware that this is particular to the capital cost only for long-distance transmission equipment. The first 10 years on this chart represent a trend for the unit cost of bandwidth transmitted over single-mode fiber (SMF). It is clear that it has dropped quite dramatically since 1990. The chart also shows where DWDM systems and nonzero (NZ)–dispersion-shifted fibers (DSFs) were introduced to support multiple wavelengths. Corning's LEAF® further helped to reduce the nonlinear effects and allowed for improvements in network design.

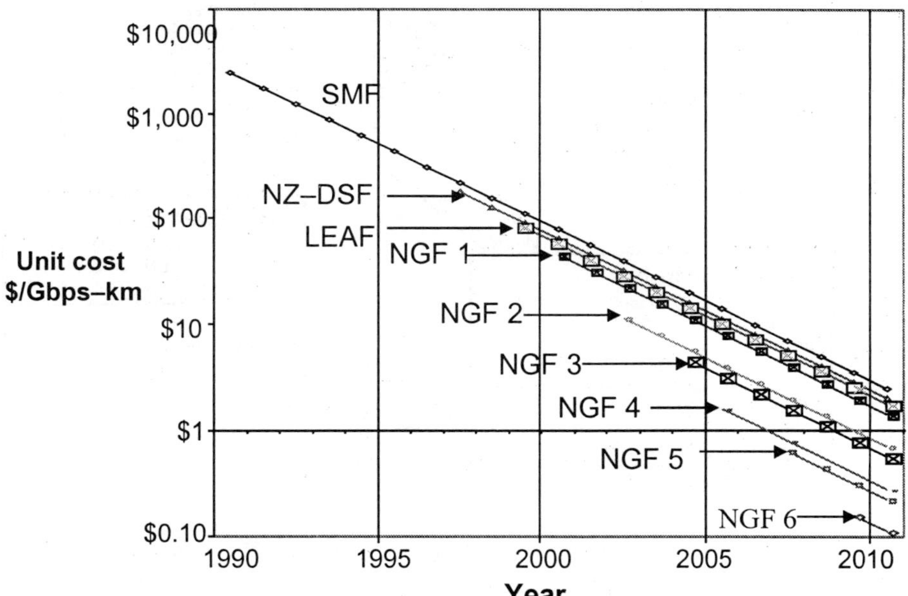

Figure 3: *What If Capital Cost Dropped Dramatically?*

This is all a wish list, but what is really being said is that as new fibers are designed, there will be improvements in the way networks can be implemented in terms of the number and packing of wavelengths and the distance that can be achieved before dispersion and other transmission effects ruin the signal. The chart is only a reasonable guess as to what sort of price reductions might be expected. Even if none of these next-generation fibers (NGFs) come into being, an almost-two-orders-of-magnitude decrease is still expected in the cost of transmission equipment.

The Penultimate Network—The Customer View

Why is the penultimate network referred to as such? The industry will never build the ultimate network. There will always be something better out there; the penultimate is the best that can be hoped for. However, from the customer point of view, any protocol, any bit rate, anywhere, anytime, and any quality of service (QoS) are important factors. Customers want to be able to sit at their computers, have some administrator dial up and request a certain amount of bandwidth from here to there with this type of QoS set up tomorrow at 3:00 pm. The industry is a long way from this point, but that is what the customers want. Because carriers want to satisfy customers, they will aim for this network flexibility as well.

Figure 4 shows a potential generic architecture that is already in development. It will possibly have some sort of wavelength or optical switch and will have hundreds of interfaces. Most commercial systems today support 32 wavelengths. Basically, one fiber pair gives 32 wavelengths. When a few more fiber pairs are added—and again, people are putting hundreds of fibers in the ground—one cable could have 100 or more fibers. This means hundreds upon hundreds of channels, so something that can support that and have that many interfaces is needed.

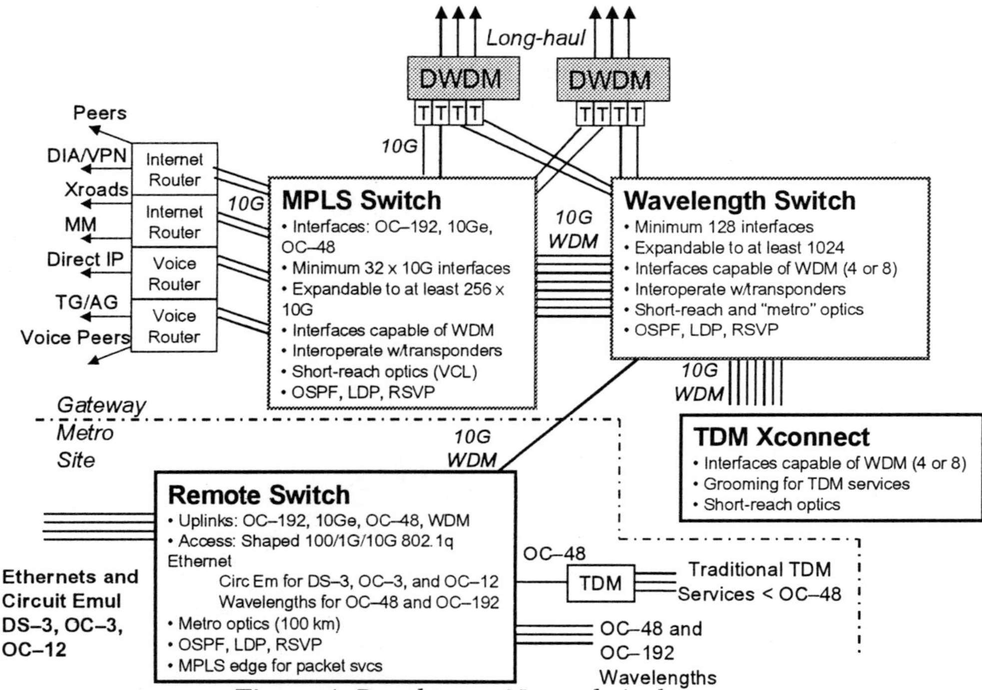

***Figure 4:** Penultimate Network Architecture*

In MPLS, some of the QoS features of asynchronous transfer mode (ATM) have been added to allow for use of Internet protocol (IP), thus avoiding the overhead of processing ATM cells, which are fairly small. The result, of course, is MPLS with IP. IP packets can be several thousand bytes—much larger than the 53-byte ATM cells. This size differential reduces the overhead on the router and the switch, which means that MPLS will be implemented and that it can support all kinds of services. Effectively, there exist numerous data interfaces, voice interfaces, VoIPs, and circuit emulations.

Time division multiplexing (TDM) services should also be supported. Some people absolutely demand a TDM circuit with 99.999 percent availability—basically five minutes of outage per year. They require protected services. On the figure, TDM cross-connects, instead of data services, would come through the MPLS switch where a mesh-routed network would be used to allow for protection in that fashion.

There is no guarantee, however, of the 99.999 percent to the same extent, but that will be discussed later in the paper. The gateway, a carrier's main switching center, is shown in *Figure 4*. Remote switches or aggregators collect various circuits. Fiber channels, 100 Mbps Ethernet up to Gigabit Ethernet, or 10-Gigabit Ethernet, enterprise system connection (ESCON), or DS–3s are collected and sent to the gateway where they are either routed around the local metropolitan area or go out on the long-distance network.

It is clear that the architecture must support quick provisioning. Bandwidth trading between service providers would be a nice feature. For example, there may be only a certain amount of capital budget, with only so much money allocated to equipment. That is put out there, and 500 requests for an OC–192 are suddenly made and cannot be satisfied. The company must purchase spare capacity from another bandwidth supplier to resell to customers.

Obviously, low unit cost is highly desired and is constantly driven into any vendors. High scalability is also necessary. Because bandwidth demands are expected to grow exponentially, the equipment purchased must be scalable. If there are 64 ports today, tomorrow there may be a need for 512 ports, and next year 4,096 may be needed. Thus, the equipment's capacity for growth is a key consideration.

Long-Distance Transport Technology Today

Most networks in today's long-distance transmission technology were designed as "industry-best" *(see Figure 5)*. Level 3's network was designed for six spans of 100 km before a regeneration of the signal. Nortel uses a bidirectional system, and they also have a unidirectional system. It matters little whether the system is bi- or unidirectional. It remains possible to have the 6x100 with erbium amplifiers providing compensation for the loss in the fiber. Dispersion compensation devices are also located at each amplifier. This supports 16 bidirectional OC–192 channels over Corning's ELEAF®; there can be optical add/drop of one to eight channels, if so desired, and an extremely low–bit-error rate (BER) of 10^{-15}.

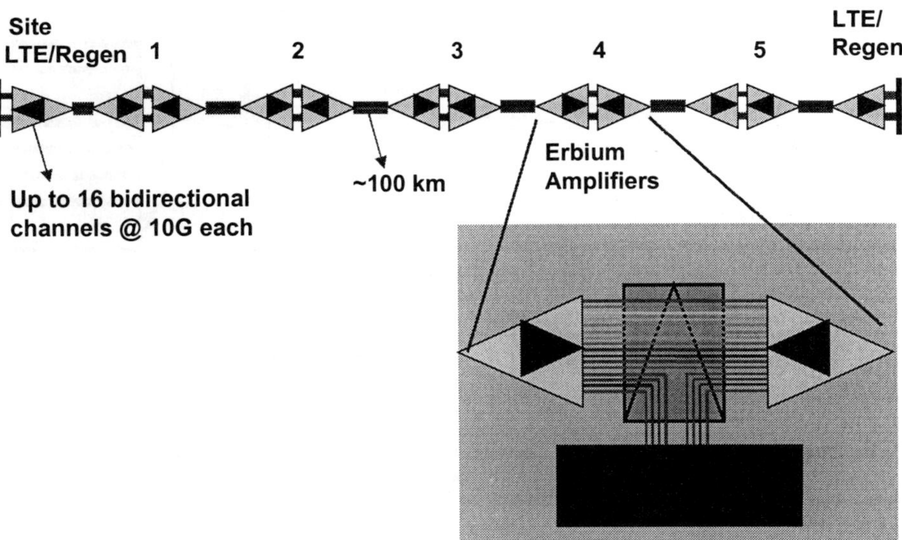

Figure 5: Long-Distance Transport Technology Today

However, the young dot-com companies are developing new applications that burn bandwidth faster than venture capitalists burn money. This is really wonderful because these companies burn the venture capitalists' money at the same time. Again, though, this is the expectation. All the new start-up companies—10 to 20 dot-coms start up each week—have great ideas for networking applications that simply burn up the bandwidth.

Remember that the 6x100 design requires regeneration of the signal at some point in its life cycle. Unfortunately, a regenerator consists of an east-facing OC–192 card and a west-facing OC–192 card. Thus, if there are 32 wavelengths, there are 32 on each card for a total of 64—without any protection. In addition, each card costs roughly $50,000. Adding the numbers, the cost is roughly $6 million per regenerator with protection.

Needless to say, numerous companies have asked the vendors for alternatives to regenerators. Qtera, Corvis Corporation, Sycamore, and others have come up with what they call the ultra-long-haul systems. Instead of having 6x100 km spans between regenerations, the ultra-long-haul systems can have, depending on the particulars of the system design and the particular manufacturer, somewhere between 2,000 and 4,000 km between regenerations. That means the system can range from New York to Los Angeles with perhaps only one regeneration. Again, because regenerators can cost $6 million each, carriers are indebted to these vendors for developing this idea, which will help push the cost curve down. Unit cost in dollars per Gbps per km is lowered.

The other features are fairly similar to the 6 x 100 *(see Figure 6)*. Dispersion compensation probably remains in each amplifier along the way. Most will support 50 to 100 percent channel add/drop, and there are products that have been announced that can support this today. There will be no limit on adding or dropping only a few channels. It is possible that as many channels as needed can be added or dropped at each amplifier site along the way. This is important to support a market such as a T-2 or T-3 city that needs only one OC–192 service and that happens to be along a 2,500 km span. An optical add/drop multiplexer (OADM) at the amplifier can support that city with an OC–192; all the other signals, which may be going from Chicago to Los Angeles, for example, do not have to be dropped. The signals can go straight through and stay optical the entire way.

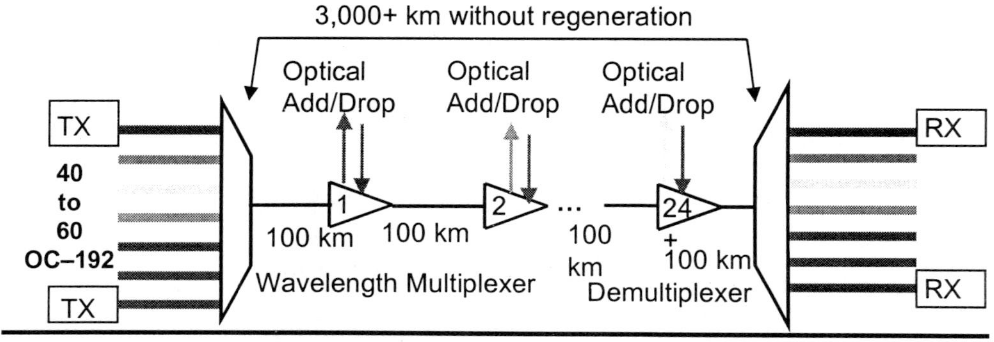

Figure 6: Next-Generation Transport Technology

What is desired in the far future? Much of what will be discussed is still research lab work, such as all-optical switching of packets. Today, routing is done electronically. The packet headers are read, which means they are detected and converted to an electrical signal; the router looks up where it is going and sends it out the right board and on its way. XRos, which was purchased by Nortel, has one; Lucent also has one that uses what are called micro-electromechanical systems (MEMSs).

MEMSs are little mirrors that are roughly a quarter of a millimeter in diameter and are controlled by electrical signals. Optical circuits can actually be switched this way. However, the important thing to remember is that the switching itself is slow—on the order of milliseconds. When transmitting 10 Gbps, milliseconds take forever. It does not involve routing the packets; instead, it is an optical circuit switch, which is useful for provisioning and protection switching. It is not designed for directing where the packets go. That still will have to be done electronically.

The term "optical switching" means that the signal remains optical and is optically switched on a packet-by-packet basis. Obviously, there will have to be detection of some header bits to determine how to set the switch. This procedure is approximately 10 years in the future, and in this business, 10 years is the far-distant future.

Because regenerators are so expensive, it would be ideal if regeneration could be performed all-optically. There have been some demonstrations of this in laboratories, although it is not yet ready for the market. As new next-generation fibers are introduced, the technology and the transmission equipment will change. This change will lead to lower costs and will help to reduce nonlinear optical effects. It probably will not reduce dispersion, but vastly improved dispersion compensation devices will be developed. Numerous companies are working on this and have tunable variable compensation devices for dispersion. Unfortunately, today they do not compensate for the dispersion slope. The amount of compensation can be changed, but it must be the same at every wavelength. That is inadequate for these ultra-long-haul systems.

Companies constantly seek ways to reduce the amount of space used in their networks. Most have facilities around the country with 50,000, 60,000, or even 70,000 square feet, and most of the space is allocated for co-location. They support their customers with space where customers can connect up to their networks, and this space is almost sold out. If the space needed for equipment can be cut, customers can be provided with more space for their equipment. Consequently, it is desirable for the density of the equipment to increase to allow more circuits in the same space. That again will help reduce costs and will cut down on power requirements.

Tunable lasers and transmitters are certainly available and useful. Today, on a 32-wavelength system, a company does not want to keep 32 wavelengths on hand as 32 separate lasers for sparing purposes. This adds cost. If there can be two or four lasers that are tunable to cover the entire range of wavelengths, it reduces sparing requirements.

Protection and Restoration

Figure 7 is a good illustration of protection and restoration. On one side there is a SONET ring. The underlined numbers on the figure represent the demand for bandwidth in bandwidth units; OC–192s will be used as an example. Because this is a SONET ring, and because of the way the vendors build SONET

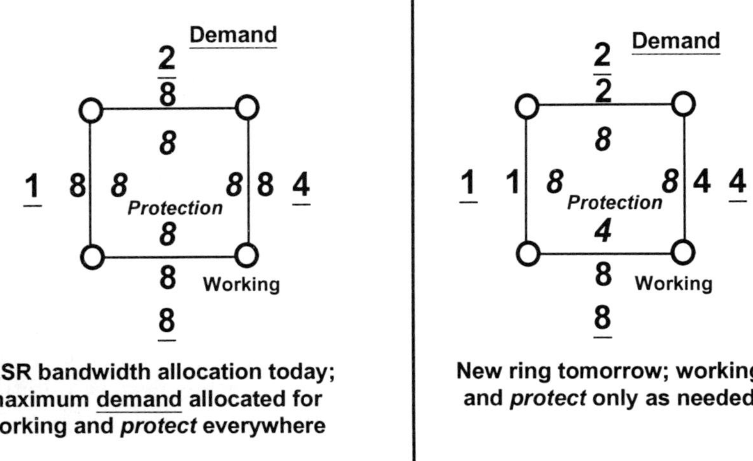

Figure 7: Ring Protection Protocol

today, whatever the maximum requirement is on the ring, that same number of units of protection and working capacity is needed. In *Figure 7*, this requirement happens to be eight units, which means that eight units of protection are needed around the entire ring. In addition, there must be eight units of working capacity around the entire ring. This is extremely wasteful and expensive.

Many vendors want to adopt an approach where the working capacity will exactly match the demand and where, if the link is cut, there is another way around. Eight units of protection are needed on one side, but only four are needed on the other to protect the next lowest maximum capacity. The total number of transmitter units is 16 x 4 for the working and protect. There are 64 units of transmission capacity on the left in today's SONET system.

In tomorrow's SONET system, there are 24, 32, 36, and 43 versus 64. Twenty units are saved. Each one of those corresponds to a pair of OC–192s, which translates into a savings of $2 million in hardware providing the same functionality. Numerous vendors are working on this, and it should be available, if not by the end of 2000, certainly in 2001. Carriers could save $2 million in hardware if the vendors change from in-band signaling to out-of-band signaling for the SONET protection switching.

In a mesh network, there are multiple connections, meaning that there are multiple ways to get between any pair of nodes (see *Figure 8*). For example, one pair of nodes may offer the direct route, which would be the preferred route for transferring information between these two nodes. If the fiber link is cut, there are four other connections and multiple ways for data to get from node to node. However, mesh gets very complicated because these nodes could be 200 miles apart and routes could be 400, 5,000, and 3,000 miles long. Many different geographic distances are involved and, again, there are regenerators and amplifiers.

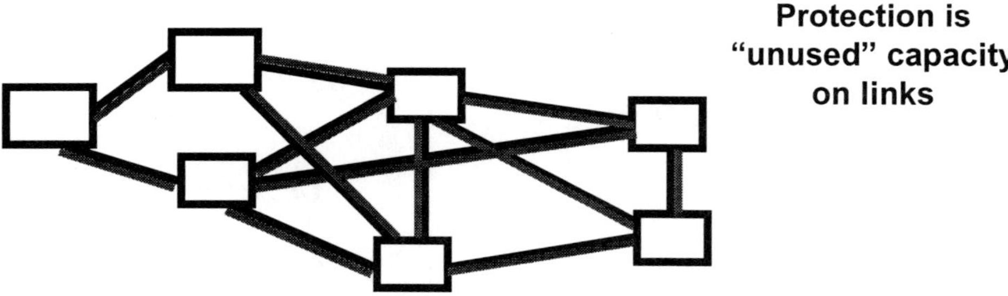

Figure 8: Protection and Restoration

Some excess capacity is on these links to protect against, say, a fiber getting cut. However, if one route is 3,000 miles and another is 300 miles, does it pay to put any protection capacity here? This is a simple example, but mesh is an extremely complicated approach to protection in addition to the process of allocating bandwidth. In other words, if fiber gets cut, is all the protection put on one link, or some on one and some on another? How is the allocation determined? What happens when there are multiple cuts?

Mesh has many advantages compared to the SONET approach of full protection capacity everywhere. With the mesh scheme, potentially one quarter of the bandwidth requirement could be allocated to one link, one quarter to another, and so on. That can be shared against multiple links, and calculations are done to determine the availability. Whatever the availability requirements, enough bandwidth to cover them can be allocated. However, by sharing protection across the multiple possible paths, protection capacity is greatly reduced, which means hardware is reduced. Mesh has a big payoff in terms of operability and cost reduction, so there is a lot of work going into it. Companies will provide mesh protocol routing and protection in their equipment.

It is important to also discuss SONET protection. *Figure 9* shows a unidirectional path-switched ring (UPSR) and a bidirectional line-switched ring (BLSR). In the UPSR, the data is duplicated and transmitted in both directions. If the fiber is cut at the end, there will be a switch to select the data from the other direction. However, the signal path uses up the entire ring to transmit the information simultaneously. A BLSR gets around that because the signal goes only from one point to the other. If there is a cut, of course, it will go around the other way. A BLSR provides greater utilization of the ring capacity.

Today's Architecture:
SONET Rings (UPSR, BLSR) are simple
- Provides <50 ms restoration time
- **However**, rings require 100 percent protection capacity
  >> Low capacity utilization

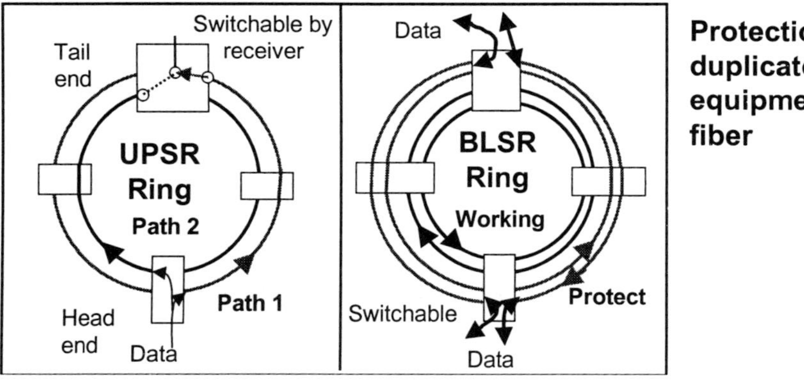

Figure 9: Protection and Restoration

Generally speaking, UPSR systems are used in metropolitan-sized networks, and the BLSR systems are used for long-distance transmission systems. An important point is that protection is done by basically completely duplicating the transmission system.

The right half of *Figure 7* shows the new ring protection protocol. Carriers hope to see this in all vendors' equipment in the near future. Again, there is wonderful reduction in protection equipment and, therefore, a lower cost associated with the new ring protection. However, it still provides one of the key benefits of SONET protection, which is the ability to recover from a fiber cut or another disruption in 50 milliseconds or fewer—a very fast guaranteed recovery.

In the future, the industry expects to see distributed mesh restoration despite the problems. Algorithms will certainly be developed to calculate where to put the bandwidth to protect against this or that cut. What is changed as cuts occur? How is reallocation to protection capacity achieved? Tools based on some clever, patented algorithms will be available, and it is certainly something that everyone anticipates, based on what the vendors are saying.

Some applications, such as voice, obviously are much more sensitive to the amount of time needed to restore service than is data. Most people will not notice if it takes a few extra seconds to download a Web page while searching the Internet, but if it is a voice call, and the caller disappears for a few seconds, someone will definitely notice that. Therefore, restoration has to be sensitive to the type of information that is being restored, especially in mesh where some traffic can perhaps be delayed while providing guaranteed restoration for other traffic.

In the open systems interconnection (OSI) model, there exist numerous layers. There is much discussion in the field as to where the protection switch should occur and in what order the routers should protect. Is there an optical switch that does protection? The best approach is unclear and may remain so for some time.

If a customer wants a cheap price and is willing to tolerate an average of 24 hours of blackout per year, the customer might agree to pay half price and simply expect that there will be some times during the year when it will be impossible to send information over the company's network. The fact that the customer is charged half the price may be more important than the blackout periods. However, other customers, such as financial institutions or air traffic controllers, run critical applications that cannot tolerate five minutes of blackout. Obviously, these customers will get bulletproof protection, but they will pay much more for it because the equipment should be able to support multiple layers of protection. The service-level agreements (SLAs) with these customers will be written appropriately, and the price will reflect the service they receive.

If the fiber is cut, the signals disappear. Thus, a protection switch is needed. However, what happens if the BER changes or the power level fluctuates, or there is some variation in signal quality? Switching solely on the basis of a loss of signal is inadequate. More sensitive protection switching than that is required, regardless of the system. Obviously, SONET has many built-in features, which is only part of the reason SONET transceivers are so expensive. They provide sensitivity to variation of signal quality, not merely to loss of signal.

The New Ultra-Broadband Switch?

What will the new ultra-broadband switch look like? *Figure 10* is one potential view. There is an IP–MPLS router with services connected to it. Services could be wavelength services, because some

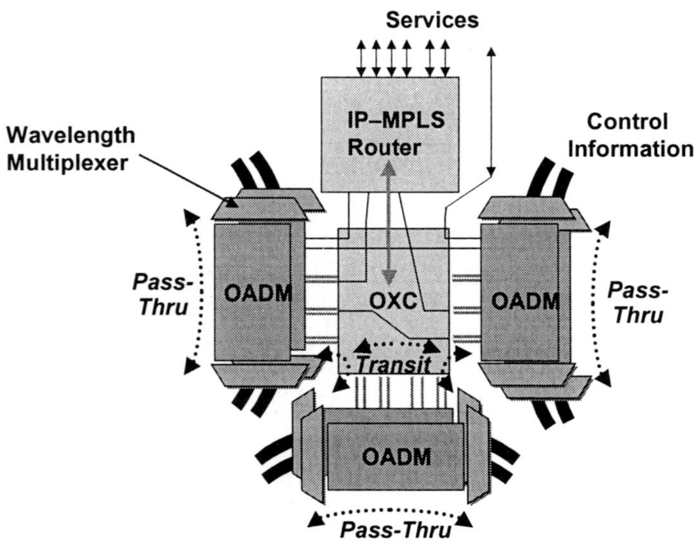

Figure 10: The New Ultra-Broadband Switch?

customers want only a wavelength. They will put it on whatever they can. That comes out of the OADM and is sent directly to the customer.

Then there is the optical cross-connect (OXC), which could be one of the Lucent micro-mirror switches or the XRos switch or another similar developing technology. The OXC is more of a provisioning tool. It takes milliseconds to set up a circuit and to move and stabilize the mirrors before the circuit can be used, but there are many applications where this is useful. This would obviously replace patch panels, which are highly prone to human error. Unfortunately, technicians are not perfect; it is a big problem in operating networks. This procedure could be entirely controlled by computer with screens showing circuit connections and alternatives. There is much less likelihood for error with the computer than with a person, who might misread the number, pull the wrong fiber out, and disconnect the customer. It has some advantages but, again, it takes milliseconds to set up the switch.

Another alternative could be the Cisco alternative, where the switch does it all. The switch is connected directly to the transport network, and it does protection, routing, etc. Wavelength services could still come directly out. They would not have to go through an IP–MPLS switch because it is not packetized or, if it is, the customer handles it.

One of the challenges is interoperability. Some have talked about how SONET is a standard, but it has also been brought up that two vendors' SONET equipment cannot be put together. This is an issue that will certainly be raised with vendors. Carriers' equipment must be able to interoperate.

Routing and protection switching flexibility is needed to support bulletproof to blackout protection options. The equipment must be sufficiently flexible to support that. Multiple protection protocols are an advantage so that the equipment could support rings as well as mesh; the carrier can simply decide at the time which of those it will use. Part of the network could possibly be ring because it supports customers who demand the best availability. Those customers are all on one ring with the equipment partitioned so that it supports them. The rest of the network perhaps is mesh.

Again, variable QoS attributes would have to be spelled out in the SLAs. What is the availability? What sort of latency is there? Perhaps the customer will have some real-time video services or voice services. Obviously, higher QoS is needed for those services than for data. If it takes a few extra minutes to complete an end-of-the-month transfer, it will make little difference.

Scaling is also important; remember that a single fiber pair today is 32 channels with at least 100 fibers in the ground per cable. In some places, there are many, many more than that. One fiber cable involves thousands of wavelength ports from one direction. If there are three or four cables coming into one site, the number of ports can rise to 10,000 very quickly. However the new equipment looks, it must be scalable and able to grow.

In terms of density, equipment should take up as little space as possible to allow more space for customers' equipment. The ideal is two bays supporting 256x256 ports. This does not yet exist, but it is anticipated.

Some customers want services where they use the SONET overhead bytes. These customers do not want the carrier to touch them, so there is the option of using the SONET overhead bytes or not. This presumes remaining in the SONET world. It is possible that SONET will go away; many people talk about SONET lite, which is also a possibility.

There is a need for upgradability in addition to scalability. Obviously, OC–768 or 40 Gbps have been demonstrated in lab systems and have been used in some very high-capacity transmission hero experiments. The equipment should at least be designed with the thought in mind of going from OC–192 to OC–768 ports.

MPLS is a Layer-2 protocol; it sticks on a four-byte header (see *Figure 11*). It can be thought of as taking MPLS and sticking IP on it; this combination gives the same features or almost all the same features as ATM. It gives QoS, virtual private networks (VPNs), and limited latency guarantees to support voice services, which cannot be done over an IP network today. With MPLS, it can be done. Bandwidth can also be reserved.

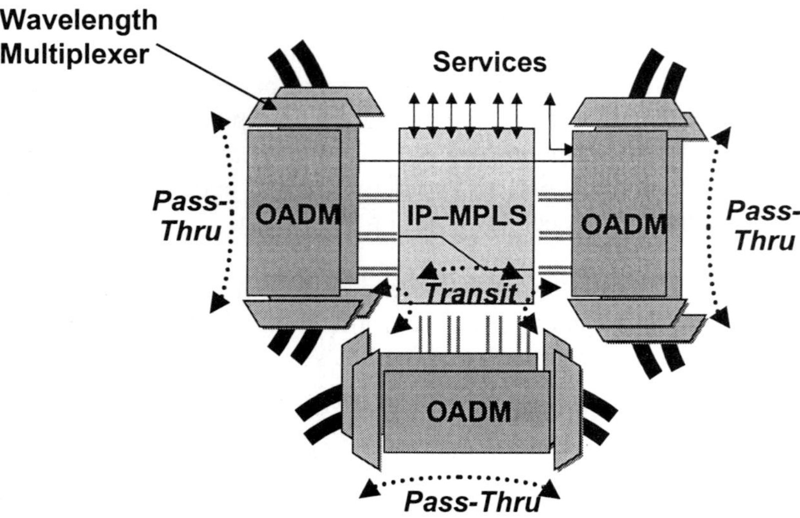

Figure 11: The New Ultra-Broadband Switch?

It supports VPNs and fast reroutes; it isolates customers. That is the "virtual" of private network service. MPLS gives the best of both worlds. It gives privacy and the QoS features of ATM. It is flexible and scalable and uses IP routing. IP is the way to go because it will help to reduce costs, and it scales to large or small networks.

Conclusions

In 1999, the wide-ranging communications industry recorded $386 billion in revenue, according to the U.S. Department of Commerce. However, this wide range includes cable television companies and television networks such as NBC, ABC, and CBS. Regardless, most of this revenue was in the telecommunications and data communications carrier business. If the same growth rate (as 1998 to 1999) continues over the next few years, revenues will reach $700 billion—and, of course, this includes only the United States. Some of these things are happening around the world, which means that there is an enormous marketplace. By 2005, more than half of the marketplace is expected to be data, not voice.

In summary, it is assumed that the demand for bandwidth is absolutely insatiable. Thus far, this has proved correct. Disruptive technological advances will continue to lower unit costs. Mesh provisioning and protection will improve. There will be better algorithms and better control over protection, which will help to reduce dramatically the cost of protection equipment. New architectures with better tools will be developed to help increase capacity utilization.

Ethernet interfaces are much cheaper than SONET, so either Ethernet will add a few features and take over SONET, or SONET must somehow show a dramatic reduction in the cost of its transceivers. Level 3 is betting on MPLS; the company expects MPLS to provide the necessary components to supply the products that its customers need.

Dynamic Control of SONET/SDH/DWDM Transport Networks to Carry Data Traffic More Efficiently

S. Makam
Staff Engineer
Tellabs

D. Levandovsky
Lead Engineer
Tellabs

C. Parollina
Staff Engineer
Tellabs

Introduction

This paper proposes ways in which the ever-growing Internet data traffic (Internet protocol [IP] and asynchronous transfer mode [ATM]) can be carried more efficiently through the synchronous optical network (SONET)/synchronous digital hierarchy (SDH)/dense wavelength division multiplexing (DWDM) infrastructure. The paper demonstrates how the capabilities of fast signaling protocols and virtual concatenation of time division multiplex (TDM) and wavelength division multiplexing (WDM) channels, when used judiciously, substantially reduce operational costs, decrease circuit provisioning time, and increase revenues by making low-cost upgrades to the existing infrastructure.

The first part of the discussion focuses on how fast-signaling and routing protocols can be utilized to set up communication channels through the TDM or WDM network instead of through static (manual) provisioning. The dynamic channel setup can be extended for fast provisioning of digital signal (DS)–1 and DS–3 private-line services across SONET/SDH as well as of optical wavelengths across WDM networks.

While dynamic protocols are used to establish paths through the TDM and WDM networks, it is important to consider the physical constraints affecting the quality of service (QoS) of these paths. Flexibility is needed in the signaling protocols to take into account the physical constraints so that the suitable paths can be determined en route.

When referring to SONET/SDH/DWDM networks, generic terms such as "transport networks" will be used in the remainder of the paper.

Limitations of the Static Transport Networks

Ever since the first intelligent transport network devices were installed in a service-provider network, there has been a vision of creating a robust, remote/automatically controlled network. The vision was to create a network that permitted remote maintenance and provisioning to improve time to repair and service-installation intervals.

The initial intelligent devices were electronic DS cross-connects (EDSXs) and certain versions of DS–1 digital terminating banks (D-banks). These devices by themselves offered some level of remote control, but the network was still saturated with devices that required on-site, manual effort for maintenance and provisioning, which prevented any real progress toward the vision.

SONET/SDH was the first real opportunity to make major steps toward achieving the vision. This new technology included broadband digital cross-connect systems (BDCSs), wideband digital cross-connect systems (WDCSs), and intelligent fiber-optic terminal equipment with varying degrees of cross-connect (XC) capabilities. The intelligence capabilities of these devices, along with the ability to make cross-connections remotely, were the catalysts needed in the transport network to achieve the vision. However, more than the network infrastructure was required to complete the package. Operations support systems (OSSs) must be in place to provide the administrative and control capabilities necessary for this automatically controlled network.

At the same time that national and international standards were being developed for SONET/SDH, efforts to define a set of standards for such an operational support infrastructure were also undertaken. These efforts led to the development of the telecommunications management network (TMN). The TMN is based on a top-down approach to network management. The TMN infrastructure includes a layered approach to solving the problem of operational support in an automatically controlled, multivendor network. The proposed layers within the TMN include the following:

- The element layer, which is the actual network element (NE) (for example, BDCS, WDCS SONET terminal)
- The element management layer (EML), which provides control and coordination of a subset of NEs and provides the mediation function between the elements and the network management layer (NML)
- The NML, which controls and coordinates the network view of all elements within a domain; this layer provisions, connects, disconnects, and modifies network capabilities for service to the customer, among other functions
- The service management layer (SML), which provides the point of contact for the customers, management of the services provided, billing, service integration, and other important functions
- The business management layer (BML), which provides for the total enterprise responsibility, along with strategic planning functions

This work to develop all that is required for the TMN has been very complex and has moved forward slowly. One of the areas of most contention is the connection between the EML and the NML. This should probably be expected because this is the boundary point of responsibilities between the transport NE, the element-manager vendor's proprietary domain, and the OSS vendor's domain. The problems encountered at this interface have contributed to the service providers' inability to achieve the vision. Some OSS vendors have attempted to alleviate this problem by incorporating the EML into the NML.

This is difficult at best. First, obtaining the necessary agreements between the OSS vendor and the element vendor to facilitate the necessary development work is difficult. Designing the function for a set of elements is a lengthy process, and each vendor's equipment may require a customized solution because the interface between the element and the enterprise messaging server (EMS) is proprietary. In addition, this solution does not promote the timely introduction of new features and functions into NEs because of the delay between the development in the elements and the developments needed to support the features and functions in the EMS.

In today's networks, multiple solutions exist. Some regional Bell operating companies (RBOCs) rely on a set of tools, of which many were developed prior to any intelligent network devices. These tools have been modified to accommodate some of the capabilities of the intelligent devices, but it is a slow process that results in only a limited set of functions. Some network providers have designed their own in-house systems. These may provide a good solution for any given point in time, but given the different element vendors that exist and the speed at which technology changes, it is a monumental task to keep the systems current.

As the network evolves, increasingly more reliance will be placed on the optical network. This optical network will include DWDM, which already has a major impact on the InterLATA (local access and transport area) carrier market and optical XCs (OXCs) and switches. The problems in achieving the service-provider vision will be the same, but the service-level demand will be much greater.

The remainder of this paper is dedicated to discussing a possible solution that will alleviate some, if not all, of the problems in the area of provisioning, including bandwidth-on-demand requirements. This solution assumes that development efforts for the TMN–like solutions will continue for all the other functions necessary in a service-provider network.

First, the provisioning and bandwidth-on-demand problem must be split into two functions, internal control and external control. Internal control is defined as the provisioning that is done as a part of the day-to-day activity of the service provider. An example would be the addition of DS–1–level systems that are provisioned to augment existing voice network trunk groups. This type of provisioning has the following characteristics:

- It is solely within the control of the service provider.
- No billing arrangements are required.
- Bandwidth on demand is not an issue.
- By definition, this is not a service offered by the service provider; it is simply a more efficient way of provisioning the network.

The external control function exists when the customer actually initiates the service or bandwidth-on-demand request through some interface to the service-provider network. The general traits are as follows:

- This is, by definition, a service offered by the service provider.
- Billing is an issue.
- Bandwidth on demand must be accommodated.

One very important part of the solution required for both the internal and external control situations is the creation of a special network control plane. This control plane will facilitate the ability to contact the NEs directly and to have the NEs automatically discover a routing path for the service in question across the network. Once the path is determined, the element would notify the EMS, which would notify the network management system (NMS).

In the case of the internal control functions, no billing issues would exist, and all OSS levels would be updated in a bottom-up approach. In the case of the external function set, the service provider offers a service, a key part of which is automatic provisioning and bandwidth on demand. In this situation, the control plane for routing discovery and service provisioning would be essentially the same; however, additional capabilities would be required. First, billing information must be provided to the OSSs. Second, an approach must be developed that will allow the service provider to control the network capacity that is accessible as a part of the service. If the characteristics of service were compared with that of the voice-switched network, a possible solution to this second problem may be found.

In the voice network a matrix of trunks is established between the various voice switches. These trunks are not dedicated to any specific customer. The trunks are shared by all customers making a call that traverses the involved trunks and switches at the ends of the trunk. These calls are set up and taken down under the control of a signaling network. At any given time one might view this trunk network and find substantial traffic carried across the network (peak hour) or little to no traffic carried (late night). Two key concepts can be taken from the voice network and applied to the external control situation. First, assume some level of capacity is made available between the NEs for the use of automatic provisioning and bandwidth on demand, similar to the trunks in the voice network. Second, establish an interface from the customer to the NE that extends some facet of the above-mentioned control plane to the customer; then such a service could be offered. Establishing the capacity is the easy part of the problem; the creation of the control plane is more difficult.

Envisioned Benefits of the Dynamic Transport Networks

Before discussing how to create the control plane it might be beneficial to discuss why the service provider and its customers might want such a service. One major point important to this discussion is the area of network trends. The trend that will provide the most incentive for such a service is the growth of the Internet and other data traffic. Considering this, both the customer and the service provider would see a major benefit in terms of the time it takes to establish service. The customers would benefit because they would need to pay only for the level of service required. The service provider would also benefit because the capacity required in the network can be maximized for efficiency. What does this mean? The data user requirements will have varying periods where the capacity demands are at a peak. Because these peak periods will vary across all users, capacity that serves one user with a particular peak period can be utilized to satisfy other users who have different peak periods. A second advantage for the service provider is that, because this is a shared-bandwidth network, like the voice network, certain pricing strategies can be implemented that could potentially increase revenues. Finally, depending on how the service provider decides to implement the capability, other private-line services can be provisioned using this control plane. This would substantially reduce the provisioning time required for the involved private-line services.

Even though the current SONET/SDH networks excel in the performance monitoring (PM) and fault management (FM) areas, they fall short in automatic neighbor/resource discovery and topology distribution. If the resource utilization information is disseminated along with the link states, a very complete picture of network status is available to a network operator for use in planning, provisioning, and operations.

A common signaling protocol, such as signaling system 7 (SS7), private network-to-network interface (PNNI), resource reservation protocol–traffic engineering (RSVP–TE) [7], or constraint-based routing–label distribution protocol (CR–LDP) [8], that is properly extended for circuit switching applications can solve the (management) interoperability problem for provisioning of circuits.

Another major benefit is the dynamic bandwidth management of circuit connections. The ability to modify the bandwidth resources of existing connections to carry bursty data traffic in a TDM network can create efficiencies.

Requirements for Dynamic Transport Networks

Walker, Mack-Crane, and Steele's article identifies a set of requirements for using a dynamic control plane in the new generation of optical transport networks [4]. To achieve a fully dynamic transport network the following functions must be available:

- Automatic neighbor discovery and link-state maintenance mechanism
- Topology distribution with link and node constraints
- Route computation (for each circuit or path) satisfying a set of constraints
- Automatic path or circuit setup and teardown procedures, along with ability to modify and abort connections
- Automatic shared mesh protection switching to handle node or link faults
- Automatic synchronization of dynamic provisioning of circuits with EMS/NMS inventory systems

In addition, to transport IP data traffic more efficiently, more granularity in SONET/SDH networks is desired. The following capabilities, when provided, will improve the bandwidth efficiency and handling:

- Contiguous concatenation
- Virtual concatenation
- Arbitrary concatenation
- Bundled connections
- Stacking and tunneling of lower-order TDM channels in higher-order channels

Dynamic Control Architectures

In the attempt to define a dynamic control architecture, the large installed base of transport networks must be considered. There is no economic driver for a wholesale redesign of SONET NEs that would make them behave like label switch routers. The task is to find an economical way to upgrade the existing NEs to provide a more dynamic establishment of n*VT1.5, synchronous transport signal (STS)–N and n*STS–1 paths and to carry the data traffic more efficiently.

In typical data networks, a portion of the link bandwidth is used for the purpose of neighbor discovery, link maintenance, topology distribution, session establishment, and path setup and rerouting or path protection. In SONET networks, the STS synchronous payload envelope (SPE) is typically unavailable for network control and management. Any spare capacity available in line or section overhead bytes of the SONET frame must be used. Some existing systems in the field may not permit the use of this spare overhead capacity and as a result, this may be a constraint in implementing dynamic control for these networks.

Some possible overhead fields that can be used to carry control information are J0 (trace byte to carry node identification), section data communications channel (DCC) (unused bandwidth for control), and line DCC (to carry new IP–based control protocol). Bidirectional control channels must be established to maintain correct adjacency/peer relationships and also for link maintenance.

There are two ways to exchange the control information between the neighboring nodes:

- Link-associated (in-band) signaling using SONET/SDH overhead bytes, optical transport network (OTN) optical supervisory channel, or OTN dedicated wavelength
- Separate (out-of-band) IP network—e.g., the data communications network (DCN) used for network management

A reference SONET/SDH transport network architecture is outlined for discussion in the remainder of the paper.

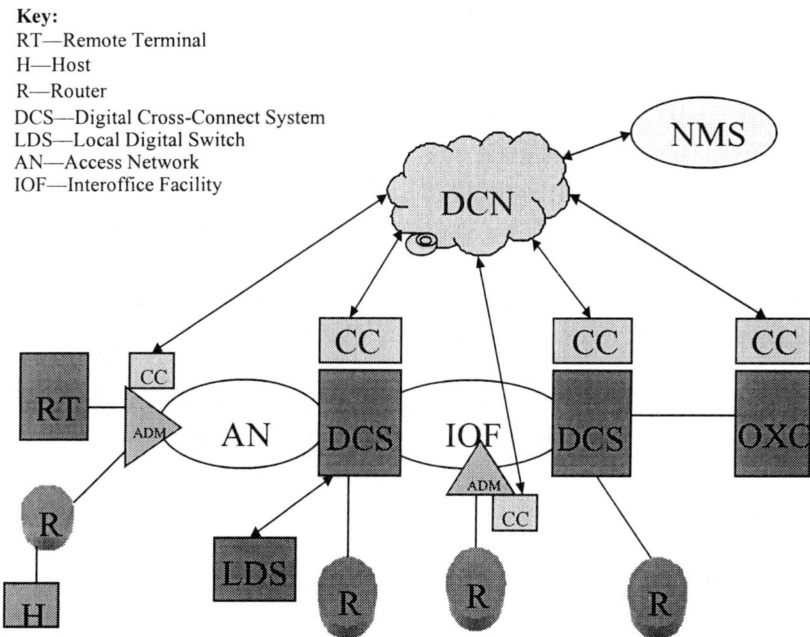

Figure 1: *Example Dynamic Transport Network Architecture*

Neighbor Discovery

A prerequisite for automatic operation of the network is the NEs' ability to recognize their neighbors and the links interconnecting them. Each NE must have a physical address, which is important for inventory control, and a logical address, which is essential for locating the element within the overall topology of the network. The link interface address (or ID) may be useful for link-specific communication and for link resource identification.

A link management protocol is needed to maintain an up-to-date working status of the links between the adjacent nodes. This should be a lightweight protocol preferably running on the link-associated control channel, such as the SONET DCC.

The NE's control component (CC) in *Figure 1* can use a dynamic routing protocol such as open shortest path first (OSPF) to acquire neighbors using its "hello" protocol [5]. The CC also runs a link management protocol and a signaling protocol.

Depending on the level of support that the SONET links provide in terms of ability to transport control and signaling information, the following alternatives can be specified:

Intact SONET Transport Links

These typically exist on SONET regenerators, optical DWDMs, and OXCs. Very basic link integrity monitoring capabilities may exist. In this case, the NMS should configure each NE with some criteria to indicate neighbor and link identity.

Section-Terminating SONET Transport Links

These typically exist on SONET add/drop multiplexers (ADMs), wideband cross-connects (XCs), and broadband XCs supporting bidirectional line-switched ring (BLSR) protection. In this case, a section overhead byte, such as J0, can be used to transmit node identification. The section DCC bytes are used for sending operations, administration, maintenance, and provisioning (OAM&P) information for managing the SONET system. It may be feasible to tunnel the node identification as well as to link maintenance information via the section DCC protocol currently defined.

Line-Terminating SONET Transport Links

These are used on SONET ADMs, wideband XCs supporting unidirectional path-switched ring (UPSR) protection or automatic protection switching (APS). Here, J0 can be used to transmit the node identification, or the line DCC can be used to support a new IP–based control protocol for neighbor discovery and link maintenance. Currently, some systems may not support access to line DCC bytes, and this may present a problem.

Path-Terminating SONET Transport Links

In path-terminating equipment, such as a wideband XC or a SONET access mux, the SPE is processed to extract virtual tributaries (VTs). Section or line overhead bytes can be used for neighbor discovery and link maintenance.

Topology Distribution

A dynamic routing protocol such as OSPF [5] or intermediate system–to–intermediate system (IS–IS) can be used to flood and exchange the link states, resource information, and adjacency information to build a database of network topology. The routing protocol needs sufficient bandwidth available in the control network. If the SONET line DCC is available, an OSPF/IP/point-to-point protocol (PPP) stack can easily be supported. Otherwise, an out-of-band control network such as the DCN, which is typically used for network management, can be used. It is important to note the distinction that a neighbor on the DCN is not necessarily a neighbor on the SONET system. A separate instance of the routing protocol here processes strictly SONET link states and not DCN link states.

After acquiring neighbors using lower-level protocols, each NE will attempt to form adjacencies with some of its newly acquired neighbors. Link-state databases are synchronized between pairs of adjacent nodes. Routing updates are sent and received only on adjacencies. A node periodically advertises its link state when the link state changes. From the link-state database each node calculates a shortest-path tree with itself as the root.

Path or Route Computation/Selection

Although a shortest path first (SPF) route database can be built using OSPF, the shortest paths are not always used as SONET paths. Additional constraints, such as node/link capability to transport a desired signal, physical constraints [6], maximum number of regenerated hops, total propagation delay, link cost, bandwidth, resource class, priority, preemption, resilience (protection and restoration), and network load balance must be used. All the nodes must be able to run basic path computation algorithms.

Explicit routes can be dictated by the NMS when it initiates a circuit-provisioning request to an NE. A client device requesting a circuit connection may also specify a constrained route to satisfy its service-level requirements. When it receives path setup request from a client or the NMS, the NE computes the complete path from itself to destination node because it does not know the full complement of constraints of a specific path.

Path Setup and Teardown

Information that is needed between two adjacent NEs for the purposes of connection establishment is appropriate for distribution via an LDP. In SONET/SDH networks, this "virtual" label corresponds to the "time slot" or STS number agreed to be on a link between two adjacent nodes for a circuit. This "virtual" label is not appended to the payload signal but rather used to identify the signal local to the link. Similarly, "virtual" labels can be used to identify wavelength/optical channel in a fiber link.

Signaling protocols are used to set up, maintain, modify/renegotiate, and tear down these connections. There are several protocols available in the industry such as SS7, used in the public switched telephone network (PSTN); PNNI, used in the ATM networks; RSVP–TE [7] and CR–LDP [8], used in the IP–multiprotocol label switching (MPLS) networks. PNNI allows traffic parameters, such as bandwidth, to be specified for a connection. Both RSVP–TE and CR–LDP support constraint-based, routed, label-switched paths.

Dynamic Modification of Path Resources
Transport networks must be optimized to support data traffic patterns that currently account for more than 50 percent of the total traffic. Dynamic bandwidth management using both NMS–initiated and client-initiated signaling will enable intervening NEs to increase or decrease bandwidth of a connection as needed. If a SONET circuit connection uses a 2*STS–1 capacity, a signaling request to modify it to 3*STS–1 capacity can be sent from the source to the destination. A later section of the paper will discuss how TDM channels can be concatenated for a connection.

Automatic Path Protection Switching
The dynamic control architecture should support end-to-end SONET path protection and shared mesh protection as well as existing linear APS, UPSR, and BLSR recovery actions. The distributed routing protocol used for topology distribution can disseminate information on different protection methods enabled on different links. Use of that information will enable the setup of an "extra traffic" connection over unprotected channels and idle protection channels.

Enhancements for Carrying Data Traffic

The second discussion topic is the use of virtual concatenation of TDM channels such as VT1.5 and STS–1 to build different bandwidth pipes dynamically to carry data traffic more efficiently. The functions needed at the transport network access points where the data traffic is mapped/extracted from the TDM channels are discussed.

Concatenation of Payload Channels
The standard SONET concatenated STS signals are restricted to multiples of three (at STS–3) and four (at STS–12 and higher). This leads to bandwidth fragmentation and regrooming requirements in the network as well as bandwidth wastage when carrying finer-grained traffic such as local-area networks (LANs) and the Internet.

Contiguous Concatenation
In SDH, the concatenation of contiguous signals in the multiplexer is called the "contiguous concatenation" [2]. At present, contiguous concatenation caters for 4 or 16 virtual circuit (VC)–4s. To transport services, such as transparent LAN at 10 Mbps, 100 Mbps, or 1 Gbps, there is no match in the bandwidth granularity. There is a single path overhead (POH) column in a contiguous concatenation and

one given payload type. In a signaling message to establish a connection, a contiguously concatenated signal can be specified by identifying the first signal and by giving a subsequent number of signals to concatenate. All intermediate nodes must support contiguous concatenation. Many installed NEs in SONET/SDH networks cannot support contiguous concatenation. Implementing this capability would require a hardware upgrade and would be prohibitively expensive.

Arbitrary Concatenation

Some SONET/SDH framer device manufacturers now support "arbitrary" concatenation with no restriction on the size of an STS–Mc (M<=N). Arbitrary concatenation includes arbitrary byte interleaving, which avoids the need for link regrooming between pairs of nodes supporting this feature [1]. This is a network service potentially available in new deployments but not a part of the existing installed infrastructure.

Virtual Concatenation

Two end systems (path terminating equipment) supporting virtual concatenation would essentially "inverse multiplex" two STS–1s into a virtual STS–2c for efficient transport of 100 Mbps Ethernet traffic [1]. It is transparent to the intermediate SONET/SDH NEs. At the sending end it will be necessary to provide each SPE/VC with information about its concatenated group identification and its position/sequence within the group, and to give each its own POH for processing in the intermediate nodes in the network [3]. At the receiving end the equipment must be capable of identifying an SPE/VC as belonging to a particular concatenated group. Because of the likelihood of differential delays, it will be necessary for the receiving-end equipment to provide buffers to store the incoming data until the latest SPE/VC arrives, when re-alignment can be performed.

The major objective of virtual concatenation is to provide multiple correctly sized channels over a SONET/SDH network for data traffic. Another benefit is the ability to add or remove an SPE/VC from the group without taking the connection down, allowing dynamic modification of bandwidth resources.

Connection Bundling

This is the process of connecting a set of nonconcatenated STS–1s together as a group [1]. All the STS–1s in the group are within the same SONET link or WDM signal and experience the same delay. This simplifies connection establishment of a batch of DS–3s that can be wholesaled. Such bundling may be combined with virtual concatenation by the end systems to aid in handling protection. The "virtual" labels used for signaling could incorporate connection bundling.

Stacking and Tunneling

In IP/MPLS networks, one label-switched path (LSP) can be tunneled through another LSP using label stacking. This concept can be extended to SONET/SDH transport networks, in which low-order channels, such as VT1.5s, can be tunneled through high-order channels, such as STS–1s. In a scenario where two broadband XCs are used as border nodes, a connection of VT1.5s must be established across these border nodes between two wideband XCs. The client device requests the wideband XC for a VT1.5 connection, and the wideband XC will map this connection to an STS–1 tunnel between itself and another wideband XC via the broadband XCs.

Summary

Some limitations of the existing SONET/SDH transport networks can be overcome with Internet-style dynamic routing and signaling protocols. By using dynamic path setup procedures and virtual concatenation of SONET/SDH payload signals, the efficiency of the transport networks can be improved. In addition, using available SONET overhead bytes and the DCN enables the cost-effective upgrade of existing networks.

References

1. Bernstein, G. Mar. 2000. Some comments on the use of MPLS traffic engineering for SONET/SDH path establishment. Internet Draft, draft-bernstein-mpls-sonet-00.txt, Work In Progress.
2. Mannie, E. Mar. 2000. MPLS for SDH control. Internet Draft, draft-mannie-mpls-sdh-control-00.txt, Work In Progress.
3. Jones, N., and C. Murton. Dec. 1999. Extending PPP over SONET/SDH, with virtual concatenation, higher-order and low-order payloads. Internet Draft, draft-ietf-pppext-posvcholo-01.txt.
4. Walker, W., B. Mack-Crane, and S. Steele. Aug. 2000. Control plane integration of multi layer networks. NFOEC 2000.
5. Moy, J. April 1998. OSPF Version 2. RFC 2328, Internet Engineering Task Force.
6. Levandovsky, D., S. Makam, and V. Sharma. Aug. 2000. Physical constraints affecting connectivity in the dynamic optical network. NFOEC 2000.
7. Awduche, D.O., et al. Feb. 2000. Extensions to RSVP for LSP tunnels. Internet Draft, draft-ietf-mpls-rsvp-lsp-tunnel-05.txt.
8. Jamoussi, B., et al. Sept. 1999. Constraint-based LSP setup using LDP. Internet Draft, draft-ietf-mpls-cr-ldp-03.txt.

Operations Support for Optical Networks Present and Future: A Network Management System Provider's Perspective

Dean Rader
Senior Systems Engineer
Telcordia Technologies

Gary Wood
TIRKS System Architect

Introduction

The public network, historically driven by voice traffic, is changing into a data-driven network. Experts predict that data traffic will dwarf voice traffic within the first decade of the next millennium. The growth of the Internet and the emergence of e-commerce are examples of this evolving trend. Bandwidth demand for data continues to increase exponentially, while voice traffic increases linearly. Dense wavelength division multiplexing (DWDM) technology can provide a means for meeting these demands.

DWDM can combat fiber exhaust in areas where spare fiber facilities are scarce. In most cases, the cost of deploying a DWDM system will be significantly less than laying new cable and can be realized in a fraction of the time. In addition, DWDM can add flexibility to the network. DWDM systems can be transparent to signal rate and format. This flexibility should make the network simpler in the long run by allowing a single transport platform at the bottom of the physical layer, which supports a diversity of services. DWDM can also add scalability to the network. Wavelengths can be added gradually as needed without interruption to services, which allows a carrier to react quickly to unexpected bandwidth demands within various regions of its network.

Initially, point-to-point DWDM applications were deployed to provide fiber exhaust relief in certain areas of the network. As the technology evolved, applications that utilize intermediate nodes and network elements (NEs) with add/drop capability became available. Optical networking capabilities also emerged to achieve dedicated protection of services at the single wavelength level (optical channel [OCh] layer) and at the aggregate signal level (optical multiplex section [OMS] layer). The technology continues to evolve, and new technological advances such as dynamic wavelength routing and optical cross-connects (OXCs) are now emerging. DWDM deployment and development are still in the early stages.

Current OSS DWDM Support

Through the end of 1998 and the beginning of 1999, Telcordia Technologies worked with a group of DWDM suppliers under the supplier-funded generic feature (SFGF) process to create a solution that supports DWDM systems within the operations support systems (OSSs) of the regional Bell operating companies (RBOCs). The solution was completed in December 1999 to support DWDM technology with the following goals:

- Support linear chain and ring topologies that utilize the optical add/drop multiplexer (OADM)
- Support optical networking technologies with dedicated protection schemes and the ability to enable and disable protection on a wavelength-by-wavelength basis
- Create enhanced processing and data structures to support multiwave and single-wave DWDM equipment

FEPS/TIRKS® System Processing

The DWDM solution was designed to fit efficiently within the existing processing operations to support the massive and rapid deployment of DWDM systems. Most of the DWDM–specific processing can be performed a minimal number of times by a small portion of the workforce. Once these processes are in place they can be invoked repeatedly through automation. This small portion of the workforce can receive necessary training on the DWDM systems and the DWDM–specific processing steps. Steps that require repetition can be performed as always with a minimal amount of new process techniques required to support DWDM. These steps are performed by a large portion of the workforce and should require minimal DWDM–specific training. *Figure 1* highlights some of the significant processing steps involved in the

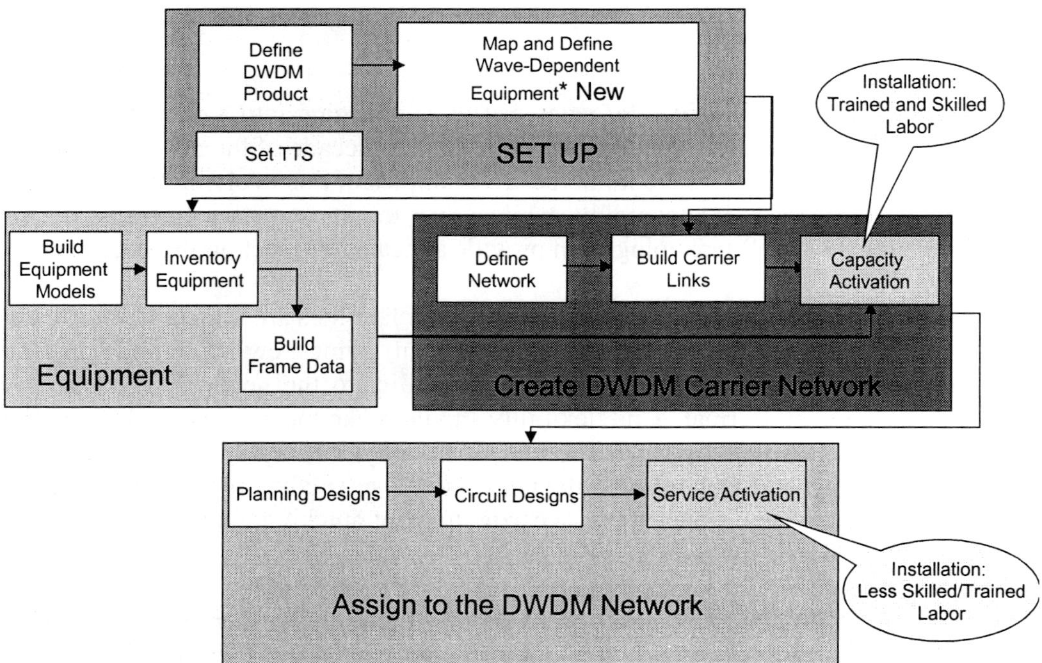

Figure 1: General Processing Flows to Inventory and Provision DWDM Systems

planning, provisioning, and inventory of DWDM systems within the facility and equipment planning system (FEPS)/TIRKS® system environment. The processing parts have been broken into four separate steps:

1. Setup—DWDM products and wave-dependent equipment are defined, and TIRKS® table system (TTS) settings are made to support DWDM deployment. Several new tables and new values to existing tables were created to support DWDM systems.
2. Equipment—equipment models are built, and equipment is inventoried. When an equipment model is created for a DWDM product, it may be reused repeatedly, as that equipment product is inventoried and placed in central offices (COs) through an RBOC region.
3. Creation of DWDM carrier networks—DWDM networks are defined, and optical carrier (OC) links that support the OMS or aggregate DWDM signal are created in inventory. Capacity is then activated.
4. Assignment to the DWDM network—planning designs are created to route services over DWDM networks. Circuit design models can be created in advance to automate and simplify the provisioning effort for a given circuit. Services are assigned to wavelengths within the DWDM network, and the circuit designs are completed. Services are then activated.

Capacity and Service Activation

Installation of equipment needed for capacity activation can be performed by a specially trained task force or by the equipment suppliers themselves. In general, service-activation installations are less predictable than capacity activations and must be performed more frequently. A wider portion of the workforce will perform the installation and software provisioning involved with service activation. These individuals often have little training with regard to DWDM systems and the processes that support them. In addition, service activation is typically a revenue-generating process that directly affects the end user. It is critical that services be turned up as quickly as possible. To support wide and rapid deployment of DWDM systems, the provisioning and installation of services should be kept as simple as possible. In keeping service activation simple, there will often be trade-offs with the flexibility of the system and the efficient use of bandwidth and equipment. Telecommunications providers should weigh these factors when creating methods and procedures to deploy and provision DWDM systems. DWDM equipment suppliers should weigh these factors when engineering and designing their products. DWDM systems designed to fit smoothly within the constraints associated with capacity and service activation will be handled more efficiently within the OSS of a telecommunications provider.

Wave-Dependent Equipment and Products

One of the key elements of the setup process is defining the frequencies of a DWDM system that can be supported and assigned to client signal (CS) services. Only International Telecommunications Union (ITU)–compliant frequencies may be entered into the TIRKS® system database for DWDM processing. While the ITU sets a standard for what wavelengths may and may not be used, the particular wavelengths supported may vary from system to system. For example a 16-wave DWDM system from one supplier may use an entirely different set of frequencies from a 16-wave system from another supplier. Consequently, the spectrum of frequencies must be defined specifically for each DWDM product within the Telcordia operations systems (OSs).

Digital carrier systems deployed for telecommunications rely on multiplexing that involves protocols and logical or time division multiplex (TDM)–based electronic connections. Unlike these other technologies, DWDM is determined by physics. The frequency of a transponder at one end of a circuit must match the frequency of an optical filter at the other end for the transmission to be complete. New data structures and databases were created to define these DWDM wave-dependent components. The heritage OSs use the concept of function codes to describe a group of equipment components associated to a certain function. These function codes can drive processes within equipment inventory, circuit design, and the allocation of bandwidth. To support the wave-dependent components, a new concept of "generic function codes" was introduced. Generic function codes are used to define equipment and optical ports that represent all the information about the equipment, with the exception of the frequencies it supports. The generic function code does not represent physical equipment; it is a placeholder for a set of completed function codes, which represent the physical wave-dependent equipment. *Figure 2* shows some examples of wave-dependent equipment within a network. Within this solution, single-wave components such as optical transponders and optical ports are defined separately from multiwave components, such as multiwave add/drop units (MADUs) and multiwave transponders. Multiwave transponders are plugs that integrate more than one laser on a single card and thus support more than one service.

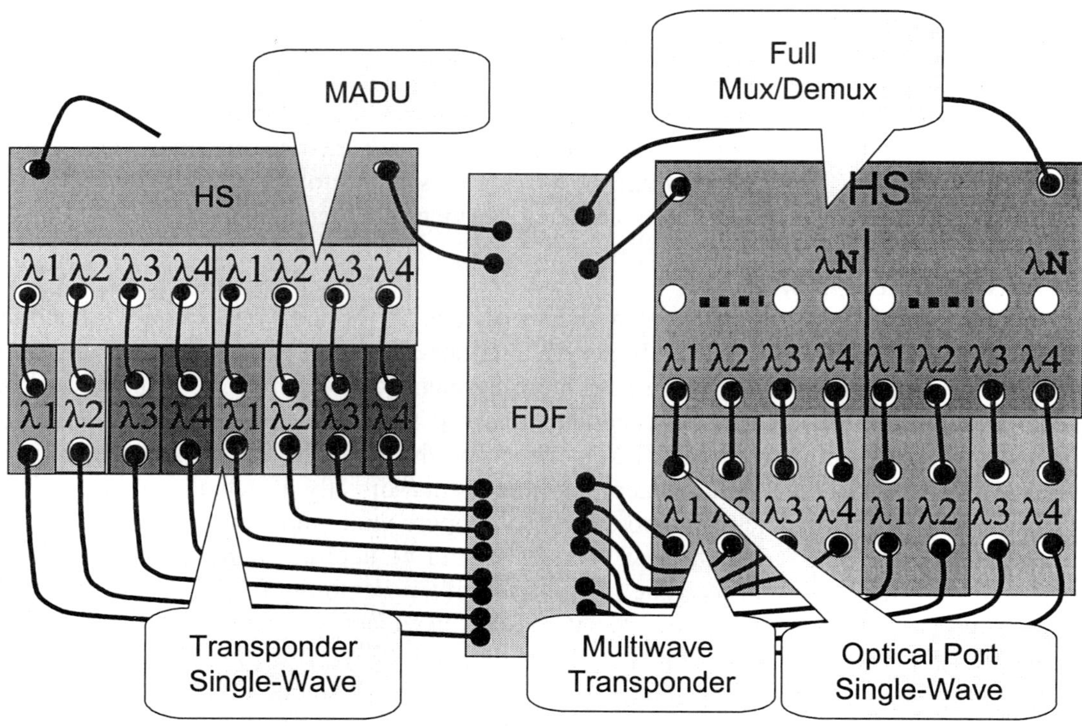

Figure 2: *Basic DWDM Configurations with Wave-Dependent Components*

New processing software was created to relate wave-dependent equipment to the frequency assignments that must be made to complete a circuit. The end user may work through processes using generic function codes. Before the circuit design is complete, the TIRKS® system will use its wavelength selection algorithm to assign a specific wavelength automatically. At this point the generic function code is replaced with a completed function code, and the specific wave-dependent components are ordered and assigned such that equipment and capacity are used efficiently. Once an optical wavelength has been selected, the TIRKS® system will check the plug inventory to ensure that the assigned equipment plugs are in stock. If any of the needed plugs are not in stock, the next best wavelength will be assigned. The process continues until a valid wavelength with available equipment is found (see *Figure* 3). To accomplish this, the TIRKS® system must interface with the plug-in inventory control system (PICS). PICS is an OS that supports the inventory and management of plug-ins within an RBOC. PICS and the TIRKS® system maintain a dynamic link during operating hours. The two systems interact to share plug-in demand and plug-in availability information.

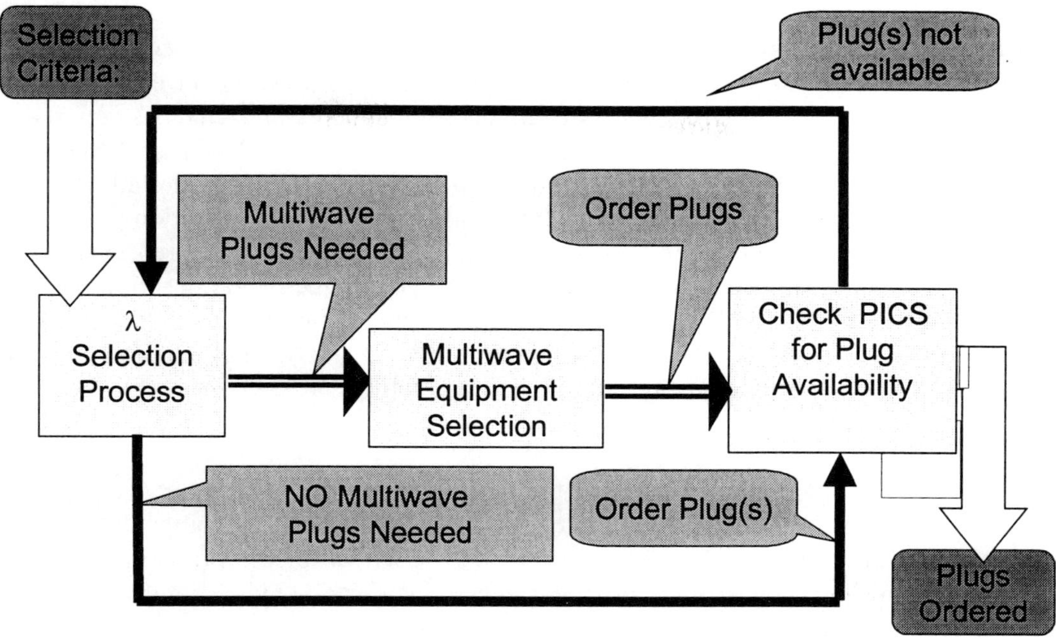

Figure 3: *Flow of the TIRKS® System Automated Wavelength Selection Algorithm*

OSS Architecture for the Current DWDM Solution

The TIRKS®/PICS interface is one example of how the heritage OSs work together to provide a complete flow-through solution for DWDM. *Figure 4* shows systems involved in this solution. Service orders can flow into the TIRKS® system through the service order analysis and control system (SOAC), which is the service order controller for the RBOC telephony network. SOAC receives data from a service order processor (SOP) and passes the order information to the TIRKS® system. The TIRKS® system provides order scheduling and tracking, provisioning and inventory of equipment, wavelength capacity management, and circuit design.

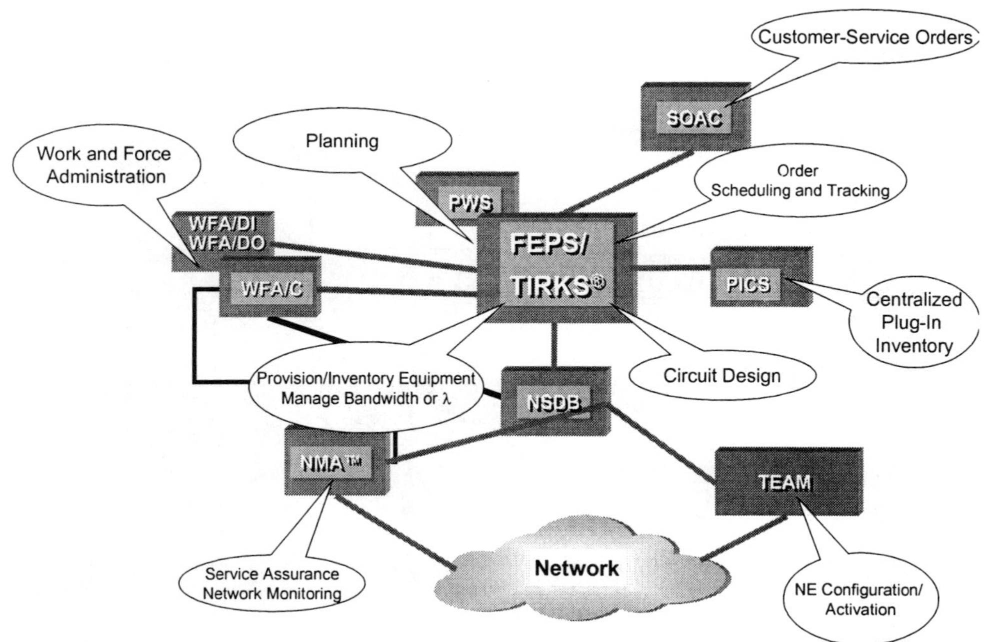

Figure 4: *Activation Architecture Heritage OSS*

FEPS has a direct interface with the TIRKS® system and provides mechanized software tools to aid in the planning of network facilities and transmission equipment. The software tools perform many functions, including routing strategies and the allocation and distribution of bandwidth over networks and facilities. The planning workstation (PWS) provides a set of decision-support and analysis reports that aid in network planning and management analysis. The network and services database (NSDB) stores data provided by the TIRKS® system for in-effect and pending circuits. NSDB provides an interface between the TIRKS® system and the downstream systems (the work and force administration [WFA], the NMA™ system, and transport element activation manager [TEAM]), so data can be shared in the flow-through installation and service assurance processes. WFA systems provide work descriptions, balance the workload between personnel, and automate the work assignments required to install and repair telecommunications equipment. TEAM interfaces on the software side with NSDB and on the equipment side with the DWDM NE. It sends messages to the NE to complete cross-connects (XCs) and to activate equipment.

The NMA™ system will continuously monitor the network for facility faults and equipment failures. When a fault occurs, this system can correlate the alarms sent by various equipment components and perform a root-cause analysis to discover and report on the source of the problem. In doing this, the NMA™ system will suppress any alarms that were triggered incidentally by the actual cause of the failure. In addition, the system can provide performance analysis of the network.

Future Goals for OSS

The initial DWDM–OSS solution supported key DWDM functionality that was available and well defined when the project was initially scoped. As time passes, the technology continues to evolve, with advances in DWDM components, NEs, and network architecture. These new developments will create new challenges for OSS. Some examples of new DWDM technologies being considered for future OSSs are discussed below.

Components

Transponders with enhanced functionality are becoming available. Transponders with tunable lasers allow a service to be assigned to one of many possible frequencies. Transponders that multiplex services during the optical-to-electrical-to-optical (O–E–O) process allow more than one service to be assigned to a single wavelength (see *Figure 5*).

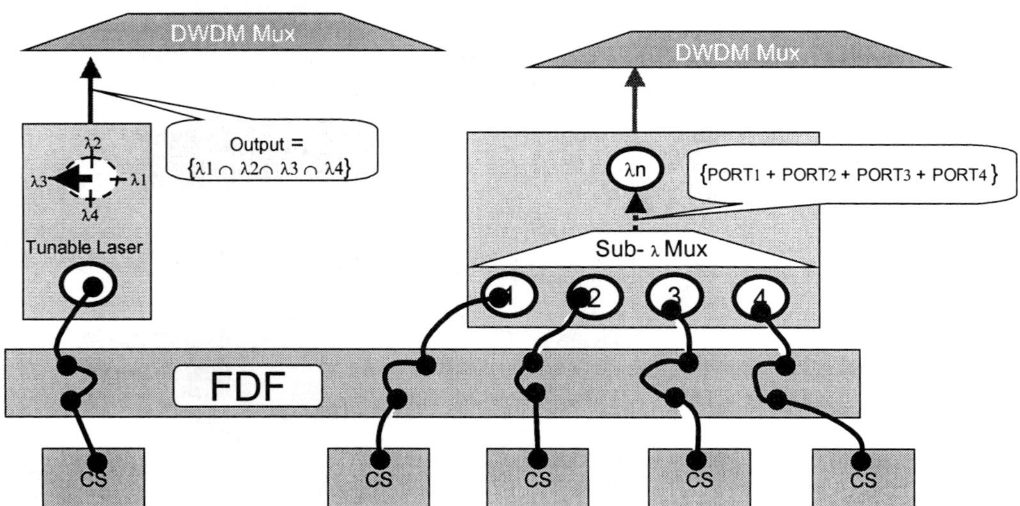

Figure 5: Transponders with Tunable Laser (Left) and Transponder with Subwave Multiplexing (Right)

NEs

DWDM NEs are evolving beyond the traditional OADM. *Figure 6* (left) shows a typical way that a DWDM NE is connected to the CS through manual fiber distribution frame (FDF) XCs. One evolving trend in NE design will integrate this functionality into a single box as shown on the right in *Figure 6*. The number of manual FDF XCs will be reduced in favor of automated XCs internal to the NE. These XCs can occur between the CS and transponder in the electrical domain, or between the frequency output of the transponder and the frequency-dependent input of the DWDM demux.

In contrast, other equipment will allow the partitioning of DWDM functionality within more than one system. An example of this would be equipment that provides centralized wavelength distribution (CWD). Such equipment would contain multitudes of transponders that interface between a CS and a passive DWDM system. The transponders can interface with several different DWDM systems as needed [1]. This equipment will need to be managed and controlled separately from the CS and DWDM systems with which it interfaces; it will be treated as its own NE.

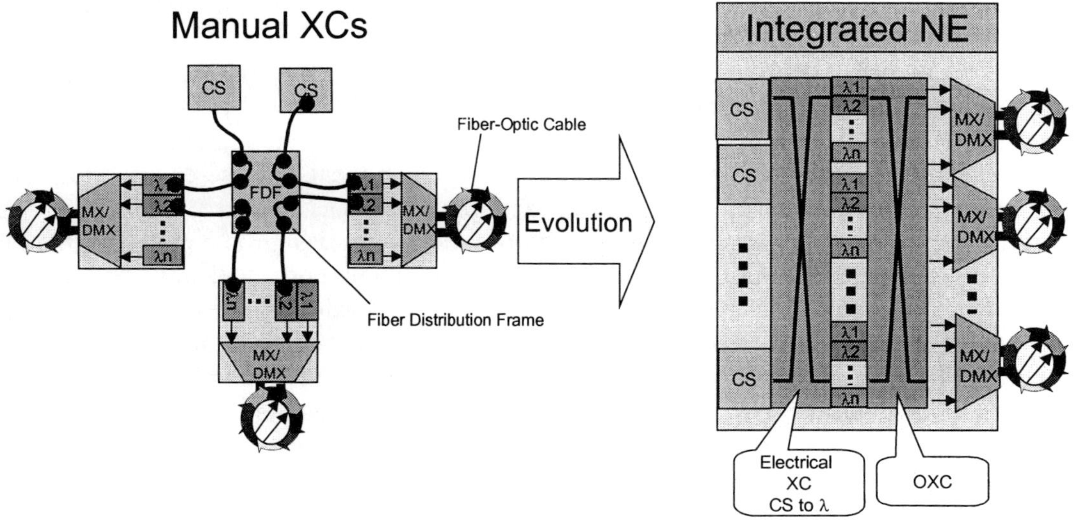

Figure 6: Discrete Systems Connected via FDF XC (left) May Evolve into a Single Integrated NE with Automated Internal Switching (right)

Optical Network Architecture

Additional applications such as unidirectional broadcast transmission with drop and continue are available for DWDM systems and may be considered for deployment in the RBOC environment. New optical protection schemes that involve switching at both ends of the circuit (i.e., at the transmission end and the receive end) are also being developed.

As demand for these new components, NEs, and DWDM applications increases, the applications will ultimately need to be supported within the OSS. In some cases, the new technology can be supported through the OS modifications for the integration of NEs (OSMINE) process with little to no software enhancements. In other cases, software enhancements that build on the initial DWDM solution will be required to support these products adequately in the OSS. The items discussed above are not to be considered as a finite list for new features being considered as potential software enhancements to the DWDM OSS.

Intelligent NEs

One of the goals for future OSS support of optical networking is to leverage off the intelligence of today's NEs. Achieving this within the current DWDM–OSS solution will require modifications to the OSS software architecture. As next-generation network (NGN)–OSS solutions begin to support DWDM technology, the intelligence within the NE will be part of the initial solution. Standard interfaces such as common object request broker architecture (CORBA) or transition language 1 (TL1) should be employed for communication between the OSS and intelligent NEs (INEs).

The auto-discovery capabilities embedded within the NE can lead to more accurate record keeping within the network management system (NMS). The NE can inform the NMS of what equipment is actually in effect and what equipment is spare. Depending on the circumstance, the NMS can handle data discrepancies in one of the following ways:

1. Automatically correct its database to match what is actually deployed
2. Invoke processes to correct automatically what is deployed in the field so that it matches the inventory
3. Identify the discrepancies to be resolved through human intervention

To achieve this, a direct or indirect interface must be established between the NE and the NMS.

Intelligent networks can help to enable efficient implementation of OSS in support of new equipment features, new services, and new network technologies. Many DWDM suppliers have engineering tools that can be used off-line to optimize product performance. These tools can address complex engineering rules that are specific to their own systems, products, and components. These engineering tools could interface with the NMS to assist in circuit design and wavelength assignment. The NMS would pass network information to the engineering tools. Just before the circuit design is completed, the NMS will invoke the engineering tools to return data that determines equipment, ports, and wavelengths that need to be assigned (see *Figure 7*). With the proper interface, a large portion of the software enhancements needed to support changes and new releases of DWDM products could be handled through the engineering tools with minimal OSS impacts. This would require a significant effort to create an effective interface that could achieve this type of cooperative assignment. To handle the assignments effectively, the engineering tools and NMS would need to have databases that contain matching records on data involving network parameters and assignment status of equipment and capacity.

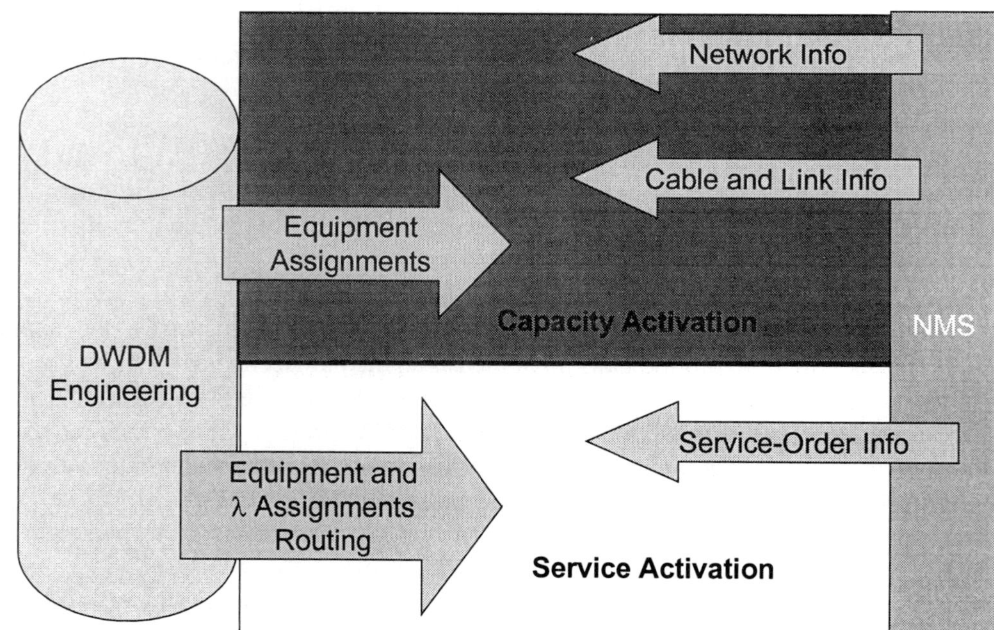

Figure 7: Cooperative Assignment between an Engineering Tool and NMS for Circuit Design

Dynamic Wavelength Routing

A significant challenge for the future will be the OSS support of DWDM systems with dynamic wavelength routing. Through protocol recognition and advanced OXC technology, services can be automatically routed through a mesh of DWDM NEs controlled by an element management system (EMS). The capacity and optical links of these dynamic networks will need to be provisioned through an NMS in a manner similar to that of static DWDM networks. The routing of services through these optical links will be handled by the NEs and their supporting EMSs. It is likely that services will sometimes need to be routed through multiple EMS environments, which do not share routing information. Dynamic wavelength routing of services could be handled for the future in a manner similar to how packet services are managed today in an NGN NMS from Telcordia. In this scenario, the NMS would manage entry and exit points between different EMS environments (DWDM subnetwork "clouds") and the links between them. The EMSs or NEs establish connections and perform routing between entry and exit points of each cloud (see *Figure 8*). Quality of service (QoS) data is sent into each subnetwork cloud so that the appropriate routing rules are applied to the service. Performance data is sent from the cloud to the OSS.

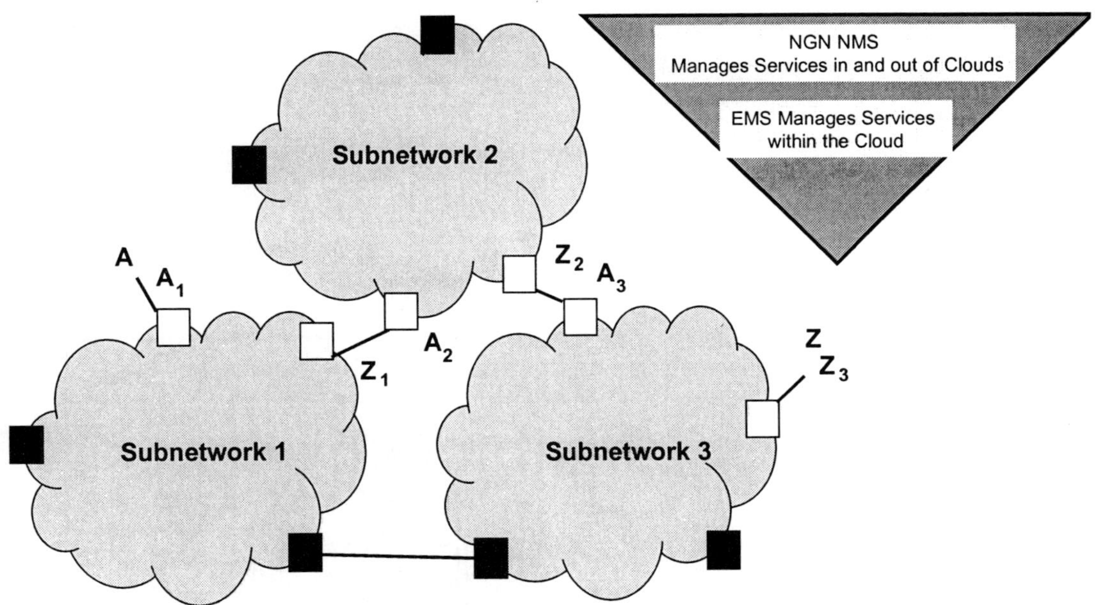

Figure 8: Interdomain Management

NGN Flow-Through

Figure 9 shows an example of how flow-through between various OSSs and telecommunications management network (TMN) layers can be achieved for the provisioning and activation process. The four TMN layers from top to bottom are as follows: The service management layer (SML) is the top layer responsible for billing and negotiation; the network management layer (NML) is responsible for provisioning, activation, and network monitoring; the element management layer (EML) serves as an interface between the NML and the NE layer (NEL) and shares in provisioning and activation; and the NEL represents the physical NEs that are deployed in the field [2].

The service activation process starts with a customer-service request that leads to an order entry, which flows from the SML to the NMS in the NML. The NML will perform processing to send installation orders to the work and force management, send service and circuit data to the service-assurance system, and send activation orders down to the EML. The EML translates the activation orders into equipment-specific messages to send down to the NEL so that the appropriate commands can be invoked. An example of processing flows associated with fault and service-level agreement (SLA) assurance is also shown in *Figure 10*. If a customer reports a problem, the trouble report will flow into the SML, where customer trouble tickets are issued and sent to the service-assurance system. Trouble-related dispatch is sent to work and force for troubleshooting and repair.

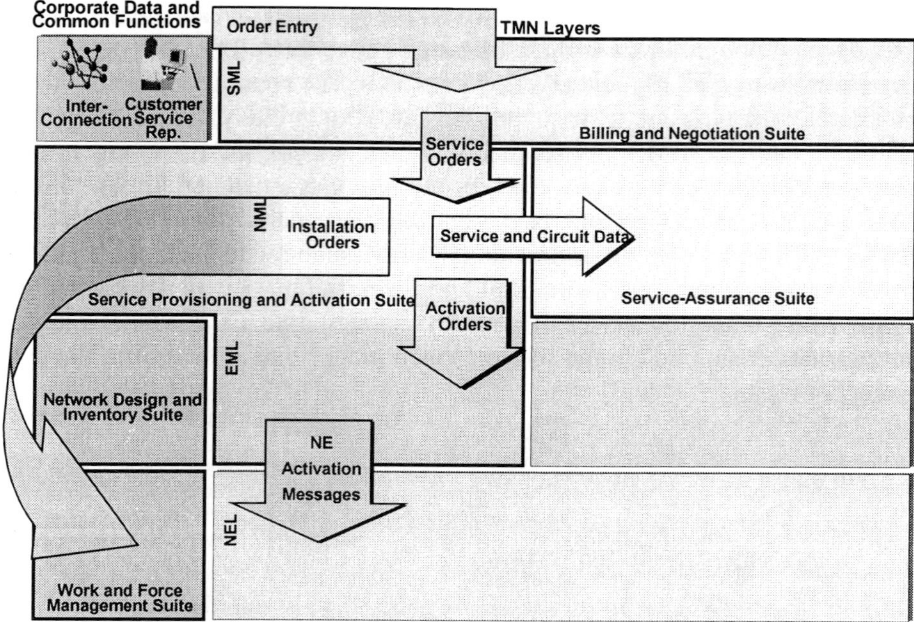

Figure 9: *Flow-Through for Provisioning and Activation*

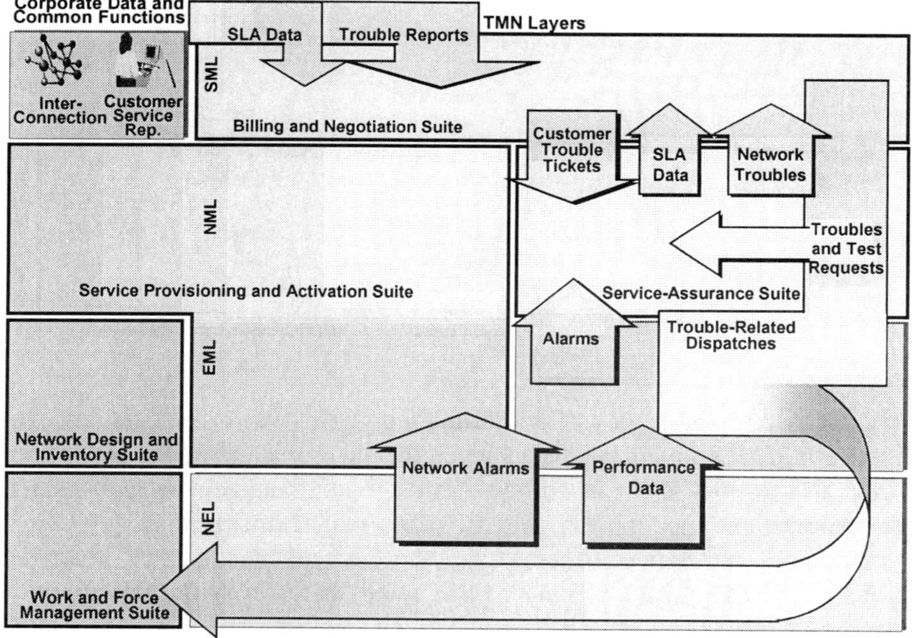

Figure 10: *Flow-Through for Fault and SLA Assurance*

It is expected that, with the constant monitoring of the network, the OSS will often discover a network failure before a customer reports it. In this scenario, the NE sends alarms to the EML, which in turn sends alarm information to the service-assurance system in the NML. Trouble information is sent to the SML to notify customer service that a problem has occurred. Trouble-related dispatch is sent to work and force to correct the problem.

In addition to field failure response, the OSS should monitor the performance of the system. The NEL will send performance data to the service-assurance system via the EML. Service assurance will take the performance information and send SLA data to the SML. Parameters such as load balancing and the degree to which SLAs are met can be reported and analyzed. These reports can be automatically generated and distributed to the appropriate management personnel.

With dynamic routing optical networks, there are additional performance parameters that are not applicable to the electronic, packet-based routing systems of today, which may be important to track. Transparent paths that route through intermediate nodes with passive optics are more cost-effective than opaque paths that have O–E–O conversions between each link. Express routing of multiple wavelengths collectively through wideband filters is more cost-efficient than routing through narrowband filters that separate each wavelength into individual components (see *Figure 11*). Metrics that determine the ability of an optical system to route a service through transparent and express paths will give the service provider a measure of the hardware costs per circuit. Such metrics can be an important factor in analyzing the quality of DWDM systems.

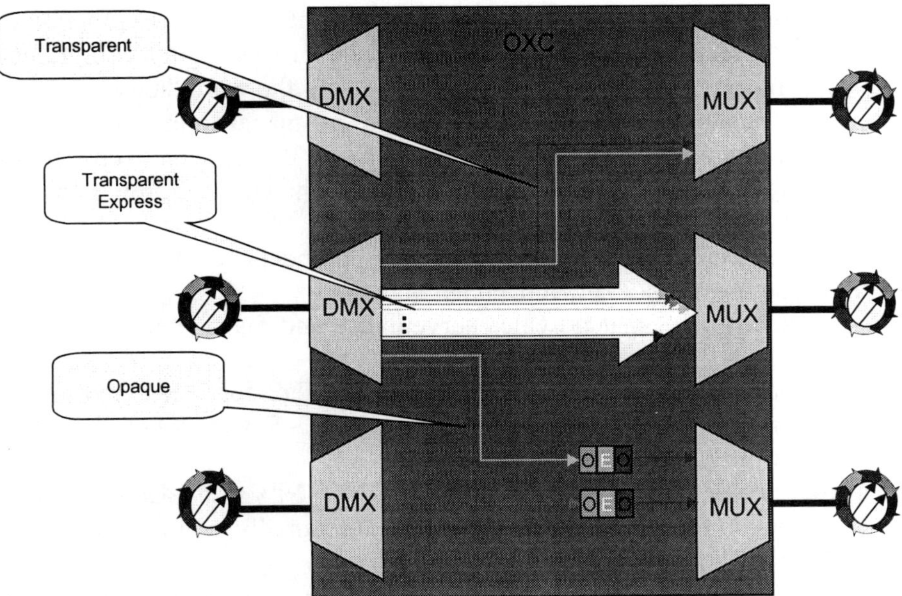

Figure 11: Routing through an Intermediate Node

SLA data can also be used to trigger the equipment-planning modules of an OS to "scale up" the drop capacity of an OXC to meet throughput expectations. Additional processing flows could be introduced to drive the ordering and installation of wave-dependent equipment after the initial capacity activation but before service activation takes place. If planned effectively, these processes can reduce service turn-up time and still take advantage of plug deferral to control equipment costs.

Conclusion

With the evolution of the NGN, service providers are carefully considering their options for the OSS of the future. The most suitable OSS architecture will vary according to the needs and strategies of a given service provider. An RBOC (incumbent local-exchange carrier [ILEC]) with a large embedded base of inventory and an established heritage OSSs will require a different OSS architecture than that of a start-up competitive local-exchange carrier (CLEC) with a small inventory and no OSS deployed. As companies merge and new services, technologies, and federal regulations evolve, new challenges appear to define

OSS architectures that support the needs of the different service providers that exist in the telecommunications market.

Selected Definitions

Capacity Activation: With DWDM this refers to the completed deployment of the optical links within a DWDM network.

CWDF: An NE between the DWDM NE and CS NE that provides ITU–compliant signal to the DWDM system

Completed Function Code: An equipment function code within the Telcordia OSS used to describe equipment associated to a specific wavelength or a group of specific wavelengths

CS: A service that is assigned to a wavelength that rides a DWDM network or system

FDF: A patch panel that provides the connection of optical fibers with optical fiber patch cords

FEPS: A Telcordia OS that provides automated tools for planning transport facilities and transmission equipment

Generic Function Code: An equipment function code within the Telcordia OSS that is not specific to a particular wavelength but instead represents a set of wave-specific completed function codes

Heritage, a.k.a. Legacy: Refers to everything of a data processing/telecommunications nature (equipment, software, files, etc.) that the provider has on hand; the word "legacy" suggests that these items have been inherited from previous generations of management; the idea is that it is unreasonably expensive and potentially problematic to replace; thus it makes sense to preserve it

MADU: An optical component used in DWDM systems, which will add/drop a specified group of wavelengths and pass the remaining wavelengths through

NGN: Terminology that represents the convergence of multiple independent networks, including voice, video, and data, into a single, unified, broadband network

The NMA System: A Telcordia OS that provides surveillance and management; NMA is a registered trademark of Telcordia

NSDB: A Telcordia OS database that stores and distributes circuit data for a telecommunications provider

Opaque Optical Transmission: Optical systems that require an O–E–O conversion between each OMS link of an optical system

OSMINE: A process that enables supplier-specific products and NEs to be accommodated in a typical RBOC network; this includes the implementation of functionally compatible interfaces between Telcordia OSs and NEs or EMS

OSS: Methods and procedures (mechanized or not) that directly support the daily operations of the telecommunications infrastructure

PICS: A Telcordia OS that supports the inventory and management of plug-ins within a client company

PWS: A Telcordia OS, which serves as a component of FEPS, that generates reports for planning and management analysis

Service Activation: With DWDM this refers to the completed deployment of a service where a client signal is assigned to a wavelength of a DWDM system.

SFGF: A Telcordia project where NE suppliers can obtain support for generic features that require software changes to one or more Telcordia OSs

SOAC: A Telcordia OS that is the service order controller for the telephony network

SOP: Front end of an OSS typically maintained and created by the service provider

The TIRKS System: A Telcordia OS for inventory and assignment of transport telecommunication equipment, facilities and automated design of complex telecommunications circuits; TIRKS is a registered trademark of Telcordia

TMN: A concept that defines the relationship between basic network building blocks in terms of standard interfaces

Transparent Optical Path: An optical path that does not undergo O–E–O conversion at intermediate points along the route

Transponder, or Wavelength Converting Transponder (WCT): The device that performs an O–E–O conversion of an input optical signal of any frequency to a specified ITU–compliant frequency

TEAM: A Telcordia OS that supports flow-through activation and management of transport NEs in a multivendor environment

TTS: Contains numerous tables used by the TIRKS system for passback information and validation of data

WFA: A suite of Telcordia OSs that helps to manage the tasks and to coordinate personnel assignments necessary to install and maintain facilities, equipment, and services; WFA/C (control), WFA/DI (dispatch in, interoffice), WFA/DO (dispatch out, outside plant) are some of the OSs within this suite

References

1. Anderson, G. Emory. 1998. Centralized wavelength distribution for dense WDM systems. *NFOEC Conference.*
2. GR–2869-CORE, Generic requirements for operations based on the telecommunications management network (TMN) architecture.

An Optical Payload and Network Management Platform for Metro Optical Networks

Li-Ran Wu
Vice President, Management Software Engineering
Sorrento Networks

Don Jiang
Software Manager

Kevin Su
Software Manager

Introduction

Omnipresent Internet access and various e-commerce solutions have fundamentally changed network service characteristics. To interconnect these heterogeneous, high-speed data streams, installation of high-throughput optical transport networks (OTNs) in metropolitan areas is occurring rapidly. This paper will provide an overview of these metropolitan optical networks (MONs), detail the key attributes of MONs, and explain the network management need, mainly at the element management and network element (NE) layers.

DWDM

Dense wavelength division multiplexing (DWDM) is intended to serve as a bit rate–insensitive, protocol-independent, and technology-transparent carrier for technologies such as synchronous optical network (SONET)/synchronous digital hierarchy (SDH), asynchronous transfer mode (ATM), Gigabit Ethernet (GbE), etc. DWDM has been widely deployed in the long-haul network. The most familiar application for DWDM is to aid fiber relief of overlaid SONET rings, as illustrated in *Figure 1*. Initially, MONs were envisioned as a scaled-down version of a long-haul network. New services from technologies such as cable modem (CM) and digital subscriber line (DSL) make increasingly clear the distinctions between long-haul and MONs. Whereas long-haul is really a piping service, the MON offers tremendous potential for creating connectivity, added quality of service (QoS), and protocol transparency. The realization of radical differences between MONs and long-haul optical networks requires different network management strategies to ensure the successful implementation of MONs.

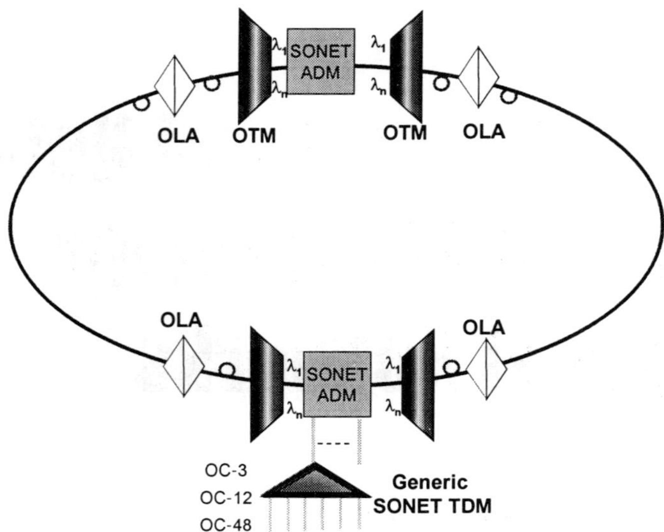

Figure 1: DWDM Piping Application for Long-Haul Overlaid SONET

With the development of sophisticated NEs such as dynamic optical add/drop multiplexers (OADMs) and optical wavelength switching routers (OWSRs), wavelength (payload) management is becoming an important task for the management platform. Network management also ensures that proper fault and state information can be propagated to the affected client layers so that appropriate action may be taken by the client layer network. To support propagation of operations, administration, and maintenance (OAM) information in the DWDM, the management model for the DWDM network has been defined based on the telecommunications management network (TMN) principles. In the TMN, the OAM and provisioning (OAM&P) functions are categorized into fault, configuration, accounting, performance, and security (FCAPS) management [1]. These functions are distributed in each management layer—NE, element management, network management, business management, and service management. The TMN–layer management model is shown in *Figure 2*. In this generic platform, the management aspects of the network are organized in an information model. Although the Q3 protocol stack is no longer the implementation choice, the TMN–layer management and information model remains the essential platform for the management of carrier-class networks [4].

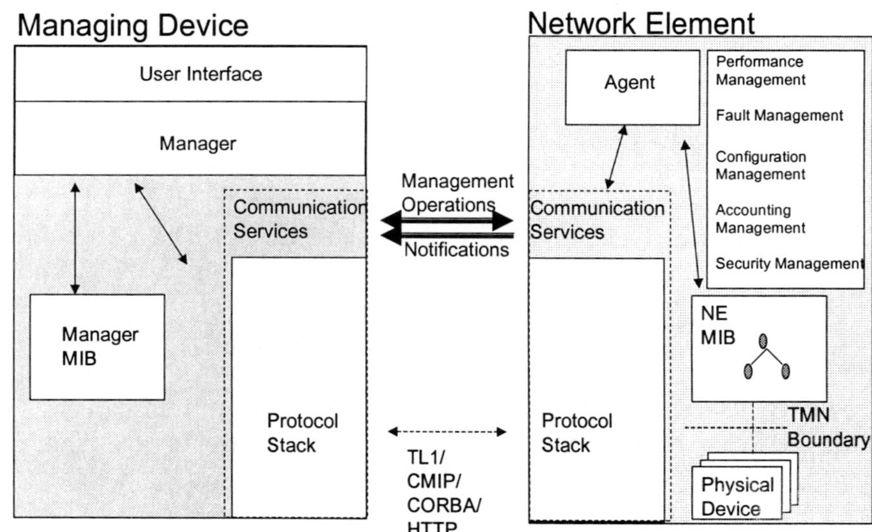

Figure 2: TMN–Layer Management Model

Overview of MONs

Figure 3 illustrates an example MON. The MON aggregates various voice, data, and video sources from various technologies and then feed into the metro inter–central office (CO) networks. The network management considerations for MON are traffic aggregation, inter–CO distribution, and long-haul backbone accessing. The MON provides the extended connectivity from the long-haul backbone to the customers. Currently, DWDM is competing with optical carrier (OC)–192 and optical fiber installation for MON deployment. Because data traffic is growing exponentially and time division multiplex (TDM) technology will reach the technology bottleneck and no longer be cost-effective, DWDM will eventually be the core component of the MON.

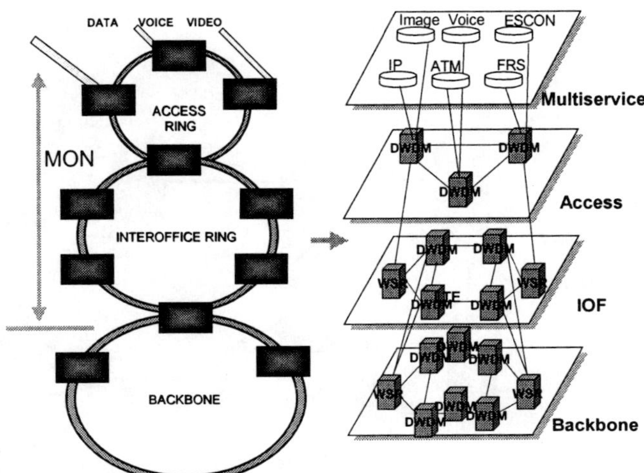

Figure 3: *Example MON*

The basic DWDM NEs are optical line amplifiers (OLAs), optical terminal multiplexers (OTMs), OADMs, and optical cross-connects (OXCs) [2, 3]. *Figure 4* gives the block diagram of each optical NE (ONE) that will be used in the later illustrations.

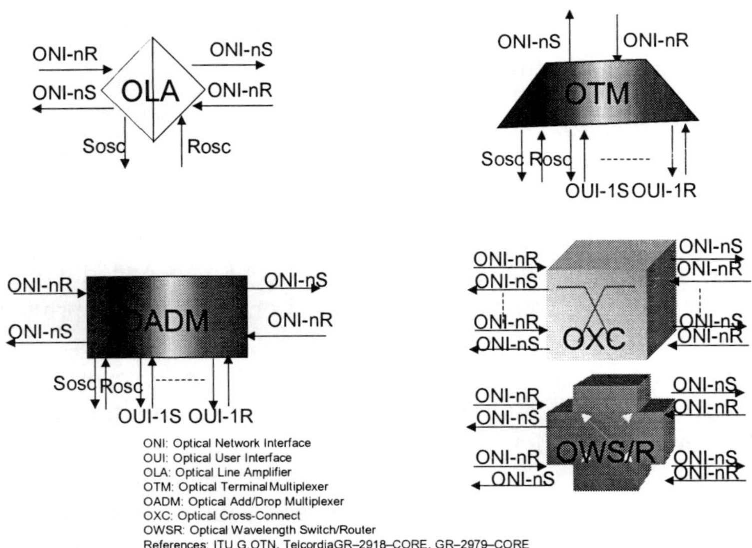

Figure 4: *WDM ONEs*

The OXC may have different capabilities, so it is necessary to define the OXC that is capable of switching or routing individual incoming wavelengths as the OWSR. The basic topologies of existing OTN are mainly point-to-point, linear, and ring, as shown in *Figure 5*. Mesh topology, which requires an OXC, remains under development. Ideally, the MON can be various combinations of point-to-point, linear, ring, and mesh subnetworks. Ring-based networks tend to have self-healing capability, while mesh-based networks require more intelligent control. Because of the dynamic bandwidth management, protection, and restoration needs, the wavelength routing mechanism is evolving and emerging in the MON. The control software for wavelength routing requires sophisticated signaling to orchestrate wavelength translation, wavelength protection, and restoration.

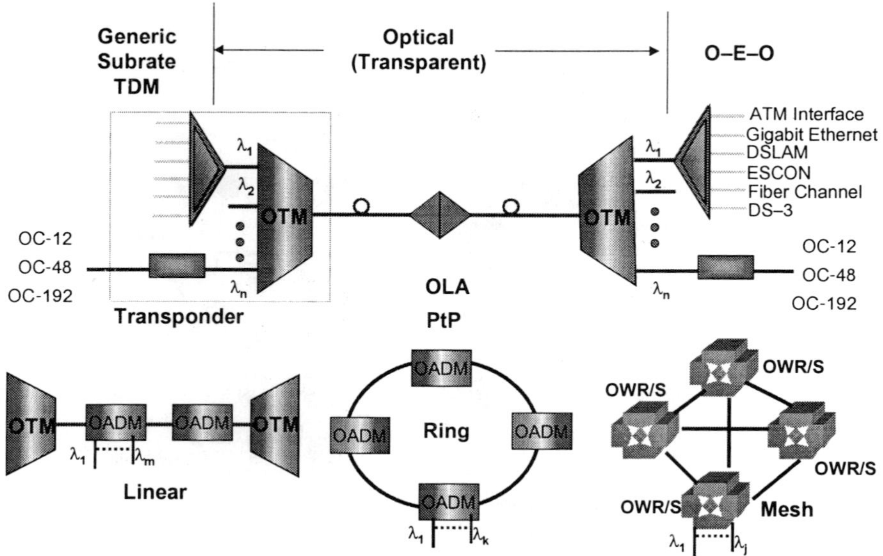

Figure 5: *Optical Network Configuration Topologies*

Key Characteristics of MONs

Fundamental changes in the upcoming telecommunications services are quick setup, lower network delay, and distance-independent pricing. Key characteristics of MONs, detailed in the following sections, would aid the advancement of telecommunications services.

Diversified Connectivity and Services

To provide connectivity is to provide a broad set of interface capabilities. SONET interface has played a significant role in the metro access edge. The SONET interface's scalability, reliability, and standardization will ensure its continued importance. The turn-up of OC–192 in the metro network will be explosive as OC–192 technology matures. To use the SONET bandwidth efficiently, multiservice platforms are employed to aggregate traffic. Services based on multiservice provisioning platforms give customers better price/performance, low latency, and flexibility. The generic time division multiplex (TDM) in *Figure 6* is designed for enterprise access and is used to aggregate from different applications into a higher-rate bit stream before feeding into wavelength division multiplexing (WDM) networks. At the same time, ATM, frame relay (FR), and native Internet protocol (IP) services are becoming the primary methods to carry packetized voice and data traffic in the metro network at optical line rates of up to OC–192 (10 Gbps). IP–based applications, such as gigabit routers, feed the wavelength directly into the DWDM network.

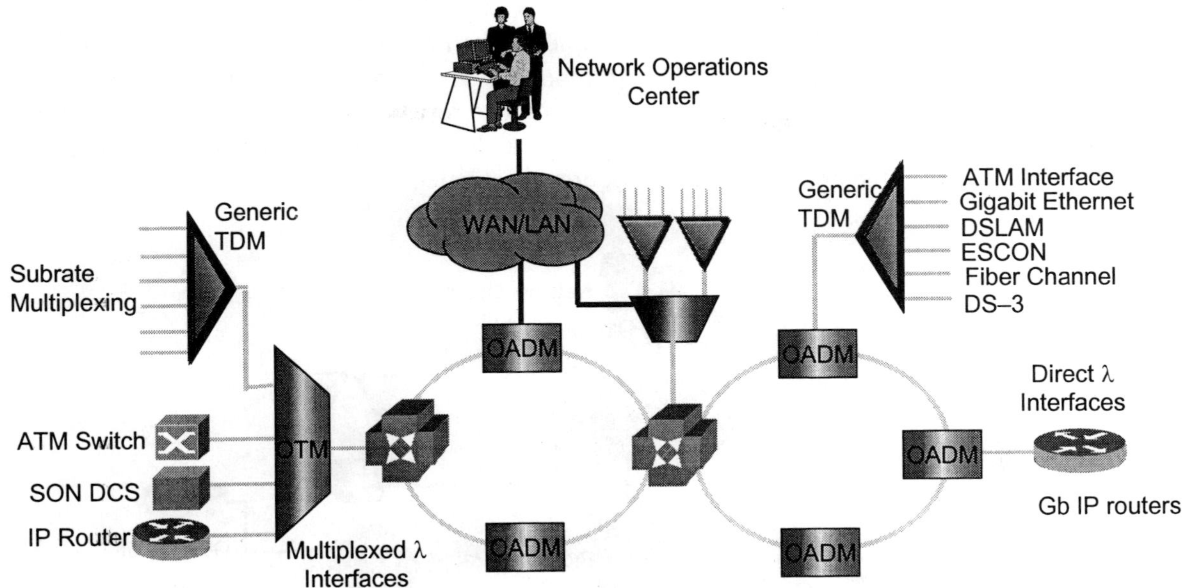

Figure 6: Ring-Interconnected MON

Transparency

In the long-haul network, traffic must be carried in a single-frame format, such as SONET. In the MON, optical transparency is important to aid the service provider's ability to provide new services. Transparency to multiservice aggregation, enterprise access, and various subrate connections are key attributes of the MON. If performance monitoring can be done at the optical domain, the next big breakthrough is the elimination of electrical interfaces in carrier backbones.

Cost-Sensitive—Pay as the Service Grows

Using existing pricing models within the metro market, deploying DWDM is not as cost-effective as deploying TDM or laying more fiber. This argument could be true for a certain installation that does not continue to grow. The advantage of the DWDM is its great scalability. The equipment designed for MON can be modular so that the user pays only to satisfy the current demand and so that it expands as the network expands without the need for major reinvestment.

Wavelength on Demand

In its infancy, point-to-point and passive linear topologies were used in the deployment of WDM networks. These networks consist of the OTMs, the OLAs, and some fixed or configurable passive OADMs. The primary purpose for such deployment was to relieve the exhausted capacity of optical fibers. With the advent of optical components and control software, other topologies and applications have begun to emerge toward a more service-oriented application.

One such service-oriented application is on-demand, wavelength-based service. This service allows wavelengths to be provisioned and assigned to different users on demand or at a prescheduled time and date. A dynamic OWSR is a key component for realizing such a service. *Figure 6* shows metro rings interconnected with an optical switch. Wavelength coming in from the OTM to the network can be switched into different rings depending on service need.

Intelligent Protection and Restoration

Dynamic OADM allows provisionable, interconnected ring deployment. The optical wavelength switching router enables the carriers to support optical-layer protection and restoration while creating new optical-layer services, such as optical virtual private networks (VPNs) and dynamic, provisionable, wavelength-based services. A fully populated OWSR network may form a mesh topology, as in *Figure 7*. The protection schemes for the all-optical ring topology are currently under standardization. The protection and restoration schemes for the mesh topology also remain under investigation. For a SONET ring, 50 ms is the required maximum restoration time for a clean ring. For the optical switch, only a few milliseconds are allowed if the optical switch protection is activated first.

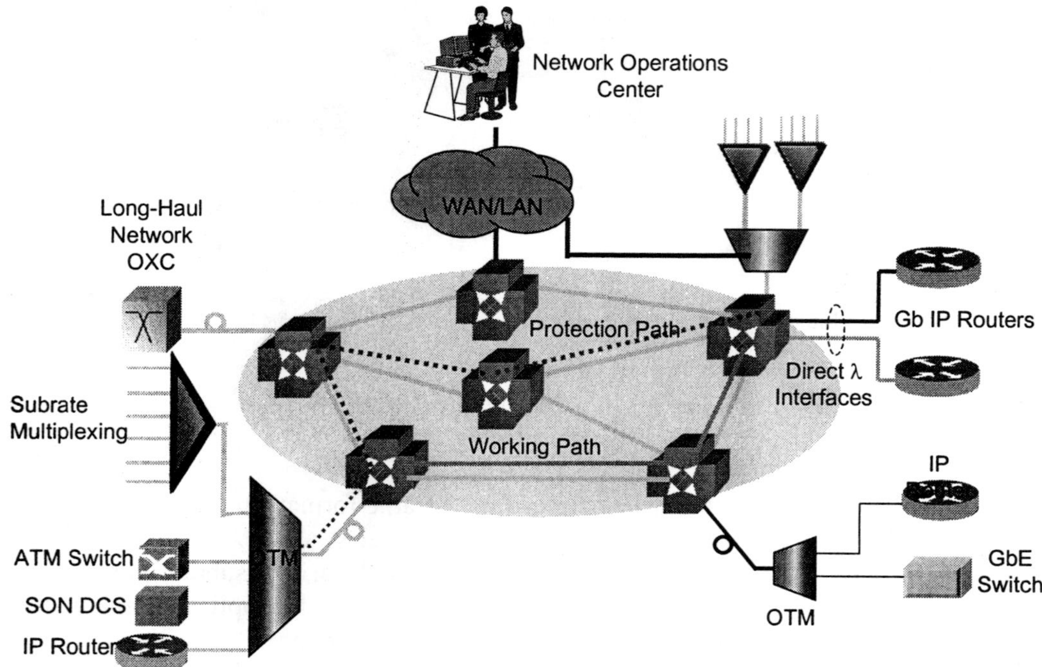

Figure 7: Mesh Topology for MON

Low protection latency, high reliability, and scalability are the favorable attributes for the mesh network over the ring network. A lambda-route control protocol based on the multiprotocol label switching (MPLS) protocol takes advantage of these key attributes for an optical mesh topology and is currently under development in the standards forum.

Management Platform for the MON

The essential features of the MON management platform are summarized in the following sections.

Management Interfaces

Of all the features in the MON, reliable services and manageable connectivity on user services matter most. The management system must accommodate the interconnection flexibility of optical data streams throughout the entire MON. Service providers such as the incumbent local-exchange carriers (ILECs), competitive local-exchange carriers (CLECs), and data local-exchange carriers (DLECs), have operations support systems (OSSs) for different technologies. The service information aggregated at the MON must be integrated into different northbound OSS platforms. To achieve multivendor interoperability, the southbound interfaces of the element manager also must adapt to various management interfaces. *Figure 8* provides an EMS platform with various management interfaces to accommodate various service providers' OSSs and vendors' NEs.

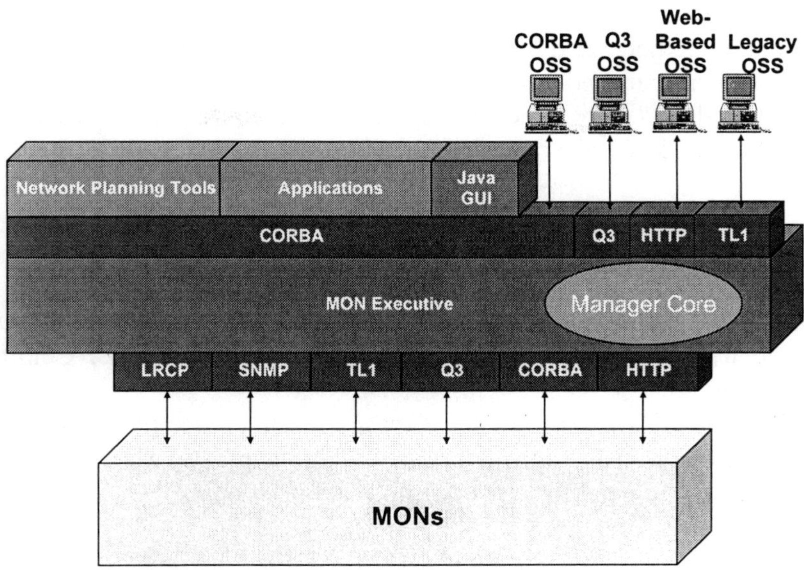

Figure 8: Unified MON Management Platform

Flexible Configuration

The MON market is cost-sensitive. To provide the lowest development/cost ratio, a single management platform is required that is capable of managing a diverse array of configurations and topologies. Such a platform should be ultra robust, modular, and scalable to accommodate the great scalability and flexibility offered by an optical network.

On-Demand Lambda Routing Management

Beside the multiservice aggregation, the unified management platform should also provide an integrated bandwidth monitoring and management function that enables a service provider to schedule wavelength allocation on demand. It should also enable the collection of accounting information for flexible billing based on bandwidth usage and QoS as defined by the carrier management application suite proposed by the TeleManagement Forum (TMF).

Intelligent Network Topology and Inventory Management

Because it adds a complex protection and restoration mechanism on top of dynamic wavelength switching, the MON requires sophisticated network topology management to keep the network topology and inventory updated. A root-cause analysis strategy is required to reflect the failure point of related alarms. With the introduction of sophisticated NEs to the networks, the complexity of the management platform is dramatically increased. New features, such as wavelength translation and dynamic wavelength assignment, require the management system to have intelligent protocols for managing restoration and protection effects on the network topology. An optical protection ring with wavelength translation (via dynamic tunable laser) is equivalent to time slot interchange (TSI) in the SONET configuration. However, the complexity of managing such a configuration would deter its implementation. The digital wrapper technology will enable the OWSR implementation, but it also gives a great challenge to the network management platform.

Conclusion

In its early deployment, the MON will require multiservice aggregation with optical-to-electrical-to-optical (O–E–O) conversion to provide some performance-monitoring capabilities. The ability to provide transparency is also essential for the deployment of just-in-time–demand–centric services.

With the introduction of more sophisticated WDM NEs (e.g., provisionable OADMs and OWSRs), coupled with an advanced method for doing on-demand wavelength conversion, the next generation of MON will likely be an interconnected network of arbitrary topologies (e.g., combinations of ring and mesh). The strategy for payload (wavelength) management is becoming a key factor to enable the MON management platform to provide service on user demand.

References:

1. ITU M.3400. Oct. 1992. TMN Management Functions.
2. Telcordia GR-2918–CORE, Issue 2. April 1998. Dense wavelength division multiplexing systems with digital tributaries for use in metropolitan-area applications: Common physical layer generic criteria.
3. Telcordia GR–2979-CORE. April 1998. Common generic requirements for optical add/drop multiplexers (OADMs) and optical terminal multiplexers (OTMs).
4. ITU M.3100. May 1995. Generic Network Information Model.